Flexible Electronics: Materials and Applications

Electronic Materials: Science and Technology

Series Editor: Harry L. Tuller
Professor of Materials Science and Engineering
Massachusetts Institute of Technology
Cambridge, Massachusetts
tuller@mit.edu

William S. Wong · Alberto Salleo

Editors

Flexible Electronics:
Materials and Applications

 Springer

Editors
William S. Wong
Palo Alto Research Center
3333 Coyote Hill Road
Palo Alto, CA 94304
wsw@parc.com

Alberto Salleo
Stanford University
Department of Materials Science &
  Engineering
Stanford, CA 94305-2205
asalleo@stanford.edu

ISBN 978-0-387-74362-2   e-ISBN 978-0-387-74363-9
DOI 10.1007/978-0-387-74363-9

Library of Congress Control Number: PCN applied for

Printed on acid-free paper.

springer.com

Preface

The advancement of flexible electronics has spanned the past forty years ranging from the development of flexible solar cell arrays made from very thin single-crystal silicon to flexible organic light-emitting diode displays on plastic substrates. The recent rapid development of this field has been spurred by the continuing evolution of large-area electronics with applications in flat-panel displays, medical image sensors, and electronic paper. Many factors contribute to the allure of flexible electronics; they are typically more rugged, lighter, portable, and less expensive to manufacture compared to their rigid substrate counterparts. Demonstrations of flexible electronics promise the availability of robust, lightweight, and low-cost electronics in the near future and this book is arranged to give the reader a survey of the materials that are used to fabricate these devices on flexible media. Subsequent chapters are organized to provide an overview of the different applications that can be created with a wide variety of materials systems. The range of polymeric to inorganic materials encompasses a wide array of performance benchmarks. It is these benchmarks of device characteristics (both electrical and mechanical) and performance, and the processes involved to make the device that will ultimately determine the suitable applications. It is not the intent of this book to give an exhaustive review of the technology but rather to provide a starting reference for a wide array of materials and applications; the chapters presented are also intended to augment existing literature in the field of flexible electronics.

The materials, processes, and applications for flexible electronics cover many different areas and we begin the book with a general overview of the field and the evolution of the technology. The mechanical stability of thin films on foil substrates is an area of active study and the understanding of these characteristics needs to be developed in order to fabricate multilayer structures with minimum offset in layer-to-layer alignment (Chapter 2). A review of processing conventional inorganic thin-film materials at low temperatures is given in Chapter 3. The chapter provides an overview of the materials and device characteristics for low-temperature silicon-based thin-film dielectrics and semiconductors. Chapter 4 describes how the understanding of the mechanical stability of thin films on foil substrates and low-temperature processing of conventional inorganic materials, such as amorphous silicon (a-Si:H), can be used for integration onto low melting point plastic platforms. These semiconductor materials have already been optimized for flat panel display applications and the transition to flexible electronics enables integration with organic light emitters to create applications in flexible emissive displays. The transition metal oxides (TMO) have the potential for creating high-performance thin-film transistor (TFT) devices beyond the performance offered by a-Si:H or poly-silicon TFTs. Chapter 5 provides an overview of the TMOs and demonstrates a novel nano-imprint lithography approach that can be used to fabricate flexible backplanes in a roll-to-roll process. Applications towards large-area flexible image sensors are another active area of research. Chapter 6 reviews ink-jet patterning techniques for fabricating a-Si:H based x-ray image sensors for medical imaging applications and ink-jet patterning for all-additive processing of backplane arrays.

The development of organic and polymeric materials for flexible electronics is progressing at a rapid rate with organic semiconductor materials producing devices that rival the performance of conventional a-Si:H TFTs. Small molecule semiconductors such as pentacene have been used to make organic TFTs and have shown performance exceeding silicon-based TFTs. Chapter 7 discusses two applications, image sensing and micro-electro-mechanical systems (MEMS), where organic materials compete well with silicon-based devices in flexible applications. In Chapter 8, the impact of materials and device physics on device and circuit design is discussed in the context of specific applications (displays and radio-frequency identification tags) based on pentacene active layers.

The transition to solution processable materials provides a potentially limitless choice of flexible electronics applications and processes (Chapter 9). These materials offer low-cost processing capabilities from simple spin casting to jet printing for device fabrication. The major limitation of the materials system is its relatively low performance for TFT applications. As these materials continue to improve, the appeal of low fabrication cost is also a catalyst for using polymeric semiconductors in photovoltaic applications. A review of excitonic solar cell properties and characteristics is presented in Chapter 11 along with an assessment of the applications for low-cost large-area power sources based on polymeric "green" technologies (Chapter 12).

Nano-scale materials are also suitable for flexible electronic applications. The size scale and electrical characteristics of randomly oriented carbon nanotubes (CNTs) mats provide a material that is highly compliant, conductive, and transparent in the visible spectrum. These three attributes give CNTs an advantage for use as transparent conductors for applications in solar cells and flexible displays (Chapter 10). Finally, none of these materials will have much functionality if a suitable substrate is unavailable. Chapter 13 reviews the characteristics required for flexible platforms for use in electronic applications and the processes and barrier materials that are required in order to make the plastic films optimal for device applications.

Lastly, we would like to thank all the contributing authors for their time and effort in preparing the manuscripts presented in this book. It is the work of these people and all the scientist, researchers, and engineers in the field that make the technology of flexible electronics exciting and challenging. We are also grateful to the Palo Alto

Research Center and Stanford University for providing support of the project and the many colleagues whom we worked with in making this book a reality.

Palo Alto, CA, USA

William S. Wong

January 2009

Alberto Salleo

Contents

Contributors

C. J. Brabec   Konarka Technologies GmbH, Altenbergerstrasse 69, A-4040 Linz, Austria, cbrabec@konarka.com

Michael L. Chabinyc   Materials Department, University of California, Santa Barbara, USA, mchabinyc@engineering.ucsb.edu

I-Chun Cheng   Department of Electrical Engineering and Graduate Institute of Photonics and Optoelectronics, National Taiwan University, Taipei, 10617 Taiwan, ichuncheng@cc.ee.ntu.edu.tw

G. Dennler   Konarka Technologies GmbH, Altenbergerstrasse 69, A-4040 Linz, Austria

Anil R. Duggal   General Electric Global Research Center, 1 Research Circle, KWD 272, Niskayuna, NY 12309, USA, duggal@research.ge.com

Ahmet Gün Erlat   General Electric Global Research Center, 1 Research Circle, KWC 331, Niskayuna, NY 12309, USA, erlat@research.ge.com

Helena Gleskova   Department of Electronic and Electrical Engineering, University of Strathclyde, Royal College Building, Glasgow, G1 1XW, UK, helena.gleskova@eee.strath.ac.uk

George Gruner   Department of Physics and Astronomy, University of California, Los Angeles, CA 90095-01547, USA, ggruner@ucla.edu

David Hecht   Department of Physics and Astronomy, University of California, Los Angeles, CA 90095-01547, USA, Dhecht99@ucla.edu

Martin Heeney   Department of Materials, Queen Mary, University of London, London, E1 4NS, UK, m.heeney@qmul.ac.uk

Warren B. Jackson   Hewlett-Packard Laboratories, Mail Stop 1198, Palo Alto, CA 94304, USA, warren.jackson@hp.com

Michael G. Kane   Sarnoff Corporation, CN5300, Princeton, NJ 08543, USA, mkane@sarnoff.com

Iain McCulloch   Department of Chemistry, Imperial College London, London SW7 2AZ, UK, i.mcculloch@imperial.ac.uk

Michael D. McGehee   Department of Materials Science and Engineering, Stanford University, Stanford, CA, USA, mmcgehee@stanford.edu

Arokia Nathan   London Centre for Nanotechnology, London, UK

Tse-Nga Ng Palo   Palo Alto Research Center, Palo Alto, CA 94304, USA

Alberto Salleo   Stanford University, Stanford, CA 94305, USA, asalleo@standford.edu

Kalluri R. Sarma   Honeywell International, Phoenix, AZ 85036, USA, kalluri.r.sarma@honeywell.com

Andrei Sazonov   The Department of Electrical and Computer Engineering, University of Waterloo, Ontario, Canada, asazonov@venus.uwaterloo.ca

P. Schilinsky   Konarka Technologies GmbH, Landgrabenstrasse 94, D-90443 Nürnberg, Germany

Shawn R. Scully   Department of Materials Science and Engineering, Stanford University, Stanford, CA, USA

Takao Someya   School of Engineering, Quantum-Phase Electronics Center, The University of Tokyo, 7-3-1 Hongo, Bunkyo-ku, Tokyo 113-8656 Japan, someya@ap.t.u-tokyo.ac.jp

Denis Striakhilev   The Department of Electrical and Computer Engineering, University of Waterloo, Canada

Zhigang Suo   Division of Engineering and Applied Science, Harvard University, Cambridge, MA 02139, USA

Sigurd Wagner   Department of Electrical Engineering, Princeton Institute for the Science and Technology of Materials, Princeton University, Princeton, NJ 08544, USA, wagner@princeton.edu

C. Waldauf   Konarka Technologies GmbH, Altenbergerstrasse 69, A-4040 Linz, Austria

William S. Wong   Palo Alto Research Center, Palo Alto, CA 94304, USA, wsw@parc.com

Min Yan   General Electric Global Research Center, 1 Research Circle, KWC 335, Niskayuna, NY 12309, USA, yanm@research.ge.com

Chapter 1
Overview of Flexible Electronics Technology

I-Chun Cheng and Sigurd Wagner

Abstract   This chapter provides an overview of the history, concepts, and possible applications of flexible electronics from the perspectives of materials and fabrication technology. The focus is on large-area capable electronic surfaces. These are made of backplane and frontplane optoelectronics that are fabricated as fully integrated circuits on flexible substrates. The discussion covers flexible electronics, and reaches back to rigid-substrate precursor technology where appropriate. Flexible electronics is a wide-open and rapidly developing field of research, development, pilot production, and field trials. The chapter puts a perspective on the technology by systematizing it and by describing representative examples.

1.1 History of Flexible Electronics

Flexible electronics has a long history. Anything thin is flexible. Forty years ago single-crystalline silicon solar cells were thinned to raise their power/weight ratio for use in extraterrestrial satellites. Because these cells were thin, they were flexible and warped like corn flakes. Today, silicon-integrated circuits are thinned to become compliant so that the owner of a smart card does not break it when he sits on it. Flexible can mean many qualities: bendable, conformally shaped, elastic, lightweight, nonbreakable, roll-to-roll manufacturable, or large-area. The field has open boundaries that move with its development and application. In this chapter we cover a newly emerging segment of flexible electronics that is largely connected with active thin-film transistor (TFT) circuits. Therefore, this survey is representative but incomplete. To the industrial community today, flexible electronics means flexible displays and X-ray sensor arrays. To researchers flexible means conformally shaped displays and sensors, electronic textiles, and electronic skin.

I-Chun Cheng (✉)
Department of Electrical Engineering, and Graduate Institute of Photonics and Optoelectronics, National Taiwan University, Taipei, 10617 Taiwan
e-mail: ichuncheng@cc.ee.ntu.edu.tw

W.S. Wong, A. Salleo (eds.), *Flexible Electronics: Materials and Applications*, Electronic Materials: Science & Technology, DOI 10.1007/978-0-387-74363-9_1,
© Springer Science+Business Media, LLC 2009

1

The development of flexible electronics dates back to the 1960s. The first flexible solar cell arrays were made by thinning single crystal silicon wafer cells to ~100 μm and then assembling them on a plastic substrate to provide flexibility [1, 2]. The energy crisis in 1973 stimulated work on thin-film solar cells as a path to reducing the cost of photovoltaic electricity. Because of their relatively low deposition temperature, hydrogenated amorphous silicon (a-Si:H) cells lend themselves to fabrication on flexible metal or polymer substrates. In 1976, Wronski, Carlson, and Daniel at RCA Laboratories reported a Pt/a-Si:H Schottky barrier solar cell made on a stainless steel substrate, which also served as the back contact [3]. In the early 1980s, n^+–i a-Si:H/Pt Schottky barrier and p^+–i–n^+ a-Si:H/ITO solar cells were made on organic polymer ("plastic") film substrates by Plattner et al. [4] and by Okaniwa et al. [5–7], respectively. Okaniwa and coworkers also studied the flexibility of their solar cells. Around the same time, the CdS that had been developed for CdS/Cu_2S solar cells on glass substrates was made by continuous deposition on a moving flexible substrate in a reel-to-reel vacuum coater [8]. Beginning in the early 1980s, the roll-to-roll fabrication of a-Si:H solar cells on flexible steel [9] and organic polymer substrates [10] was introduced. Today, a-Si:H solar cells routinely are made by roll-to-roll processes.

The first flexible TFT dates back to 1968, when Brody and colleagues made a TFT of tellurium on a strip of paper and proposed using TFT matrices for display addressing. In the following years, Brody's group made TFTs on a wide range of flexible substrates, including Mylar, polyethylene, and anodized aluminum wrapping foil. The TFTs could be bent to a 1/16″ radius and continued to function. They could be cut in half along the channel direction, and both halves remained operational [11, 12].

In the mid-1980s, the active-matrix liquid-crystal display (AMLCD) industry started in Japan by adopting the large-area plasma enhanced chemical vapor deposition (PECVD) machines that had been developed for a-Si:H solar cell fabrication. The success of the a-Si:H TFT backplane-based AMLCD industry and the demonstration of a-Si:H solar cells on flexible substrates stimulated research on silicon-based thin-film circuits on novel substrates. In 1994, Constant et al. at Iowa State University demonstrated a-Si:H TFT circuits on flexible polyimide substrates [13]. Their demonstration included two approaches to achieving overlay registration in photolithography: (1) the edge of the polyimide substrate was affixed to a rigid silicon wafer by using vacuum compatible epoxy resin and (2) a conformal coating of polyimide was applied to a silicon wafer to form a polyimide film; after the TFT circuitry had been fabricated on top of the polyimide film, it was detached from the wafer. In 1996, a-Si:H TFTs were made on flexible stainless steel foil [14]. In 1997, polycrystalline silicon (poly-Si) TFTs made on plastic substrates using laser-annealing were reported [15, 16]. Since then, research on flexible electronics has expanded rapidly, and many research groups and companies have demonstrated flexible displays on either steel or plastic foil substrates. For example, in 2005 Philips demonstrated a prototype rollable electrophoretic display [17] and Samsung announced a 7″ flexible liquid crystal panel [18]. In 2006, Universal Display Corporation and the Palo Alto Research Center presented a prototype flexible organic light-emitting diode (OLED) display with full-color and full-motion with a poly-Si TFT backplane made on steel foil [19].

1.2 Materials for Flexible Electronics

A generic large-area electronic structure is composed of (1) a substrate, (2) backplane electronics, (3) a frontplane, and (4) encapsulation. To make the structure flexible, all components must comply with bending to some degree without losing their function. Two basic approaches have been employed to make flexible electronics: (1) transfer and bonding of completed circuits to a flexible substrate and (2) fabrication of the circuits directly on the flexible substrate.

In the transfer-and-bond approach, the whole structure is fabricated by standard methods on a carrier substrate like a Si wafer or a glass plate. Then it is transferred to [20–23] or fluidic self-assembled on [24] a flexible substrate. The transfer-and-bond approach has been extended to the bonding of ribbons of Si and GaAs devices to a stretched elastomer, which upon relaxation forms a "wavy" semiconductor that can be stretched and relaxed reversibly [25, 26]. The transfer approaches have the advantage of providing high-performance devices on flexible substrates. These processes are sophisticated advances over the original flexible wafer-based solar cell arrays [1, 2]. Their drawbacks are small surface area coverage and high cost. Bonded circuits will likely be added to large-area electronic surfaces at low density for high-speed communication and computation, lasing, and similarly demanding functions.

In many applications, the majority of the surface will be covered with electronics fabricated directly on the substrate. There are many approaches to integrating disparate materials and oftentimes flexible substrates are not fully compatible with existing planar silicon microfabrication processes. Direct fabrication may require (1) relying on polycrystalline or amorphous semiconductors because these can be grown on foreign substrates, (2) developing new process techniques, (3) introducing new materials, and (4) striking a compromise between device performance and low process temperatures tolerated by polymer foil substrates. Direct fabrication on flexible substrates is a hotbed of process research. New process techniques include the printing of etch masks [27, 28], the additive printing of active device materials [29–31], and the introduction of electronic functions by local chemical reaction [32]. Nanocrystalline silicon and printable polymers for OLEDs [33] are also materials of intense research. We will concentrate on the direct fabrication on flexible substrates, as it is the most direct, and sometimes more innovative, approach to the manufacturing of large-area electronic surfaces.

1.2.1 Degrees of Flexibility

Flexibility can mean many different properties to manufacturers and users. As a mechanical characteristic, it is conveniently classified in the three categories illustrated by Fig. 1.1 : (1) bendable or rollable, (2) permanently shaped, and (3) elastically stretchable. The tools for microfabrication have been developed for flat substrates. Therefore, at present all manufacturing is done on a flat workpiece that is shaped only as late as possible in the process. This approach benefits from the tremendous technology base established by the planar integrated circuit and display industries.

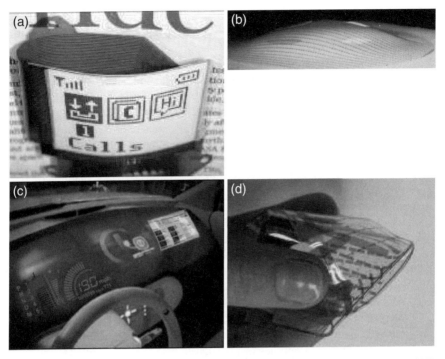

Fig. 1.1 (**a**) A bendable wristband display [Courtesy of Dr. Yu Chen, E Ink Corporation, 2001]. (**b**) Silicon islands on a spherically shaped foil substrate [34]. (**c**) Concept for a conformably shaped digital dashboard [Courtesy Professor Miltiades Hatalis, Lehigh University]. (**d**) Stretchable interconnects on an elastomer [Dr. Stéphanie P. Lacour, Princeton University]

When a mechanically homogeneous sheet of thickness d is bent to a cylindrical radius r, perpendicularly to the axis of bending, its outside surface expands and its inside surface is compressed by the bending strain $\varepsilon = d/2r$. When the sheet is not homogeneous, as is the case for an a-Si:H TFT layer on a plastic foil, the strain in the surfaces is modified from this simple expression, which however remains a useful approximation. In TFT backplanes or in entire displays, the strain ε must be kept below a critical value. The straightforward approach to keeping ε low even in sharp bending, to small r, is to make the structure thin. In this way, the strain experienced by the active devices in bendable or rollable electronics can be kept small, particularly when the devices are placed in the neutral plane.

Conformally shaped electronic surfaces are also made by existing planar fabrication techniques. Continuous, unbroken, surfaces are shaped to the desired geometry by plastic deformation [35, 36]. The extent of deformation can easily exceed the critical tensile strain, for fracture or necking, of thin-film inorganic semiconductors or metals, which typically lies between 0.1 and 1%. Therefore, the devices are protected by placement onto rigid islands. However, if only materials that can undergo large plastic deformation are employed, a flat electronic surface may be shaped without causing damage to unprotected devices [37].

Elastically stretchable electronics can undergo large and reversible deformation. When the substrate is an elastomer, devices are placed on rigid islands and are interconnected with elastically stretchable conductors [38, 39]. Alternatively, a substrate configured as a net-like open surface can be deformed reversibly without damaging the electronics that are made on the ribs of the net [40].

1.2.2 Substrates

Flexible substrates that are to serve as drop-in replacements for plate glass substrates must meet many requirements:

(1) Optical properties – Transmissive or bottom-emitting displays need optically clear substrates. In addition, substrates for LCDs must have low birefringence.

(2) Surface roughness – The thinner the device films, the more sensitive their electrical function is to surface roughness. Asperities and roughness over short distance must be avoided, but roughness over long distance is acceptable. As-received metal substrates usually are rough on both scales, while plastic substrates may be rough only over long distance.

(3) Thermal and thermomechanical properties – The working temperature of the substrate, for example the glass transition temperature (T_g) of a polymer, must be compatible with the maximum fabrication process temperature (T_{max}). Thermal mismatch between device films and substrate may cause films to break during the thermal cycling associated with fabrication. A rule of thumb for tolerable mismatch is $|\Delta CTE \cdot \Delta T| \leq 0.1$–$0.3\%$, where ΔCTE is the difference in coefficients of thermal expansion between substrate and device film, and ΔT is the temperature excursion during processing. Silicon-based circuits benefit from substrates with low CTE. High thermal conductivity may be important for the cooling of current-load circuits. Dimensional stability during processing is a concern with plastic substrates.

(4) Chemical properties – The substrate should not release contaminants and should be inert against process chemicals. Of advantage are substrates that are good barriers against permeation by atmospheric gases: for OLED application the water vapor permeation rate should lie below 10^{-6} g/m^2/day and the oxygen permeation rate below 10^{-3} to 10^{-5} cm^3/m^2/day.

(5) Mechanical properties – A high elastic modulus makes the substrate stiff, and a hard surface supports the device layers under impact.

(6) Electrical and magnetic properties – Conductive substrates may serve as a common node and as an electromagnetic shield. Electrically insulating substrates minimize coupling capacitances. Magnetic substrates can be used for the temporary mounting of the substrate during fabrication, or for affixing the finished product.

Table 1.1 Properties of substrates for flexible backplanes [after ref. 41]

Property	Unit	Glass (1737)	Plastics (PEN, PI)	Stainless steel (430)
Thickness	μm	100	100	100
Weight	g/m^2	250	120	800
Safe bending radius	cm	40	4	4
Roll-to-roll processable?	–	Unlikely	Likely	Yes
Visually transparent?	–	Yes	Some	No
Maximum process temperature	°C	600	180, 300	1,000
CTE	ppm/°C	4	16	10
Elastic modulus	GPa	70	5	200
Permeable to oxygen, water vapor		No	Yes	No
Coefficient of hydrolytic expansion	ppm/%RH	None	11, 11	None
Prebake required?	–	Maybe	Yes	No
Planarization required?	–	No	No	Yes
Buffer layer required? Why?	–	Maybe	yes: adhesion, chemical passivation	yes: electrical insulator, chemical passivation
Electrical conductivity	–	None	None	High
Thermal conductivity	W/m·°C	1	0.1–0.2	16
Plastic encapsulation to place electronics in neutral plane	Substrate thickness	5×	1×	8×
Deform after device fabrication	–	No	Yes	No

Three types of substrate materials are available for flexible applications: metals, organic polymers (plastics), and flexible glass. Properties of typical materials are listed in Table 1.1 for 100-μm-thick foils.

1.2.2.1 Thin Glass

Glass plates are the current standard substrates in flat panel display technology. Plate glass becomes flexible when its thickness is reduced to several 100 μm [42, 43]. Glass foils as thin as 30 μm can be produced by the downdraw method. Foil glass retains all advantages of plate glass: optical transmittance of >90% in the visible, low stress birefringence, smooth surface with RMS roughness of 1 nm or less, temperature tolerance of up to 600°C, high dimensional stability, a low coefficient of thermal expansion (CTE) of $\sim 4 \times 10^{-6}$/°C, which matches those of silicon device materials, resistance to most process chemicals, no outgassing, impermeability against oxygen and water, scratch resistance, and electrical insulation. However, flexible glass is fragile and difficult to handle. To reduce breakage during handling, foil glass can be made to resist crack propagation by

(1) laminating it with plastic foil, (2) applying a thin hard coat, or (3) applying a thick polymer coat.

1.2.2.2 Plastic Film

Polymer foil substrates are highly flexible, can be inexpensive, and permit roll-to-roll processing. However, they are thermally and dimensionally less stable than glass substrates and are easily permeated by oxygen and water. A glass transition temperature, T_g, compatible with the device process temperature is essential. However, a high T_g alone is not sufficient. Dimensional stability and a low CTE are also important factors. Typical polymer films are shrunk by heating and cooling cycles. They shrink less if prestabilized by prolonged annealing [44]. Because the elastic modulus of polymer substrates is a factor of 10–50 lower than that of inorganic device materials, a small thermal mismatch stress can make the free-standing workpiece curve and cause misalignment during the overlay registration of the flattened piece. A large CTE mismatch coupled with a large temperature excursion during processing can break a device film [45]. Polymer substrates with CTE below 20 ppm/°C are preferred as substrates for silicon-based device materials.

Candidate polymers for flexible substrates include (1) the thermoplastic semicrystalline polymers: polyethylene terephthalate (PET) and polyethylene naphthalate (PEN), (2) the thermoplastic noncrystalline polymers: polycarbonate (PC) and polyethersulphone (PES), and (3) high-T_g materials: polyarylates (PAR), polycyclic olefin (PCO), and polyimide (PI). PC, PES, PAR, and PCO are optically clear and have relatively high T_g compared to PET and PEN, but their CTEs are 50 ppm/°C or higher, and their resistance to process chemicals is poor. Much research has been conducted with PET, PEN, and PI, with their relatively small CTEs of 15, 13, and 16 ppm/°C (Kapton E), respectively, relatively high elastic moduli, and acceptable resistance to process chemicals. Both PET and PEN are optically clear with transmittance of >85% in the visible. They absorb relatively little water (~0.14%), but their process temperatures are only ~150 and ~200°C, even after prestabilization by annealing. In contrast, PI has a high glass transition temperature of ~350°C, but it is yellow because it absorbs in the blue, and it absorbs as much as 1.8% moisture [46, 47]. No polymer meets the extremely demanding requirement for low permeability in OLED applications. The typical water and oxygen permeation rates of flexible plastic substrates are 1–10 g/m^2/day and 1–10 cm^3/m^2/day, respectively, instead of the required 10^{-6} g/m^2/day and 10^{-5} cm^3/m^2/day [48]. Barrier layer coatings can reduce absorption and permeability by gas, raise resistance to process chemicals, strengthen adhesion of device films, and reduce surface roughness.

1.2.2.3 Metal Foil

Metal foil substrates less than ~125 μm thick are flexible and are attractive substrates for emissive or reflective displays, which do not need transparent substrates. Stainless steel has been most commonly used in research because of its high resistance to corrosion and process chemicals, and its long record of application in amorphous silicon solar cells. Stainless steel substrates can tolerate process temperature as high as 1,000°C, are dimensionally stable, present a perfect permeation barrier against moisture and oxygen, can serve as heat sink, and can provide electromagnetic shielding. Certain magnetic steels lend themselves to magnetic mounting and handling. In general, stainless steel substrates are more durable than plastic and glass foils.

A typical stainless steel foil comes with sharp rolling marks and micrometer-size inclusions, which may cause devices to fail. The most carefully rolled steel has a surface roughness of ~100 nm, in contrast to display glass with less than 1 nm. To ensure the electrical integrity of the thin-film devices made on them, steel foil substrates must be either polished well [49, 50] or planarized with a film [14, 51, 52]. The planarization layer may be organic, inorganic, or a combination. Commercially available materials for planarization include (1) organic polymers, (2) methylsiloxane spin-on-glass, and (3) silicate spin-on-glass. The higher the organic content of the planarizing material, the thicker the film that can be applied without forming cracks and the smoother the resulting surface. This is particularly important for OLEDs, which need a surface roughness of less than 5-nm RMS. However, the organic content imposes a ceiling on any subsequent, sustained process temperature. An organic polymer-planarizing layer can be protected against excessive transient heating with an inorganic overlayer by pulsed-laser processing. The typical process steps for applying the planarization layer include spin-on, hot-plate bake, and high-temperature cure. As a rule of thumb, the final curing temperature sets the upper temperature limit for the subsequent device process. Incorporation of getter materials, for example phosphorus in a spin-on silicate film, enhances the barrier properties against possible contaminants that may originate in the steel substrate.

Metal substrates are electrically conducting. In some applications they can serve as back contact, as in solar cells. For other applications metal must be coated with an insulating layer to provide circuit isolation. SiN_x or SiO_2 layers are commonly used for this purpose on steel substrates. The electrical insulation also functions as adhesion layer and as barrier against process chemicals. SiN_x:H, a standard material in the a-Si:H TFT process, is also an excellent diffusion barrier. However, the high hydrogen content of SiN_x:H may cause the film to crack or ablate during subsequent processing at high temperature or during laser annealing. In such cases, an SiO_2 film is deposited instead and provides the added advantage of a lower dielectric constant than that of SiN_x:H, which helps reduce the capacitive coupling between the substrate and the electronics. A barrier layer of 0.2–0.4 μm suffices for electrical insulation as it reduces the leakage current at a potential difference of 50 V to 1 nA/cm^2. An active-matrix circuit driven at video rate may require a thicker layer to reduce the coupling capacitance to 1 nF/cm^2.

1.2.3 Backplane Electronics

Backplanes provide or collect power and signal to or from frontplanes. Backplanes may be passive or active. The ideal flexible active-matrix backplane should be rugged, rollable or bendable, capable of CMOS operation, and should lend itself to low-cost manufacturing. Today's TFT backplane technologies are best described by their active semiconductor, which may be amorphous, nanocrystalline, or poly-crystalline silicon, a II–VI compound semiconductor, or an organic semiconductor in polymer or molecular form. As thin-film semiconductor technologies develop away from glass substrates and toward flexible substrates, the backplanes differentiate further by the type of substrate. The substrate material defines a region of TFT fabrication conditions in temperature–time space. The maximum tolerable process temperature is set by the type of substrate materials: <300, <600, and <1,000°C for organic polymer, glass, and steel substrates, respectively. We now briefly review the active materials for TFT backplanes and the associated fabrication technologies.

1.2.3.1 Silicon Thin-Film Transistors

Silicon-based materials benefit from the advantages of a well-established technology and a native oxide that is a high-quality insulator. Three approaches can be taken to preparing TFT-grade silicon on foil substrates: (1) direct deposition of the channel semiconductor, (2) deposition of a precursor film followed by crystallization, and (3) physical transfer of separately fabricated circuits. Techniques (1) and (2) are explored for large-area display applications. Direct deposition can provide the full range of devices, from low OFF current amorphous silicon (a-Si:H) TFTs to CMOS capable TFTs of nanocrystalline silicon (nc-Si:H). The highest ON current is obtained in TFTs of both polarities in polycrystalline silicon (poly-Si) made by the crystallization of an amorphous silicon precursor film.

a-Si:H made from a glow discharge was first reported in the late 1960s and the first a-Si:H TFT was demonstrated by Snell and coworkers in 1981 [53]. Since the mid-1990s, a-Si:H TFTs have been made on flexible substrates, including PI films [13] and stainless steel foil [14]. Today's baseline growth process for a-Si:H TFTs is PECVD at a substrate temperature of 250–350°C. a-Si:H is a material with reproducible optoelectronic properties that provides TFTs with electron field-effect mobilities of $\sim$0.5–1 $cm^2V^{-1}s^{-1}$. The a-Si:H growth process has been adapted to low-temperature polymer substrates with process temperatures of 150°C or less [54]. The only useful a-Si:H TFTs are n-channel devices as the hole mobility of $\sim$0.003 $cm^2V^{-1}s^{-1}$ is too low to be useful.

nc-Si:H is a promising candidate material for fully integrated electronics on plastic because it is CMOS capable and can be made by PECVD at low substrate temperature. The top-gate geometry has produced the highest mobilities to date. However, the nc-Si:H TFT fabrication process needs considerably more research. The nc-Si:H growth schedule (direct or prenucleation), the location of source/drain contacts (in-plane or staggered), efficient source/drain doping (particularly p$^+$), the type of gate dielectric (SiO_2 or SiN_x), and the techniques for post-process annealing are not settled.

Poly-Si can be either directly deposited by low-pressure chemical vapor deposition (LPCVD) or formed by the crystallization of a precursor film, typically a-Si:H. Crystallization produces the superior material, with larger grain sizes and smoother surface than obtained by LPCVD. a-Si:H can be crystallized by furnace anneal at

600°C or higher, or at ~550°C by silicide-mediated crystallization (also known as metal-induced crystallization), and by pulsed excimer laser annealing. Poly-Si TFTs have been made on steel foil substrates since the late 1990s [55, 56]. While furnace crystallization is not compatible with organic polymer substrates, the spatial and temporal confinement of heat in excimer laser annealing permits making poly-Si on plastic, as first demonstrated with poly-Si TFTs made on plastic in 1997 [16]. Controlling the geometry of the laser spot produces large grain sizes [57], and poly-Si TFTs with high performance have been made on flexible steel substrates [58, 59].

1.2.3.2 Organic Thin-Film Transistors

The first demonstration of the field effect in a small-molecule organic material dates back to 1964 [60]. In 1983, the field effect in a polymer structure made by a solution process was reported [61]. Since then the performance of organic thin-film transistors (OTFTs) has been raised impressively by the introduction of new organic channel materials and by improved fabrication [62]. In the late 1990s, OTFTs with ON currents comparable to those of a-Si:H TFT were reported [63]. In 2000, pentacene TFTs with a hole field-effect mobility of 3.2 $cm^2V^{-1}s^{-1}$ and an ON/OFF ratio of $>10^9$ were demonstrated [64]. Because in most organic materials the hole mobility is higher than the electron mobility, most OTFTs are p-channel devices.

Organic polymers are soluble, and small molecules can be derivatized to soluble precursors. Therefore, OTFTs can be fabricated by solution processing near room temperature [31], compatible with low-temperature plastic substrates. Since the late 1990s, OTFTs and circuits have been made on flexible plastic substrates, including PI [32, 65], PEN[66], PET [67–69], PC [70, 71], and on paper [72]. More recently, displays have been made on OTFT backplanes on flexible polymeric substrates. The displays include reflective LCDs [73, 74], electrophoretic displays [75], and OLED displays [76–79].

1.2.3.3 Transparent Thin-Film Transistors

TFTs made of transparent materials [80, 81] may not need shielding from visible light to suppress photoconductance, and can raise the pixel aperture of transmissive displays, for example on windscreens of cars. They have been developed from the conventional wide-bandgap compound semiconductors: GaN or SiC, and from the transparent oxide semiconductors: ZnO [82], In_2O_3 [83, 84], and SnO_2 [85]. The first flexible transparent TFT was made by Nomura et al. in 2004 from the amorphous In–Ga–Zn–O system on PET. Saturation mobilities reached 6–9 $cm^2V^{-1}s^{-2}$, and device characteristics were stable under mechanical bending [86].

More recently, flexible transparent TFTs have been demonstrated with organic channel materials, including a conductive polymer [69] and single-walled carbon nanotubes (SWNT). SWNTs have enabled flexible transparent organic TFTs with mobilities comparable to that of a-Si:H TFTs [87–89].

1.2.3.4 Materials for Interconnects and Contacts

Transparent Conductive Oxides

Typical electrode and contact materials are metallic. Displays and thin-film solar cells, however, need electrodes that are transparent electrical conductors. To date, these have been metal oxides [90–92]. The first transparent conducting oxide (TCO), CdO, was reported by Baedeker in 1907 [93]. Transparent conducting tin-doped indium oxide (ITO), originally developed for the defrosting of aircraft windshields, is the most widely used TCO. It is the transparent electrode in flat panel displays, is used for electromagnetic shielding, and is deposited on optical grade polyester (PET) substrates by roll-to-roll dc sputtering for touch screens. In the trade-off between electrical conductivity and optical transparency, ITO has reached a performance plateau at a resistivity of $\sim 10^{-4}$ $\Omega \cdot$cm. New TCO materials based on ZnO and SnO_2, and mixed ternary and quaternary oxides are under development.

Depositing ITO on organic polymer foil substrates presents two challenges: opto-electronic quality and mechanical integrity. The best-quality display grade ITO on glass is dc-magnetron sputtered at a substrate temperature of 300–400°C, or first sputtered on the cold substrate and then annealed at 200°C in a controlled oxygen environment. Because polymeric substrate materials are heat sensitive, ITO is deposited on them at low temperature as an amorphous phase. Fortunately, amorphous ITO films deposited at room temperature have an optical transparency and an electrical conductivity that is adequate for present display technologies. The mismatch in mechanical properties between the ITO and the polymeric substrate is a challenge. ITO is stiff and brittle with an elastic modulus of $\sim$118 GPa. Depositing it on a compliant substrate may cause cracks and delamination during fabrication or in use. The mechanisms of cracking and the effect of cracking on the electrical properties of ITO films on PET substrates have been investigated by Cairns et al. The critical strain for cracking is reciprocal to the square root of film thickness, similar to what observed in a typical ceramic film [94]. ITO cracks at a strain of $\sim$1%. Prior to electrical failure, small cracks form and may grow by fatigue under cyclic loading. The onset of ITO conductivity loss in tension can be delayed by reducing the ITO layer thickness, by using a high-modulus but nonbrittle undercoat for the ITO, or by introducing compressive prestress into the ITO layer. Under compressive load, the brittle ITO film may delaminate and then buckle and crack. The strain at which buckling occurs depends on the strength of adhesion between the ITO and the substrate [95].

Conducting Organic Polymers

We just described how the mismatch of mechanical properties between inorganic device films and polymeric substrates can break the films during processing and use. These risks of film fracture provide one motive for using conductive polymers as contact and interconnect materials as part of all-organic flexible electronics [96]. In 1976, Heeger, MacDiarmid, and Shirakawa made the discovery that polyacetylene exhibits a high electrical conductivity when doped with the oxidant iodine.

Today, such "self-doped" conductive polymers are available commercially. Polyaniline (PANI) and polypyrrole (PPY) are used as antistatic materials, polythiophene derivatives, e.g., polyethylene dioxythiophene (PEDOT) replace inorganic ITO in some device applications and also function as hole injectors and transport layers in OLEDs, and polyphenylene vinylene (PPV) derivatives are active materials in polymer light-emitting diodes (PLEDs). Conductive polymers need much further research as they are not yet well defined and have suboptimal device properties.

Stretchable Interconnects

Flexible electronic surfaces that undergo large plastic deformation need stretchable interconnects. One-time deformation, for example to a conformally shaped surface, needs interconnects that can be deformed plastically. Under certain conditions, metal films can be deformed once and remain electrically conducting [35–37]. Repeated deformation, as experienced by electronic skin, requires elastic interconnects. Several approaches have been demonstrated to realizing elastically stretchable interconnects. They include making "filled" elastomers conducting by blending in conductive polymers [97–99] or by embedding metal particles [100]. Conductor-filled elastomers are used mainly for forming interconnects between rigid components. [101, 102]. Often the electrical conductivities of these conducting elastomers are insufficient for electrical interconnects. Because the filled-elastomer conductors are applied with thick-film techniques, they are not directly compatible with thin-film circuits. The search for elastic thin-film interconnects has focused on metal films deposited on or encased in elastomers. Microfabricated gold-film serpentines encased in a silicone elastomer conduct up to 54% tensile strain [103]. Stripes of thin gold film patterned on prestretched or flat elastomeric membranes remain electrically conducting when the membranes are stretched up to twice their initial length. Such stretchable interconnects were used in the demonstration of an inverter circuit made of a-Si:H TFTs on a silicone membrane [38, 104, 105].

1.2.4 Frontplane Technologies

Frontplanes carry the specific optoelectronic application. The frontplane materials of displays include liquid crystals for transmissive displays, reflective-mode liquid crystals and electrophoretic foils for reflective displays, and OLEDs for emissive displays. The frontplane might also be an X-ray sensor, an image sensor, a pressure sensor, a chemical sensor, an actuator or an artificial muscle in a smart textile.

1.2.4.1 Liquid Crystal Displays

The LCD is a light intensity filter that is controlled by the electric field between two transparent electrodes. These are fabricated on glass plates that are held $\sim$5 μm apart by transparent spacers [106]. To make the LCD flexible, the liquid is encapsulated in a polymer foil. The first examples were the polymer-dispersed liquid crystal (PDLC) [107] and the nematic curvilinear aligned phase (NCAP) [108, 109]. The phase separation procedure developed for PDLCs can also be used for bistable cholesteric liquid crystal [110]. Today, the term PDLC is used to describe liquid crystal droplets in encapsulated form. The PDLC dispersion techniques do not require alignment layers or polarizers, and large devices are relatively simple to fabricate.

The first LCDs fabricated on flexible plastic substrates date back to 1981 [111, 112]. In the meantime, full-color reflective cholesteric displays on plastic substrates have been developed [113–116], and twisted nematic LCDs made on a transparent flexible plastic substrate were demonstrated with low-voltage operation and high contrast [117]. A paintable LCD technology, a sequence of coating and UV curing, is under development for making flexible LCDs [118].

1.2.4.2 Electrophoretic Displays

Electrophoretic displays are reflective-type bistable devices. Bistable displays consume power only when they are overwritten. Because they combine low-power consumption with adequate contrast and full viewing angle, they are particularly suitable for electronic books. In electrophoretic displays, the electric field across the electrophoretic material controls the location and orientation of charged objects that are suspended in a liquid. Metcalfe and Wright first used electrophoresis for creating an image in 1956 [119]. In the early 1970s, displays using electrophoresis to create reversible images were demonstrated [120, 121]. These electrophoretic displays suffered from sticking particles, discoloration of dyes, and migration of particles to electrode edges. To limit the lateral migration of the particles, microcellular electrophoretic films were developed in the late 1970s and early 1980s, by using photoresist to create a square grid of confining ribs [122, 123]. In the late 1990s, microencapsulated electrophoretic display films were developed by E-Ink Corporation [124] and by NOK Corporation [125]. Instead of filling the electrophoretic fluid into microcellular structures, the fluid is microencapsulated by means of polycondensation of urea and formaldehyde, which produces discrete, mechanically strong and optically clear, microcapsules. These microcapsules are subsequently spread on ITO-coated polyester foil, and then rear electrodes are printed. Microencapsulation not only solves the problem of particle clustering, agglomeration, and lateral migration, but also maintains the cell gap during deformation. Normally, electrophoretic displays are bicolor (black and white). Schemes for obtaining full color include (1) using three combinations of two-color electrophoretic fluids in a subpixellated scheme to form various combinations of red, green, and blue, (2) combining an external color filter over a black and white electrophoretic film, and (3) stacking three imaging films over each other [126].

The current electrophoretic film produced by E Ink Corporation is a dual-particle formulation. It contains positively charged white particles and negatively charged black particles that are suspended in a clear fluid. It has a black-to-white update time of less than 30 ms at 15 V drive voltage, and gray tones are obtained by partial addressing between black and white states [127]. Active-matrix electrophoretic displays with a-Si:H TFTs on steel foil substrates have been demonstrated by E Ink

[128, 129]. E Ink's microencapsulated electrophoretic films have been integrated with Plastic Logic Limited's active-matrix organic TFT backplane made by solution processing on a plastic substrate. Paper-like displays and prototypes of rollable displays based on E Ink's electrophoretic films have been demonstrated by Philips [130, 17].

1.2.4.3 Organic Light-Emitting Displays

Since the discovery of OLEDs in the late 1980s, the technology has developed rapidly and OLED displays encapsulated in glass are commercially available. OLEDs have wider viewing angles, faster response time, lower voltage operation, and possibly lower power consumption than backlit AMLCDs. Because of their thin-film structure, OLEDs are a natural choice for flexible displays. The two types of OLED materials are small-molecule, with the higher efficiency, and conjugated polymers. Small-molecule OLEDs are usually prepared by thermal evaporation, and polymer OLED by solution processing. The latter enables many potentially inexpensive fabrication steps, such as spin coating, ink-jet printing, and spraying, and is easily compatible with roll-to-roll manufacturing. Therefore, efforts are underway to develop solution-processable precursors for small-molecule OLED fabrication, as mentioned earlier.

Electroluminescence in organic materials was first studied in the 1960s [131–133]. Interest in organic electroluminescence was revived by Tang and VanSlyke's discovery of electroluminescence in small-molecule organic thin-film diodes made by thermal evaporation [134]. Visible light emission from polymer organic light-emitting diodes (PLEDs) reported by Burroughes et al. in 1990 attracted further attention. The PLED was deposited by spin coating a solution-processable precursor polymer, which then was converted to a conjugated polymer by heating to 250°C [135]. A year later, in 1991, Braun and Heeger reported the use of a soluble conjugated polymer that eliminated the need for high-temperature processing [136]. Soon thereafter a flexible OLED display on PET was demonstrated by Gustafsson et al., who employed solution-processable MEH–PPV as the active material and the conductive polymer PANI as the hole-injecting contact [137]. In 1997, a-Si:H TFTs on steel foil were integrated with a printed top-emitting OLED [33]. Many research groups now have demonstrated active-matrix OLEDs on flexible plastic or steel substrates [138–140]. A typical pixel cross section of an active-matrix bottom-emitting OLED on a clear plastic substrate is shown in Fig. 1.2 .

In OLEDs and PLEDs, an $\sim$100-nm-thick organic stack is sandwiched between an anode and a cathode. In bottom-emitting (through-the-substrate) OLEDs, the anode is usually ITO with high work function, and the cathode is a metal with low work function, such as Ca or Mg, which is covered by an interconnect metal like Al. Because top-emitting OLEDs need a transparent cathode, they are coated with a very thin layer of the cathode metal, which is then covered with ITO. Electrons and holes injected under forward bias combine to bound polaron–excitons. These radiatively decay via electron–hole recombination to produce the electroluminescence.

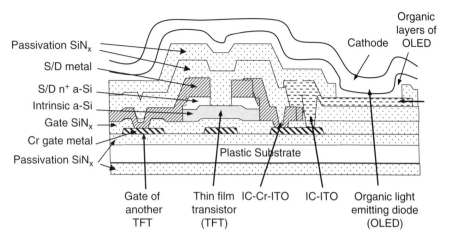

Fig. 1.2 Schematic cross section through an a-Si:H TFT and an OLED on a clear plastic substrate [Dr. Ke Long, Princeton University]

The OLED stack contains charge injection and transport layers that serve to keep the excitons away from nonradiatively recombining at the contacts. The injection and transport layers are also made of organic materials, for example PEDOT/polystyrene sulfonate or PANI [142]. Alternative OLED architectures include inverted top-emitting OLEDs that have the cathode in contact with the substrate, and transparent OLEDs whose anodes and cathodes both are transparent [143, 144]. The biggest challenge to making flexible OLED display is the demand for extremely low rates of permeation of moisture and oxygen to ensure acceptably long OLED lifetime. Oxidation at the organic–cathode interface can inhibit charge injection and results in black spots.

1.2.4.4 Sensors

Flexible sensors have captured the imagination for applications in biomedicine, artificial skin, and wearable electronics. We describe some recent examples. The shear-stress sensor skin by Xu et al. integrates microelectromechanical systems (MEMS) and ICs onto a flexible Parylene (poly-para-xylylene) "skin," which can be affixed to the human body like a Band-Aid [145]. A hand posture and gesture monitoring glove has been made by integrating strain sensors with fabric [146]. The electrical characteristics of resistive, capacitive, or field-effect devices made of conducting or semiconducting polymers change upon gas absorption. This is the operating principle of chemical sensing transistors made on fibers for wearable electronics. [147]. A conformal H_2-sensing skin for the safety monitoring of H_2 fuel tanks is based on a piezoelectrically driven sound resonance cavity. The skin itself consists of a porous PVDF polymer layer that allows leaked H_2 to diffuse freely, and an impermeable silicone rubber that serves as the seal for the porous polymer [148]. Electronic skin developed by Someya and coworkers is made flexible and conformable by a net-like structure that carries thermal and pressure sensors made of organic materials on its filaments. Conductive rubber functions as pressure sensor and organic diodes as thermal sensors [149–151, 40]. A plastic film with pentacene TFT-based electronic circuits is processed to form the net-like structure that allows the large deformation of the electronic skin. For ultra-high flexibility, the OTFTs are embedded at a neutral strain position [152]. A pressure sensor and microphone based on a ferroelectric polymer foil have been combined with an a-Si:H TFT on a PI foil substrate. Depending on the function, the TFT is used as a switch or amplifier [153].

1.2.4.5 Actuators

A soft electrically insulating polymer functions as dielectric in a capacitor with mechanically compliant electrodes. This electroactive polymer is compressed when the electrodes are charged and, by conservation of volume, expands parallel to the planes of the electrodes. Large deformation has been demonstrated in appropriately designed dielectric elastomer actuators [154, 155]. Electroactive polymer actuation has been switched by thin films on polymer foil, pointing to a possible technology for large-area actuation [156]. Potential applications include artificial muscles for biologically inspired robots, animatronics, prosthetics, conformally shaped loudspeakers, and smart skin which can modulate electromagnetic properties, control of airflow and heat transfer, and control of the texture of large surfaces.

1.2.4.6 Electronic Textiles

Low stiffness is inherent in textile fabric. E-textiles, or smart textiles, have electronic circuitry woven into or integrated onto the fabric [157, 158]. A circuit has been woven of yarns that carried a-Si:H TFTs or conductors, or functioned as insulating spacers [159]. Figure 1.3 illustrates the concept of an e-suit. E-textiles have been developed and test-marketed for wireless communication, entertainment, health, and safety. Electrically conducting fibers are a prerequisite. They can be made of wire or by coating fibers with metals, metallic salts, or conductive polymer. Input and output devices, sensors, actuators, and power supplies are integrated by application or weaving. One particular goal is a power supply obtained by integrating photovoltaics into the fabric. Nanotechnology and MEMS are expected to increasingly enable the direct integration of electronic devices. A number of practical issues that include conformal encapsulation of active-device yarn remain to be addressed.

1.2.5 Encapsulation

Thin-film encapsulation has been researched and applied for decades. Single-layer metallic barrier coatings [160–162] were followed by single-layer transparent barrier coatings [163–166] and now multilayer composite (oxide or nitride/polymer) barrier coatings. Since the early 1970s, thin aluminum films evaporated onto polymeric substrates have been applied commercially as gas barriers in food packaging.

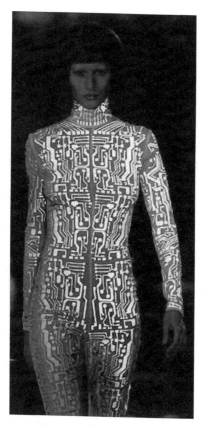

Fig. 1.3 Concept of an e-suit [Givenchy Fall 1999 collection]

Today, metallized plastic foil is widely used for packaging not only of food but also of medical supplies and sensitive electronic components. A 10- to 100-nm-thick aluminum film reduces the permeability by atmospheric gases to 1/1,000 of that of the bare polymeric foil. For microwaveability and product visibility, transparent oxide barrier coatings applied by sputtering or chemical vapor deposition entered development in the 1980s. Today, polymeric substrates are routinely coated with thin layers of brittle oxides, which become flexible when made thin.

The typical inorganic barrier materials such as SiO_2, SiN_x, and Al_2O_3 are highly impermeable to atmospheric gases. If it were perfect, a single thin layer could reduce the gas or vapor permeation to nondetectable values. However, microscopic defects, in particular granular film growth, cracks above occluded dust particles, and pinholes provide diffusion paths for gas and vapor [160, 166]. To slow down the ingress of atmospheric gas molecules, the length of their diffusion path is increased by stacking and alternating thin inorganic barrier layers and thick polymeric layers to a multilayer barrier. The polymeric layers stop the upward propagation of defects in the inorganic barrier layers, and make the composite structure more flexible than a single thick inorganic barrier structure would be. The entire multilayer barrier is typically several micrometers thick. In 1994, aluminum layers as barriers alternating with acrylate polymer layers were found to reduce the oxygen permeation rate through a polypropylene substrate by four orders of magnitude [167]. In 1996, oxygen and water permeation rates below the detection limit of <0.015 cm^3/m^2/day were obtained by using a three-layer barrier structure of 1-μm acrylate/25-nm AlO$_x$/0.24-μm acrylate on a PET substrate [168].

Making impermeable thin-film barrier coatings has become the major challenge to making flexible OLED displays. Polymeric substrates needed a barrier on the substrate side ("passivation") and on top ("encapsulation") of the device to ensure long life. The barrier through which the light is emitted needs to meet the additional criteria of high optical transparency of 85–90% over the visible spectrum, control of optical microcavity effects, and, if on top of the OLED, low stress to avoid shear damage to the OLED, dense and conformal coating to avoid through-layer and edge leakage, and low process temperature. Multilayer organic/inorganic barrier coats have been proven to extend the operating life of OLEDs [48, 169–173]. For the purpose of reducing the cost and barrier complexity, single, dense, inorganic barrier layers of aluminum oxide made by atomic-layer deposition remain under study [174, 175].

An important complement to the development of the barrier technology itself is improved techniques for the evaluation of barrier quality, and with sufficient sensitivity for relevance to OLEDs. Direct identification and characterization of the defects that cause permeation are parts of this task in the development of flexible OLEDs.

1.3 Fabrication Technology for Flexible Electronics

1.3.1 Fabrication on Sheets by Batch Processing

Electronic devices and circuits and display panels are made by batch processing. Flexible foil substrates, cut to sheets, will be the drop-in replacement for the rigid glass plates or silicon wafers.

Flexible sheet substrates may be handled in several ways during processing:

- on a rigid carrier, facing up and loose;
- on a rigid carrier, facing up or down and bonded to the carrier for the duration of processing;
- in a tensioning frame, facing up or down;
- in a frame, facing down and loose;
- electrostatically bonded to a rigid carrier;
- magnetically attached to a rigid carrier.

Rigid substrates are best suited to free standing and loose mounting. The flexibility of the substrate is given by its flexural rigidity $D = Et^3/12(1-\nu^2)$, where E is Young's modulus, t is the thickness of the substrate, and ν is its Poisson ratio.

a-Si TFTs have been made on polymeric substrates held loosely in a frame [54]. The substrate was flattened and temporarily bonded with water to plate glass for photolithography. However, holding the substrate loosely is a technique confined to the laboratory, because device films may crack or the sample may curve from mismatch strain between the deposited films and the substrate. Since varying curvature corresponds to varying size of the flattened sample, any variation of the stress-induced curvature between alignment steps is synonymous with misalignment in mask overlay [45].

Temporarily bonding the foil substrate to a rigid carrier for processing can improve the substrate's dimensional stability. Bonding may be particularly advisable if inorganic device materials are deposited on compliant polymer substrates, because of the large strain this combination may generate. The adhesive must provide sufficient shear strength between the substrate and the carrier, resist the process chemicals, degas little and release few contaminants. At the end of processing it must be removed without damaging the electronics. Thermoplastic adhesives provide the necessary resistance against solvents and can be detached by heating. However, they impose a ceiling on the process temperature that necessarily is lower than the highest working temperature of the substrate. This requirement makes the process window narrow and therefore may degrade device performance. Because the mechanical force needed for debonding may cause damage to the devices and may reduce the yield, special equipment will be needed for debonding in a manufacturing setting.

1.3.2 Fabrication on Web by Roll-to-Roll Processing

Flexible electronics are naturally associated with roll-to-roll processing [174]. Roll-to-roll fabrication of large-area electronics, including solar cells, is desirable for cost reduction. Indeed, amorphous silicon solar cells are manufactured on flexible steel [177] and PI foils [178] by roll-to-roll processes. The solar cells on PI even take advantage of the easy through-substrate connection that can be had by punching holes into the plastic foil substrate. In contrast to solar cell manufacture, making displays and other active electronic circuitry requires a large number of patterning steps. The device layers can be patterned by the additive processes of directly printing the active materials, shadow masking, or subtractive patterning by photolithography. All of these techniques can be adapted to web processing. The big challenge is that backplane circuits need high precision, accuracy, and yield. The roll-to-roll photolithography and etching tools available today are not capable of 2-μm resolution and overlay registration, particularly when combined with the tensioning applied for winding and with process cycles at elevated temperature, both of which cause substrate deformation. The goal of roll-to-roll fabrication of flexible electronics is stimulating innovations in equipment and process design [179], process recipes, and system integration. Tools for roll-to-roll processing that are available today include web cleaner, PECVD, sputtering, plasma etcher, develop etch strip line (for printed circuit boards with high-density interconnects), die punch (which can be modified for control of alignment and winding/unwinding), evaporator (with linear source to improve film uniformity in roll-to-roll applications), laser writer, inkjet printer, screen printer, and inspection devices.

1.3.3 Additive Printing

If flexible electronics could be fabricated by additive printing, their cost could become as low as a few dollars per square meter [180]. Additive printing is roll-to-roll process compatible, is a high-throughput process, uses device materials efficiently, may not require vacuum, and may provide a solution to overlay registration problem through digital compensation. Noble-metal conductors, organic conductors, semiconductors, and insulators can be printed. The printability of organic materials has stimulated experiments on the printing of TFTs [29, 31]. Masks for etching or lift-off patterns, as well as certain inorganic materials [30, 181], can be printed [27, 28]. Printing metallic conductors from nanoparticles may reduce the required sintering temperature to values acceptable for plastic substrates. The printing of high-quality gate dielectrics and the operational stability of printed devices are issues that remain to be resolved.

1.4 Outlook

Most of the present industrial development of flexible electronics is for flexible displays and X-ray sensor arrays. Conformally shaped electronics are perhaps five years away from industrial development, and elastically stretchable electronic surfaces, ten years. Research laboratories are inventing many new flexible technologies that range from elastically conforming sensor surfaces to electronic nets. The task of the researcher is to offer the industrialist a choice of new applications and show the path to them by developing the necessary architecture, circuits, materials, and fabrication technology.

Acknowledgments   The authors gratefully acknowledge support of their research by the United States Army Research Laboratory, the United States Display Consortium, Universal Display Corporation, Hewlett-Packard Laboratories, the DuPont Company, the National Institutes of Health, DARPA, the Eastman Kodak Company, and the New Jersey Commission for Science and Technology.

References

1. Crabb RL, Treble FC (Mar 25, 1967) Thin silicon solar cells for large flexible arrays. Nature 1223–1224
2. Ray KA (Jan, 1967) Flexible solar cell arrays for increased space power. IEEE Trans Aerosp Electron Syst v AES-3, n 107–115
3. Wronski CR, Carlson DE, Daniel RE (1976) Schottky-barrier characteristics of metal-amorphous-silicon diodes. Appl Phys Lett 29:602–604

4. Plattner PD, Kruhler WW, Juergens W, Moller M (1980) '80 Photovoltaic Solar Energy Conf p 121
5. Okaniwa H, Nakatani K, Asano M, Yano M, Hamakawa Y (1982) Production and properties of a-Si:H solar cell on organic polymer film substrate. In: Conference record of the sixteenth IEEE photovoltaic specialists conference – 1982, San Diego, CA, USA, Sep. 27–30, San Diego, CA, USA, 1982, pp 1111–1116
6. Okaniwa H, Nakatani K, Yano M, Asano M, Suzuki K (1982) Preparation and properties of a-Si:H solar cells on organic polymer film substrate. Jpn J Appl Phys 21:239–244
7. Okaniwa H, Nakatani K (1983) Flexible substrate solar cells. In: Hamakawa Y (ed) JARECT vol. 6. Amorphous Semiconductor Technologies & Devices. Ohusha, Tokyo, pp 239–250
8. Russell TWF, Rocheleau RE, Lutz PJ, Brestovansky DF, Baron BN (1982) Properties of continuously-deposited photovoltaic-grade CdS. In: Conference record of the sixteenth IEEE photovoltaic specialists conference – 1982, San Diego, CA, USA, Sep. 27–30, San Diego, CA, USA, pp 743–747
9. Nath P, Izu M (1985) Performance of large area amorphous Si-based single and multiple junction solar cells. In: Rec 18th IEEE Photovoltaic Specialists Conference. Las Vegas, NV, Oct 21–25, p 939
10. Yano M, Suzuki K, Nakatani K, Okaniwa H (1987) Roll-to-roll preparation of a hydrogenated amorphous silicon solar cell on a polymer substrate. Thin Solid Films 146:75–81
11. Brody TP (1984) The thin-film transistor – A late flowering boom. IEEE Trans Electron Devices ED-31:1614–1628
12. Brody TP (1996) The birth and early childhood of active matrix – A personal memoir. J Soc Inf Disp 4(3):113–127
13. Constant A, Burns SG, Shanks H, Gruber C, Landin A, Schmidt D, Thielen C, Olympie F, Schumacher T, Cobbs J (1995) Development of thin film transistor based circuits on flexible polyimide substrates. Electrochem Soc Proc 94-35:392–400
14. Theiss SD, Wagner S (1996) Amorphous silicon thin-film transistors on steel foil substrates. IEEE Electron Device Lett 17:578–580
15. Yang ND, Harkin G, Bunn RM, McCulloch DJ, Wilks RW, Knapp AG (1997) Novel fingerprint scanning arrays using polysilicon TFT's on glass and polymer substrates. IEEE Electron Device Lett 18:19–20
16. Smith PM, Carey PG, Sigmon TW (1997) Excimer laser crystallization and doping of silicon films on plastic substrates. Appl Phys Lett 70:342–344
17. http://www.research.philips.com/newscenter/archive/2005/050902-rolldisp.html
18. http://www.samsung.com/PressCenter/PressRelease/PressRelease.asp?seq=20051128_0000217879
19. http://www.universaldisplay.com/press/press-2006-2-7.htm
20. Inoue S, Utsunomiya S, Saeki T, Shimoda T (2002) Surface-free technology by laser annealing and its application to poly-Si TFT–LCDs on plastic film with integrated drivers. IEEE Trans Electron Devices 49:1353–1360
21. Lee Y, Li H, Fonash SJ (2003) High-performance poly-Si TFTs on plastic substrates using a nano-structured separation layer approach. IEEE Electron Device Lett 24:19–21
22. Asano A, Kinoshita T (2002) Low-temperature polycrystalline-silicon TFT color LCD panel made of plastic substrates. Soc Inf Display Symp Digest Tech Papers 33:1196–1199
23. Berge C, Wagner T, Brendle W, Craff-Castillo C, Schubert MB, Werner JH (2003) Flexible monocrystalline Si films for thin film device from transfer process. Mat Res Soc Symp Proc 769, H. 2.7:53–58
24. Stewart R, Chiang A, Hermanns A, Vicentini F, Jacobsen J, Atherton J, Boiling E, Cuomo F, Drzaic P, Pearson S (2002) Rugged low-cost display systems. Proc SPIE – Int Soc Opt Eng 4712:350–356
25. Khang DY, Jiang H, Huang Y, Rogers JA (2006) A stretchable form of single-crystal silicon for high-performance electronics on rubber substrates. Science 311:208–212

26. Sun Y, Choi WM, Jiang H, Huang YY, Rogers JA (2006) Controlled buckling of semiconductor nanoribbons for stretchable electronics. Nat Nanotechnol 1:201–207

27. Gleskova H, Wagner S, Shen DS (1995) Electrophotographic patterning of thin-film silicon on glass foil. IEEE Electron Device Lett 16:418–420

28. Wong WS, Ready SE, Matusiak R, White SD, Lu JP, Ho J, Street RA (2002) Amorphous silicon thin-film transistors and arrays fabricated by jet printing. Appl Phys Lett 24:610

29. Garnier F, Hajlaouir R, Yassar A, Srivastava P (1994) All-polymer field-effect transistor realized by printing techniques. Science 265:1684–1686

30. Ridley BA, Nivi B, Jacobson JM (1999) All-inorganic field effect transistors fabricated by printing. Science 286:746–749

31. Sirringhaus H, Kawase T, Friend RH, Shimoda T, Inbasekaran M, Wu W, Woo EP (2000) High-resolution inkjet printing of all-polymer transistor circuits. Science 290: 2123–2126

32. Drury CJ, Mutsaers CMJ, Hart CM, Matters M, de Leeuw DM (1998) Low-cost all-polymer integrated circuits. Appl Phys Lett 73:108–110

33. Wu CC, Theiss SD, Gu MH, Lu M, Sturm JC, Wagner S, Forrest SR (1997) Integration of organic LEDs and amorphous Si TFTs onto flexible and lightweight metal foil substrates. IEEE Electron Device Lett 18:609–612

34. Hsu PHI, Huang M, Wanger S, Suo Z, Sturm JC (2000) Plastic deformation of thin foil substrates with amorphous silicon islands into spherical shapes. Mat Res Soc Symp Proc 621:Q8.6

35. Hsu PHI, Bhattacharya R, Gleskova H, Huang M, Suo Z, Wagner S, Sturm JC (2002) Thin film transistor circuits on large-area spherical surfaces. Appl Phys Lett 81: 1723–1725

36. Bhattacharya R, Wagner S, Tung YJ, Esler J, Hack M (2005) Organic LED pixel array on a dome. Proc IEEE 93:1273–1280

37. Bhattacharya R, Wagner S, Tung YJ, Esler J, Hack M (2006) Plastic deformation of a continuous organic light emitting surface. Appl Phys Lett 88:033507-1-3

38. Lacour SP, Wagner S, Huang Z, Suo Z (2003) Stretchable gold conductors on elastomeric substrates. Appl Phys Lett 82:2404–2406

39. Gray DS, Tien J, Chen CS (2004) High-conductivity elastomeric electronics. Adv Mater 16:393–397

40. Someya T, Kato Y, Sekitani T, Iba S, Noguchi Y, Murase Y, Kawaguchi H, Sakurai T (2005) Conformable, flexible, wide-area networks of pressure and thermal sensors with organic transistor active matrixes. Proc Nat Acad Sci USA 102:12321–12325

41. Cannella V, Izu M, Jones S, Wagner S, Cheng IC (Jun, 2005) Flexible stainless-steel substrates. Inf Display24–27

42. Plichta A, Weber A, Habeck A, Glas S (2003) Ultra thin flexible glass substrates. Mat Res Soc Symp Proc 769:H9.1

43. Plichta A, Habeck A, Knoche S, Kruse A, Weber A, Hildebrand N (2005) Chapter 3: Flexible glass substrates. In: Grawford GP (ed) Flexible Flat Panel Displays. Wiley, England, pp 35–55

44. MacDonald WA (2004) Engineered films for display technologies. J Mater Chem 14:4–10

45. Cheng IC, Kattamis A, Long K, Sturm JC, Wagner S (2005) Stress control for overlay registration in a-Si:H TFTs on flexible organic-polymer-foil substrates. J Soc Inf Disp 13(7): 563–568

46. MacDonald BA, Rollins K, MacKerron D, Rakos K, Eveson R, Hashimoto K, Rustin B (2005) Chapter 2: Engineered films for display technologies. In: Grawford GP (ed) Flexible Flat Panel Displays. John, England, pp 11–33

47. http://www.Dupont.com/kapton/products/H-78305.html

48. Lewis JS, Weaver MS (2004) Thin-film permeation-barrier technology for flexible organic light-emitting devices. IEEE J Selected Top Quantum Electron 10:45–57

49. Haruki H, Uchida Y (1983) Stainless steel substrate amorphous silicon solar cell. In: Hamakawa Y (ed) JARECT, Amorphous Semiconductor Technologies & Devices. Ohusha, Tokyo, pp 216–227

50. Afentakis T, Hatalis M, Voutsas AT, Hartzell J (2006) Design and fabrication of high-performance polycrystalline silicon thin-film transistor circuits on flexible steel foils. IEEE Trans Electron Devices 53:815–822

51. Ma EY, Wagner S (1999) Amorphous silicon transistors on ultrathin steel foil substrates. Appl Phys Lett 74:2661–2662

52. Wu M, Bo XZ, Sturm JC, Wagner S (2002) Complementary metal–oxide–semiconductor thin-film transistor circuits from a high-temperature polycrystalline silicon process on steel foil substrates. IEEE Trans Electron Devices 49:1993–2000

53. Snell AJ, Mackenzie KD, Spear WE, LeComber PG, Hughes AJ (1981) Application of amorphous silicon field effect transistors in addressable liquid crystal display panels. Appl Phys 24:357–362

54. Gleskova H, Wagner S, Suo Z (1998) a-Si:H TFTs made on polyimide foil by PE-CVD at 150°C. Proc Mater Res Soc 508:73–78

55. Wu M, Pangal K, Sturm JC, Wagner S (1999) High electron mobility polycrystalline silicon thin-film transistors on steel foil substrates. Appl Phys Lett 75:2244–2246

56. Howell R, Stewart M, Karnik S, Saha S, Hatalis M (2000) Poly-Si thin-film transistors on steel substrates. IEEE Electron Device Lett 21:70–72

57. Sposili RS, Im JS (1996) Sequential lateral solidification of thin silicon films on SiO_2. Appl Phys Lett 69:2864–2866

58. Serikawa T, Omata F (1999) High-mobility poly-Si TFTs fabricated on flexible stainless steel substrates. IEEE Electron Device Lett 20:572–576

59. Afentakis T, Hatalis M, Voutsas T, Hartzell J (2003) High performance polysilicon circuits on thin metal foils. Proc SPIE 5004:122–126

60. Heilmeier GH, Zanoni LA (1964) Surface studies of α-copper phthalocyanine films. J Phys Chem Solids 25:603–611

61. Ebisawa F, Kurokawa T, Nara S (1983) Electrical properties of polyacetylene/polysiloxane interface. J Appl Phys 54:3255–3259

62. Dimitrakopoulos CD, Malenfant PRL (2002) Organic thin film transistors for large area electronics. Adv Mater 2:99–117

63. Lin YY, Gundlach DJ, Nelson S, Jackson TN (1997) Stacked pentacene layer organic thin-film transistors with improved characteristics. IEEE Electron Device Lett 18:606–608

64. Schon JH, Kloc C, Batlogg B (2000) On the intrinsic limits of pentacene field-effect transistors. Org Electron 1:57–64

65. Gelinck GH, Geuns TCT, de Leeuw DM (2000) High-performance all-polymer integrated circuits. Appl Phys Lett 77:1487–1489

66. Kane MG, Campi J, Hammond MS, Cuomo FP, Greening B, Sheraw CD, Nichols JA, Gundlach DJ, Huang JR, Kuo CC, Jia L, Klauk H, Jackson TN (2000) Analog and digital circuits using organic thin-film transistors on polyester substrates. IEEE Electron Device Lett 21:534–536

67. Rogers JA, Bao Z, Dodabalapur A, Makhija A (2000) Organic smart pixels and complementary inverter circuits formed on plastic substrates by casting and rubber stamping. IEEE Electron Device Lett 21:100–103

68. Fix W, Ullmann A, Ficker J, Clemens W (2002) Fast polymer integrated circuits. Appl Phys Lett 81:1735–1737

69. Lee MS, Kang HS, Kang HS, Joo J, Epstein AJ, Lee JY (2005) Flexible all-polymer field effect transistors with optical transparency using electrically conducting polymers. Thin Solid Films 477:169–173

70. Dimitrakopoulos CD, Mascaro DJ (2001) Organic thin-film transistors: A review of recent advances. IBM J Res Dev 45:11–27

71. Park SK, Kim YH, Han JI, Moon DG, Kim WK (2002) High-performance polymer TFTs printed on a plastic substrate. IEEE Trans Electron Devices 49:2008–2015
72. Eder F, Klauk H, Halik M, Zschieschang U, Schmid G, Dehm C (2004) Organic electronics on paper. Appl Phys Lett 84:2673–2675
73. Mach P, Rodriguez SJ, Nortrup R, Wiltzius P, Rogers JA (2001) Monolithically integrated, flexible display of polymer-dispersed liquid crystal driven by rubber-stamped organic thin-film transistors. Appl Phys Lett 78:3592–3594
74. Sherwa CD, Zhou L, Huang JR, Gundlach DJ, Jackson TN, Kane MG, Hill IG, Hammond MS, Campi J, Greening BK, Francl J, West J (2002) Organic thin-film transistor-driven polymer-dispersed liquid crystal displays on flexible polymeric substrates. Appl Phys Lett 80:1088–1090
75. Rogers JA, Bao Z, Baldwin K, Dodabalapur A, Crone B, Raju VR, Kuck V, Katz H, Amundson K, Ewing J, Drzaic P (2001) Paper-like electronic displays: Large-area rubber-stamped plastic sheets of electronics and microencapsulated electrophoretic inks. Proc Nat Acad Sci USA 98:4835–4840
76. Inoue Y, Fujisaki Y, Suzuki T, Tokito S, Kurita T, Mizukami M, Hirohata N, Tada T, Yagyu S (2004) Active-matrix OLED panel driven by organic TFTs. In: Proc Int Display Workshops (IDW), Niigata, Japan, pp 355–358
77. Zhou L, Wanga A, Wu SC, Sun J, Park S, Jackson TN (2006) All-organic active matrix flexible display. Appl Phys Lett 88:083502
78. Mizukami M, Hirohata N, Iseki T, Ohtawara K, Tada T, Yagyu S, Abe T, Suzuki T, Fujisaki Y, Inoue Y, Tokito S, Kurita T (2006) Flexible AM OLED panel driven by bottom-contact OTFTs. IEEE Electron Device Lett 27:249–251
79. Lee S, Koo B, Park JG, Moon H, Hahn J, Kim JM (2006) Development of high-performance organic thin-film transistors for large-area displays. MRS Bull 31:455–459
80. Thomas G (1997) Invisible circuits. Nature 389:907–908
81. Wager JF (2003) Transparent electronics. Science 300:1245–1246
82. Hoffman RL, Norris BJ, Wager JF (2003) ZnO-based transparent thin-film transistors. Appl Phys Lett 82:733–735
83. Seager CH, McIntyre DC, Warren WL, Tuttle BA (1996) Charge trapping and device behavior in ferroelectric memories. Appl Phys Lett 68:2660–2662
84. Lavareda G, de Caryalho CN, Fortunato E, Ramos AR, Alves E, Conde O, Amaral A (2006) Transparent thin film transistors based on indium oxide semiconductor. J Non-Crystalline Solids 352:23–25
85. Prins MWJ, Gross-Holz KO, Muller G, Cillessen JFM, Giesbers JB, Weening RP, Wolf RM (1996) A ferroelectric transparent thin-filn transistor. Appl Phys Lett 68:3650–3652
86. Nomura K, Ohta H, Takagi A, Kamiya T, Hirano M, Hosono H (2004) Room-temperature fabrication of transparent flexible thin-film transistors using amorphous oxide semiconductors. Nature 432:488–492
87. Hur SH, Park OO, Rogers JA (2005) Extreme bendability of single-walled carbon nanotube networks transferred from high-temperature growth substrates to plastic and their use in thin-film transistors. Appl Phys Lett 86:243502
88. Artukovic E, Kaempgen M, Hecht DS, Roth S, Gruner G (2005) Transparent and flexible carbon nanotube transistors. Nano Lett 5:757–760
89. Takenobu T, Takahashi T, Kanbara T, Tsukaqoshi K, Aoyaqi Y, Iwasa Y (2006) High-performance transparent flexible transistors using carbon nanotube films. Appl Phys Lett 88:33511
90. Ginley DS, Bright C (2000) Transparent conducting oxides. MRS Bull 25:15–18
91. Lewis BG, Paine DC (2000) Applications and processing of transparent conducting oxides. MRS Bull 25:22–27
92. Paine DC, Yeom HY, Yaglioglu B (2005) Chapter 5: Transparent conducting oxide materials and technology. In: Grawford GP (ed) Flexible Flat Panel Displays. Wiley, England, pp 80–98

93. Baedeker K (1907) Über die elektrische Leitfähigkeit und die thermoelektrische Kraft einiger Schwermetallverbindungen. Ann Phys 22:749–766
94. Cairns DR, Witte RP II, Sparacin DK, Sachsman SM, Paine DC, Crawford GP, Newton RR (2000) Strain-dependent electrical resistance of tin-doped indium oxide on polymer substrates. Appl Phys Lett 76:1425–1427
95. Bouten PCP, Slikkerveer PJ, Leterrier Y (2005) Chapter 6: Mechanics of ITO on plastic substrates for flexible displays. In: Grawford GP (ed) Flexible Flat Panel Displays. Wiley, England, p 117
96. "Bert" Groenendaal L (2005) Chapter 8: Conductive polymers. In: Grawford GP (ed) Flexible Flat Panel Displays. Wiley, England, p 157
97. Rubner M, Lee K, Tripathy S, Morris P, Georger J Jr, Jopson H (1984) Electrically conductive polyacetylene/elastomer blends. Mol Crystals Liquid Crystals 106:408
98. Chiang LY, Wang LY, Kuo CS, Lin JG, Huang CY (1997) Synthesis of novel conducting elastomers as polyaniline-interpenetrated networks of fullerenol-polyurethanes. Synth Met 84:721–724
99. EI-Tantawy F (2005) Development of novel functional conducting elastomer blends containing butyl rubber and low-density polyethylene for current switching, temperature sensor, and EMI shielding effectiveness applications. J Appl Polym Sci 97:1125–1138
100. Xie J, Pecht M, DeDonato D, Hassanzadeh A (2001) An investigation of the mechanical behavior of conductive elastomer interconnects. Microelectron Reliab 41:281–286
101. Tamai T (1982) Electrical properties of conductive elastomer as electrical contact material. IEEE Trans Compon Hybrids Manuf Technol 5:56–61
102. Lanotte L, Ausanio G, Barone AC, Campana C, Lannotti V, Luponio C, Pepe GP (2006) Giant resistivity change induced by strain in a composite of conducting particles in an elastomer matrix. Sens Actuators A (Phys) 127:56–62
103. Gary DS, Tien J, Chen CS (2004) High-conductivity elastomeric electronics. Adv Mater 16:393–397
104. Lacour SP, Tsay C, Wagner S (2004) An elastically stretchable TFT circuit. IEEE Electron Device Lett 25:792–794
105. Lacour SP, Wagner S (2005) Thin film transistor circuits integrated onto elastomeric substrates for elastically stretchable electronics. IEEE IEDM 2005 Tech. Digest, IEEE, New York, paper 5.2
106. Crawford GP (2005) Chapter 16: Encapsulated liquid crystal materials for flexible display applications. In: Grawford GP (ed) Flexible Flat Panel Displays. Wiley, England, pp 313–330
107. Doane JW, Vaz NA, Wu BG, Zumer S (1986) Field controlled light scattering from nematic microdroplets. Appl Phys Lett 48:269–271
108. Fergason JL (1985) Polymer encapsulated nematic liquid crystals for display and light control applications. SID Digest Tech Papers XVI:68–71
109. Drzaic PS (1986) Polymer-dispersed nematic liquid crystal for large area displays and light valves. J Appl Phys 60:2142–2148
110. Doane JW (1990) Chapter 14: Polymer dispersed liquid crystal displays. In: Bahadur B (ed) Chapter 14, Liquid Crystals: Applications and Uses. World Scientific, Singapore, pp 361–395
111. Penz PA, Surtani K, Wen W, Johnson MR, Kane D, Sanders L, Culley B, Fish J (1981) Plastic substrate LCD. Proc SID XII:116–117
112. Takahashi S, Shimokawa O, Inoue H, Uehara K, Hirose T, Kikuyama A (1981) A liquid crystal display panel using polymer films. SID Digest Papers XII:86–87
113. Okada M, Hatano T, Hashimoto K (1997) Reflective multicolor display using cholesteric liquid crystals. SID Digest 28:1019–1022
114. Hashimoto K, Okada M, Nishiguchi K, Masazumi N, Yamakawa E, Taniguchi T (1998) Reflective color display using cholesteric liquid crystals. SID Digest 29:897–900

115. Khan A, Huang XY, Doane JW (2004) Low power cholesteric LCD and electronic book. Proceedings of the SPIE Defence and Securities Symposium, Orlando, FL
116. Stephenson S (2004) Development of flexible displays using photographic technology. SID Digest 35:774–777
117. Kim J, Vorflusev V, Kumar S (1999) Flexible display prepared using phase separated composite organic films of liquid crystals. Proc SID 30:880–885
118. Vogels JPA, Klink SI, Penterman R, de Koning H, Huitema EEA, Broer DJ (2004) Robust flexible LCDs with paintable technology. J SID 12:411–416
119. Metcalfe KA, Wright RJ (1956) Fine grain development in xerography. J Sci Instrum 33:194–195
120. Evans PF, Lees H, Maltz M, Dailey J (1971) Color display devices. US Patent 3,612,758
121. Ota I, Ohnishi J, Yoshiyama M (1973) Electrophoretic image display panel. Proc IEEE 61:832–836
122. Hopper MA, Novotny V (1979) An electrophoretic display, its properties, model and addressing. IEEE Trans Electron Devices 26:1148–1151
123. Blazo SF (1982) High resolution electrophoretic display with photoconductor addressing. SID Digest 82:92–93
124. Comiskey B, Albert JD, Yoshizawa H, Jacobson J (1998) An electrophoretic ink for all-printed reflective electronic displays. Nature 394:253–255
125. Kawai H, Kanae N (1999) Microencapsulated electrophoretic rewritable sheet. SID Digest 30:1102–1105
126. Amundson K (2005) Chapter 19: Electrophoretic imaging films for electronic paper display. In: Grawford GP (ed) Flexible Flat Panel Displays Wiley, England, pp 381–382
127. Whitesides T, Walls M, Paolini R, Sohn S, Gates H, McCreary M, Jacobson J (2004) Towards video-rate microencapsulated dual-particle electrophoretic displays. SID Symp Digest Tech Papers 35:133–135
128. Chen Y, Au J, Kazlas P, Ritenour A, Gates H, Coodman J (2002) Ultra-thin, high-resolution, flexible electronic ink displays addressed by a-Si active-matrix TFT backplanes on stainless steel foil. In: Technical Digest of IEDM, pp 389–392
129. Chen Y, Au J, Kazlas P, Ritenour A, Gates H, McCreary M (2003) Flexible actrive-matrix electronic ink display. Nature 423:136
130. www.research.philips.com/technologies/display/ovˈelpap.html
131. Pope M, Kallmann HP, Magnante P (1963) Electroluminescence in organic crystals. J Chem Phys 38:2042–2043
132. Mehl W, Buchner W (1965) Durch elektrochemische Doppelinjektion angeregte Elektrolumineszenz in Anthracen-Kristallen. Z. Krist Phys Chem 47:76
133. Helfrich W, Schneider WG (1965) Recombination radiation in anthracene crystals. Phys Rev Lett 14:229–231
134. Tang CW, VanSlyke SA (1987) Organic electroluminescent diodes. Appl Phys Lett 51:913–915
135. Burroughes JH, Bradley DDC, Brown AR, Marks RN, Mackay K, Friend RH, Burns PL, Holmes AB (1990) Light-emitting diodes based on conjugated polymers. Nature 347:539–541
136. Braun D, Heeger AJ (1991) Visible-light emission from semiconducting polymer diodes. Appl Phys Lett 58:1982–1984
137. Gustafsson G, Cao Y, Treacy GM, Klavetter F, Colaneri N, Heeger AJ (1992) Flexible light-emitting diodes made from soluble conducting polymers. Nature 357:477–479
138. Sarma KR, Schmidt J, Roush J, Chanley C, Dodd SR (2004) AMOLED using a-Si TFT backplane on flexible plastic substrate. Proc SPIE – Inter Soc Optical Eng 5443:165–176
139. Chuang TK, Roudbari AJ, Troccoli MN, Chang YL, Reed G, Hatalis M, Spirko J, Klier K, Preis S, Pearson R, Najafov H, Biaggio I, Afentakis T, Voutsas A, Forsythe E, Shi J, Blomquist S (2005) Active-matrix organic light-emitting displays on flexible metal foils. Proc SPIE – Inter Soc Optical Eng 5801:234–248

140. Chwang AB, Hack M, Brown JJ (2005) Flexible OLED display development: Strategy and status. J SID 13(6):481–486

141. Long K (2005) Flexible full-color active matrix organic light-emitting displays: Dry dye printing for OLED integration and 280°C amorphous-silicon thin-film transistors on clear plastic substrates. Princeton University Ph. D. Thesis

142. Hildner ML (2005) Chapter 15: OLED displays on plastic. In: Grawford GP (ed) Flexible Flat Panel Displays. Wiley, England, pp 285–312

143. Bulovic V, Gu G, Burrows PE, Forrest SR, Thompson ME (1996) Transparent light-emitting devices. Nature 380:29

144. Lee KH, Ryu SY, Kwon JH, Kim SW, Chung HK (2003) QCIF full color transparent AMOLED display. Proc Soc Inform Display Symp Dig Tech Papers 34:104–107

145. Xu Y, Tai YC, Huang A, Ho CM (2002) IC-integrated flexible shear-stress sensor skin. Solid-State Sensor, Actuator and Microsystems Workshop, Hilton Head Island, SC, pp 354–357

146. Lorussi F, Scilingo EP, Tesconi M, Tognetti A, Rossi DD (2005) Strain sensing fabric for hand posture and gesture monitoring. IEEE Trans Inf Technol Biomed 9:372–381

147. Collins GP (Aug, 2004) Next stretch for plastic electronics. Sci Am 291(2):76–81

148. Dong S, Cao H, Bai F, Yan L, Li JF, Viehland D (2004) Conformal sensor skin approach to the safety-monitoring of H_2 fuel tanks. Appl Phys Lett 84:4153–4154

149. Someya T, Sakurai T (2003) Integration of organic field-effect transistors and rubbery pressure sensors for artificial skin applications. 2003. IEEE International Electron Devices Meeting (IEDM), 8.4, Washington DC, Dec 8–10, pp 203–206

150. Someya T, Sekitani T, Iba S, Kato Y, Kawaguchi H, Sakurai T (2004) A large-area, flexible pressure sensor matrix with organic field-effect transistors for artificial skin applications. Proc Nat Acad Sci USA 101(27):9966–9970

151. Someya T, Kawaguchi H, Sakurai T (2004) Cut-and-Paste Organic FET Customized ICs for Application to Artificial Skin. 2004. IEEE International Solid-State Circuits Conference (ISSCC 2004), 16.2, San Francisco Marriott, San Francisco, CA, Feb 14–19, pp 288–289

152. Sekitani T, Iba S, Kato Y, Noguchi Y, Someya T, Sakurai T (2005) Ultraflexible organic field-effect transistors embedded at a neutral strain position. Appl Phys Lett 87:173502

153. Graz I, Keplinger Ch, Schwödiauer R, Bauer S, Lacour SP, Wagner S (2006) Flexible ferroelectret field-effect transistor for large-area sensor skins and microphones. Appl Phys Lett 88:073501-1-3

154. Pelrine R, Sommer-Larsen P, Kornbluh R, Heydt R, Kofod G, Pei Q, Gravesen P (2001) Applications of dielectric elastomer actuators. Proc SPIE 4329:335–349

155. Ashley S (Oct, 2003) Artificial muscles. Sci Am 289:52–59

156. Lacour SP, Prahlad H, Pelrine R, Wagner S (2004) Mechatronic system of dielectric elastomer actuators addressed by thin film photoconductors on plastic. Sensors Actuators A Phys 111:288–292

157. Moeli D, May-Plumlee T (2002) Interactive electronic textile development. J Textile Apparel, Technol Manage 2:1–12

158. Gould P (Oct, 2003) Textiles gain intelligence. Mater Today 10:38–43

159. Bonderover E, Wagner S (2004) A woven inverter circuit for e-textile applications. IEEE Electron Device Lett 25:295–297

160. Prins W, Hermans J (1959) Theory of permeation through metal coated polymer films. J Phys Chem 63:716–719

161. Jamieson EHH, Windle AH (1983) Structure and oxygen-barrier properties of metallized polymer film. J Mater Sci 18:64–80

162. Weiss J, Leppin C, Mader W, Salzberger U (1989) Aluminum metallization of polyester and polypropylene films: Properties and transmission electron microscopy microstructure investigations. Thin Solid Film 174:155

163. Klemberg-Sapiepha JE, Martinu L, Kuttel OM, Wertheimer M (1993) Transparent gas barrier coatings by dual-frequency PECVD. 36th Annual Technical Conference Proceedings of the Society of Vacuum Coaters, p 445

164. Philips RW, Markantes TM, LeGallee C (1993) Evaporated dielectric colorless films on PET and OPP exhibiting high barriers towards moisture and oxygen. 36th Annual Technical Conference Proceedings of the Society of Vacuum Coaters, pp 293–301

165. Brody AL (Feb, 1994) Glass-coated flexible films for packaging: An overview. Packaging Technol Eng 44

166. Chatham H (1996) Review oxygen diffusion barrier properties of transparent oxide coatings on polymeric substrates. Surf Coatings Technol 78:1–9

167. Shaw DG, Langlois MG (1994) Use of vapor deposited acrylate coatings to improve the barrier properties of metallized film. 37th Annual Technical Conference Proceedings of the Society of Vaccum Coaters, p 240

168. Affinito JD, Gross ME, Coronado CA, Graff GL, Greenwell EN, Martin PM (1996) A new method for fabricating transparent barrier layers. Thin Solid Films 290/291:63–67

169. Weaver MS, Michalski LA, Rajan K, Rothman MA, Silvernail JA, Burrows PE, Graff GL, Gross ME,. Martin PE, Hall M, Mast E, Bonham C, Bennett W, Zumhoff M (2002) Organic light-emitting devices with extended operating lifetimes on plastic substrates. Appl Phys Lett 81:2929–2931

170. Chwang AB, Rothman MA, Mao SY, Hewitt RH, Weaver MS, Silvernail JA, Rajan K, Hack M, Brown JJ, Chu X, Moro L, Krajewski T, Rutherford N (2003) Thin film encapsulated flexible OLED display. Appl Phys Lett 83:413–415

171. Yoshida A, Fujimura S, Miyake T, Yoshizawa T, Ochi H, Sugimoto A, Kubota H, Miyadera T, Ishizuka S, Tsuchida M, Nakada H (2003) 3-inch Full-color OLED display using a plastic substrate. Proc Soc Inform Display Symp Dig Tech Papers 34:856–859

172. Graff GL, Williford RE, Burrows PE (2004) Mechanisms of vapor permeation through multilayer barrier films: Lag time versus equilibrium permeation. J Appl Phys 96:1840–1849

173. Kim TW, Yan M, Erlat AG, McConnelee PA, Pellow M, Deluca J, Feost TP, Duggal AR, Schaepkens M (2005) Transparent hybrid inorganic/organic barrier coatings for plastic organic light-emitting diode substrates. J Vac Sci Technol A 23:971–977

174. Ghosh AP, Gerenser LJ, Jarman CM, Fornalik JE (2005) Thin-film encapsulation of organic light-emitting devices. Appl Phys Lett 86:223503

175. Carcia PF, McLean RS, Reilly MH, Groner MD, George SM (2006) Ca test of Al_2O_3 gas diffusion barriers grown by atomic layer deposition on polymers. Appl Phys Lett 89:031915

176. Crawford GP (2005) Chapter 21: Roll-to-roll manufacturing of flexible displays. In: Crawford GP (ed) Flexible Flat Panel Displays. Wiley, England, pp 409–445

177. Yang J, Banerjee A, Guha S (2003) Amorphous silicon based photovoltaics – From earth to the "final frontier". Solar Energy Mater Solar Cells 78:597–612

178. Takano A, Tabuchi K, Uno M, Tanda M, Wada T, Shimosawa M, Sakakibara Y, Kiyofuji S, Nishihara H, Enomoto H, Kamoshita T (2006) Production technologies of film solar cell. Mater Res Soc Symp Proc 910:0910-A25-04

179. Mei P, Jackson WB, Taussig CP, Jeans A (Jun 6, 2006) Forming a plurality of thin-film devices using imprint lithography. US Patent 7056834

180. Wagner S, Gleskova H, Sturm JC, Suo Z (2000) Novel processing technology for macroelectronics. In: Street RA (ed) Technology and Applications of Hydrogenated Amorphous Silicon. Springer, Berlin, pp 222–251

181. Shimoda T, Matsuki Y, Furusawa M, Aoki T, Yudasaka I, Tanaka H, Iwasawa H, Wang D, Miyasaka M, Takeuchi Y (2006) Solution-processed silicon films and transistors. Nature 440:783–786

Chapter 2
Mechanical Theory of the Film-on-Substrate-Foil Structure: Curvature and Overlay Alignment in Amorphous Silicon Thin-Film Devices Fabricated on Free-Standing Foil Substrates

Helena Gleskova, I-Chun Cheng, Sigurd Wagner, and Zhigang Suo

Abstract   Flexible electronics will have inorganic devices grown at elevated temperatures on free-standing foil substrates. The thermal contraction mismatch between the substrate and the deposited device films, and the built-in stresses in these films, cause curving and a change in the in-plane dimensions of the workpiece. This change causes misalignment between the device layers. The thinner and more compliant the substrate, the larger the curvature and the misalignment. We model this situation with the theory of a bimetallic strip, which suggests that the misalignment can be minimized by tailoring the built-in stress introduced during film growth. Amorphous silicon thin-film transistors (a-Si:H TFTs) fabricated on stainless steel or polyimide (PI) (Kapton E®) foils need tensile built-in stress to compensate for the differential thermal contraction between the silicon films and the substrate. Experiments show that by varying the built-in stress in just one device layer, the gate silicon nitride (SiN_x), one can reduce the misalignment between the source/drain and the gate levels from $\sim$400 parts-per-million to $\sim$100 parts-per-million.

2.1 Introduction

Flexible displays built on metallic or plastic foil substrates are becoming a reality. Many flat panel display companies around the world have manufactured flexible display prototypes using a variety of thin-film technologies [1].

The inorganic device films are typically grown at an elevated temperature while the substrate is held flat. Upon cooling, a stress field arises due to the difference

H. Gleskova (✉)

Department of Electronic and Electrical Engineering, University of Strathclyde, Royal College Building, Room 3.06A, 204 George Street, Glasgow, G1 1XW, UK

e-mail: helena.gleskova@eee.strath.ac.uk

W.S. Wong, A. Salleo (eds.), *Flexible Electronics: Materials and Applications*, Electronic   29
Materials: Science & Technology, DOI 10.1007/978-0-387-74363-9_2,
© Springer Science+Business Media, LLC 2009

in thermal contraction between the film and the substrate. If the substrate is thick and stiff, for example a silicon wafer or a plate of glass, the film-on-plate structure remains almost flat (it forms a spherical cap with a very large radius of curvature), even when the stress is large. The stress in the film is much bigger than in the substrate. If the stress in the film becomes too large or the bond between the film and the substrate is weak, the film may crack or peel off the substrate [2].

When the substrate's thickness is reduced, its strength, the product of thickness with Young's modulus, may become comparable to that of the deposited device film. Upon cooling, such a film-on-substrate-foil structure rolls to a cylinder with the film facing outward (film in compression) or inward (film in tension). When flattened for circuit fabrication, the workpiece has a different size from before film deposition. This change in size leads to an error in mask overlay alignment.

In flexible electronics, the rigid plate glass substrate of current liquid crystal displays is replaced with plastic or metallic foils. Among metallic materials, stainless steel, molybdenum, and aluminum foils have been utilized as substrates in the fabrication of thin-film transistors and solar cells. A number of plastic (organic polymers) substrates have also been tested successfully in a variety of thin-film applications. Table 2.1 lists the thermo-mechanical properties of the common foil substrates and of silicon nitride (SiN_x) and hydrogenated amorphous silicon (a-Si:H) deposited by plasma enhanced chemical vapor deposition (PECVD).

Foil substrates with thicknesses ranging from a few to several hundred micrometers were either free-standing during the whole fabrication process, or temporarily

Table 2.1 Thermo-mechanical properties of substrates used for flexible electronics, and of some TFT materials

Material	Brand name	Upper working temperature (°C)	Young's modulus (GPa)	CTE (K^{-1})	Source
Stainless steel	AISI 304	~1,400	190–210	18×10^{-6}	Goodfellow
Polyethylene naphthalate (PEN)	Kaladex, Kalidar	155	5–5.5 (biaxial)	21×10^{-6}	Goodfellow
Polyethylene terephthalate (PET)	Arnite, Dacron, Mylar, etc.	115–170	2–4	30–65×10^{-6}	Goodfellow
Polyimide (PI)	Kinel, Upilex, Upimol, etc.	250–320	2–3	30–60×10^{-6}	Goodfellow
Polyimide (PI)	Kapton E®	>300	5.2	16×10^{-6}	DuPont
PECVD silicon nitride	–	–	183	2.7×10^{-6}	[3, 4]
PECVD amorphous silicon	–	–	140	3.0×10^{-6}	[5]

attached to a carrier during the fabrication and separated after the fabrication was completed. Each approach faces different challenges. We focus on the mechanics of thin-film devices fabricated on free-standing foils because of our experience with them, and because of their relevance to roll-to-roll processing. The simple thermo-mechanical theory introduced in this chapter addresses two important issues that arise when the device films are deposited on the foil substrates at an elevated temperature: (1) the change in the dimensions of the flattened workpiece and (2) the curving of the workpiece.

Mechanical strain is defined as a relative change in dimensions. Therefore, the terms *change in dimensions* and *strain* are interchangeable and they both require a reference point. For any given temperature, the substrate and the film have different reference points: they are their respective in-plane dimensions, were they relaxed instead of bonded together. For example in Fig. 2.1 , $\varepsilon_s(T_d) < 0$, $\varepsilon_f(T_d) > 0$,

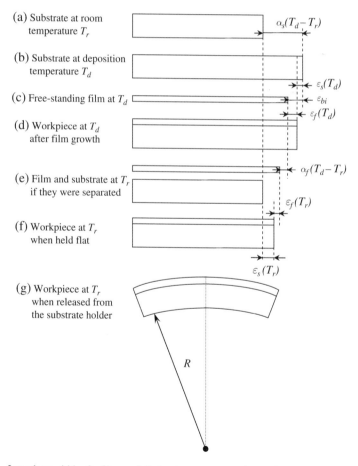

Fig. 2.1 Length or width of a film-on-foil structure at room and growth temperatures

$\varepsilon_s(T_r) > 0$, and $\varepsilon_f(T_r) < 0$, where the subscripts s and f are for the substrate and film, respectively, and T_d and T_r are for deposition and room temperatures, respectively.

Figure 2.1d represents a film grown on a free-standing foil substrate at an elevated temperature. The substrate is held flat during film growth. It expands when heated to the deposition temperature but remains stress-free (Fig. 2.1b). The growing film may develop a built-in strain ε_{bi} (Fig. 2.1c) that causes a built-in stress σ_{bi}. The built-in strain arises from atoms deposited in out-of-equilibrium positions. If the substrate is not constrained, the built-in strain ε_{bi} in the film forces the substrate to shrink or expand. Figure 2.1d depicts film with a built-in tensile stress ($\sigma_{bi} > 0$), which forces the substrate into compression ($\varepsilon_s(T_d) < 0$). Next, the unconstrained workpiece is cooled to room temperature. Since the coefficient of thermal expansion (CTE) of the device film, α_f, is typically much smaller than that of the substrate, α_s, the substrate wants to shrink more than the film, forcing the workpiece to roll into a cylinder with the film on the outside (Fig. 2.1g). When flattened for photolithographic alignment, the film and the substrate have the same in-plane dimensions. These dimensions are, however, different from those that the substrate and the film would have, were they not bonded together. Figure 2.1f shows that the substrate is in tension ($\varepsilon_s(T_r) > 0$), while the film is in compression ($\varepsilon_f(T_r) < 0$). The cause of alignment error, $\varepsilon_s(T_r)$, and the radius of curvature, R, of the workpiece are correlated.

For this chapter, we adapt in Section 2.2 the theory of a bimetallic strip to calculate the strains in the film and the substrate during film growth and after cooling to room temperature. The formulae we develop can be used for any combination of film/substrate materials. In Section 2.3, we provide a specific, quantitative guide to controlling $\varepsilon_s(T_r)$ and R in amorphous silicon thin-film transistors (a-Si:H TFTs) fabricated on free-standing foils of stainless steel and the polyimide (PI), Kapton E.

2.2 Theory

Several recent papers have adapted the classical theory of bimetallic strips to integrated circuits on crystalline substrates [6]. We focus on aspects specific to the film-on-foil structure [7]. The stress field due to misfit strains is biaxial in the plane of the film and the substrate. A small, stiff wafer bends into a spherical cap with an equal and biaxial curvature. However, a film-on-foil structure bends into a cylindrical roll. The transition from cap to roll as the substrate becomes thinner and larger has been studied extensively [6]. Foil substrates are on the roll side well away from the cap-to-roll transition point. They bend into a cylindrical shape, as shown in Fig. 2.2.

The strain in the axial direction, ε_A, is independent of the position throughout the sheet – a condition known as generalized plane strain. Let z be the through-thickness coordinate, whose origin is placed arbitrarily. The geometry dictates that the strain in the bending direction, ε_B, be linear in z, namely:

$$\varepsilon_B = \varepsilon_0 + \frac{z}{R} \tag{2.1}$$

where ε_0 is the strain at $z = 0$.

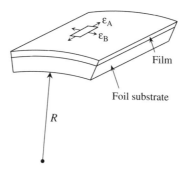

Fig. 2.2 A film-on-foil structure bends into a cylindrical roll [7]. This illustration corresponds to Fig. 2.1(**g**)

If the film and the substrate were separated, they would strain by different amounts but develop no stress (Fig. 2.1e). Let e be the strain developed in an unconstrained stress-free material. For example, thermal expansion produces a strain $e = \alpha \, \Delta T$, where α is the CTE, and ΔT, the temperature change (Fig. 2.1b). One may also include in e the built-in strain ε_{bi} (Fig. 2.1c). Now, if the film and the substrate are bonded to each other and cannot slide relative to each other, a stress field arises (Fig. 2.1d, f). The two stress components in the axial and the bending directions, σ_A and σ_B, are both functions of z (Figs. 2.1g and 2.2). Each layer of the material is taken to be an isotropic elastic solid with Young's modulus Y and Poisson ratio v, and obeys Hooke's law $\sigma = Y\varepsilon$. (Uniaxial tensile stress along the x-axis of a freestanding sample causes stretch along the x-direction coupled with shrinkage along the unconstrained y and z directions, thus $v = -\varepsilon_y/\varepsilon_x = -\varepsilon_z/\varepsilon_x$.) The film and the substrate are dissimilar materials: e, Y, and v are known functions of z. Hooke's law relates the strains to the stresses as:

$$\varepsilon_A = \frac{\sigma_A}{Y} - v\frac{\sigma_B}{Y} + e \tag{2.2a}$$

$$\varepsilon_B = \frac{\sigma_B}{Y} - v\frac{\sigma_A}{Y} + e \tag{2.2b}$$

The first term on the right-hand side is the strain caused by stress applied in the same direction in which the strains ε_A and ε_B are measured, the second term is the strain due to stress applied in the perpendicular in-plane direction, and e is the strain developed in a stress-free material. Any changes in z-direction are neglected. Equation (2.2) can be modified to:

$$\sigma_A = \frac{Y}{1-v}\left(\frac{\varepsilon_A + \varepsilon_B}{2} - e\right) + \frac{Y}{1+v}\left(\frac{\varepsilon_A - \varepsilon_B}{2}\right) \tag{2.3a}$$

$$\sigma_B = \frac{Y}{1-v}\left(\frac{\varepsilon_A + \varepsilon_B}{2} - e\right) - \frac{Y}{1+v}\left(\frac{\varepsilon_A - \varepsilon_B}{2}\right) \tag{2.3b}$$

During film growth, the substrate is held flat but it is let loose upon cooling. No external forces are applied and the substrate is allowed to bend. Force balance requires that:

$$\int \sigma_A dz = 0, \quad \int \sigma_B dz = 0, \quad \int \sigma_B z dz = 0. \tag{2.4}$$

By inserting Eqs. (2.1) and (2.3) into Eq. (2.4) and integrating, we obtain three linear algebraic equations for three constants ε_A, ε_0, and R. The procedure outlined here is applicable to any number of layers and arbitrary functions $e(z)$, $Y(z)$, and $v(z)$.

When the film and the substrate are bonded together and are forced to be flat, the strains in the substrate and the film are equal at all times, and $\varepsilon_A = \varepsilon_B = \varepsilon$ (see Fig. 2.3). The corresponding stress components are also equal, $\sigma_A = \sigma_B$, but the stress in the substrate, σ_s, is different from that in the film, σ_f. In this case, Eq. (2.2) lead to:

$$\varepsilon = \frac{(1 - v_s) \cdot \sigma_s}{Y_s} + e_s = \frac{\sigma_s}{Y_s^*} + e_s \tag{2.5a}$$

$$\varepsilon = \frac{(1 - v_f) \cdot \sigma_f}{Y_f} + e_f = \frac{\sigma_f}{Y_f^*} + e_f \tag{2.5b}$$

where ε, σ, v, and Y are the strain, stress, Poisson ratio, and Young's modulus, respectively. $Y_f^* = \frac{Y_f}{1-v_f}$ and $Y_s^* = \frac{Y_s}{1-v_s}$ are the biaxial strain moduli of the film and the substrate, respectively.

For a flattened workpiece with no external in-plain force applied, Equations (2.4) lead to:

$$\sigma_f d_f + \sigma_s d_s = 0 \tag{2.6}$$

Equations (2.5) and (2.6) lead to an expression for the strain ε of the workpiece at an arbitrary temperature T:

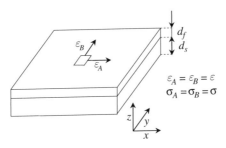

$$\varepsilon_A = \varepsilon_B = \varepsilon$$
$$\sigma_A = \sigma_B = \sigma$$

Fig. 2.3 Flattened film-on foil structure. This illustration corresponds to Fig. 2.1(**d**) and (**f**)

$$\varepsilon(T) = \frac{e_s(T) + \frac{Y_f^* d_f}{Y_s^* d_s} \cdot e_f(T)}{1 + \frac{Y_f^* d_f}{Y_s^* d_s}} \tag{2.7}$$

where $e_f = \alpha_f (T - T_d) + \varepsilon_{bi}$ and $e_s = \alpha_s (T - T_d)$. T_d is the deposition temperature, and α_f and α_s are the CTEs of the film and the substrate, respectively.

The in-plane strain in the substrate coated with a film (strain in the substrate of a flattened workpiece) with respect to the uncoated substrate at the same temperature, at any temperature T, can be expressed as:

$$\varepsilon_s(T) = \varepsilon(T) - e_s(T) = \frac{\left[(T_d - T)(\alpha_s - \alpha_f) + \varepsilon_{bi}\right]}{1 + \frac{Y_s^* d_s}{Y_f^* d_f}} \tag{2.8a}$$

Similarly, the in-plane strain in the film (strain in the film of a flattened workpiece) with respect to the free-standing film at the same temperature, at any temperature T, can be expressed as:

$$\varepsilon_f(T) = \varepsilon(T) - e_f(T) = -\frac{\left[(T_d - T)(\alpha_s - \alpha_f) + \varepsilon_{bi}\right]}{1 + \frac{Y_f^* d_f}{Y_s^* d_s}} \tag{2.8b}$$

2.2.1 The Built-in Strain ε_{bi}

The built-in strain ε_{bi} that develops in the film during its deposition manifests itself by the built-in stress σ_{bi}. At $T = T_d$, the only stress in the film is the built-in stress σ_{bi} and Eqs. (2.5) and (2.6) lead to:

$$\varepsilon_{bi} = -\frac{\sigma_{bi}}{Y_f^*} \cdot \left(1 + \frac{Y_f^* d_f}{Y_s^* d_s}\right) \tag{2.9}$$

The minus sign indicates that, at the deposition temperature, a film with built-in tensile stress ($\sigma_{bi} > 0$) forces the substrate to shrink, while a film with built-in compressive stress ($\sigma_{bi} < 0$) forces the substrate to expand. (If the deposition is done at room temperature, the film with built-in tensile stress forces the substrate into compression. If released, it will curve such that the film lies on the concave side.) The term in the parentheses is the modification for a compliant substrate: when the film strength, $Y_f^* d_f$, becomes comparable to the substrate strength, $Y_s^* d_s$, the substrate changes its dimension in response to film deposition. For stiff substrates when $Y_f^* d_f \ll Y_s^* d_s$, this term equals 1.

Both ε_{bi} and σ_{bi} can be complicated functions of the film's through-thickness coordinate z. Typically, neither $\varepsilon_{bi}(z)$ nor $\sigma_{bi}(z)$ is known. Therefore, an average strain value $\overline{\varepsilon_{bi}}$ is extracted from experiments for a given film/substrate couple, for example from the radius of curvature of the relaxed workpiece [8].

For the remainder of this chapter, we will use the built-in strain ε_{bi}. Remember that, following Eq. (2.9), the built-in strain is *negative if tensile built-in stress* is grown into the film, and it is *positive for compressive built-in stress*.

2.3 Applications

2.3.1 Strain in the Substrate, $\varepsilon_s(T_d)$, and the Film, $\varepsilon_f(T_d)$, at the Deposition Temperature T_d

A growing film typically builds in strain ε_{bi}. This built-in strain arises from atoms deposited in out-of-equilibrium positions. If the substrate is thin and/or compliant and unconstrained, the growing film will relieve some of its built-in strain/stress by forcing the substrate to shrink (tensile built-in stress in the film) or expand (compressive built-in stress in the film). The change in the substrate's in-plane dimensions (strain in the substrate of a flattened workpiece) at the deposition temperature T_d can be calculated from Eq. (2.8a) by setting $T = T_d$:

$$\varepsilon_s(T_d) = \frac{\varepsilon_{bi}}{1 + \frac{Y_s^* d_s}{Y_f^* d_f}} \tag{2.10}$$

Equation (2.10) compares the in-plane substrate's dimensions at the deposition temperature T_d after and before film growth. Tensile built-in strain ($\varepsilon_{bi} < 0$) leads to a negative $\varepsilon_s(T_d)$, indicating substrate shrinkage during film growth (see Fig. 2.1d). Compressive built-in strain ($\varepsilon_{bi} > 0$) leads to a positive $\varepsilon_s(T_d)$, indicating substrate expansion during film growth.

Setting $T = T_d$ in Eq. (2.8b) expresses the change in the in-plane film's dimensions (strain in the film of a flattened workpiece) at the deposition temperature T_d:

$$\varepsilon_f(T_d) = -\frac{\varepsilon_{bi}}{1 + \frac{Y_f^* d_f}{Y_s^* d_s}} \tag{2.11}$$

Equation (2.11) describes the in-plane strain in the film at the deposition temperature T_d when the built-in strain ε_{bi} is present (compared to the absence of ε_{bi}). Tensile built-in strain ($\varepsilon_{bi} < 0$) leads to a positive $\varepsilon_f(T_d)$, indicating that the film is stretched (see Fig. 2.1d). Compressive built-in strain ($\varepsilon_{bi} > 0$) leads to a negative $\varepsilon_f(T_d)$, indicating compression in the growing film.

$\varepsilon_s(T_d)$ and $\varepsilon_f(T_d)$ both equal zero if the film has no built-in strain/stress. For $\varepsilon_{bi} \neq 0$ and a thick and rigid substrate ($Y_f^* d_f \ll Y_s^* d_s$), all strain/stress is taken up by the film, $\varepsilon_f(T_d) = -\varepsilon_{bi}$, leaving the substrate free of strain and stress, $\varepsilon_s(T_d) \to 0$. For compliant substrates, the denominator in Eqs. (2.10) and (2.11) becomes a finite number larger than 1, indicating that the built-in strain is now divided between the film and the substrate depending on their relative strengths, $\frac{Y_f^* d_f}{Y_s^* d_s}$.

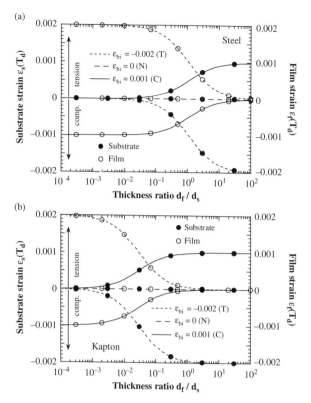

Fig. 2.4 Strain in the substrate, ε_s (T_d), and the film, ε_f (T_d), of a flattened workpiece at deposition temperature after deposition of SiN_x, as a function of film/substrate thickness ratio. (**a**) The upper frame is for steel ($Y_f^*/Y_s^* \cong 0.92$) substrate and (**b**) the lower frame for Kapton E ($Y_f^*/Y_s^* \cong 35$) substrate. The parameter ε_{bi} represents the built-in strains measured in SiN_x films deposited by PECVD. T stands for built-in tensile stress, C for built-in compressive stress, and N for no built-in stress in the film

Figure 2.4 shows examples of the calculated strain in the substrate and the film at the deposition temperature T_d as a function of the film/substrate thickness ratio. The upper frame is for steel, and the lower is for Kapton E. Three different values of the built-in strain ε_{bi}, −0.002 (T for tensile built-in stress), 0 (N for no built-in stress), and +0.001 (C for compressive built-in stress) were chosen to match the typical values of the built-in strain in SiN_x films deposited by PECVD [8]. Other parameters used in the calculation are $Y_f^* = 183/(1-\nu)$ GPa, $Y_s^* = 200/(1-\nu)$ GPa (AISI 304 steel) or $5.2/(1-\nu)$ GPa (Kapton E), and $\nu = 0.3$ for both steel and Kapton E.

Strain develops in the film and the substrate at the deposition temperature only under built-in stress. When the film is very thin compared to the substrate, all strain/stress is taken up by the film and the substrate's dimensions do not change. With increasing film thickness, the strain in the film decreases and the strain in the substrate increases. If a 1-μm-thick SiN_x film is deposited on a 100-μm-thick steel foil, almost all strain is taken up by the film. However, if a 1-μm-thick SiN_x film is deposited on a 100-μm-thick Kapton E foil, approximately one-fourth of the built-in strain is taken up by the substrate, which reduces the strain in the film.

2.3.2 Strain in the Substrate, $\varepsilon_s(T_r)$, and the Film, $\varepsilon_f(T_r)$, at Room Temperature T_r

Precise overlay alignment is crucial for the fabrication of any integrated circuit. Alignment errors between device layers can affect critical device tolerances, for example, the overlap between the source/drain and the gate electrodes in a TFT. Increasing the tolerances to avoid open circuits caused by misalignment raises parasitic capacitances and reduces circuit speed. When fabricating devices on a rigid plate of glass, the alignment tolerances must increase as the exposure frame size increases. When working with thinner (steel) and/or more compliant (organic polymer) substrates, the alignment tolerances must be increased even further. As a result, one must either raise the tolerances at the cost of reduced circuit speed, or use self-alignment, or find means to maintain the substrate dimensions unchanged throughout the fabrication process. This section analyzes how the dimensions of the substrate at room temperature are changed after film growth at elevated temperature.

By substituting $T = T_r$ in Eq. (2.8) we obtain:

$$\varepsilon_s(T_r) = \frac{\left[(T_d - T_r)\left(\alpha_s - \alpha_f\right) + \varepsilon_{bi}\right]}{1 + \frac{Y_s^* d_s}{Y_f^* d_f}} \tag{2.12}$$

$$\varepsilon_f(T_r) = -\frac{\left[(T_d - T_r)\left(\alpha_s - \alpha_f\right) + \varepsilon_{bi}\right]}{1 + \frac{Y_f^* d_f}{Y_s^* d_s}} \tag{2.13}$$

Equation (2.12) describes the strain of the substrate at room temperature, after film deposition at elevated temperature (Fig. 2.1f). It compares the in-plane dimensions of the flattened substrate/workpiece after and before film growth. For accurate overlay alignment, $\varepsilon_s(T_r)$ should be close to zero. $\varepsilon_s(T_r)$ can be made small by minimizing the numerator $\left[(T_d - T_r)\left(\alpha_s - \alpha_f\right) + \varepsilon_{bi}\right]$ and maximizing the denominator $1 + \frac{Y_s^* d_s}{Y_f^* d_f}$. This can be achieved by three ways:

(1) Choosing a substrate with a CTE close to that of the device layers. This is challenging for plastic substrates whose CTE usually is much larger than those of silicon device films.
(2) Minimizing $(T_d - T_r)$ by lowering the deposition temperature. In a-Si:H thin-film technology, lower deposition temperature leads to worse electronic properties. Here, organic electronics, for example polymer light-emitting diodes and thin-film transistors, have an advantage.

(3) Compensating the CTE mismatch with strain built into the device films, such that $\left[(T_d - T_r)(\alpha_s - \alpha_f) + \varepsilon_{bi} = 0\right]$. The built-in strain depends on the deposition conditions and sometimes can be easily adjusted (see Fig. 2.12).

One can also (1) choose a thicker substrate, which however is not desirable for flexing or "shaping" applications, (2) keep the device structure very thin, which however may reduce the device yield, and (3) choose substrate with a large Young's modulus (not possible for plastic substrates). These measures may be combined.

Equation (2.13) describes the strain in the film after the flattened workpiece has been cooled to room temperature. The larger the thermal misfit strain between the substrate and the film, $(T_d - T_r)(\alpha_s - \alpha_f)$, the larger the strain in the substrate and the film. When $\alpha_s > \alpha_f$, a tensile built-in strain in the film ($\varepsilon_{bi} < 0$) helps to compensate the thermal misfit and to reduce the strain in the substrate, which causes alignment error, and in the film, which causes film fracture.

Figure 2.5 depicts the strain of the substrate at room temperature, $\varepsilon_s(T_r)$, after film deposition, calculated from Eq. (2.12). Positive strain (tension) means that at room temperature the substrate is elongated after film growth, while negative strain (compression) indicates shrinkage. Three different values of the built-in strain in the film ε_{bi}, −0.002 (tensile built-in stress), 0 (no built-in stress), and +0.001 (compressive built-in stress), were chosen to match typical values of the built-in strain in SiN$_x$ films deposited by PECVD [8]. The calculations are done for two different deposition temperatures $T_d = 150$ and $250°C$ for three types of substrates: steel, Kapton E, and a high-CTE plastic. The Young's modulus and CTE for each substrate are given below:

Young's modulus
Steel: $Y_s^* = 200/(1-\nu)$ GPa
Kapton E: $Y_s^* = 5.2/(1-\nu)$ GPa
High-CTE plastic: $Y_s^* = 3/(1-\nu)$ GPa
CTEs
Steel: $\alpha_s = 18 \times 10^{-6}$ K^{-1}
Kapton E: $\alpha_s = 16 \times 10^{-6}$ K^{-1}
High-CTE plastic: $\alpha_s = 40 \times 10^{-6}$ K^{-1}

In all cases, $Y_f^* = 183/(1-\nu)$ GPa, $\nu = 0.3$, $\alpha_f = 2.7 \times 10^{-6}$ K^{-1}, and $T_r = 20°C$. For the values of ε_{bi} used here, the steel substrate is seen to remain dimensionally stable ($\varepsilon_s(T_r) \cong 0$) up to $d_f/d_s \cong 0.01$, Kapton E up to $d_f/d_s \cong 0.0003$, and high-CTE plastic for $d_f/d_s < 0.0001$. It is clear that steel foil is dimensionally most stable during TFT processing. For all substrates, built-in tensile stress is preferable to compressive or no built-in stress because it can compensate differential thermal contraction. However, deposition on a high-CTE plastic foil causes a substantial elongation of the substrate regardless of the built-in stress in the film, even if the film is very thin ($d_f/d_s \sim 10^{-3}$). Note that brittle films withstand higher compressive strain than tensile strain, because tensile strain lower than $\sim 5 \times 10^{-3}$ may fracture the films. Yet fracture of device films is commonly observed on high-CTE plastics.

Figure 2.6 depicts the strain of the SiN$_x$ film at room temperature, $\varepsilon_f(T_r)$, after its deposition at elevated temperature T_d, calculated from Eq. (2.13). Since all

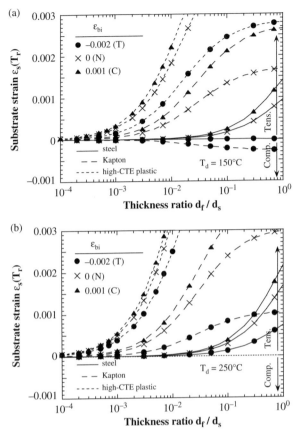

Fig. 2.5 Strain of the substrate at room temperature, ε_s (T_r), after deposition of SiN$_x$, as a function of film/substrate thickness ratio. The three substrates illustrated are steel ($Y_f^*/Y_s^* \cong 0.92$, $\alpha_s = 18 \times 10^{-6}$ K^{-1}), Kapton E ($Y_f^*/Y_s^* \cong 35$, $\alpha_s = 16 \times 10^{-6}$ K^{-1}), and high-CTE plastic ($Y_f^*/Y_s^* \cong 45$, $\alpha_s = 40 \times 10^{-6}$ K^{-1}). The calculations are done for two different deposition temperatures: (**a**) $T_d = 150°$C and (**b**) $T_d = 250°$C. The parameter ε_{bi} is the built-in strain measured in SiN$_x$ films deposited by PECVD. Other parameters are listed in the text. T stands for built-in tensile stress, C for built-in compressive stress, and N means no built-in stress in the film

substrates presented here have CTEs substantially larger than that of SiN$_x$ film, when the flattened workpiece is cooled from the deposition temperature, the film is typically under compression. The compressive strain approaches 1% if the film is grown on a high-CTE plastic at 250°C. The strain in the film remains constant for steel up to $d_f/d_s \cong 0.03$, Kapton E up to $d_f/d_s \cong 0.001$, and high-CTE plastic for $d_f/d_s \cong 0.0001$, and it is partially relieved if thick films are grown.

The effect of the film's built-in strain ε_{bi} on substrate dimensions at room temperature, ε_s (T_r), is shown in Fig. 2.7 for 100-μm-thick foil substrates after a 1-μm-thick SiN$_x$ deposition at the temperatures of 150, 200, and 250°C. Positive strain (tension) means that at room temperature the substrate is elongated after film growth, negative strain (compression) indicates shrinkage. ε_s (T_r), given by

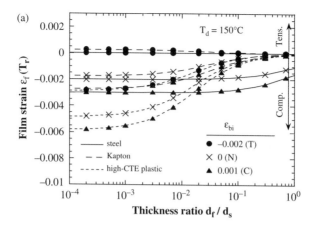

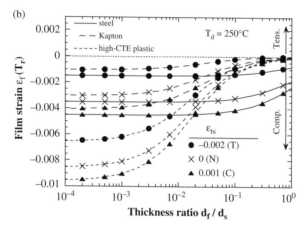

Fig. 2.6 Strain of the SiN$_x$ film at room temperature, ε_f (T_r), after its deposition at elevated temperature T_d, as a function of film/substrate thickness ratio. (**a**) $T_d = 150°C$ and (**b**) $T_d = 250°C$. Parameters are the same as in Fig. 2.5

Eq. (2.12), is plotted as a function of the built-in strain in the film for steel (upper frame) and Kapton E (lower frame) substrates. As mentioned above, the film/substrate couple is held flat, as required for the alignment procedure. Because the CTE of steel and Kapton E are larger than that of SiN$_x$, both substrates are elongated after deposition and cooling if no built-in stress is grown into the film. The steel substrate is elongated by ∼18 ppm for $T_d = 150°C$, by ∼25 ppm for 200°C, and by ∼32 ppm for 250°C; Kapton E by ∼450 ppm for $T_d = 150°C$, by ∼620 ppm for 200°C, and by ∼800 ppm for 250°C. When the SiN$_x$ film is grown with built-in *tensile* stress, ε_s (T_r) is reduced. For $T_d = 250°C$ for Kapton E, a built-in tensile strain of ∼-3×10^{-3} completely compensates the CTE mismatch between the substrate and the film, leaving the dimensions of the workpiece unchanged. Therefore, by tailoring the built-in strain/stress in the TFT layers, one can keep the film/substrate couple dimensionally stable for accurate photomask overlay alignment. Built-in

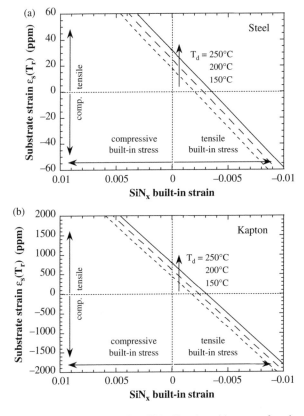

Fig. 2.7 Strain of the substrate ε_s (T_r) after SiN$_x$ film deposition, as a function of the built-in strain in a 1-μm-thick SiN$_x$ film deposited on 100-μm-thick (**a**) steel or (**b**) Kapton foil substrates at three deposition temperatures $T_d = 150°$C, $200°$C, and $250°$C

tensile strain in the film is needed to compensate for the CTE mismatch between the substrate and the film.

2.3.3 Radius of Curvature R of the Workpiece

When $Y_f\, d_f \ll Y_s\, d_s$, the substrate dominates and the film complies with it, as a TFT does on a plate glass substrate. The stress in the substrate is negligible, and the film/substrate couple curves only slightly, even when the film is highly stressed. The strain is biaxial in the plane of the film, and the structure forms a spherical cap with the radius of curvature given by:

$$R = \frac{d_s}{6\frac{Y_f^* d_f}{Y_s^* d_s}\left(e_f - e_s\right)} \cdot \frac{\left(1 - \frac{Y_f^* d_f^2}{Y_s^* d_s^2}\right)^2 + 4\frac{Y_f^* d_f}{Y_s^* d_s}\left(1 + \frac{d_f}{d_s}\right)^2}{1 + \frac{d_f}{d_s}} \qquad (2.14)$$

where $(e_f - e_s)$ is the mismatch strain between the film and the substrate. The mismatch strain has two dominant components. One is the thermal mismatch strain caused by the difference between the CTE of the substrate, α_s, and that of the film, α_f. The other is the built-in strain ε_{bi} grown into the film during its deposition. Therefore,

$$e_f - e_s = \alpha_f \cdot (T_r - T_d) + \varepsilon_{bi} - \alpha_s \cdot (T_r - T_d) = (\alpha_f - \alpha_s) \cdot (T_r - T_d) + \varepsilon_{bi} \quad (2.15)$$

where $(T_r - T_d)$ is the difference between the room and the deposition temperatures. $R > 0$ when the film is on the convex side and $R < 0$ when the film is on the concave side of the film-on-foil structure. When $e_f - e_s = 0$, $R \to \infty$, and the structure becomes flat.

Since $Y_f \, d_f \ll Y_s \, d_s$, the second fraction in Eq. (2.14) is typically neglected, leading to the Stoney formula:

$$R = \frac{d_s}{6 \frac{Y_f^* d_f}{Y_s^* d_s} \left(e_f - e_s\right)} \quad (2.16)$$

In such cases, R is very large.

A stiff film and a compliant substrate, for example an a-Si:H layer on an organic polymer foil, may have similar products of elastic modulus and thickness, $Y_f \, d_f \approx Y_s \cdot d_s$. The structure rolls into a cylinder instead of forming a spherical cap. An example is shown in Fig. 2.8, for a 500-nm-thick layer of SiN$_x$ deposited on two different plastic substrates by PECVD at 150°C.

In such a case, the Stoney formula is no longer valid. The radius of curvature R is now given by [9]:

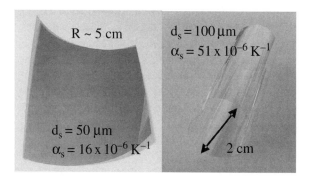

Fig. 2.8 500-nm-thick PECVD SiN$_x$ film deposited on two different plastic substrates. The substrate thickness and the CTE are shown for each substrate. The film is deposited on the top and the structures roll into cylinders. The *left* picture shows the SiN$_x$ film in tension, and the *right* one in compression [9]

$$R = \frac{d_s}{6\frac{Y_f' d_f}{Y_s' d_s}(e_f - e_s)} \cdot \left\{ \frac{\left[\left(1-\frac{Y_f' d_f^2}{Y_s' d_s^2}\right)^2 + 4\frac{Y_f' d_f}{Y_s' d_s}\left(1+\frac{d_f}{d_s}\right)^2\right]\left[(1-v_s^2)+\left(\frac{Y_f' d_f}{Y_s' d_s}\right)^2(1-v_f^2)\right]}{\left(1+\frac{d_f}{d_s}\right)\left(1+\frac{Y_f' d_f}{Y_s' d_s}\right)\left[(1-v_s^2)(1+v_f)+\frac{Y_f' d_f}{Y_s' d_s}\left(1-v_f^2\right)(1+v_s)\right]} + \right.$$
$$\left. + \frac{3\left(\frac{Y_f' d_f}{Y_s' d_s}\right)^2\left(1+\frac{d_f}{d_s}\right)^2\left[(1-v_s^2)+(1-v_f^2)\right]+2\frac{Y_f' d_f}{Y_s' d_s}(1-v_s v_f)\left(1+\frac{Y_f' d_f}{Y_s' d_s}\right)\left(1+\frac{Y_f' d_f^3}{Y_s' d_s^3}\right)}{\left(1+\frac{d_f}{d_s}\right)\left(1+\frac{Y_f' d_f}{Y_s' d_s}\right)\left[(1-v_s^2)(1+v_f)+\frac{Y_f' d_f}{Y_s' d_s}\left(1-v_f^2\right)(1+v_s)\right]} \right\}$$

(2.17)

Here, $Y_f' = \frac{Y_f}{1-v_f^2}$ and $Y_s' = \frac{Y_s}{1-v_s^2}$ are the plane strain moduli of the film and the substrate, respectively. (During the derivation of Eq. (2.17), the $1/(1-v^2)$ factor accompanies each corresponding Young's modulus. Therefore, the introduction of the plane strain modulus $Y' = \frac{Y}{1-v^2}$ simplifies the expressions.)

If the Poisson ratios v of the film and the substrate are identical, Eq. (2.17) simplifies to the form [7]:

$$R = \frac{d_s}{6\frac{Y_f d_f}{Y_s d_s}(e_f - e_s)(1+v)} \cdot \frac{\left(1-\frac{Y_f d_f^2}{Y_s d_s^2}\right)^2 + 4\frac{Y_f d_f}{Y_s d_s}\left(1+\frac{d_f}{d_s}\right)^2}{\left(1+\frac{d_f}{d_s}\right)}$$

(2.18)

The mismatch strain $(e_f - e_s)$ is again given by Eq. (2.15). The first fraction in Eq. (2.18) is the Stoney formula divided by $(1+v)$, a factor arising from the generalized plane strain condition (cylindrical shape). The $(1+v)$ factor in the denominator of Eq. (2.18) reflects the observation that a compliant substrate rolls into a cylinder, rather than the spherical cap assumed in the Stoney formula. The second fraction is identical to that of Eq. (2.14) and constitutes the deviation from the Stoney formula for compliant substrates. When the mechanical properties of the film and the substrate and the radius of curvature R are known, one can extract the built-in strain ε_{bi} of the film [8].

The normalized radius of curvature, calculated from Eqs. (2.16) and (2.18), is plotted as a function of d_f/d_s in Fig. 2.9. The Stoney formula, Eq. (2.16), which is an approximation for $Y_f d_f \ll Y_s d_s$, becomes invalid when the film thickness approaches the substrate thickness, regardless of materials used. For typical TFT materials on a steel substrate, $Y_f/Y_s = 0.92$, and the Stoney formula with the $(1+v)$ factor included is a good approximation for $d_f/d_s \leq 0.01$. For Kapton substrate, $Y_f/Y_s = 35$, and the Stoney formula is useful only for $d_f/d_s \leq 0.001$. In the specific case of a 100-μm-thick Kapton E substrate, Eq. (2.18) must be used for $d_f > 100$ nm.

Figure 2.10 shows the film-on-foil curvature $1/R$ as a function of the built-in strain in the film calculated using Eq. (2.18). The calculation is done for a 1-μm-thick PECVD SiN$_x$ film deposited on two different substrates, stainless steel and Kapton E, at three deposition temperatures $T_d = 150°C$, $200°C$, and $250°C$. The following parameters were used in the calculations: $d_f = 1$ μm, $d_s = 100$ μm, $Y_f = 183$ GPa, $Y_s = 200$ GPa (steel) or 5.2 GPa (Kapton E), $v = 0.3$, $\alpha_f = 2.7 \times 10^{-6}$ K^{-1}, $\alpha_s = 18 \times 10^{-6}$ K^{-1} (steel) or 16×10^{-6} K^{-1} (Kapton E), and $T_r = 20°C$.

Fig. 2.9 Normalized radius of curvature as a function of film/substrate thickness ratio. *Full lines* are the solution for a cylindrical roll, Eq. (2.18), with $\nu = 0.3$. *Dashed* lines are the Stoney formula, Eq. (2.16), for a spherical cap formed by a thick, stiff wafer

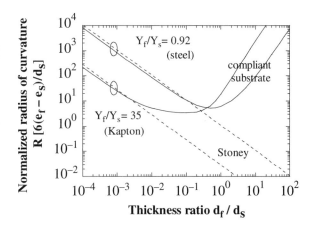

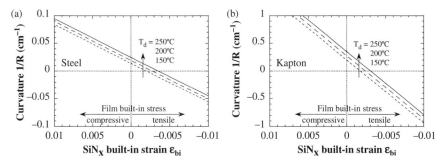

Fig. 2.10 Curvature $1/R$ of the film-on-foil structure at room temperature as a function of the built-in strain imposed by the film grown at an elevated temperature, calculated using Eq. (2.18). A 1-µm-thick PECVD SiN_x film is deposited on two different substrates of a 100-µm-thick (**a**) steel or (**b**) Kapton E foil. The calculations are done for three different deposition temperatures T_d of 150, 200, and 250°C. The remaining parameters are listed in the text

A 100-µm-thick Kapton foil curves substantially more after the growth of a 1-µm-thick PECVD SiN_x film than a 100-µm-thick stainless steel foil coated with an identical film. If the SiN_x film is grown without built-in stress ($\sigma_{bi} = 0$ and $\varepsilon_{bi} = 0$), the structure curves with the film on the outside. The room-temperature curvature $1/R$ of the SiN_x-on-Kapton structure is 0.19 cm^{-1} ($R = 5.3$ cm) with a film grown at 150°C, 0.26 cm^{-1} ($R = 3.8$ cm) at 200°C, and 0.34 cm^{-1} ($R = 2.9$ cm) at 250°C. The room-temperature curvature $1/R$ of the SiN_x-on-steel structure is 0.013 cm^{-1} ($R \sim 77$ cm) with a film grown at 150°C, 0.019 cm^{-1} ($R \sim 53$ cm) at 200°C, and 0.025 cm^{-1} ($R \sim 40$ cm) at 250°C.

Both Kapton and steel have larger thermal expansion coefficients than the SiN_x film. Upon cooling from the growth temperature, the substrate shrinks more than the SiN_x film. The SiN_x is forced to occupy the longer, outside perimeter of the cylindrical roll. Built-in compressive stress in the film forces the film-on-foil structure to curve more, while built-in tensile stress in the film makes the structure more flat. When the built-in strain fully compensates the differential thermal expansion between the film and the substrate, i.e. $e_f - e_s = 0$, the denominator in Eq. (2.18) becomes zero. $R \to \infty$, and the film-on-foil structure becomes flat. For each film/substrate couple and each deposition temperature, there is an optimal amount of built-in tensile strain, as can be seen from Fig. 2.10. The higher the deposition temperature, though, the higher this compensating built-in tensile strain must be. If the built-in tensile strain is increased beyond this optimal point, the structure bends the other way, and the film faces inward.

2.3.4 Strain of the Substrate and the Curvature of the Workpiece for a Three-Layer Structure

The preceding theory can be extended easily to any number of layers. The layer design will benefit from a brief discussion of the strain of the substrate at room temperature, $\varepsilon_s(T_r)$, and the radius of curvature, R, for a three-layer structure.

The strain of the substrate at room temperature (change in substrate dimensions) is:

$$\varepsilon_s(T_r)$$
$$= \frac{Y_{f1}^* d_{f1}\left[(T_d - T_r)(\alpha_s - \alpha_{f1}) + \varepsilon_{bi1}\right] + Y_{f2}^* d_{f2}\left[(T_d - T_r)(\alpha_s - \alpha_{f2}) + \varepsilon_{bi2}\right]}{Y_s^* d_s + Y_{f1}^* d_{f1} + Y_{f2}^* d_{f2}}$$
(2.19)

Here, $f1$ and $f2$ denote the first and second films, respectively. The other symbols were introduced earlier. Equation (2.19) is valid whether each side of the substrate is coated with one film or both films are deposited on the same side. Similarly to a two-layer structure, the higher the deposition temperature and the larger the CTE mismatch between the substrate and either of the films, the larger must be the compensating built-in tensile strain. $\varepsilon_s(T_r) \neq 0$ unless the built-in strains induced by the growing films are tailored appropriately.

If the Poisson ratio v is the same for all layers, the radius of curvature R of the relaxed three-layer structure (see Fig. 2.11) at room temperature is given by:

$$R = \frac{d_1}{6(1+v)} \cdot$$
$$\left\{ \frac{\left(1 - \frac{Y_2 d_2^2}{Y_1 d_1^2} - \frac{Y_3 d_3^2}{Y_1 d_1^2}\right)^2 + 4\frac{Y_2 d_2}{Y_1 d_1}\left(1 + \frac{d_2}{d_1}\right)^2 + 4\frac{Y_3 d_3}{Y_1 d_1}\left[\left(1 + \frac{d_3}{d_1}\right)^2 + 3\frac{d_2}{d_1}\left(1 + \frac{d_2}{d_1} + \frac{d_3}{d_1}\right) + \frac{Y_2 d_2}{Y_1 d_1}\left(\frac{d_2^2}{d_1^2} + \frac{d_2 d_3}{d_1^2} + \frac{d_3^2}{d_1^2}\right)\right]}{\left(1 + \frac{d_2}{d_1}\right)\left[\frac{Y_2 d_2}{Y_1 d_1}(e_2 - e_1) + \frac{Y_3 d_3}{Y_1 d_1}(e_3 - e_1)\right] + \frac{Y_3 d_3}{Y_1 d_1}\left(\frac{d_2}{d_1} + \frac{d_3}{d_1}\right)\left[(e_3 - e_1) + \frac{Y_2 d_2}{Y_1 d_1}(e_3 - e_2)\right]} \right\}$$
(2.20)

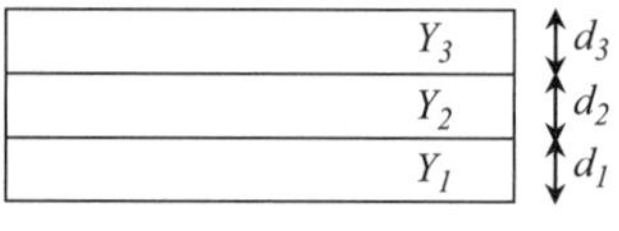

Fig. 2.11 Three-layer structure

If both layers are deposited on the same side of the substrate, then the Subscript 1 denotes the substrate and the Subscripts 2 and 3 denote the films. Equation (2.20) can also be used if one film is deposited on each side of the substrate. Then the Subscript 2 denotes the substrate and the Subscripts 1 and 3 denote the films. From analyzing Eq. (2.20), one can easily conclude that the deposition of identical layers on both sides of the substrate always leads to a flat structure ($R \rightarrow \infty$). In this case, it is not necessary to tailor the built-in strain in the films to achieve the flat structure. However, it is essential to tailor the built-in strain in the films, if the three-layer structure should have the same dimensions as the initial substrate (see Eq. (2.19)).

If two dissimilar layers are deposited on the same side of the substrate, the only way to keep the workpiece flat and at its initial size is to tailor the built-in strains in the films to compensate the thermal contraction mismatches. The benefit of a three-layer structure is that the two films work together to compensate the thermal mismatch. It becomes possible to achieve $\varepsilon_s(T_r) = 0$ and $1/R = 0$, even if the deposition of each film separately leads to $\varepsilon_s(T_r) \neq 0$ and $1/R \neq 0$.

2.3.5 Experimental Results for a-Si:H TFTs Fabricated on Kapton

As concluded above, to achieve flat flexible electronic surfaces that have identical dimensions at each mask alignment step, one must tailor the built-in strains introduced by film growth. While this can be easily done in some materials, it is not trivial in others. Figure 2.12a shows the curvature of a 50-μm-thick Kapton E foil after PECVD deposition of SiN$_x$ at 150°C. The change in curvature was obtained by varying the rf deposition power from 5 to 25 W. The SiN$_x$ film, which faces left, is under tension for low deposition power and under compression for the highest deposition power. Figure 2.12b shows two other materials, Cr and a-Si:H, whose built-in strains are difficult to vary. Cr is typically under tension and a-Si:H under compression.

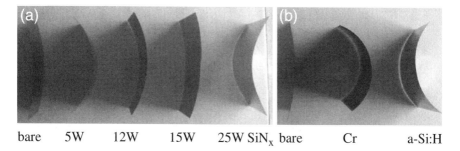

bare   5W   12W   15W   25W SiN$_x$ bare   Cr   a-Si:H

Fig. 2.12 Curvature induced by (**a**) SiN$_x$ deposited over a range of rf power and (**b**) Cr and a-Si:H, all on 50-μm-thick Kapton E polyimide substrates. All films are facing *left*. The 300–500-nm-thick SiN$_x$ and the 250-nm-thick a-Si:H films were deposited at 150°C, and the 80-nm-thick Cr was deposited by thermal evaporation without control of substrate temperature. The bare substrate is also shown for comparison [8]

1. front SiN$_x$ passivation
2. back SiN$_x$ passivation

3. 80-nm sputtered Cr gate metal

4. bottom Cr gate pattern by wet etch

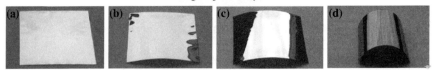

5. ~300-nm **5W** SiN$_x$ 5. ~300-nm **12W** SiN$_x$ 5. ~300-nm **20W** SiN$_x$ 5. ~300-nm **25W** SiN$_x$

6. 200-nm 6W (i)a-Si:H
7. 50-nm 4W (n$^+$)a-Si:H

8. 80-nm evaporated Cr

Fig. 2.13 Curvature variation in a-Si:H TFT fabrication on Kapton foil at five steps from substrate passivation to evaporation of the top Cr source/drain contact metal. For all four samples, the entire process is identical except for the deposition of the gate SiN$_x$, where the rf power was varied from 5 to 25 W. The square substrate is 7.5×7.5 cm^2 [8]

The ability to vary the built-in strain in SiN$_x$ over a wide range was put to use in TFT fabrication such that the gate and source/drain patterns aligned. Complete a-Si:H TFTs were fabricated at 150°C in a non-self-aligned, back-channel-etched geometry. We began by passivating a 50-μm-thick Kapton E substrate foil with an ~0.45-μm-thick PECVD SiN$_x$ on both sides (see Fig. 2.13). Following substrate passivation, the TFT fabrication process was carried out. An ~80-nm-thick Cr film was sputtered at room temperature and wet etched to create the first device pattern, the bottom gate electrode. All samples are flat after this first Cr evaporation. Next, a silicon stack composed of ~300-nm SiN$_x$ gate dielectric layer, ~200-nm (i)a-Si:H channel layer, and ~50-nm (n$^+$)a-Si:H source/drain layer was deposited by PECVD, followed by ~80-nm-thick thermally evaporated Cr for the source/drain metal contacts. The details of fabrication are described elsewhere [8]. Note that only sample (a) is flat after the $SiN_x/(i)a$-Si:H/(n^+)a-Si:H stack deposition. If a mask were aligned to the gate pattern at this step, sample (a) would experience the smallest misalignment error. However, the situation changes after the source/drain Cr is evaporated, which makes samples (b) and (c) flat instead. This sequence shows that all four device layers work together to compensate the CTE mismatch.

We measured the misalignment between the first and second photolithography levels, i.e. the bottom gate and the top source/drain. Steps 5–8 of Fig. 2.13 fall between these two photolithographic steps. In all four samples of Fig. 2.13, we kept these layers the same, except for intentionally varying the built-in stress in the gate

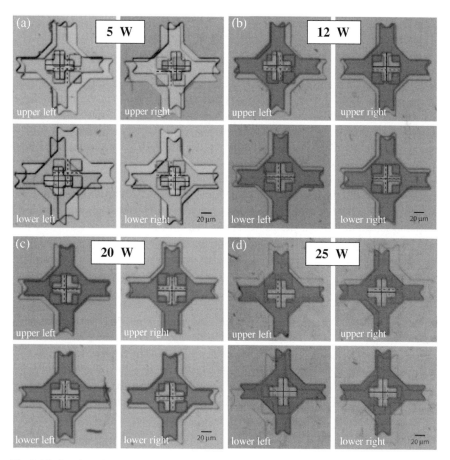

Fig. 2.14 Overlay misalignment between the first, gate, and the second, source/drain, photolithography levels in the back-channel-etched a-Si:H TFT process with (**a**) 5 W, (**b**) 12 W, (**c**) 20 W, and (**d**) 25 W gate SiN_x. The frames lie 52 mm apart near the corners of the substrate. The *dashed* crosses mark the centers of the alignment marks at the gate level and the *solid black* crosses, at the source/drain level [8]

SiN_x by tuning the rf deposition power from low to high, 5, 12, 20, and 25 W (22, 53, 89, and 111 mW/cm^2).

Figure 2.14 shows the corresponding alignment marks for the gate and the source/drain for all four samples. Both the gate and the source/drain patterns were aligned at the center of the substrate, and the misalignment was measured on alignment marks in four corners of a 52-mm × 52-mm square. These marks are shown in Fig. 2.14. The centers of the alignment marks are indicated by dashed crosses for the gate level and solid crosses for the source/drain. On average, sample (a) shrunk substantially, by ~480 ppm, sample (b) shrunk slightly, by ~96 ppm, sample (c) expanded slightly, by ~115 ppm, and sample (d) expanded substantially, by ~385 ppm. As expected, the dimensions for the flat samples (b) and (c) changed the least. As fabricated, TFTs of all samples had similar electrical performance.

2.4 Conclusions

When inorganic devices are grown at elevated temperature on a rigid plate substrate, the device films' built-in stresses may affect the device performance, but have little or no effect on the alignment error between device layers. The device films simply are too weak compared to the substrate. However, if the devices are grown on a flexible free-standing foil, built-in stresses in the films may cause substantial misalignment between device layers and unmanageable curving of the workpiece. The thinner and more compliant the substrate, the larger the misalignment and the tighter the curvature. With the help of theory of a bimetallic strip, the misalignment can be minimized by adjusting the built-in strain in the device films during growth. a-Si:H TFTs fabricated on foil substrates of stainless steel or Kapton E need tensile built-in strain to compensate for the differential thermal contraction between the device films and the substrate.

Misalignment and curvature obtained in a-Si:H TFTs fabricated on 50-μm-thick Kapton E foil prove that this approach is effective. By varying the built-in strain in only the gate SiN_x layer, one can reduce the misalignment from ~400 to ~100 ppm. Further optimization of the a-Si:H TFT growth process is possible for lowering this value even further.

Future work will address second-order misalignment effects caused by viscoelastic flow and nonreversible shrinkage of plastic substrate, and by modulation of local mechanics caused by device patterns. To which extend the electrical and the mechanical properties may be varied independently of each other also remains to be explored.

Acknowledgment   The work at Princeton University was supported by the United States Display Consortium.

References

1. Crawford GP (ed) (2005) Flexible Flat Panel Displays. Wiley, Chichester
2. Hutchinson JW, Suo Z (1991) Mixed-mode cracking in layered materials. Adv Appl Mech 29:63–191

3. Jansen F, Machonkin MA (1988) Thermomechanical properties of glow discharge deposited silicon and silicon oxide films. J Vac Sci Technol A6:1696–1698
4. Maeda M, Ikeda K (1998) Stress evaluation of radio-frequency-biased plasma-enhanced chemical vapor deposited silicon nitride films. J Appl Phys 83:3865–3870
5. Witvrouw A, Spaepen F (1993) Viscosity and elastic constants of amorphous Si and Ge. J Appl Phys 74:7154–7161
6. Freund LB, Suresh S (2003) Thin film materials. Cambridge University Press, New York
7. Suo Z, Ma EY, Gleskova H, Wagner S (1999) Mechanics of rollable and foldable film-on-foil electronics. Appl Phys Lett 74:1177–1179
8. Cheng I-C, Kattamis A, Long K, Sturm JC, Wagner S (2005) Stress control for overlay registration in a-Si:H TFTs on flexible organic polymer foil substrates. J SID 13:563–568
9. Gleskova H, Cheng I-C, Wagner S, Sturm JC, Suo Z (2006) Mechanics of thin-film transistors and solar cells on flexible substrates. Solar Energy 80:687–693

Chapter 3
Low-temperature Amorphous and Nanocrystalline Silicon Materials and Thin-film Transistors

Andrei Sazonov, Denis Striakhilev, and Arokia Nathan

Abstract   Low-temperature processing and characterization of amorphous silicon (a-Si:H) and nanocrystalline silicon (nc-Si) materials and devices are reviewed. An overview of silicon-based low-temperature thin-film dielectrics is given in the context of thin-film transistor (TFT) device operation. The low-temperature growth and synthesis of these materials are also presented and compared to conventionally fabricated high-temperature processed devices. The effect of using nc-Si contacts on a-Si:H TFTs and the stability of nc-Si TFTs is reviewed.

3.1 Introduction

Electronic devices and systems fabricated on flexible substrates are the subjects of growing attention within both the research and the industrial community [1]. Rapid development of wireless telecommunications increases demand in portable, low-power, and inexpensive electronics. Here, the use of flexible substrates enables the development of such new products as rollable light-weight displays, flexible solar cells integrated into clothing, or cylindrical and spherical cameras for security applications. Furthermore, the fabrication cost of electronic devices on flexible substrates can be reduced compared to existing planar or flat-panel technology due to implementation of high-throughput roll-to-roll technology. Here, a roll of thin plastic or metal foil used as the substrate can be kilometers long and meters wide compared to a silicon wafer diameter of 10–12″ in integrated circuit technology, or a glass sheet size of about 2 m×2 m in a flat-panel display (FPD) manufacturing process [2, 3]. Based on this technology, a number of commercial products are already being produced (such as solar cells [4]), and several others are in the prototyping stage

A. Sazonov (✉)

The Department of Electrical and Computer Engineering, University of Waterloo, Ontario, Canada
e-mail: asazonov@venus.uwaterloo.ca

W.S. Wong, A. Salleo (eds.), *Flexible Electronics: Materials and Applications*, Electronic
Materials: Science & Technology, DOI 10.1007/978-0-387-74363-9_3,
© Springer Science+Business Media, LLC 2009

(active-matrix organic light-emitting diode (AMOLED) displays, thin-film batteries, imaging arrays) [5–7], with more products to be developed.

Unlike traditional electronics, where the value is added by downscaling the devices, the value is added by increasing the fabrication area in flexible electronics. The device size requirements are more relaxed. Therefore, the capital equipment cost in flexible electronics is lower, and a larger part of the product costs falls onto the materials with a wide variety of electronic materials that can be used in flexible electronics: organic semiconductors, high-temperature inorganic semiconductors, and low-temperature materials [8] in addition to the substrate materials that replace conventional glass substrates.

Organic semiconductors are attractive due to low-temperature and large-area fabrication process compatibility. In particular, polymers can be deposited at room temperature by low-cost spin coating, or by roll-to-roll technology compatible with ink-jet printing. The carrier mobility in organic semiconductors is already comparable to that of amorphous silicon (a-Si:H) ($\sim$1 cm^2 V^{-1} s^{-1}) [7]. These materials, however, degrade in air and, therefore, need encapsulation, which is inorganic thin-film based [9]. Thus, inorganic thin-film technology still has to be used in organic electronics fabrication process, increasing its cost. Furthermore, the carrier transport in organic semiconductors, such as pentacene, is sensitive to contamination and is strongly interface-dependent, thus demanding very smooth substrates and various interface preparation procedures [10]. The need for low-cost fabrication processes and the nonuniformity of materials and device characteristics over large substrate areas make the applications for organic semiconductors in large-scale flexible electronics a matter for the future. These areas of development are discussed in separate chapters.

Another option is to transfer the existing high-temperature semiconductor technology to flexible substrates with high maximum working temperatures (e.g., high-temperature plastics such as polyimide (PI) [11], flexible glass, and metal foils [12, 13]) or by transferring the devices prefabricated on rigid substrates onto plastic foils [14, 15]. This approach facilitates the utilization of high fabrication temperature and electronic materials with improved device performance (e.g., polycrystalline Si (poly-Si) with the carrier mobility exceeding 100 cm^2 V^{-1} s^{-1}). However, high substrate and process costs restrict this technology to a limited number of applications, mainly very high added-value products (such as high-resolution flexible displays and high-end radio frequency (RF) ID tags).

A more direct approach to integrate high-performance devices on low-melting point platforms is to reduce the maximum fabrication temperature of inorganic thin-film transistors (TFTs) to a level compatible with the thermal budget of the low-cost substrates. This approach has several advantages:

- wider variety of substrate materials available, including low-cost plastics, paper, or tissue;
- lower thermal budget materials can be integrated in the process, such as adhesives, polymers, and biomaterials;

- thermal deformations of the substrate are reduced, and so is mechanical stress occurring due to mismatch between thermal expansion coefficients of the substrate and the films;
- materials science, device physics, fabrication process, and equipment are already well established, for example, in a-Si:H technology.

Therefore, accommodation of existing inorganic amorphous, nanocrystalline, or polycrystalline thin-film technology to flexible substrates achievable by the reduction of deposition temperature seems to be the most promising approach to enable flexible electronics in the near future.

3.2 Low-temperature Amorphous and Nanocrystalline Silicon Materials

Due to the low maximum working temperatures of most low-cost plastics (the glass transition temperatures, T_g, in the range of 80–150°C), the thermal budget in the fabrication process is limited by 100–150°C. Process temperatures are too low to enable the manufacturing techniques developed for crystalline silicon, such as thermal oxidation, diffusion, or epitaxy. Moreover, low thermal budget severely restricts the process parameters of plasma deposition, sputtering, and photolithography.

Consider an example of hydrogenated a-Si:H. a-Si:H films with predominantly monohydride bonding have been deposited at substrate temperature as low as 35°C by use of hydrogen or helium dilution followed by postdeposition annealing at 150°C [16]. More recently, there have been reports of TFT fabrication at 150°C on Kapton® E [17] and at 110°C on polyethylene terephthalate (PET) [18], showing performance characteristics close to those fabricated at 250–300°C.

3.2.1 Fundamental Issues for Low-temperature Processing

In traditional plasma enhanced chemical vapor deposition (PECVD), thin-film silicon process used in FPD manufacturing, the substrate temperature, T_s, is in the range of 200–300°C. Its reduction below 150°C leads to the change in deposition mechanisms. The concentration of defects (dangling Si bonds) increases, the concentration of di- and polyhydride-bonded hydrogen in the films increases, the mass density decreases, and the doping efficiency drops [19]. The films have low mass density and high charge trapping, and their electronic properties are generally poor. This change is usually attributed to reduced surface mobility of the film-building radicals due to lower thermal energy on the growth surface [20]. The thermal energy loss can be compensated by using "soft" ion bombardment by light ions (H^+, He^+) with the energy less than 50 eV, or by producing exothermic chemical reactions [20, 21]. Following this approach, the deposition parameters can be adjusted at low substrate temperature to increase the surface energy and thus to provide the growth

Table 3.1 The properties of a-Si:H deposited at 300, 120, and 75°C

Parameter	Deposition temperature		
	300°C [16]	120°C [17]	75°C [28]
Dark conductivity (σ_{300}), Ω^{-1} cm^{-1}	10^{-10}	4×10^{-11}	9×10^{-11}
Optical gap (E_g), eV	1.75	1.92	1.90
Urbach slope (E_0), meV	–	50	–
Hydrogen concentration (C_H), %	~10	10.9	9.5
Microstructure parameter $R = $ SiH$_n$/(SiH+SiH$_n$)	<0.1	~0	~0

conditions similar to those at higher T_s but without overheating the substrate. More thin-film deposition details are reported elsewhere [22–27].

3.2.2 Low-temperature Amorphous Silicon

In our experiments, we optimized the deposition process for a-Si:H films at 120°C [28] and 75°C. The a-Si:H properties are presented in Table 3.1 together with those of "device quality" material deposited by standard PECVD at higher temperature (180°C). The films have dark conductivity below 10^{-10} Ω^{-1} cm^{-1}, optical bandgap between 1.7 and 1.9 eV, Urbach parameter of 50 meV, hydrogen concentration of about 10 at.%, and microstructure parameter R about zero, which is indicative of predominant hydrogen monohydride bonding. As seen, these a-Si:H films deposited at low temperatures have the properties suitable for electronic device applications [22, 23].

3.2.3 Low-temperature Nanocrystalline Silicon

By comparison, the properties of nc-Si deposited at 75°C are presented in Table 3.2 together with those of "device quality" material deposited at higher temperature (260°C) and those of highly doped n$^+$ nc-Si contact layer deposited at 75°C. It is seen that the low-temperature nc-Si is comparable with its high temperature counterpart. Both materials have average crystal size of 15–25 nm. The 100-nm-thick films

Table 3.2 The properties of undoped nc-Si deposited at 260 and 75°C and of highly doped n$^+$ nc-Si deposited at 75°C

Parameter	Undoped 250°C [43]	Undoped 75°C [18]	n$^+$-doped 75°C [17]
Film thickness, nm	100	100	70
Dark conductivity (σ_{300}), Ω^{-1} cm^{-1}	10^{-6}	3×10^{-7}	0.3
Crystallinity, % (Raman)	82	75	72
Crystal grain size, nm (Raman, XRD)	~30	~20	~15

have the crystallinity of 75% and above, and low dark conductivity. Thus, as it follows from Tables 3.1–3.2, device quality a-Si:H and nc-Si films can be deposited at a temperature as low as 75°C using existing industrial plasma deposition equipment.

3.3 Low-temperature Dielectrics

The gate dielectrics in a-Si:H TFTs and passivation dielectrics are usually based on plasma deposited amorphous silicon nitride (a-SiN$_x$), whereas for poly-Si TFTs, amorphous silicon oxide (a-SiO$_x$) gate dielectric is preferred [17]. In order to meet the requirements for a good gate dielectric, the gate dielectric material has to withstand electric fields of about 2 MV/cm without breakdown, must have good insulating properties and low charge-trapping rate at lower electric fields, and should form a high-quality interface with the a-Si:H semiconductor layer. While PECVD a-SiN$_x$ gate dielectric technology at ~300°C is well established, the reduction of deposition temperature to 150°C for compatibility with flexible plastic substrates and organic light-emitting diodes (OLEDs) often leads to a material with high hydrogen concentration and poor dielectric performance.

3.3.1 Characteristics of Low-temperature Dielectric Thin-film Deposition

At low deposition temperatures (below 200°C), the leakage current through a-SiN$_x$ and a-SiO$_x$ films goes up rapidly due to low film mass density [29, 30]. The high-density a-SiN$_x$ films can be obtained by RF PECVD [31]. The oxide mass density can be increased using high-density plasma systems (e.g., electron cyclotron resonance [32]). The maximum deposition area in such systems, however, is limited by 10–12″, which is rather insufficient for large-area fabrication. Despite this, there are research groups who have recently reported successful low-temperature fabrication of a-Si:H TFTs using PECVD silicon nitride (SiN$_x$) [31, 33, 34] and one can observe a gradual progress in the performance of the devices [35].

3.3.2 Low-temperature Silicon Nitride Characteristics

The properties of our a-SiN$_x$ deposited by RF PECVD at 120 and 75°C are presented in Table 3.3 together with those of "device quality" material deposited at higher temperature (260°C). As it is seen from Table 3.3, the electrical and mechanical properties of a-SiN$_x$ films (mass density, mechanical stress, electrical resistivity, and breakdown voltage) can be well controlled at low deposition temperature, and high device quality can be achieved by adjusting the deposition conditions (RF power and gas mixture composition). In Table 3.3, the film (a) deposited at 75°C under high power density (100 mW/cm^2) and high hydrogen dilution of silane and ammonia mixture yielded the compressive stress of –221 MPa accompanied by high

Table 3.3 The properties of a-SiN$_x$ deposited at 260, 120, and 75°C

	260°C	120°C [20]	75°C (a) [28]	75°C (b) [28]
Mass density, g/cm^3	~2.7	2.1	2.57	2.44
Stress, MPa	220	−173	−221	−44
Dielectric constant	6.5	6.2	5.6	5.8
Resistivity at 1 MV/cm, Ω·cm	~10^{15}	2×10^{15}	2×10^{15}	2×10^{14}
Breakdown field at leakage current of 10^{-6} A/cm^2, MV/cm	3−8	5.5	10.8	7.4

resistivity (2×10^{15} Ω cm) and high breakdown field (10.8 MV/cm). In film (b), deposited at 75°C under lower power density (50 mW/cm^2) and lower hydrogen dilution, the compressive stress goes down, achieving −44 MPa, whereas the resistivity and the breakdown field are still high (2×10^{14} Ω cm and 7.4 MV/cm, respectively). Thus, good electronic properties can be combined with low mechanical stress in the film, which is important for mechanical integrity of devices fabricated on flexible substrates.

3.3.3 Low-temperature Silicon Oxide Characteristics

The properties of our PECVD SiO$_x$ films deposited by conventional 13.56 MHz glow discharge decomposition of silane and nitrous oxide mixture at 75 and 120°C are presented in Table 3.4 together with those deposited at higher temperature (250°C). Helium (He), nitrogen (N$_2$), and argon (Ar) were used as diluent gases.

Table 3.4 Properties of RF PECVD SiO$_x$ films deposited at different conditions

T_s, C	Diluent	Mass density, g/cm^3	Compressive stress, GPa	ε	Refractive index	Deposition rate, nm/min	Break down field, MV/cm	Leakage @ 5 MV/cm, A/cm^2
250	He	1.70 ± 0.17	0.118 ± 0.006	2.03	1.32 ± 0.17	7.17	>9	3.7×10^{-10}
120	None	1.77 ± 0.18	0.061 ± 0.012	–	1.44 ± 0.06	9.25	>9	10^{-8}
120	He	1.71 ± 0.17	0.097 ± 0.001	2.75	1.27 ± 0.20	7.20	8.4	9×10^{-9}
120	N$_2$	1.63 ± 0.16	0.134 ± 0.007	–	1.29 ± 0.17	6.93	4.8	10^{-8}
120	Ar	1.66 ± 0.16	0.064 ± 0.002	3.57	1.15 ± 0.06	7.43	>9	10^{-8}
75	He	1.69 ± 0.17	0.086 ± 0.004	2.45	1.19 ± 0.16	7.57	>9	2.2×10^{-8}

As one can see, all samples are characterized by low mechanical stress (below 100 MPa). The SiO_x mass density is also lower in comparison with thermally grown SiO_2 (2.2–2.3 g/cm^3) [36]. Moreover, poor thickness uniformity (more than 5%), which is a well-known problem of PECVD SiO_x [37], was observed in the film deposited with no dilution. We used He, N_2, and Ar dilution to improve the film uniformity.

The refractive index, which is related to the film density, increases with the deposition temperature. However, it is lower than it was reported for PECVD SiO_x deposited at $T_s > 250°C$, which was in 1.47–1.60 range [37–39], again indicating a lower film density. The SiO_x films fabricated with no dilution are characterized by the highest value of refractive index, which is close to thermally grown SiO_2 films ($n = 1.47$). This refractive index corresponds to the highest film density. The dielectric permittivity deduced from high-frequency (1 MHz) CV measurements on metal–oxide–semiconductor (MOS) capacitors was in the range between 2.03 and 3.57, which is lower than the value of 4.1 observed for SiO_x films deposited at $T_s > 200°C$ [40].

In the investigated deposition conditions range, no strong correlation between film density and mechanical stress was found. However, the film uniformity was improved by using dilution, and the best uniformity was obtained with He dilution. It was also reported that He or Ar dilution can decrease the concentration of undesirable Si–OH and Si–H bonds, which are responsible for the lower film density and poorer electrical properties of these materials [38]. In our case, no presence of SiH stretching at 2,000 cm^{-1} or NH-stretching vibrations at 3,320 cm^{-1} was found in the Fourier Transformed Infra-Red (FTIR) spectra.

Dielectric breakdown at 8.4 MV/cm was observed for the SiO_x film deposited at 120°C with He dilution and at 4.8 MV/cm for the film deposited with N_2 dilution. The films deposited at higher temperatures are characterized by lower leakage current density, which was 3.7×10^{-10} A/cm^2 for the sample deposited at 250°C, 9×10^{-9} A/cm^2 at 120°C, and 2.2×10^{-8} A/cm^2 at 75°C (all values are given for 5 MV/cm). Comparing the influence of different diluent gases on leakage currents of the MOS capacitors, it should be noted that the best results are achieved for SiO_x films deposited from silane mixtures with He dilution. Thus, our SiO_x films fabricated at low temperatures with He dilution are characterized by the lowest leakage current, best uniformity, and reduced mechanical stress.

3.4 Low-temperature Thin-film Transistor Devices

In order to develop a reliable and industrially relevant TFT technology for flexible electronics applications, the following goals have to be achieved:

– Maximum deposition temperature should not exceed the thermal budget of the low-cost plastic substrates (~80°C for PET substrates, 150°C in the case of polyethylene naphthalate (PEN) substrates, and 300°C for PI substrates).

- TFT technology should be compatible with existing FPD manufacturing process (i.e., a thin-film silicon process is preferable).
- The use of existing fabrication equipment (large-area 13.56-MHz parallel plate PECVD reactors) without major modifications is desirable.

3.4.1 Device Structures and Materials Processing

Pursuing these goals and based on the above described thin-film silicon-based materials, TFTs were fabricated at maximum process temperatures of 120 and 75°C. Both bottom-gate (inverted-staggered) and top-gate (staggered) TFTs were designed and fabricated on plastic foils. For a-Si:H TFTs, the inverted-staggered structure is preferential due to lower defect density on top surface of SiN_x gate dielectric. For nc-Si devices, however, staggered structure is preferable to take advantage of better electronic properties on top of nc-Si film.

The bottom-gate TFT fabrication process flow on plastic substrate is shown in Fig. 3.1a [22]. After cleaning, the substrate was double-side coated by a 0.5-μm-thick a-SiN_x film. The purpose of this coating is to encapsulate the plastic foil and protect it from aggressive chemicals, as well as to improve adhesion to the substrate. Furthermore, due to low a-SiN_x thermal expansion coefficient, the alignment error in the photolithography processes is reduced. On top of a-SiN_x, a 100–120-nm aluminum film was deposited at room temperature by RF magnetron sputtering and patterned by wet etching to form the TFT gate electrodes. Then, a trilayer consisting of a-SiN_x (300 nm)/a-Si:H(50 nm)/a-SiN_x (300 nm) was deposited by PECVD, and source/drain contact windows were opened in the top nitride by wet etching. Next, we deposited a 50-nm-thick n^+-doped nc-Si-based film and a 300-nm thick a-SiN_x passivation layer. After etching off the n^+ layer around the source and drain contacts, the a-Si:H layer underneath them was etched off to electrically isolate the devices from each other. KOH solution was used for the etching. Finally, an ∼1-μm-thick aluminum film was deposited at room temperature by RF magnetron sputtering, and top contact pads were then formed by wet etching.

In comparison, a top-gate TFT fabrication process flow is shown in Fig. 3.1b. After the substrate was cleaned, a 60-nm-thick Cr film was deposited by RF magnetron sputtering at room temperature, followed by a 70-nm-thick n^+ nc-Si film deposited by PECVD. Next, both films were patterned by a combination of dry and wet etching, forming the source/drain contacts. After patterning, a bilayer consisting of a 100-nm-thick undoped nc-Si film and a 150-nm-thick gate dielectric was deposited by PECVD and then dry etched to electrically isolate single devices and get access to the source/drain contacts. Next, a 150-nm-thick dielectric film was deposited by PECVD to passivate the exposed TFT sidewalls. The source/drain contact windows were then opened in the dielectric by wet etching. Finally, a 1-μm-thick aluminum film was deposited by RF magnetron sputtering at room temperature and patterned by wet etching to form the source/drain electrode contacts and the gate electrode. TFTs were fabricated with various channel lengths (from 25 to 200 μm) and widths (from 100 to 200 μm).

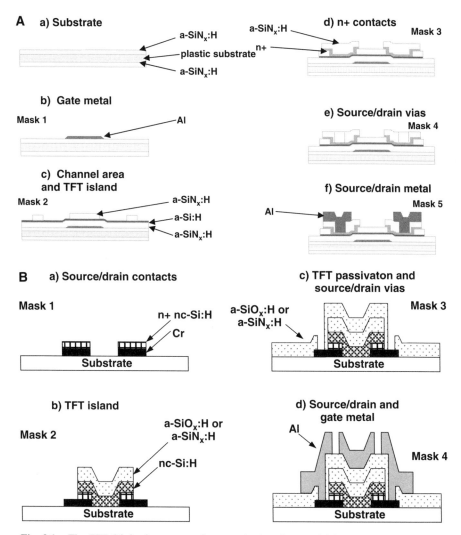

Fig. 3.1 The TFT fabrication process flow on plastic substrate: (**a**) bottom-gate structure and (**b**) top-gate structure

3.4.2 Low-Temperature a-Si:H Thin-Film Transistor Device Performance

Figure 3.2a,b shows the output and transfer characteristics, respectively, of a typical a-Si:H TFT fabricated at 300°C on glass substrates [42]. For TFT applications as switching devices, a step-like TFT transfer characteristics is desired: if the gate voltage, V_G, is below the threshold voltage, V_T, (i.e., $V_G < V_T$), then the drain current, I_D, is ideally zero; at $V_G > V_T$, the drain current, I_D, is limited by the load resistance.

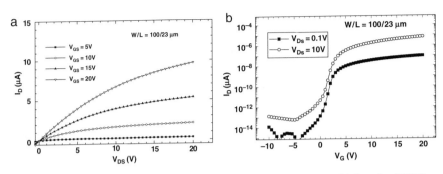

Fig. 3.2 The output (**a**) and transfer (**b**) characteristics of an a-Si:H TFT fabricated at 300°C on glass substrate [41]

In real TFTs, the leakage current, (I_{off}), flows through the TFT at $V_G < V_T$; the on-current, I_{on}, is limited by the channel conductivity, and the switching occurs within the range of V_G and is characterized by the subthreshold slope, $S = \left[\frac{\partial(\log I_D)}{\partial V_G}\right]^{-1}$ [43].

The TFT drain current depends on the gate voltage as

$$I_D = \mu_{FE} \cdot C \cdot (W/2L) \cdot (V_G - V_T)^2, \qquad (3.1)$$

where μ_{FE} , C, W, and L are the TFT field-effect mobility, the gate capacitance per unit area, the channel width, and the channel length, respectively. Note that the TFT mobility extracted from the transfer characteristics using Eq. (3.1) may be significantly affected by the source/drain contact resistance and thus is lower than the field-effect mobility in the channel.

The device in Fig. 3.2 is characterized by low I_{off} (<0.1 pA), the I_{on} above 10 μA, low threshold voltage ($V_T \sim 2.5$ V), and the subthreshold slope of 0.3 V/dec. The field-effect mobility extracted from these characteristics using Eq. (3.1) is about 1 cm² V⁻¹ s⁻¹.

3.4.3 Contacts to a-Si:H Thin-film Transistors

In Fig. 3.3a,b, the transfer characteristics of an a-Si:H TFT fabricated at 120°C with n⁺ nc-Si:H contacts is shown along with the dependencies of extracted μ_{FE} and V_T on the channel length [27]. The I_{off} and V_T values are similar to those at 300°C. However, the S value increased to 0.5 V/dec, which is attributed to higher interface defect density, and μ_{FE} decreased to 0.8 cm² V⁻¹ s⁻¹. The I_{on} decreased, and the resulting I_{on}/I_{off} ratio is about $10^6 - 10^7$, lower than at 300°C. As seen from Fig. 3.3b, the extracted μ_{FE} goes down in the shorter channel TFTs ($L < 75$ μm), which can be explained by increased contribution of the source/drain contact series resistance [42]. In its ON state, a TFT can be represented by a series of the channel resistance

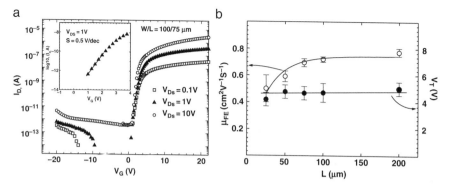

Fig. 3.3 (a) The transfer characteristics of an a-Si:H TFT fabricated at 120°C and (b) the dependencies of μ_{FE} and V_T on the channel length [20]

(R_{ch}) and the source/drain resistances (R_{DS}):

$$V_{DS} = I_D(R_{ch} + R_{DS}) = I_D R_m. \tag{3.2}$$

Here, R_{ch} is the function of the channel length, whereas R_{DS} does not depend on the channel length, L:

$$R_m = R_{DS} + R_{ch} = R_{DS} + AL, \tag{3.3}$$

where

$$A = \frac{2}{k} \cdot \frac{V_{ch}}{(V_G - V_T)} \cdot \frac{1}{W}, \tag{3.4}$$

V_{ch} is the voltage drop across the channel, and $k = \mu_{FE} \cdot C$.

In the case of a long TFT channel or low R_{DS}, we have $R_{ch} \gg R_{DS}$ and the μ_{FE} value is not affected. If R_{ch} and R_{DS} are comparable, then I_D is affected by R_{DS}, and the extracted μ_{FE} value is lower than the true field-effect mobility. From the $R_m(L)$ dependence extrapolated to zero channel length, R_{DS} can be extracted. The results are given in Table 3.5. It is seen that at lower deposition temperatures, the R_{DS} increases, which is attributed to a lower doping efficiency in n$^+$-doped layer. High R_{DS} also explains the I_{on} reduction at lower deposition temperatures.

In Fig. 3.4a,b, the transfer characteristics of an a-Si:H TFT fabricated at 75°C with n$^+$ a-Si:H contacts is shown along with the dependencies of μ_{FE} and V_T on the channel length [25]. The I_{off} is still below 1 pA, and so is the gate leakage current, I_g, indicating the high quality of the bulk gate dielectric. The I_{on}, however, decreased by

Table 3.5 The a-Si:H TFT source/drain contact series resistance, R_{DS}, at different fabrication temperatures

Deposition temperature	260°C	75°C
R_{DS} W, kΩ·cm	0.229	932

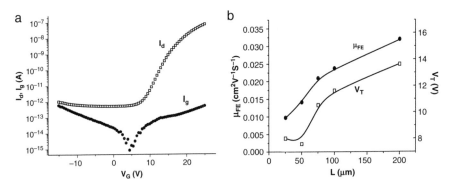

Fig. 3.4 (a) The transfer characteristics of an a-Si:H TFT fabricated at 75°C and (b) the dependencies of μ_{FE} and V_T on the channel length [26]

two orders of magnitude, V_T increased to 10 V, and the S value went up to 2.1 V/dec. Consequently, the extracted μ_{FE} dropped down to 0.03 cm^2 V^{-1} s^{-1}. High V_T value is usually attributed to increased built-in charge in the gate dielectric and high trap density in a-Si:H layer, and a large S value typically stems from high defect density at the interface between the gate dielectric and the channel [44]. As seen from Fig. 3.4b, μ_{FE} increases with the channel length up to $L = 200\,\mu$m, which means that the source/drain series resistance is comparable with the channel resistance even in long channel TFTs. The calculated $R_{DS} = (93.2 \pm 3.7)$ MΩ (see Table 3.4) is four orders of magnitude higher than at 260°C [44]. Therefore, extracted μ_{FE} value is well underestimated in this case.

3.4.4 Low-temperature Doped nc-Si Contacts

From the results in Figs. 3.2–3.4, the a-Si:H TFTs fabricated at low deposition temperature have increased source/drain series resistance and more defects at the interface between the gate dielectric and the channel. The series resistance can be reduced by using highly doped n$^+$ nc-Si thin films [45–47]. In Fig. 3.5, a TFT transfer characteristics with n$^+$ nc-Si contacts is shown in comparison with that of a TFT with n$^+$ a-Si:H contacts. In TFT with n$^+$ nc-Si contacts, the I_{on} is one order of magnitude higher, whereas I_{off} is one order of magnitude lower; thus, I_{on}/I_{off} ratio is two orders of magnitude higher.

Improving the interface between the gate dielectric and the channel can be achieved by the plasma treatments of the gate dielectric surface prior to the channel layer deposition [48] or by using a-SiO$_x$ as the gate dielectric (in a-SiO$_x$, the interface states density is reported to be one order of magnitude lower than in a-SiN$_x$) [26]. In Fig. 3.6a,b, the transfer characteristics of an a-Si:H TFT fabricated at 75°C with a-SiO$_x$ gate dielectric is demonstrated along with the dependencies of μ_{FE} and V_T on the channel length [26]. Despite high I_{off} originating from the high oxide

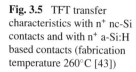

Fig. 3.5 TFT transfer characteristics with n⁺ nc-Si contacts and with n⁺ a-Si:H based contacts (fabrication temperature 260°C [43])

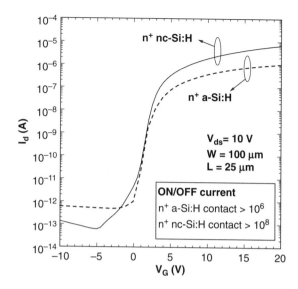

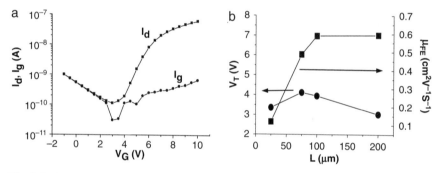

Fig. 3.6 (a) The transfer characteristics of an a-Si:H TFT fabricated at 75°C with a-SiO$_x$ gate dielectric and (b) the dependencies of μ_{FE} and V_T on the channel length [19]

porosity [26], the S value improved to 1 V/dec, and V_T went down to 3 V. From Fig. 3.6b, the contact resistance is much lower than R_{ch} at $L > 100$ µm, which was achieved by using n⁺ nc-Si contacts. Therefore, in this case, the contribution of R_{DS} into the estimated value of μ_{FE} is low. The calculations give us $\mu_{FE} = 0.6$ cm² V⁻¹ s⁻¹ – a dramatically improved result comparable to that of an a-Si:H TFT fabricated entirely at 300°C.

Figure 3.7a,b shows the output and transfer characteristics of a top-gate nc-Si TFT fabricated at 75°C on glass substrate with a-SiN$_x$ gate dielectric. No current crowding is visible on the output characteristics, which demonstrates good quality of the source/drain contacts. The device transfer characteristics show strong dependence on the gate voltage, with the I_{off} varying between 10^{-9} A and 10^{-12} A, and the I_{on} between 10^{-8} A and 10^{-6} A. The V_T is below 10 V, and the subthreshold slope

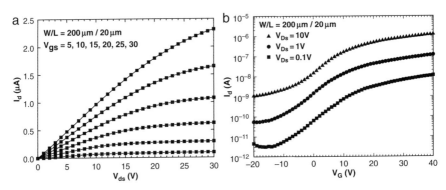

Fig. 3.7 The output (**a**) and transfer (**b**) characteristics of a top-gate nc-Si TFT fabricated at 75°C with a-SiN$_x$ gate dielectric

is about 4 V/dec. Both values are very high, indicating poor interface between the channel and the gate dielectric. The μ_{FE} is 0.02–0.05 cm^2 V^{-1} s^{-1}. While further device optimization is needed, the use of nc-Si as an active layer is advantageous due to much higher μ_{FE} (about 40 cm^2 V^{-1} s^{-1}) that can be achieved compared to a-Si:H TFTs [11].

Thus, high-quality TFTs can be fabricated at substrate temperatures below 150°C by using standard industrial plasma-deposition equipment. Such a reduction in the deposition temperature leads to increased series resistance of the source/drain contact layer, and to higher defect density at the interface between the channel layer and the gate dielectric. To eliminate the high series resistance, n$^+$ nc-Si has been used to form the source and drain contacts, and the use of oxide-based gate dielectrics can improve the interface. TFTs fabricated at 120C demonstrate the same performance as their high temperature counterparts (μ_{FE} of 0.8 cm^2 V^{-1} s^{-1} and V_T of 4.5 V), and can be used in AMOLED displays. The deposition temperature can be further reduced to 75°C, and similar performance (including μ_{FE} of 0.6 cm^2 V^{-1} s^{-1} and V_T of $\sim$ 4 V) can be achieved.

3.4.5 Low-temperature nc-Si TFTs

Although low-temperature processed a-Si:H TFT technology is highly compatible with flexible plastic substrates, high throughput and hence low cost, and large-area processing capability, the circuit performance of a-Si:H is limited when it comes to application areas, which go beyond current flat-panel imagers and displays that demand high speed and system-on-panel integration. In particular, there is a strong quest for RF and higher frequency applications involving reduced power, reduced bandwidth, and higher information content. While RF performance may potentially be achieved with laser-recrystallized a-Si:H to yield high-mobility poly-Si, several issues remain with that technology, notably scalability, throughput and cost, nonuniformity, and more importantly, plastic substrate compatibility. Other silicon-based technologies have also been reported, including TFTs made from highly crystalline silicon thin films or nanowires using a variety of solution and dry processing, and including transfer techniques with low thermal budget and with performance matching that of single crystal devices (see [8] and references therein). While these approaches offer potentially low cost and highly promising performance, they are still very much at the research phase and have yet to demonstrate the reliability and stability specifications dictated by the industry.

A possible technological solution that is economically viable, plastic-compatible, and which can meet the immediate requirements of FPDs (liquid crystal and OLED) is nc-Si. In particular, with AMOLED displays, a TFT field-effect mobility of at least 5 cm^2 V^{-1} s^{-1} can provide significant design leverage and considerably enhances lifetime under low supply voltages [49]. A higher mobility also decreases the settling time and/or the compensation error to meet frame rate requirements for large-area displays. But can such TFT mobilities be achieved with nc-Si? Preliminary results are promising and show that it is indeed possible to obtain high field-effect mobilities using fine grain nc-Si with high crystallinity and passivated grain boundaries. TFTs fabricated at plastic-compatible temperatures of 150°C show n- and p-channel maximum mobilities of 450 cm^2 V^{-1} s^{-1} and 150 cm^2 V^{-1} s^{-1}, respectively [50]. However, the question remains on whether industrial scale RF PECVD can be employed for fabrication of nc-Si TFT backplanes and at the same time maintains the same cost position as a-Si:H.

3.5 Device Stability

Thin-film silicon-based TFTs exhibit threshold voltage (V_T) shift under bias stress that causes the drain current to decrease with time. There are two main mechanisms of V_T shift: (1) defect states generation in the channel layer and (2) charge trapping in the gate dielectric [51]. Defect generation dominates in a-Si:H TFTs; in that case, threshold voltage shift is temperature- and bias-dependent and irreversible upon stress removal. Charge trapping prevails in poly-Si TFTs; here, ΔV_T saturates once the trap states at the gate dielectric interface are filled up, and returns back to its initial state upon charge detrapping.

A decrease in deposition temperature down to the values compatible with low-cost plastic foils results in poorer electronic properties in a-Si:H and in higher charge-trapping rate in the gate dielectric (see Section 3.3 in this chapter). This behavior, in turn, causes the threshold voltage shift to increase. It was reported that in a-Si:H TFTs fabricated at the substrate temperature of 110°C, a-Si:H defect generation increased fivefold, whereas the charge trapping in the nitride gate dielectric went up by one order of magnitude [33]. Therefore, optimization of both low-temperature a-Si:H and a-SiN$_x$ is required in order to reduce ΔV_T. Bias stressing performed on a-Si:H TFTs fabricated at 150°C using optimized a-Si:H and a-SiN$_x$ gate dielectric shows that ΔV_T decreased to the values similar to those measured in high-temperature devices [52].

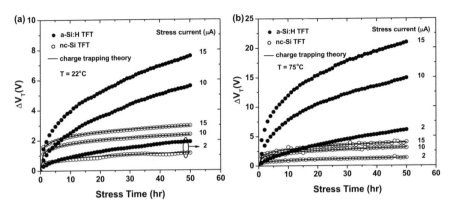

Fig. 3.8 Threshold voltage shift as a function of time for different stress currents and temperatures: (**a**) 22°C and (**b**) 75°C. *Filled circles* – a-Si:H TFT; *open circles* – nc-Si TFT; and *lines* – calculations using stretched-exponential Eq. (3.1) for charge trapping. The a-Si:H data are from [55]

Further ΔV_T reduction can be achieved by using nc-Si TFTs. It was reported that ΔV_T in bottom-gate nc-Si TFTs is remarkably lower than in a-Si:H counterparts [53]. Esmaeili Rad et al. have shown that ΔV_T in nc-Si TFTs is controlled by the charge trapping in the gate dielectric [54]. Figure 3.8a,b, shows the V_T shift in bottom-gate nc-Si TFTs [54]. Here, the TFTs were subject to 50 h of continuous constant current applied to the drain, while keeping the drain and gate terminals shorted in a diode-connected configuration with the source grounded. This configuration allows on-line monitoring of the gate voltage and ΔV_T extraction. Figure 3.8a,b shows ΔV_T for stress currents of 2, 10, and 15 µA at 22 and 75°C, respectively. The data previously reported on a-Si:H TFTs are also shown for comparison [55]. Here, two fundamental differences are noticeable. First, ΔV_T in nc-Si TFT saturates at prolonged stress times, but does not in a-Si:H TFT. Second, ΔV_T in nc-Si TFT is weakly temperature-dependent, in contrast to ΔV_T in a-Si:H-based device. From Fig. 3.8, after 50 h stressing at 15 µA, ΔV_T in nc-Si TFT is 3 V and 4 V at 22 and 75°C, respectively, whereas ΔV_T in a-Si:H TFT is 7.6 V and 21 V at 22 and 75°C, respectively. This behavior suggests a difference in the ΔV_T mechanisms between a-Si:H and nc-Si TFTs. It should be noted that the initial threshold voltage values are comparable in both devices ($\sim$4 V for nc-Si and $\sim$2.6 V for a-Si:H), and the initial applied gate bias is also comparable and less than 25 V [55]. In this range of gate bias stress, the V_T shift in a-Si:H TFTs is commonly attributed to defect state creation in the channel [55]. Since the band tail carrier density (n_{BT}) remains unchanged during constant current stress, V_T of a-Si:H TFT may increase indefinitely until the gate voltage reaches the supply voltage or the density of weak bonds (N_{WB}) becomes a rate limiting factor [55]. At higher stress voltages, however, defect state creation in a-Si:H TFTs is no longer dominant [51]. In that case, ΔV_T is controlled primarily by charge injection from the channel into the gate dielectric interface without subsequent penetration in the bulk, and this process is temperature-independent. Our measurement results show that ΔV_T is only weakly temperature-dependent, which is consistent with the results reported by Powell et al. [51].

ΔV_T can be quantitatively described by considering the stretched-exponent model for the charge-trapping kinetics in a-Si:H TFTs:

$$\Delta V_T = C \left(1 - \exp\left[-\left(\frac{t}{\tau}\right)^{\beta} \right] \right) \tag{3.5}$$

where C, τ, and β are the fitting parameters [51]. ΔV_T calculations using Eq. (3.1) are shown by solid lines in Fig. 3.8a,b. As seen, the stretched-exponential time dependence fits well the measurement data. Values of the extracted parameters are given in Table 3.6. The parameter C was set to $(2/3)(V_{GS}-V_T)$ in our calculations, as opposed to $(V_{GS}-V_T)$ in [51]. Here, the difference is due to device operation regime, which is saturation in our case and linear in the latter. ΔV_T follows the channel charge concentration, which at pinch-off ($V_{DS} = V_{GS}-V_T$) is about two-third of that in linear regime ($V_{DS} = 0$). This relationship is nearly valid although $V_{DS} = V_{GS}$ in this series of experiments.

When the temperature increases from 22 to 75°C, the parameter τ changes in the range 10^8–10^7 s in nc-Si TFTs and 10^8–10^5 s in a-Si:H TFTs [51]. Smaller τ leads to larger ΔV_T for a given stress time. One can see that an increase in temperature reduces τ in nc-Si/a-Si:H TFTs by one order of magnitude, whereas in the case of a-Si:H, its reduction is by three orders of magnitude. This result implies that nc-Si TFTs have higher stability and long-term reliability even at high operation temperatures. From Table 3.6, one may also find that the parameter β is fairly constant at various stress currents and temperatures.

Similar observations have been made for poly-Si TFTs [56] and even for nc-Si TFTs made on PET foil by transfer process [57]. Thus, the threshold voltage shift in nc-Si TFTs is controlled by the charge trapping in the gate dielectric, and a control over the interface defect density is particularly important at low deposition temperatures.

Table 3.6 Parameter values for Eq. (3.1) resulting from fit to measured data indicated in Fig. 3.8a,b

Temperature (°C)	Stress current (μA)	C (V)	β	τ (h)
22	2	5	0.25 ± 0.01	10,000 ± 2,549
22	10	10	0.2018 ± 0.001	32,422 ± 1,894
22	15	14	0.187 ± 0.001	42,730 ± 2,496
75	2	5	0.268 ± 0.0098	3,791 ± 701
75	10	10	0.2374 ± 0.006	3,262 ± 446
75	15	12	0.247 ± 0.009	2,099 ± 345

3.6 Conclusions and Future Prospective

A number of parameters for low-temperature growth of silicon materials have been presented. Conventional thin-film silicon devices offer a wide range of advantages including entrenched fabrication facilities, known properties, and low barrier for entry into large-area electronics applications. Here, a number of issues need to be addressed with implications on growth rate, gate dielectric quality and/or pertinent interface modification, and device stability, in the low-temperature range, all of which require significant further research. One of the key issues that must be addressed here is the industrial implementation of nc-Si TFTs, in which a high growth rate is imperative. nc-Si films can be deposited at 0.01–0.1 nm/s using high dilution of silane with hydrogen; high dilution and high flow rate tend to limit the growth rate. However, it has been shown [58] that the growth rate can be increased to 2 nm/s at higher pressures (7–9 Torr) by increasing the very-high frequency power and moving into the silane depletion regime. Films deposited at higher pressure have been found to consist of denser grains, limiting the oxidation to yield higher short-circuit current in solar cells. Growth rates of 1.5–2 nm/s would be acceptable by the industry; however, further work is needed to achieve the desired growth rates in 13.56-MHz plasma while maintaining optimal nc-Si microstructure and transport property.

References

1. Fruehauf N, Chalamala BR, Gnade BE, Jang J (2004) Flexible Electronics 2004 – Materials and Device Technology, vol 814. Materials Research Society Symposia Proceedings, Pittsburgh, PA, USA
2. Sheats JR (2002) Roll-to-roll manufacturing of thin film electronics, Proc SPIE 4688: 240–248
3. Allen K (2004) Reel to real: Prospects for flexible displays, In: Fruehauf N, Chalamala BR, Gnade BE, Jang J (eds) Flexible Electronics 2004 – Materials and Device Technology, vol 814. Materials Research Society Symposia Proceedings, Pittsburgh, PA, USA, pp I1.1.1–I1.1.3
4. Ichikawa Y et al. (2001) Production technology for amorphous silicon-based flexible solar cells, Solar Energy Mater Solar Cells 66:107–115
5. Kim BS et al. (2004) Developments of transmissive a-Si TFT-LCD using low temperature processes on plastic substrate, In: SID 2004 International Symposium Digest, pp 19–21
6. McDermott J, Brantner PC (2003) Thin-film solid-state lithium battery for body worn electronics, In: Shur MS, Wilson PM, Urban R (eds) Electronics on Unconventional Substrates – Electrotextiles and Giant-Area Flexible Circuits, vol 736. Materials Research Society Symposia Proceedings, Pittsburgh, PA, USA, pp D5.1.1–D5.1.9
7. Kelley TW et al. (2003) High performance organic thin film transistors, In: Blom PWM, Greenham NC, Dimitrakopoulos CD, Frisbie CD (eds) Organic and Polymeric Materials and Devices, vol 771. Materials Research Society Symposia Proceedings, Pittsburgh, PA, USA, pp L6.5.1–L6.5.11
8. Reuss R et al. (2005) Macroelectronics: Perspectives on technology and applications, Proc IEEE 93:1239
9. Lewis JS, Weaver MS (2004) Thin-film permeation-barrier technology for flexible organic light-emitting devices, IEEE J Selected Top Quantum Electron 10:45–57
10. Kelley T (2006) High-performance pentacene transistors, In: Klauk H (ed) Organic Electronics, Wiley-VCH, Weinheim, pp 35–57

11. Wagner S, Gleskova H, Cheng IC, Wu M (2003) Silicon for thin-film transistors, Thin Solid Films 430:15–19
12. Plichta A, Weber A, Habeck A (2003) Ultra thin flexible glass substrates, In: Fruehauf N, Chalamala BR, Gnade BE, Jang J (eds) Flexible Electronics – Materials and Device Technology, vol 769. Materials Research Society Symposia Proceedings, Pittsburgh, PA, USA, pp H9.1.1–H9.1.10
13. Afentakis T, Hatalis MK, Voutsas AT, Hartzell JW (2003) High performance polysilicon circuits on thin metal foils, Proc SPIE 5004:122–126
14. Inoue S, Utsunomiya S, Saeki T, Shimoda T (2002) Surface-free technology by laser annealing (SUFTLA) and its application to poly-Si TFT-LCDs on plastic film with integrated drivers, IEEE Trans Electron Dev 49:1353–1360
15. Jongerden GJ (2003) Monolithically series integrated flexible PV modules manufactured on commodity polymer substrates, Proceeding of Third World Conference on Photovoltaic Energy Conversion, pp 2109–2111
16. Srinivasan E, Lloyd DA, Parsons GN (1997) Dominant monohydride bonding in hydrogenated amorphous silicon thin films formed by plasma enhanced CVD at room temperature, J Vac Sci Technol A 15:77
17. Gleskova H, Wagner S, Suo Z (1998) a-Si:H TFTs made on polyimide foil by PECVD at 150°C. Mater Res Soc Symp Proc 508:73
18. Parsons GN, Yang CS, Klein TM, Smith L (1999) Surface reactions for low temperature (110°C) amorphous silicon TFT formation on transparent plastic, Mater Res Soc Symp Proc 507:19
19. Robertson J (2000) Deposition mechanism of hydrogenated amorphous silicon, J Appl Phys 87:2608–2617
20. Perrin J (1995) Reactor design for a-Si:H deposition, In: Bruno G, Capezzuto P, Madan A (eds) Plasma Deposition of Amorphous-Based Materials, Academic Press, San Diego, CA, pp 177–241
21. Sazonov A, Nathan A (2000) A 120°C fabrication technology for a-Si:H thin film transistors on flexible polyimide substrates, J Vac Sci Technol A 18:780–782
22. Nathan A, Striakhilev D, Servati P, Sakariya K, Sazonov A, Alexander S, Tao S, Lee CH, Kumar A, Sambandan S, Jafarabadiashtiani S, Vygranenko Y, Chan IW (2004) a-Si AMOLED display backplanes on flexible substrates, In: Fruehauf N, Chalamala BR, Gnade BE, Jang J (eds) Flexible Electronics 2004 – Materials and Device Technology, vol 814. Materials Research Society Symposia Proceedings, Pittsburgh, PA, USA, pp I3.1.1–I3.1.12
23. Sazonov A, Striakhilev D, Nathan A (2000) Materials optimization for TFTs fabricated at low temperature on plastic substrate, J Non-Cryst Solids 266–269:1329–1334
24. Lee CH, Sazonov A, Nathan A (2004) Low temperature (75°C) hydrogenated nanocrystalline silicon films grown by conventional plasma enhanced chemical vapor deposition for thin film transistors, In: Ganguly G, Kondo M, Schiff EA, Carius R, Biswas R (eds) Amorphous and Nanocrystalline Silicon Science and Technology – 2004, vol 808. Materials Research Society Symposia Proceedings, Pittsburgh, PA, USA, pp A4.17.1–A4.17.6
25. McArthur C, Meitine M, Sazonov A (2003) Optimization of 75°C amorphous silicon nitride for TFTs on plastics, In: Fruehauf N, Chalamala BR, Gnade BE, Jang J (eds) Flexible Electronics – Materials and Device Technology, vol 769. Materials Research Society Symposia Proceedings, Pittsburgh, PA, USA, pp 303–308
26. Meitine M, Sazonov A (2003) Low temperature PECVD silicon oxide for devices and circuits on flexible substrates, In: Fruehauf N, Chalamala BR, Gnade BE, Jang J (eds) Flexible Electronics – Materials and Device Technology, vol 769. Materials Research Society Symposia Proceedings, Pittsburgh, PA, USA, pp 165–170
27. Stryahilev D, Sazonov A, Nathan A (2002) Amorphous silicon nitride deposited at 120°C for OLED-TFT arrays on plastic substrates, J Vac Sci Technol A 20:1087–1090
28. Sazonov A, Nathan A (2000) 120°C Fabrication technology for a-Si:H thin film transistors on flexible polyimide substrates, J Vac Sci Technol A18:780–782

29. Kuo Y (2004) Deposition of dielectric thin films for a-Si:H TFT. In: Kuo Y (ed) Thin Film Transistors, Materials and Processes, vol I. Kluwer Academic Publishers, Boston, MA, pp 241–271

30. Meitine M, Sazonov A (2003) Low temperature PECVD silicon oxide for devices and circuits on flexible substrates, In: Fruehauf N, Chalamala BR, Gnade BE, Jang J (eds) Flexible Electronics – Materials and Device Technology, vol 769. Materials Research Society Symposia Proceedings, Pittsburgh, PA, USA, pp 165–170

31. Kattamis A, Cheng IC, Allen S, Wagner S (2004) Hydrogen in ultralow temperature SiO_2 for nanocrystalline silicon thin film transistors, In: Fruehauf N, Chalamala BR, Gnade BE, Jang J (eds) Flexible Electronics 2004 – Materials and Device Technology, vol 814. Materials Research Society Symposia Proceedings, Pittsburgh, PA, USA, pp I10.14.1–I10.14.6

32. Rashid R, Flewitt AJ, Grambole D, Kreibig U, Robertson J, Milne WI (2001) High quality growth of SiO_2 at 80°C by electron cyclotron resonance (ECR) for thin film transistors, In: Im JS, Werner JH, Uchikoga S, Felter T, Voutsas T, Kim HJ (eds) Advanced Materials and Devices for Large-Area Electronics, vol 695E. Materials Research Society Symposia Proceedings, Pittsburgh, PA, USA, pp D13.1.1–D13.1.6

33. Yang CS, Smith LL, Artur CB, Parsons G (2000) Stability of low-temperature amorphous silicon thin film transistors formed on glass and transparent plastic substrates, J Vac Sci Tech B18:683.

34. Sazonov A, Nathan A, Striakhilev D (2000) Materials optimization for thin film transistors fabricated at low temperature on plastic substrate, J Non-Cryst Solids 266–269:1329

35. Gleskova H, Wagner S, Gašparik V, Kováč P (2001) 150°C Amorphous silicon thin-film transistor technology for polyimide substrates, J Electrochem Soc 148:G370

36. Revesz AG, Anwand W, Brauer G, Hughes HL, Skorupa W (2002) Density gradient in SiO_2 films on silicon as revealed by positron annihilation spectroscopy, Appl Surf Sci 194:101

37. Hsieh SW, Chang CY, Hsu SC (1993) Characteristics of low-temperature and low-energy plasma-enhanced chemical vapor deposited SiO2. J Appl Phys 74:2638

38. Martinu L, Poitras D (2000) Plasma deposition of optical films and coatings: A review, J Vac Sci Technol A18:2619

39. Pereyra I, Alayo MI (1997) High quality low temperature DPECVD silicon dioxide, J Non-Cryst Solids 212:225

40. Han SS, Ceiler M, Bidstrup SA, Kohl P (1994) Modeling the properties of PECVD silicon dioxide films using optimized back-propagation neural networks, IEEE Trans Comp Pack Manufac Technol A17:174

41. Street RA (1991) Hydrogenated Amorphous Silicon, Cambridge University Press, Cambridge, pp 18–61

42. Nathan A, Servati P, Karim KS, Striakhilev D, Sazonov A (2004) Device physics, compact modeling, and circuit applications of a-Si:H TFTs, In: Kuo Y (ed) Thin Film Transistors, Materials and Processes, vol I. Kluwer Academic Publishers, Boston, MA, pp 79–181

43. Streetman BG, Banerjee S (2000) Solid State Electronic Devices, Prentice Hall, Upper Saddle River, NJ

44. Sazonov A, McArthur C (2004) Sub-100C a-Si:H TFTs on plastic substrates with silicon nitride gate dielectrics, J Vac Sci Technol A22:2052–2055

45. Charania T, Sazonov A, Nathan A (2000) Use of 120°C n^+-μc-Si:H in low temperature TFT fabrication, In: Kuo J (ed) Thin Film Transistor Technologies V, vol 2000-31. Proceeding of the Electrochemical Society, Pennington, NJ, pp 4–62

46. Meitine M, Sazonov A (2004) Top gate TFT for large area electronics, In: Fruehauf N, Chalamala BR, Gnade BE, Jang J (eds) Flexible Electronics 2004 – Materials and Device Technology, vol 814. Materials Research Society Symposia Proceedings, Pittsburgh, PA, USA, pp I6.12.1–I6.12.6

47. Lee CH, Stryahilev D, Nathan A (2004) Intrinsic and doped μc-Si:H TFT layers using 13.56 MHz PECVD at 250C. In: Ganguly G, Kondo M, Schiff EA, Carius R, Biswas R (eds)

Amorphous and Nanocrystalline Silicon Science and Technology – 2004, vol 808. Materials Research Society Symposia Proceedings, Pittsburgh, PA, pp A4.14.1–A4.14.6

48. Umezu I, Kuwamura T, Kitamura K, Tsuchida T, Maeda K (1998) Effect of plasma treatment on the density of defects at an amorphous Si:H-insulator interface, J Appl Phys 84:1371–1377

49. Nathan A, Striakhilev D, Chaji R, Ashtiani S, Lee CH, Sazonov A, Robertson J, Milne W (2006) Backplane requirements for active matrix organic light-emitting diode displays, MRS Symp Proc 910:373–387

50. Lee CH, Sazonov A, Nathan A, Robertson J (2006) Directly deposited nanocrystalline silicon thin-film transistors with ultra high mobilities, Appl Phys Lett 89:252101–252103

51. Powell MJ, van Berkel C, Hughes JR (1989) Time and temperature dependence of instability mechanisms in amorphous silicon thin-film transistors, Appl Phys Lett 54:1323

52. Gleskova H, Wagner S (2001) DC-gate-bias stressing of a-Si :H TFTs fabricated at 150°C on polyimide foil, IEEE Trans Electron Dev 48:1667–1671

53. Lee CH, Striakhilev D, Nathan A (2007) Stability of nc-Si:H TFTs with silicon nitride gate dielectric, IEEE Electron Dev Lett 54:45–51

54. Esmaeili-Rad MR, Sazonov A, Nathan A (2007) Absence of defect state creation in nanocrystalline silicon thin film transistors deduced from constant current stress measurements, Appl Phys Lett 91:113511

55. Jahinuzzaman SM, Sultana A, Sakariya K, Servati P, Nathan A (2005) Threshold voltage instability of amorphous silicon thin-film transistors under constant current stress, Appl Phys Lett 87:023502

56. Jeong Y, Nagashima D, Kuwano H, Nouda1 T, Hamada H (2002) Effects of various hydrogenation processes on bias-stress-induced degradation in p-channel polysilicon thin film transistors, Jpn J Appl Phys 41:5048–5054

57. Yuan HC, Celler GK, Ma Z (2007) Observation of threshold-voltage instability in single-crystal silicon TFTs on flexible plastic substrate, IEEE Electron Dev Lett 28:590–592

58. Matsui T, Matsuda A, Kondo M (2004) High-Rate Plasma Process for Microcrystalline Silicon: Over 9% Efficiency Single Junction Solar Cells, In: Ganguly G, Kondo M, Schiff EA, Carius R, Biswas R (eds) Amorphous and Nanocrystalline Silicon Science and Technology – 2004, p.A8.1.1. Materials Research Society Symposia Proceedings 808, Warrendale, PA

Chapter 4
Amorphous Silicon: Flexible Backplane and Display Application

Kalluri R. Sarma

4.1 Introduction

Advances in the science and technology of hydrogenated amorphous silicon (a-Si:H, also referred to as a-Si) and the associated devices including thin-film transistors (TFT) during the past three decades have had a profound impact on the development and commercialization of major applications such as thin-film solar cells, digital image scanners and X-ray imagers and active matrix liquid crystal displays (AML-CDs). Particularly, during approximately the past 15 years, a-Si TFT-based flat panel AMLCDs have been a huge commercial success. a-Si TFT-LCD has enabled the note book PCs, and is now rapidly replacing the venerable CRT in the desktop monitor and home TV applications. a-Si TFT-LCD is now the dominant technology in use for applications ranging from small displays such as in mobile phones to large displays such as in home TV, as well-specialized applications such as industrial and avionics displays. a-Si TFT-LCDs as large as $108''$ diagonal for TV applications have been demonstrated (by Sharp Corporation at CES, Las Vegas, NV, in January 2007). While this flat rigid glass substrate-based a-Si TFT backplane and the associated electronics and display technologies continue to mature, recently over the past several years, there is a growing interest and investments in the development of flexible a-Si TFT backplanes for large-area electronics and display applications.

The high level of current interest in large-area flexible electronics and displays fabricated using flexible plastic or metal foil substrates is due to their potential for being very thin, light weight, highly rugged with greatly minimized propensity for breakage, and amenable to low-cost RTR manufacturing, in comparison to devices built on the conventional rigid glass substrates. In addition, flexible electronics and displays can enable a variety of new applications due to their ability to have unique form factors and to be curved and conformable. Further, flexible displays can be rollable or foldable when not in use during storage or transportation. During the

K.R. Sarma (✉)
Honeywell International, 21111 N. 19th Avenue, Phoenix, AZ 85036, USA
e-mail: kalluri.r.sarma@honeywell.com

W.S. Wong, A. Salleo (eds.), *Flexible Electronics: Materials and Applications*, Electronic
Materials: Science & Technology, DOI 10.1007/978-0-387-74363-9_4,
© Springer Science+Business Media, LLC 2009

recent years, significant progress has been made in the flexible display development and the interest and R&D investments in flexible display development continue to grow significantly. At the time of this writing, market studies [1] project that flexible display market will have annual revenue of over $1.9 billion per year by the year 2015. Initial flexible display products using monochrome, bistable, electrophoretic display (EPD) media have already been announced for market introduction in 2007 by companies such as Polymer Vision [2].

Development of a flexible TFT backplane is a crucial enabler for realizing flexible electronic systems and displays properties. In this chapter, we will discuss the recent progress in the development of flexible a-Si TFT backplanes and displays with a focus on the approach using flexible plastic substrates and organic light-emitting diode (OLED) display media. Flexible OLED displays are believed to be the holy grail of the flexible display development efforts. In this chapter, we will first discuss the required enabling technologies of flexible backplanes including flexible substrates, barrier layers, and TFT technologies and processes that are compatible with flexible substrates. This will be followed by a discussion of the display media for flexible substrates. We will then discuss the requirements of the flexible a-Si TFT backplanes for OLED displays. This will be followed by the progress in the fabrication of low-temperature a-Si TFT backplanes and displays. For comparison purposes, the recent progress on flexible plastic backplanes with a-Si TFTs for the EPD media will also be discussed. Finally, we will discuss the outlook for future low-temperature a-Si TFT backplane and display manufacturing.

4.2 Enabling Technologies for Flexible Backplanes and Displays

Several enabling technologies must be developed to realize flexible TFT backplanes and displays. These enabling technologies include flexible substrates with the required characteristics, flexible substrate compatible (low-temperature) TFT designs and fabrication processes, display media, and appropriate barrier layers for the protection of the display media from the ambient (oxygen and moisture).

4.2.1 Flexible Substrate Technologies

Thin metal foils such as stainless steel, and thin polymer substrate materials are the leading candidates for use as flexible substrates. In the following, we will discuss the relative advantages and issues in these two approaches.

4.2.1.1 Flexible Stainless Steel Substrates

Metal foil substrates offer the advantages of higher process temperature capability (for TFT fabrication), dimensional stability (no shrinkage of the substrate during high-temperature processing associated with the TFT fabrication), and being impervious to oxygen and moisture (inherent barrier for the ambient oxygen and moisture). The disadvantages and limitations of the metal foil substrates include the following:

(1) Being opaque, it cannot be used for transmissive displays or bottom emission OLED displays. It can only be used for reflective displays and top emission OLED displays.
(2) Poor surface smoothness characteristics.
(3) Compatibility issues with the TFT process chemicals.

The smoothness of the metal foil substrates must be increased by polishing (for example, by using chemical-mechanical polishing CMP), and/or by applying additional surface smoothing (planarization) layers, to achieve acceptable yield of the TFT and OLED devices to be fabricated. Compatibility with the TFT process chemicals can be addressed by using an appropriate protective film at the backside of the stainless steel substrate. Note that a metal foil substrate, by itself, is a good barrier (for oxygen and moisture) and thus it does not require an additional barrier layer. However, the display fabricated using the metal foil substrate would still require a good barrier (encapsulation) layer to be applied on top of the fabricated TFT and the display media such as OLED. Another consideration in the use of metal foil substrate is the parasitic coupling capacitance due to coupling of the backplane electronics to the conductive substrate. Stainless steel is being actively investigated as a substrate for the flexible backplanes using low-temperature polysilicon (LTPS) TFT as well as a-Si TFT for reflective [e.g., 3–5] and top emission mode OLED [e.g., 6, 7] display applications.

4.2.1.2 Flexible Plastic Substrates

A transparent plastic substrate has the advantage of being compatible with transmissive as well as reflective displays. Thus it is compatible with both top- and bottom-emitting OLED device architectures, thereby making it suitable for a broader range of applications.

Table 4.1 shows the properties of some of the common candidate plastic substrate materials for flexible backplane and display fabrication. These candidate substrates include polyethylene terephthalate (PET – e.g. Melenix® from DuPont Teijin Films), polyethylene naphthalate (PEN, e.g., Teonex® Q65 from DuPont Teijin Films), poly carbonate (PC, e.g., Lexan® from GE), polyethersulfone (PES, e.g., Sumilite® from Sumitomo Bakellite), and polyimide (PI, e.g., Kapton® from DuPont). While Kapton has high glass transition temperature (Tg), it absorbs in the visible (yellow colour), and thus is not suitable for transmissive displays or bottom emission OLED displays. Higher process temperature ($>350°C$) capable clear plastic substrates are being developed and investigated [8, 9] for use as a drop-in replacement for glass with conventional (high-temperature) a-Si TFT fabrication process. However, as these high-temperature clear plastic substrates are not commercially available at this time, we will not discuss this further.

Table 4.1 Available candidate plastic substrates

	PET (Melinex)	PEN (Teonex)	PC (Lexan)	PES (Sumilite)	PI (Kapton)
T_g, °C	78	121	150	223	410
CTE (−55 to 85°C), ppm/°C	15	13	60–70	54	30−60
Transmission (400–700 nm), %	89	87	90	90	Yellow
Moisture absorption, %	0.14	0.14	0.4	1.4	1.8
Young's modulus, Gpa	5.3	6.1	1.7	2.2	2.5
Tensile strength, Mpa	225	275	–	83	231
Density, gcm^{-3}	1.4	1.36	1.2	1.37	1.43
Refractive index	1.66	1.5–1.75	1.58	1.66	–
Birefringence, nm	46	–	14	13	–

Some of the important limitations of the available plastic substrates include limited process temperature capability, lack of dimensional stability (during TFT processing involving high temperatures), and significant differences in the linear thermal coefficient of expansion (TCE) between the plastic substrate and the TFT thin films. Plastic substrates are believed to have a lower cost potential compared to the metal foil substrates. Based on the availability and the broad range of desirable film properties (in comparison to the other candidate polymer substrate materials), DuPont Teijin Film's (DTF) PEN substrates are being widely used in the development of flexible TFT backplanes for flexible OLED display [e.g., 10, 4, 5] and EPD [11]

Table 4.2 shows a comparison of the properties of the stainless steel and PEN substrates with the standard rigid glass substrates for use in TFT backplane applications. Major advantages of the stainless steel in comparison with PEN plastic films include higher process temperature capability (allowing direct fabrication of conventional a-Si or LTPS TFT arrays) and lower TCE mismatch with the TFT thin films. In comparison with the TCE values for stainless steel (10) and PEN (16),

Table 4.2 Comparison of PEN and stainless steel substrates with glass substrates

	Glass	PEN	Stainless steel
Weight, gm/m^2 (for 100-μm-thick film)	220	120	800
Transmission in the visible range (%)	92	90	0
Maximum process temperature (°C)	600	180	>600
TCE (ppm /°C)	5	16	10
Elastic modulus (Gpa)	70	5	200
Permeability for oxygen and moisture	No	Yes	No
Coefficient of hydrolytic expansion (ppm/%RH)	0	11	0
Planarization necessary	No	No	Yes
Electrical conductivity	None	None	High
Thermal conductivity (W/m.°C)	1	0.1	16

the TCE values for the common TFT materials are in the following range: glass = 5 ppm/°C, SiN_x = 1.5 ppm/°C, Si = 3.4 ppm/°C, Cr = 6.5 ppm/°C, Mo = 5 ppm/°C, and Al = 24 ppm/°C.

The oxygen and moisture barrier properties of stainless steel (while being excellent) are not believed to be compelling, as they do not obviate the need for an additional effective barrier layer for encapsulating the top side of the display media built on the substrate.

Based on availability and long-term potential, PEN plastic substrate approach continues to be a viable candidate for flexible OLED display development. In the next section, we will discuss the characteristics of the PEN plastic substrates in detail as they relate to a-Si TFT backplane fabrication.

4.2.1.3 Flexible PEN Plastic Substrates

Polyester films (e.g., PET and PEN from DTF) are well-known substrates for a wide range of electronic applications such as membrane touch switches and flexible circuitry [12]. New developments in polyester film substrates are contributing to the successful development of plastic substrates for use in flexible active matrix display applications. In this section, we will discuss the characteristics of the PEN films [13] as they are more widely used in the flexible a-Si TFT backplane development. These PEN-based substrates offer a unique combination of excellent dimensional stability, low moisture pickup, good solvent resistance, high clarity, and very good surface smoothness. This combination of attributes makes PEN a promising substrate (among the available plastic substrates) for subsequent vacuum and other coating processes for potential use in flexible active matrix backplane fabrication.

The technical challenges in the plastic substrates for active matrix display application are, however, extremely demanding. The plastic substrates, while being flexible, need to offer glass-like properties and must, therefore, have high clarity, smoothness of surface, excellent dimensional and thermal stability, and low thermal coefficient of thermal expansion (CTE) mismatch with the a-Si TFT thin films, coupled with excellent oxygen and moisture barrier properties. In the following, we will discuss the characteristics of the presently available PEN substrates, in relation to the requirements of TFT backplane and display applications.

Optical Properties

Good optical properties are achieved with Teonex® Q65 films by close control of the polymer recipe [14, 15]. Typically, Teonex® Q65 has a total light transmission (TLT) of 87% over 400–700 nm coupled with a haze of less than 0.7%. The substrate is optically clear and colorless, and thus can be used for transmissive or reflective displays and bottom- as well as top-emitting OLED displays. They are not, however, suitable for the liquid crystal displays (LCDs) because of their birefringence. As PEN is a semicrystalline biaxially oriented thermoplastic material, it is birefringent, and thus is not a suitable substrate for the LCD media that depend on the polarization control of the propagated light. Amorphous polymer substrates are not birefringent, and thus are suitable for the LCD media. Birefringence is not an issue for the OLED and EPD media.

Surface Quality

Surface smoothness and cleanliness are essential to prevent pinpricks in subsequent barrier coatings and to ensure that the defects from the substrate do not deleteriously affect the active matrix TFT manufacturing yield. Industrial grade PEN typically has a rough surface with a large (unacceptable) concentration of peaks of up to 0.1 μm high. By control of recipe and film process optimization, DTF achieves a much smoother surface without any 0.1-μm high peaks, but only a small concentration of 0.05-μm high peaks in Teonex® Q65. These remaining surface defects are still detrimental to the performance of the thin films deposited on top and are then removed by the application of a coating layer, typically comprised of a scratch resistant material. This coating layer acts to smooth over all the underlying PEN surface defects and additionally helps to prevent surface scratches during the subsequent substrate handling operations. This surface is now smooth without any peaks or protrusions and is acceptable for the TFT active matrix backplane fabrication using thin films with a thickness of the order of 100 nm.

Dimensional Stability

Dimensional stability during TFT array processing (involving temperature cycles between the room temperature and the TFT process temperatures) is extremely critical to ensure that the features in each layer of the TFT device structure align properly with the corresponding features in the previous layers. Plastic films undergo a variable and undesirable change in dimensions at the glass transition temperature, Tg, due to both molecular relaxation events associated with the increased mobility of the polymer chains and shrinkage or expansion associated with the relaxation of residual strain within the oriented parts of the film structure. This is an artifact of the film manufacturing conditions [14]. The dimensional stability of Teonex® Q65 produced by biaxial stretching is enhanced by a heat stabilization process where the internal strain in the film is relaxed by exposure to high temperature while under minimum line tension. Shrinkage at a given temperature is measured by placing the sample in a heated oven for a given period of time. The percent shrinkage is calculated as the percent change of dimension of the film in a given direction before and after heating. Heat-stabilized films exhibit shrinkage of the order of <0.1% and typically <0.05% when exposed to temperatures of up to 180°C for 5 min. Once Teonex® Q65 is heat stabilized, it remains a dimensionally reproducible substrate up to 200°C. Its improved thermal resistance provides a dimensionally reproducible substrate over this temperature range and permits a continuous use at a temperature of about 180°C. It should be noted that shrinkage of 0.05% is not acceptable for fabricating the TFT backplanes. In Section 4.4, we will discuss this further and describe a prestabilization process to reduce the shrinkage to manageable levels to fabricate low-temperature a-Si TFT backplanes and OLED displays.

Table 4.3 CTE of PEN (Teonex® Q65) as a function of temperature and orientation

	CTE (ppm/°C)			
	−50 to 0°C	0–50°C	50–100°C	100–150°C
Machine direction	13	16	18	25
Transverse direction	8	11	18	29

TCE, and more particularly the difference in the TCE of the plastic substrate and the TFT thin film materials, is an important factor in the backplane fabrication, due to the deleterious effect of the thermally induced strain (in the TFT thin films) during cooldown to room temperature from the process temperatures. The CTE in the heat-stabilized Teonex films varies with temperature and orientation (machine direction versus transverse direction) as shown in Table 4.3. Excessive strains/stresses result in film cracking, delamination, and substrate curling/buckling problems.

Solvent and Moisture Resistance

The Teonex® Q65 brand has excellent solvent resistance to most acids and organic solvents and will typically withstand the solvents used in AMOLED display fabrication. Indeed no specific issues of significance are observed using the Teonex® Q65 substrate during the fabrication of the a-Si TFT backplanes and AMOLED test displays. While the PEN substrate does not react with moisture, it does absorb moisture and results in a dimensional change. Figure 4.1 shows the moisture absorption in the

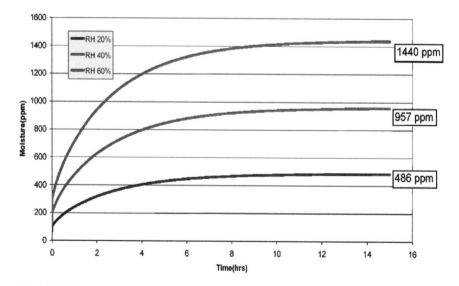

Fig. 4.1 Moisture absorption in PEN plastic substrates [15]

PEN substrate as a function of relative humidity (RH) and time [15]. At 40% RH, the equilibrium moisture concentration in the film is expected to be about 957 ppm which is very high since for every 100 ppm of moisture absorbed, the film is estimated to expand by ∼45 ppm. This is a very significant dimensional change and can deleteriously affect the TFT backplane process if it is not managed. Moisture absorption is reversible by heating the substrate in vacuum or in an inert atmosphere. Uncontrolled moisture absorption/desorption during the TFT backplane fabrication can potentially have far more impact on the substrate dimensional stability than the dimensional instability due to the inherent PEN substrate shrinkage. We will discuss this further in Section 4.4 on backplane fabrication.

Barrier Properties

The inherent barrier properties of PEN films are typically of the order of ca. 1 g/m^2/day for water vapor transmission rates and an equivalent ca of 3 mL/m^2/day for oxygen transmission rates. This is still a long way from the levels required for the protection of OLED displays, which require water vapor transmission rates of $<10^{-6}$ g/m^2/day and oxygen transmission rates of $<10^{-5}$ mL/m^2/day. No polymer substrate meets these requirements, and the flexible substrates currently being developed need to use an effective barrier layer to encapsulate the OLED devices for protection against oxygen and moisture ingression to enhance the OLED lifetime. Note that the EPDs have far less sensitive to moisture, and thus do not impose such stringent requirements on the barrier layer performance. The progress in the barrier layer development is discussed in Section 4.2.1.3 in more detail.

4.2.2 TFT Technologies for Flexible Backplanes

Flexible substrate compatible TFT backplane technology development is a critical item for the fabrication of flexible active matrix displays. Both the well-established TFT technologies, namely a-Si TFT and LTPS TFT, are being considered for flexible display applications. In addition, emerging TFT technologies, namely O-TFT (organic TFT) and oxide semiconductor (such as ZnO) TFT technologies are also being developed for flexible backplane applications. The TFT processes developed for the flat rigid glass substrates cannot readily be applied for use with the flexible plastic substrates, due to reasons such as lower process temperature constraints, thermal stress limitations due to a larger CTE mismatch, and dimensional stability issues.

Currently, there are two main approaches being considered for producing flexible plastic backplanes:

(1) Standard (high temperature) TFT fabrication on a conventional display glass substrate, followed by transfer of the TFT circuit (backplane) on to a flexible plastic substrate by adhesive bonding at a lower temperature (e.g., <150°C). This process is referred to as Device Layer Transfer (DLT) process.
(2) Fabrication of TFT array directly on the flexible plastic substrate using lower (plastic substrate compatible) temperature processes.

The DLT process [16, 17] can be a viable approach for flexible displays, when low cost is not a consideration. This approach can provide the most optimum TFT device performance with respect to mobility, leakage current, stability, and uniformity as the TFTs are fabricated using conventional LTPS (low-temperature polysilicon at ~450°C) process, and then transferred on to the flexible plastic substrate at a low temperature. DLT will be a higher cost approach compared to the direct fabrication approach, due to wastage of the glass substrate and additional cost of the transfer process. While several companies have demonstrated this approach, it is not believed to be solution for the desired low-cost flexible displays ultimately.

Table 4.4 shows a comparison of the various TFT approaches. The conventional LTPS process used in the current AMLCDs uses a typical process temperature in the range of ~450°C using a polysilicon film produced by Excimer laser recrystallization of an a-Si film. Due to the high process temperature requirement, the conventional LTPS TFT approach may be appropriate for use with the stainless steel substrates, but not with plastic substrates with a process temperature limitation of less than 200°C. To overcome this problem using a plastic substrate, ultra low-temperature (<200°C) polysilicon (ULTPS) TFT processes are being developed [e.g., 18]. The ULTPS TFT approach has the potential for providing high mobility CMOS TFT devices suitable for driving the OLED pixels, as well as for fabricating the row and column drivers directly on the plastic substrate. Good progress has been made in producing TFTs with high mobility and satisfactory threshold voltages for the n- and p-channel devices. However, the leakage currents need to be reduced further for fabricating high-quality active matrix displays. The ULTPS technology continues to be developed to reduce the leakage currents and to further enhance its maturity and increase process yield.

a-Si TFT is currently the workhorse of the well-established AMLCD technology, and thus it will have several advantages if it can be adapted for use in flexible electronics and display manufacturing applications. However, for AMOLED applications, currently, a-Si TFT does have some issues that include the following:

(1) Low TFT mobility ($\mu_{fe} \sim 1$ cm^2 V^{-1} s^{-1}) which requires use of TFTs with a large W/L (W and L are the TFT channel width and length, respectively) ratio to achieve the desired pixel drive current. This results in a reduced pixel aperture ratio, and thus reduced pixel luminance in bottom emission OLED displays. As OLED materials and devices with improved efficiency are developed, this will be less of an issue. Top emission OLED pixel architecture alleviates this problem completely.

(2) The low mobility does not allow integration of the row and column drivers on the display glass.

(3) Only NMOS TFTs are available in a-Si, which restrict the choice of pixel circuit designs.

(4) The TFT stability with respect to gate bias is poor. The impact of this is discussed in more detail below.

Table 4.4 TFT technology options for flexible displays

| | Poly-Si | | a-Si | | | |
	DLT	LTPS	ULTPS	Conventional a-Si	Low-temperature a-Si	Organic TFT	Oxide (e.g., ZnO)
Process temperature (°C)	~450°C	~450°C	<200°C	~300°C	<200°C	<150°C	<200°C
Circuit type	CMOS	CMOS	CMOS	NMOS	NMOS	PMOS	NMOS
Device performance	–	–	–	–	–	–	–
– Mobility ($cm^2 V^{-1} . s^{-1}$)	~100	~100	~100	~1	~1	~1	~40
– Off-current	Excellent	OK	Issue	Excellent	Excellent	OK	OK?
– Uniformity	Good	Issue	Issue	Good	Good	OK	OK?
– Stability	Excellent	OK	Issue	Issue	Issue!	Issue !!	OK?
Cost	High	Medium	Medium	Low	Low	Very low?	Low
Maturity	Low	High	Low	High	Low	Very low	Very low

The conventional a-Si TFTs used in the current AMLCDs are fabricated at a typical process temperature of 300°C. Again for the obvious reason of high process temperature requirement, the conventional a-Si TFT process could not be used with the candidate plastic substrate with a process temperature limitation of less than 200°C. In recent years, significant advances have been made in the process temperature reduction, and a-Si TFTs have been successfully fabricated using low process temperatures of about 150°C [e.g., 19–21, 10], with a performance comparable to the 300°C process with respect to mobility, threshold voltage, and leakage current. However, the device stability under gate bias stress remains to be one item that needs to be improved particularly for the low-temperature processed a-Si TFTs. a-Si TFTs are known to exhibit threshold voltage, V_t, shifts under prolonged positive gate bias, particularly under higher operating temperature conditions. In AMOLED display operation, the drive TFT is typically subjected to positive gate voltage bias for the entire frame time (as opposed to only during the row address time as in an AMLCD). We will discuss this further under the next section on pixel drive electronics. One approach for increasing the a-Si TFT stability and decreasing the propensity for V_t shift is to reduce the display operating voltages and pixel current drive requirements. The recent progress in OLED materials toward lower drive voltages and pixel current requirements is very encouraging.

As opposed to fabricating the TFT backplane on a self-supporting flexible substrate, an alternate method involves bonding the flexible substrate to a rigid carrier substrate such as glass using a temporary adhesive, prior to the TFT array fabrication, for ease of handling during TFT array fabrication. After the TFT array is fabricated, the flexible substrate with the backplane circuit is separated from the temporary adhesive (and the carrier substrate). The issues in this approach include (1) temperature constraints imposed by the temporary adhesive, (2) potential chemical contamination by the temporary adhesive during the TFT processing, (3) yield of the bonding and debonding (of the flexible substrate/backplane from the rigid carrier substrate) operations with complete removal of the temporary adhesive, (4) cost of the bonding and debonding operations, and (5) cost of the temporary substrate if it is not reuseable.

An interesting variation of this approach is being developed by Philips, [22] which involves a-Si TFT backplane fabrication on a polyimide foil (10-μm-thick flexible substrate) that is spin coated on a rigid glass substrate. This process is named as the EPLaR (Electronics on Plastic by Laser Release) process. This process involves two extra process steps compared to a conventional a-Si TFT process on a rigid glass substrate. The first is an additive process of spin coating a 10-μm-thick polyimide layer (which subsequently becomes the self-supporting flexible substrate/backplane). The second is the laser release process. The temperature capability of this polyimide layer exceeds the requirements of the conventional a-Si TFT process; thus it can be processed in conventional a-Si TFT backplane fabrication facilities using standard processes. EPD media is then laminated to the TFT backplane, and the resulting display on the polyimide foil is then separated from the rigid carrier glass substrate, by a laser release process, which relies on the appropriate glass surface treatments prior to the polyimide spin coating, and use of the appropriate type of polyimide. Flexible EPDs have been demonstrated using this process. This process has a potential to be adopted for the LTPS backplanes as well, and for other display media such as an OLED as well.

Direct fabrication on the self-supporting flexible substrate (i.e., without the use of a carrier substrate) is expected to have a lower cost, and has the advantage of a relatively easier transition to a roll-to-roll (RTR) process. However, due to the limited process temperature capability of the PEN substrate, development of a low-temperature plastic compatible a-Si TFT processes are required, and this is discussed in Section 4.2.2.1.

Currently, there is a lot of interest in the development of organic electronics involving TFTs fabricated using organic semiconductors. O-TFTs have the advantage of very low process temperature, and they could be fabricated using low-cost solution processing methods (e.g., spin coating, ink-jet printing, etc.) instead of the more expensive vacuum-based thin-film deposition methods. To date OTFTs fabricated using vacuum-deposited pentacene as the organic semiconductor have shown the best performance [e.g., 23] with a field-effect mobility of over 2 cm^2 V^{-1} s^{-1}, near zero V_t, and "on"–"off" current ratio of over 10^8. However, the solution processable organic semiconductor-based O-TFTs have shown a mobility in the range of about 0.05 cm^2 V^{-1} s^{-1}. One major advantage of O-TFTs with complete solution processing (e.g., ink-jet printing) is that it is easier to compensate for dimensional instability of the plastic substrate during backplane processing [24]. O-TFT backplanes are beginning to be commercialized using the EPD media for some initial applications. While impressive progress is being made on this approach, this technology is still developing and not yet satisfactory for flexible OLED displays. This topic "Organic and Polymeric TFTs for Flexible Displays and Circuits" is discussed in detail in Chapter 8.

Transparent oxide semiconductors such as zinc oxide (ZnO) are beginning to be actively investigated for use in low-temperature TFT backplanes for displays [25]. ZnO is a wide bandgap (-3.3 eV at 300 K) semiconductor and has the advantage of being deposited directly in a polycrystalline phase even at room temperature (such as by RF magnetron sputtering), and thus is compatible with the currently available flexible plastic substrates. Hirao et al. achieved a field-effect mobility and threshold voltage of 50.3 cm^2 V^{-1} s^{-1} and 1.1 V, respectively, for ZnO TFTs and demonstrated an AMLCD display. An additional feature of ZnO TFT is its transmission in the visible range. It has an average optical transmission (including the glass substrate) of about 80% in the visible part of the spectrum. The combination of transparency, high channel mobility, and room temperature processing makes the ZnO TFT very promising for flexible and transparent electronics and display applications.

In the following, we will discuss the recent progress in low-temperature a-Si TFT technology for flexible plastic backplanes and displays.

4.2.2.1 Low-temperature a-Si TFT

The conventional a-Si TFTs used in fabricating the backplanes on glass substrates for the AMLCD displays use a typical process temperature of about 300°C. The deposition conditions such as process gas flow rates, gas pressure, and RF power for producing the SiN_x gate dielectric and the a-Si semiconductor films are optimized at this high ($\sim$300°C) temperature to achieve high-quality SiN_x and a-Si films for the TFT devices. This $\sim$300°C processed a-Si TFT is ideally suited for use for a rigid-glass substrate based AMLCD display as a pixel switch because it has adequate mobility (on-current, I_{on}) and low leakage current (off-current, I_{off}), with an I_{on}/I_{off} ratio of over 10^7. The TFT performance characteristics are adequate for driving an OLED pixel as well using the currently available OLED material and device architectures to achieve the desired pixel luminance. However, the issue at hand is achieving this performance in a-Si TFTs fabricated at low process temperatures that are compatible with the available flexible plastic substrates. A straightforward reduction of the TFT process temperature (without changing the other process conditions) results in degraded mobility, and higher threshold voltage, leakage current, and sensitivity to gate bias stress-induced instability.

To achieve good quality SiN_x gate dielectric film and a-Si:H film at low deposition temperatures, and thereby to fabricate a-Si TFTs at these low temperatures without degrading the mobility, threshold voltage, and leakage currents, helium (He) and/or hydrogen (H^2) dilution of the process gases has been used [e.g., 21, 20, 10]. The process gas dilution coupled with optimization of the other process conditions such as gas flow rates, pressure, and RF power density of the plasma, was found to improve the electronic properties of the SiN_x gate dielectric and the a-Si semiconductor films deposited at low temperatures ($\sim$150°C). For example, Table 4.5 shows the recent literature on low-temperature ($\sim$150°C) a-Si TFT fabricated on

Table 4.5 Typical reported characteristics of low-temperature a- Si TFTs on plastic substrates

Reference	Plastic substrate	Processs temperature (°C)	TFT structure	Performance
Gleskova et al. [20]	PI	150	BCE	$\mu = 0.45 \text{ cm}^2\text{V}^{-1}\text{s}^{-1}$ $V_{th} = 3 \text{ V}$ SS = 0.5 V/dec
Sazanov et al. [26]	PI	120	CHP	$\mu = 0.0.7{-}0.8 \text{ cm}^2\text{V}^{-1}\text{s}^{-1}$ $V_{th} = 4{-}5 \text{ V}$ SS = 0.5 V/dec
Won et al. [27]	PES	150	BCE	$\mu = 0.4 \text{ cm}^2\text{V}^{-1}\text{s}^{-1}$ $V_{th} = 0.7 \text{ V}$ SS = 0.82 V/dec
Sarma et al. [28]	PEN	150	CHP	$\mu = 0.863 \text{ cm}^2\text{V}^{-1}\text{s}^{-1}$ $V_{th} = 2.5 \text{ V}$

Fig. 4.2 CHP-type TFT
structure fabricated on
Teonex® Q65 PEN plastic
substrates [10]

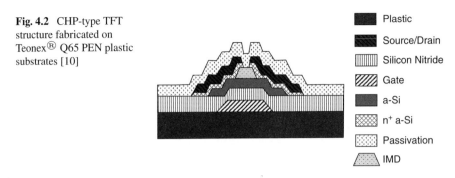

Plastic

Source/Drain

Silicon Nitride

Gate

a-Si

n⁺ a-Si

Passivation

IMD

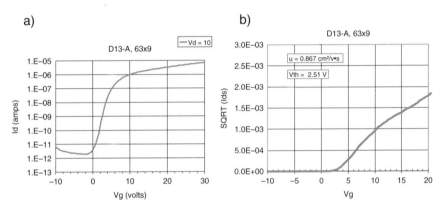

Fig. 4.3 Typical TFT characteristics in a fabricated backplane: (**a**) transfer characteristics and (**b**) Square I_{ds} versus V_{gs} ($=V_{ds}$) for a TFT in saturation

plastic substrates. Sarma et al. developed a 150°C a-Si TFT process for fabricating flexible PEN plastic backplanes for flexible OLED displays, using a 6-mask channel passivated (CHP) structure as shown in Fig. 4.2. The 150°C process is optimized by diluting the process gases SiH₄ (for a-Si), and SiH₄ and NH₃ (for SiN$_x$) with He and optimizing the other process conditions of gas flow rates, gas pressure, and RF power. Figure 4.3 shows the typical performance of a 150°C TFT in a fabricated backplane. Figure 4.3a shows the transfer characteristics for a TFT with a W/L ratio of 63 μm/9 μm. This device shows an on-current, I_{on}, of 7 μA, and an off-current, I_{off}, in the pico-ampere range. Figure 4.3b shows a plot of I_{off} versus V_{gs} ($=V_{ds}$) for the TFT in saturation. This TFT shows a mobility of 0.87 cm² V⁻¹ s⁻¹ and a threshold voltage, V_t, of 2.5 V. These performance characteristics are sufficient to drive an AMOLED pixel.

In addition to acceptable drive current and low leakage current, TFT device stability under gate bias stress is important for use in OLED display applications in particular. The low-temperature a-Si TFTs developed to date were found to have somewhat higher drifts in the device threshold voltage compared to the devices fabricated at higher process temperatures [8, 9]. Further work is required to improve the stability of a-Si TFT against the gate bias stress in the OLED display applications. In parallel, OLED pixel circuit designs that result in minimizing gate bias stress, and that compensate for small V_t drifts on the display performance are being developed [e.g., 29, 30]

4.2.3 Display Media for Flexible Displays (LCD, Reflective-EP, OLED)

For active matrix flexible displays the popular display media being considered includes liquid crystal display (LCD), electrophoretic display (EPD), and organic light-emitting diode (OLED) display. These display media also happen to be most popular ones in use or under development using flat rigid glass substrates.

4.2.3.1 LCD Media

One of the significant issues with the use of LCD media for flexible displays is the LC (liquid crystal) cell gap control. LC cell gap value has a significant effect on the display optical performance (luminance, contrast ratio, viewing angle, etc.), and maintaining this cell gap is difficult as the display is bent or flexed. Second, the high performance transmissive LCD mode displays require a backlight that needs to be flexible and color filter array that needs to be fabricated on a flexible substrate as well. Reflective LCD mode does not require a backlight, and offers the opportunity for lower power operation for daytime use. However, it would need to be frontlit for night-time viewing, thereby giving up the advantage of lower power mode of operation. Some reflective LCD modes such as cholesteric mode [31] are bistable and do not require any power except when updating the image on the display, and thus have the ultimate low-power potential. However, the bistable LCDs are typically slow responding with a response time of the order of ~100 ms, and thus are not suitable for video applications

4.2.3.2 Electrophoretic Display Media

EPD is a reflective bistable (low-power) display that does not have the cell gap control issues as in LCDs. Recently, significant advances are made in the EPD [e.g., 32] technology in reducing the drive voltage, and improving the response time. However, response time in the present practical devices is still in the ~100 ms range and does not support video applications. Further, additional work remains to be done for realizing a viable color capable EPD. However, because the requirements of a barrier layer (for protection of the EPD media) and the requirements of the active matrix TFT backplane (for driving the EPD pixel) are not stringent, and the simplicity of the monochrome reflective, bistable EPD technology, currently several companies are actively commercializing flexible displays using this display media, for applications such as e-books. We will discuss the recent status of these types of flexible displays in Section 4.5.

4.2.3.3 OLED Display Media

AMOLED display technology offers significant advantages over the current well-entrenched AMLCD with respect to superior image quality with wide viewing angle and fast response time, being lighter and thinner, lower cost (does not need backlight or color filters), and lower power. Because of this, many companies are actively developing and beginning to commercialize the AMOLED displays built using rigid flat glass substrates. Also, OLED media is believed to be natural choice for use in a flexible display as it represents the ultimate flexible display with a rugged solid-state display media and other attributes including full color, superior image quality, full-motion video, and low power. However, the OLED display media has very stringent requirements with respect to both the barrier layer specifications and the active matrix TFT backplane performance. Due to the long-term potential of flexible OLED displays, there is considerable interest in developing this technology. Also, OLED display application for flexible a-Si TFT backplanes is the focus of this chapter.

4.2.4 Barrier Layers

Lack of impermeability to moisture and oxygen is a serious deficiency of all the plastic substrates for the flexible display applications. All display media including LCDs, EPDs, and OLEDs degrade when exposed to oxygen and moisture in the ambient, even though at different rates with OLED having the most sensitivity to moisture and oxygen. For example, for the protection of an OLED display the plastic substrate (barrier layer) must have a permeability of less than 10^{-6} gm/m^2/day for moisture and 10^{-5} mL/m^2/day for oxygen. In comparison, LCD displays have a barrier requirement of 10^{-2} gm/m^2/day for both oxygen and moisture, which is significantly less stringent compared to OLEDs. The base plastic substrates typically have about 10 gm/m^2/day transmission rates for both oxygen and moisture, implying the need for incorporating a separate barrier layer.

In principle, a thin layer of an inorganic film such as SiO_2, SiN_x, and Al_2O_3 deposited on the flexible plastic substrate can serve as a barrier layer with the required impermeability to oxygen and moisture. However, in practice multilayer barrier film structures are believed to be required to counter the effects of the pinholes/cracks in single layer deposited barrier layers. Several organizations are currently developing optically transparent multilayer barrier coatings for flexible OLED displays [33]. Vitex Systems [34] uses such kind of an approach for their barrier film called Barix^{TM}, which employs alternating layers of a polymer and a ceramic inorganic barrier coatings applied in vacuum, as shown in Fig. 4.4. The inorganic films serve as barrier films for oxygen and moisture, organic layers serve the planarization/smoothing function, and multilayers provide redundancy against

Fig. 4.4 Vitex barrier comprising a multilayer stack of organic and inorganic films [34]

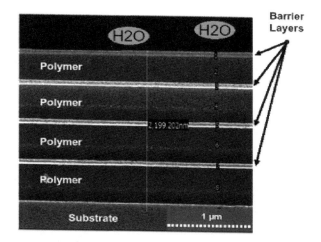

pinhole defects in the barrier films. The Barix™ layer is found to be an effective barrier layer, by minimizing the detrimental effects of pinholes and diffusion at grain boundaries. The film is about 3 μm thick and consists of about 500 Å thick Al_2O_3 ceramic films sandwiched between UV-cured acrylate films. Barix™ films were found to have water permeability in the range of 10^{-6} gm m^{-2} day^{-1}.

Note that whether using a plastic substrate or a stainless steel substrate, the top side of the OLED must be protected with an impermeable encapsulation (barrier) layer.

4.3 Flexible Active Matrix Backplane Requirements for OLED Displays

As with LCDs, OLEDs can be addressed by either the passive matrix or the active matrix scheme. Passive matrix displays are the simplest to manufacture. The row and column addressing buses are connected directly to the cathode and anode terminals of the OLED pixel at the intersection of each row and column. Because the duty cycle of the pixel scales inversely with the number of rows, N, in the display to be addressed, the peak luminance, L_{peak}, has to increase to achieve the desired average pixel luminance, L_{av} ($L_{av} = L_{peak}/N$) as the display resolution increases. This results in a higher current density (higher voltage) operation for the OLED device, and a consequent decrease in OLED efficiency (and increase in power consumption) and shorter lifetime. In addition, the large currents cause significant voltage drops along the addressing bus lines resulting in display luminance nonuniformities. These issues limit the display resolution to perhaps less than QVGA resolution and maximum size achievable in passive matrix addressed OLEDs to less than 5 in. in size. Significant improvements in OLED efficiency and lifetime, at high luminance levels can somewhat alleviate but not eliminate the limitations to the passive matrix addressed OLED displays.

Active matrix addressing scheme removes the limitations to the resolution and display size associated with passive matrix OLEDs. With active matrix addressing scheme, the pixel is driven with 100% duty cycle, regardless of the number of rows in a display. Active matrix addressing allows operation of the OLED devices (pixels) in the efficient regime (in the current density versus optical efficiency curve) and provides the best image quality. Other advantage of active matrix is that the cathode (which is typically a low-work function reactive material such as Li and Ca) does not need to be patterned in the active area of the display. A simple shadow mask can be used during deposition for patterning the common cathode.

4.3.1 Active Matrix Addressing

LCDs are voltage-controlled devices and involve varying the voltage (data) across the LC pixel to control transmission through the pixel (gray level). On the other hand, OLEDs are current-controlled devices involving control of the current through the OLED pixel to control pixel luminance (gray level). The pixel circuit in an AMLCD typically involves only a switching TFT and a storage capacitor to hold the pixel electrode node at the data voltage during the frame time. However, in case of AM OLED, the pixel circuit is required to control the current flow through the OLED pixel, and involves at least two TFTs and a storage capacitor. In the following section, we will discuss the two general implementations of a current driven pixel circuit in an AMOLED display.

4.3.1.1 Voltage Programming

Figure 4.5 shows the basic current driven pixel circuit for an AMOLED. This circuit consists of a select transistor, T1, a drive transistor (current source element), T2, and a storage capacitor, Cs. During the row select period, select transistor T1 turns on and transfers the voltage (data) signal from the column electrode to the gate of the drive transistor T2. After the addressing period, T1 is switched off, and the programmed voltage (data) is held on the gate of T2 for the rest of the frame time. The storage capacitor Cs prevents discharge of the T2 gate node (by leakage through T1) by any appreciable amount. Thus, Cs allows continuous driving of the OLED by T2, through the common power supply, L_{DD}, while the other rows in the display are addressed sequentially. When T2 is biased in saturation ($|V_d| > |V_{gs} - V_t|$), it behaves as a constant current source, with the current, I_{sd}, given by:

$$I_{sd} = K \cdot \mu_{fe} \cdot (V_{gs} - V_t)^2$$

where K is a constant based on the transistor size (channel length, L, and channel width, W) and the gate capacitance per unit area, C, μ_{fe} is the field effect mobility, V_{gs} is the gate-source voltage, and V_t is the threshold voltage.

Fig. 4.5 Typical current drive
pixel circuit design utilizing
two TFTs and one capacitor

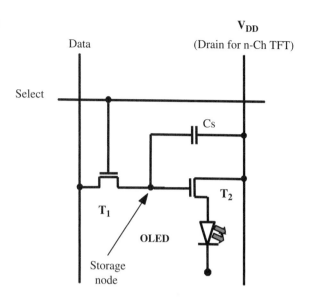

K is given by:

$$K = C \cdot W/L$$

Thus, the OLED pixel is driven by a constant current provided by T2 operating in the saturation regime with the programmed gate (data) voltage. This addressing scheme works well when the TFT performance is uniform across the display surface. Because the pixel current (current through T2) is proportional to $(V_{gs}-V_t)^2$ and μ_{fe}, any nonuniformities in the threshold voltage and mobility will result in pixel luminance variations.

A PMOS transistor is more suited for driving the AMOLED pixel, with the OLED pixel constructed with the ITO pixel electrode (which is part of the backplane fabrication) serving as the anode at the bottom and a low work function common cathode layer at the top. (Deposition of an ITO anode layer on top of the organic layers generally damages the organic due to exposure to oxygen in the RF sputter deposition atmosphere.) Thus, connecting the OLED to the drain side of the TFT as in Fig. 4.5 ensures that the gate-source (power supply bus) bias across the drive TFT is held constant to achieve a constant current drive. If an NMOS drive transistor is used, then power supply bus will be connected to the TFT drain, and OLED will be on the source side of the TFT. With this configuration, the drive TFT gate voltage is divided between the TFT gate-source voltage, V_{gs}, and the OLED, V_{OLED}. Any variations in the OLED devices across the display will result in variation in V_{gs}, and thus variations in the drive current and the luminance. Thus, an n-channel drive TFT requires more uniform OLED device (pixel) performance across the display surface to achieve uniform display luminance.

While a simple current source circuit consisting of two TFTs and one capacitor, C, (as shown in Fig. 4.5) is sufficient when the TFT mobility and threshold voltage are uniform and stable, additional compensation circuitry is required at each pixel when the TFT performance is not uniform or stable [29, 30].

4.3.1.2 Current Programming

There are many current-programmed pixel drive circuits reported in the literature [35]. While the basic operating principle in all these circuits is similar, the detailed architectures are different. Figure 4.6 shows an example of a current programmed pixel drive circuit using n-channel TFTs. Here, the data signal is provided as a current rather than a voltage. This pixel circuit employs four TFTs, three control buses (two row/gate buses and one column/data bus), a V_{DD} supply bus, and a storage capacitor, C. The circuit requires two sets of row (gate) drivers "V_row" and "SW".

During the pixel programming time, M1, M2, and M3 are turned on, and M4 is turned off. The programmed pixel current flows through M2 and M3. Current also flows through M1 to charge the storage capacitor, C, to a voltage that will sustain the programmed pixel current through M3. After the pixel has been programmed to the desired current during the row address time, M2 and M1 are turned off, and M4 is turned on. Assuming that negligible voltage drop occurs across C when M1 and M2 are turned off, M3 will continue to flow the programmed pixel current through M4 during the video hold (frame) time ($\sim$16.5 ms for a 60-Hz display refresh rate). During the row programming time, the row (gate) driver V_row is enabled and the row driver SW is disabled. After the pixel has been programmed, V_row is disabled and SW is enabled at the end of the row time. The V_{DD} supply bus provides a source

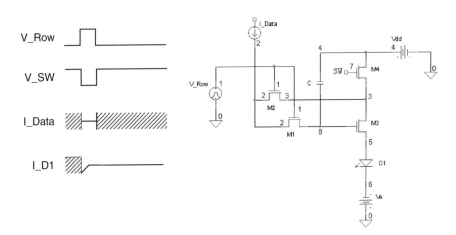

Fig. 4.6 Schematic of current-programmed AMOLED pixel circuit

of current for pixel operation during the hold time. This circuit compensates for the pixel-to-pixel variation in mobility and V_{th} of the TFT across the display surface.

In case of LTPS, the TFT performance (threshold voltage, V_{th}, and mobility, μ) is generally not very uniform due to the randomness of the grain boundaries, and thus it can benefit from a current programmed current source pixel circuit design. Compared to LTPS TFTs, a-Si TFTs are more uniform with respect to μ and V_t, and hence a simple voltage-programmed current source pixel design with 2T+1C architecture should be adequate. However, in practice, due to gate bias stress, the threshold voltage of the a-Si TFTs can vary, requiring compensation for the threshold voltage variations [e.g., 29, 30].

One consequence of having many TFTs in a pixel circuit is the reduction of pixel aperture ratio in case of bottom emission pixel architecture. Thus, top emission has the advantage of not being impacted by the pixel circuit real estate.

4.4 Flexible AMOLED Displays Using a-Si TFT Backplanes

4.4.1 Backplane Fabrication Using PEN Plastic Substrates

Dimensional stability issues arising from the substrate shrinkage and moisture absorption/desorption are a major consideration in the successful fabrication of the a-Si backplanes on PEN substrates. Sarma et al. developed a prestabilization process involving annealing of the PEN plastic substrates in vacuum at 160°C for 4 h to increase the dimensional stability against shrinkage. The need for dimensional stability of the plastic substrate can be illustrated when we consider the typical design rules used in the TFT backplane fabrication. For a typical 3-μm design rule used (for a contact via, as an example), a shrinkage (misalignment) of more than 1.5 μm is problematic. The as received "heat stabilized PEN substrate" shrinks about 0.05% during TFT backplane processing. This translates to a misalignment of 250 μm over a span of 50 mm (for a 2″ display). Clearly, this level of shrinkage (dimensional instability) is not acceptable. With the developed prestabilization process, the shrinkage during TFT backplane processing is reduced to 1.5 μm over a 60-mm span (~25 ppm or 0.0025%). Also, as all other plastic materials, Q65 PEN substrate absorbs moisture resulting in a dimensional change [15]. Figure 4.1 shows the moisture absorption in PEN with time as a function of RH at 20°C ambient temperature. Note that every 100 ppm of moisture absorption results in a dimensional change of about 45 ppm, and this level of dimensional change is very inconsistent with the dimensional stability requirements for backplane fabrication. To eliminate or greatly minimize the dimensional changes associated with moisture absorption/desorption during TFT processing, the PEN substrate is coated with 3,000-A-thick plasma CVD deposited SiN_x moisture barrier on both top and bottom surfaces, after the substrate prestabilization and prior to the TFT array fabrication. While the single layer SiN_x film does not eliminate the moisture absorption completely, it greatly minimizes it, thereby enhancing the substrate's dimensional stability during the TFT processing steps. Also, as an additional precaution, prior to any new film deposition step during the TFT array fabrication, the substrate is prebaked under standard conditions to restore the baseline dimensional stability.

Using the 150°C a-Si TFT process discussed in Section 4.2.2.1, Sarma et al. developed and implemented a backplane process using the PEN plastic substrate that was prestabilized and coated with 3,000 Å of SiN_x barrier layer on both top and bottom. The process sequence for the backplane fabrication uses six masks and includes the following steps:

(1) Deposit, pattern, and etch gate metal layer
(2) Deposit SiN_x gate insulator and a-Si layer, pattern, and etch
(3) Deposit, pattern, and etch SiN_x intermetal dielectric layer
(4) Deposit, pattern, and etch source-drain layer
(5) Deposit pattern and etch passivation layer
(6) Deposit contact plug layer
(7) ITO pixel electrode layer deposition and pattern.

The above process sequence is similar to that of conventional high-temperature CHP-type a-Si TFT process. However, the process recipes for the TFT thin-film depositions, particularly for the a-Si and SiN_x dielectric layers, are optimized for 150°C process to achieve the mobility and leakage current characteristics comparable to the high-temperature processed TFTs as discussed in Section 4.2.2.1. Further, the mask and process design details are optimized by taking into consideration the expected level of plastic substrate shrinkage during the TFT process and thin-film stresses due to CTE mismatch. Four-in. diameter, 125-μm-thick PEN plastic substrates are utilized for fabricating the backplanes for the test displays. Two backplane (test display) designs were developed. The first design involved a 64×64 pixel monochrome display with an 80 dpi (dots per inch.) resolution to demonstrate the basic feasibility with proof of concept test displays. The second design involved a 160×160(×3) pixel display to demonstrate a display with a larger size and higher resolution and to determine display size and resolution limits. Both designs were successfully fabricated and functional displays were built using these backplanes as discussed in the next section.

Figure 4.7a shows a photograph of a fully processed (with three 64×64 pixel backplanes on each wafer) 4″ diameter PEN plastic substrates. Figure 4.7b shows a photograph of a fabricated pixel in the 64×64 pixel larray. Figure 4.8 shows photographs of a 160×160(×3) pixel backplane fabricated on a 4″ diameter PEN plastic substrate and illustrates the flexural capabilities of the fabricated backplane. Each design included various test structures and process control monitors placed at the periphery of the pixel arrays.

One of the critical requirements for successful backplane fabrication is maintaining registration of various TFT mask levels during processing as the substrate dimension changes due to shrinkage and the level of moisture (absorbed in the plastic substrate). Using the substrate prestabilization process, and SiN_x barrier layers, we achieved acceptable dimensional stability and layer-to-layer alignment accuracy

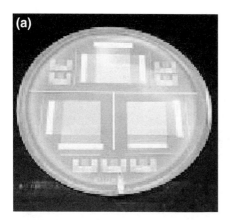
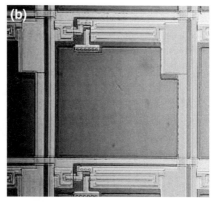

Fig. 4.7 Photographs of a-Si TFT backplanes processed on a 4″ diameter Teonex® Q65 flexible plastic substrate: (**a**) Substrate with three 64×64 pixel arrays and (**b**) Fabricated pixel in the array

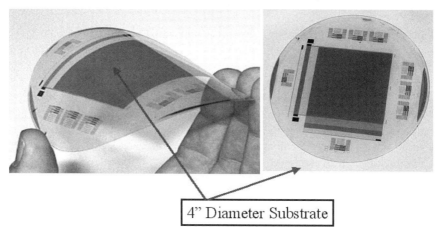

Fig. 4.8 Photographs of a 160×160(×3) pixel backplane fabricated on a flexible PEN plastic substrate illustrating its flexural capability

sufficient for fabricating functional backplanes and displays. Figure 4.9 shows photographs of a pixel in the 160(×3)×160 pixel backplane at the four extreme regions of a fabricated array (UL, upper left; UR, upper right; LL, lower left; and LR, lower right) in a 160×160(×3) pixel array with a pixel pitch of 100 μm×300 μm. The photographs and electrical testing of the TFTs at these pixels indicated that we achieved sufficient registration accuracy to ensure functioning backplane, and thus a functioning test display.

Fig. 4.9 Photograph illustrating that sufficient alignment accuracy is achieved at the four extreme regions of the fabricated 160×160(×3) pixel backplane for proper functioning of the backplane

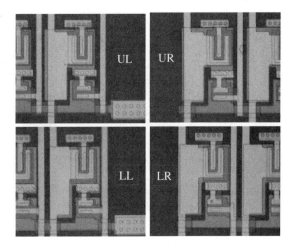

4.4.2 Flexible OLED Display Fabrication

Table 4.6 shows the salient design features of the 64×64 pixel test display designed and fabricated. The basic design features of the 160×160(×3) pixel test display are similar to the ones in Table 4.6 except for a subpixel pitch of 100 μm×300 μm and an active display area of 48 mm×48 mm with an 80 cgpi resolution.

The OLED device fabrication involves first spin coating of a hole transport layer (PDOT) on the backplane. This is followed by spin coating of a yellow polymer emitter layer. Then a low work function cathode metal is evaporated through a shadow mask to achieve the desired cathode pattern on the active matrix substrate. This structure is then laminated to a cover glass with an adhesive to protect the OLED from the ambient oxygen and the moisture. Starting from coating of the fully processed active matrix backplane with PDOT until the completion of the display including the glass substrate lamination, all the processing is done inside a glove box with an inert ambient to exclude exposure of oxygen and moisture from the OLED structure.

Table 4.6 Specifications for the 64×64 pixel test display

Display resolution	64×64 pixels
Color	Monochrome yellow
Pixel pitch	80 dpi – 300 μm×300 μm
Active area	19.2 mm×19.2 mm
Active matrix element	150°C a-Si TFT, inverted-staggered structure, CHP type
AMOLED pixel design	Two TFTs/pixel, voltage-programmed current drive
Backplane processing	Seven mask process
Display driving	External row/column drivers
Gray scale	Analog

Fig. 4.10 160×160(×3)
pixel display connected to the
COF column drivers and flex
cable connected to a row
driver board

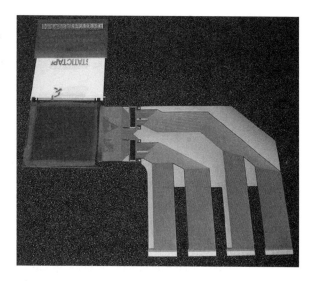

Appropriate drive electronics are designed to exercise and evaluate each test display. Both the test displays utilized a conventional AMLCD COTS row driver. Since no suitable commercial grayscale driver was available for a display of only 64 columns, a simple 64-channel sample and hold (S/H) circuit was devised for the column driver. For the 160×160(×3) pixel test display, an AMLCD-type column driver is used with a chip-on-flex (COF) implementation. The source drivers provide 8-bit voltage control to the display. The COF column drivers and a flex cable for the row driver board are then bonded to the column and row bus pads on the display with heat seal connections, to complete the display assembly as shown in Fig. 4.10. The display assembly is then connected to the display electronics system for test and evaluation.

Figure 4.11 shows the photographs of test images in the 64×64 pixel displays fabricated. As seen in Fig 4.11, while these display have some pixel and line defects, they are found to function as designed, validating the 150°C a-Si TFT process and the backplane design. The fabricated displays are found to be capable of displaying

Fig. 4.11 Photographs of test images being displayed on a 64 x 64 pixel AMOLED fabricated using a flexible PEN plastic backplane built with low-temperature a-Si TFTs

Fig. 4.12 Photograph of a text image being displayed on a 160($\times$3)$\times$160 pixel AMOLED fabricated using a flexible PEN plastic backplane built with low-temperature a-Si TFTs

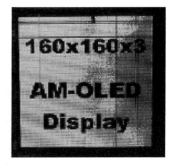

grayscale images and full-motion video. In general, the control displays fabricated using glass substrates were found to perform similarly except for having fewer pixel and line defects. The surface quality of the PEN plastic substrate was found to have a significant impact on the quality of the displays fabricated with respect to pixel and line defects observed. Displays fabricated on PEN substrates with improved surface quality exhibited significantly fewer display defects.

Similarly, Fig 4.12 shows a photograph of a text image on a 160($\times$3)$\times$160 pixel display. While this backplane was designed for a full-color display with R, G, and B pixels, only a monochrome (yellow) OLED media is used for the display fabrication to verify the design and fabrication of the backplane with 80 cgpi resolution. Complete functionality of this display does verify the feasibility of fabrication flexible OLED displays with this resolution.

4.4.3 Flexible AMOLED Display Fabrication with Thin-film Encapsulation

After verifying the feasibility of the low-temperature a-Si TFT backplanes for driving OLED displays using the displays encapsulated using glass substrates, Sarma et al. fabricated flexible AMOLED test displays using the same backplane designs, and OLED processing but using Barix thin-film encapsulation [34] for the OLED device protection. Figure 4.13 shows [36] a schematic cross section of the AMOLED display fabricated. The Barix (barrier film) is of the order of only a few microns. Thus, the thickness of the flexible display fabricated is about $\sim$130 μm. Figure 4.14 [36] shows photographs of a flexible AMOLED test display fabricated along with its flexural capabilities. The fabricated displays were found to function essentially similar to the test displays in Figs. 4.11 and 4.12 using the glass substrate lamination. While systematic long-term lifetime measurements were not made, the general lifetime observations on the fabricated flexible OLED displays indicated the effectiveness of the thin-film encapsulation on the display lifetime.

The flexible display evaluations highlighted handling issues for the backplanes and displays during fabrication and subsequent use. The flexible displays can

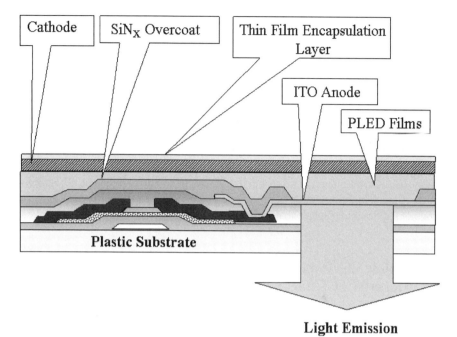

Fig. 4.13 Schematic cross section through the AMOLED test display with thin-film encapsulation

Fig. 4.14 Photographs of (**a**) a checker board image being displayed on a 64×64 pixel AMOLED fabricated using a flexible PEN plastic backplane built with low-temperature a-Si TFTs and thin-film encapsulation and (**b**) its flexural capability

tolerate flexing up to a critical minimum bend radius. However, when the back-planes and display get bent with a bend radius smaller than the critical value, it leads to cracking of the overcoat layer on the plastic substrate, and this crack can propagate through the address bus lines. The cracked bus structures cause the displays to exhibit intermittent line failures, gross line failures, or large regions of nonfunctioning areas in the display including complete display failure. The cracking can occur during the backplane or display fabrication if the substrate gets bent by more than the critical radius. Thus, it is essential to have appropriate tooling and fixturing during fabrication to ensure successful fabrication.

The results on the backplane and OLED display fabrication and evaluation indicate the feasibility of fabricating $160 \times 160 (\times 3)$ displays with an 80-cgpi resolution and 48 mm $\times$ 48 mm active area, using low-temperature a-Si TFT backplanes on PEN plastic substrates. Using this approach, it should be possible to fabricate large-size flexible AMOLED displays with higher resolutions (e.g., $\sim$120 cgpi) by further improvements in the dimensional stability of the plastic substrates and by using a top emission OLED device architecture. In addition, further reductions in a-Si TFT backplane fabrication temperature and use of next generation alignment/lithography tools with substrate distortion compensation capability can further extend the size range of the flexible displays that can be fabricated using the PEN plastic substrates.

4.5 Flexible Electrophoretic Displays Fabricated using a-Si TFT Backplanes

While the focus of this chapter has been on low-temperature a-Si TFT backplanes on flexible plastic substrates for fabricating flexible OLED displays, for the sake of completeness, we will discuss status of the application of these same backplanes for fabricating EPDs.

Recently, significant progress has been made in adapting EPD media to the low-temperature a-Si TFT fabricated on flexible plastic substrates. EPD media is generally viewed as the logical first choice for the development and commercialization of flexible displays, as the requirements of the TFT backplane performance as well as the barrier layer performance are not stringent compared to the OLED display media. Figure 4.15 shows a photograph of a flexible Active Matrix Electro-Phoretic Display (AMEPD) [11] fabricated using 120°C a-Si TFT backplane on a flexible PEN plastic substrate. This is a 14.3″ (A4 size) display with a 1280×900 pixel resolution with a $\pm$ 15 V drive. While this is a monochrome display, it is targeted for the e-book applications.

4.6 Outlook for Low-Temperature a-Si TFT for Flexible Electronics Manufacturing

The advances in computers, communications, and displays, and the systems and applications enabled by these advances in recent years have been nothing short of revolutionary. Display is a central part and a significant enable of these revolutionary advances. The recent developments in flat panel displays, and more particularly the active matrix TFT displays, have been extraordinary, having enabled applications that would not have been possible without them. Now active matrix TFT-LCDs are ubiquitous. The value proposition is compelling for replacing the

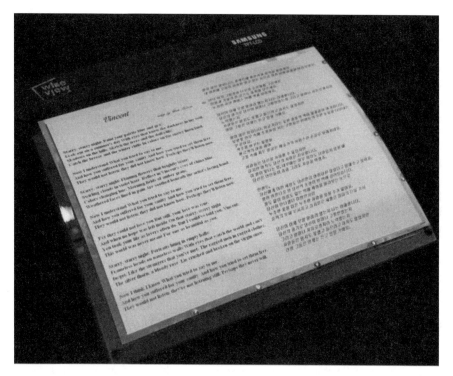

Fig. 4.15 Photograph of an AMEPD display fabricated using a flexible PEN plastic backplane built using low-temperature a-Si TFT [11]

dominant (and bulky) CRT display with a TFT-LCD with its attributes of being significantly thinner, lighter, and having superior image quality and consuming less power. The TFT technology is the heart of the active-matrix displays, and more particularly the advances in a-Si TFT technology have been central to the success of the TFT-LCD industry.

Having experienced the phenomenal successes in the active-matrix flat-panel display developments, the display industry is now fascinated at the prospects of repeating that success with the development of flexible displays. Flexible displays represent a new paradigm in display technology development. First, compared to the rigid flat glass-based displays, flexible displays have a potential for being significantly lighter, thinner, more rugged, and rollable or foldable for transportation and storage. Second, flexible displays have a potential for RTR manufacturing for significant cost reduction.

The recent progress in flexible backplane electronics and display development to date clearly shows the feasibility of fabricating a-Si TFT backplanes and displays directly on available flexible plastic substrates. Direct fabrication of the TFT backplane on the flexible plastic substrate (as opposed to using a device layer transfer process, or by means of a carrier substrate with bond/debond operations) represents an ultimate low-cost manufacturing process that can be adapted in a RTR operation. However, there are several challenges in accomplishing this objective.

First, the flexible plastic substrates must be improved with respect to dimensional stability (against shrinkage due to high-temperature processing and moisture absorption/desorption), and surface quality (with respect to particulates and other surface imperfections), and by reducing the TCE mismatch with the TFT thin films. These substrate advances will allow fabrication of large-size, high-resolution displays.

Second, improving the low-temperature (<200°C) a-Si TFT process with respect to device stability against the gate bias stress would be a significant help in the development of flexible OLED displays. Development and optimization of the OLED pixel circuits that minimize gate bias stress and provide compensation for some level of V_t drift are expected to help achieve this objective as well. TFT gate bias stress is much more of an issue for the OLED display media compared to the EPD media.

Finally, cost-effective multilayer barrier film technology must be advanced and integrated with the backplane and display fabrication processes to protect the backplane and the display media such as OLED from the ambient air. Again, the barrier requirements for the OLED display are much more stringent than for the EPD media.

Very impressive flexible EPDs built using low-temperature a-Si TFT backplanes have been demonstrated. An example of these demonstrations includes Samsung's 14.3″, 1280×900 pixel EPD. These type of flexible displays are expected to be commercialized soon for the e-book and other very low-power display applications. Because a flexible OLED is considered an ultimate display, further development of various technology elements are expected to continue in order to realize this potential.

Compared to LTPS (or ULTPS) technology, low-temperature a-Si TFT technology has made more progress to date, toward demonstrating proof-of-concept flexible backplanes and demonstration display using direct fabrication approach. If plastic substrate with higher process temperature capability (e.g., that would allow use of conventional ~300°C a-Si TFT process) are developed, that would lower the barrier for adaption of a-Si TFT backplanes for flexible displays. However, the flexible plastic substrate industry is not expecting the prospect of developing any new flexible substrates with a significantly higher process temperature capability to be very high. While the emerging O-TFT and ZnO TFT technologies are making significant progress, and continue to show potential for use in flexible displays, it is expected that they need more time to mature with respect to technology, process equipment (such as for printing), and infrastructure needed for fabricating the TFT devices and backplanes for high-resolution displays involving small design rules, and OLED display media with stringent requirements. The big advantage of low-temperature a-Si technology is having the required infrastructure in place and large-area capability.

Acknowledgments   The contributions of Charles Chanley, Sonia Dodd, Jerry Roush, John Schmidt, and other members of the Honeywell Displays and Graphics Group in the development of flexible AMOLED displays using low-temperature a-Si TFT Backplanes are acknowledged.

References

1. Young B (2007) A roadmap for flexible displays. In: Proceeding of USDC Flexible Displays & Microelectronics Conference & Exhibit, 5–8 Feb 2007

2. McGoldrick K (2007) A rollable display in every device. In: Proceeding of USDC Flexible Displays & Microelectronics Conference & Exhibit, 5–8 February

3. Paek SH, Kim KL, Seo HS, Jeong YS, Yi SY, lee SY, Choi NB, Kim SH, Kim CD, Chung IJ (2006) 10.1 inch SVGA ultra thin and flexible active matrix electrophoretic display. SID 06 Digest, p 1834

4. Raupp GB, Colaneri N, O'Rourke SM, Kaminski J, Allee DR, Venugopal SM, Bawolek EJ, Loy DE, Moyer C, Angeno SK, O'Brien BP, Bottesch D, Rednour S, Blanchard R, Marrs M, Dailey J, Long K (2006) Flexible display technology in a pilot line manufacturing environment. Army Science Conference, Orlando, FL, 27–30 November

5. Raupp G (2007) Active matrix TFT technology development and pilot line manufacturing for reflective and emissive flexible displays. In: Proceeding of USDC Flexible Displays & Microelectronics Conference & Exhibit, 5–8 February

6. Jin DU, Jeong JK, Shin HS, Kim MK, Ahn TK, Kwon SY, Kwack JHo, Kim TW, Mo YG, Chung HK (2006) 5.6-inch Flexible full color top emission AM OLED display on stainless steel foil. SID 06 Digest, p 1855

7. Chwang A, Hewitt R, Urbanik K, Silvernail J, Rajan K, Hack M, Brown J, Lu JP, Shih C, Ho J, Street R, Ramos T, Moro L, Rutherford N, Tognoni K, Anderson B, Huffman D (2006) Full color 100 dpi AM OLED displays on flexible stainless steel substrates. SID 06 Digest, p 1858

8. Long K, Kattamis AZ, Cheng IC, Gleskova H, Wagner S, Sturm JC, Stevenson M, Yu G, O'Regan M (2006) Active matrix amorphous silicon TFT arrays at 180°C on clear plastic substrates for organic light emitting displays. IEEE Trans Electron Dev 53(8):1789

9. Long K, Kattamis AZ, Cheng IC, Gleskova H, Wagner S, Sturm JC (2006) Stability of amorphous silicon TFTs deposited on clear plastic substrates at 250°C to 280°C. IEEE Electron Dev Lett 27(2):111

10. Sarma KR, Chanley C, Dodd S, Roush J, Schmidt J, Srdanov G, Stevenson M, Yu G, Wessel R, Innocenzo J, O'Regan M, MacDonald WA, Eveson R, Long K, Gleskova H, Wagner S, Sturm JC (2003) Flexible active matrix OLED using 150°C a-Si TFT backplane. Cockpit Displays X, Proceedings of the SPIE 2003, pp 180–191

11. Hwang TH et al. (2007) 14.3 inch Active matrix-based plastic electrophoretic display using low temperature processes. SID 07 Digest, p 1684

12. MacDonald WA, Mace JM, Polack NP (2002) 45th Annual Technical Conference Proceedings of the Society of Vacuum Coaters, p 482

13. Teonex® is registered trademark of Teijin DuPont Films Japan Limited and licensed to DuPont Teijin Films U.S. Limited Partnership

14. MacDonald WA, Rollins K, Rustin RA, Handa M (2003) Plastic displays – New developments in polyester films for plastic electronics. SID 03 Digest, p 264

15. MacDonald WA, Eveson R, MacKerron D, Adam R, Rollins K, Rustin R, Looney MK, Hashimoto K (2006) The impact of environment on dimensional reproducibility of polyester film during flexible electronics processing. SID 06 Digest, p 414

16. Utsunomiya S et al. (2003) Flexible color AM OLED display fabricated using surface free technology by laser ablation/annealing (SUFTLA™). SID 03 Digest, pp 864–867

17. Asano A, Kinoshita T, Otani N (2003) A plastic 3.8-in low-temperature polycrystalline silicon TFT color LCD panel. SID 03 Digest, pp 988–991

18. Kwon JY, Jung JS, Park KB, Kim JM, Lim H, Lee SY, Kim JM, Noguchi T, Hur JH, Jang J (2006) 2.2 inch qqVGA AM OLED driven by Ultra Low Temperature Poly Silicon (ULTPS) TFT direct fabricated below 200°C. SID 06 Digest, pp 1358–1361

19. Gleskova H, Wagner S (1999) Amorphous silicon thin-film transistors on compliant polyimide foil substrates. IEEE Electron Dev Lett 20(9):473

20. Gleskova H, Wagner S (2001) DC-gate bias stressing of a-Si:H TFTs fabricated at 150°C on polyimide foil. IEEE Trans Electron Dev 48(8):1667

21. He S, Nishiki H, Hartzell J, Nakata Y (2000) Low temperature PECVD a-Si TFT for plastic substrate. SID 2000 Digest, pp 278–281

22. French I, George D, Boerefijn I, Chuiton E, Gomez G, Mazel F, Trestour C, Kretz T (2007) Flexible displays made by the EPLARTM process in a factory. In: Proceeding of USDC Flexible Displays & Microelectronics Conference & Exhibit, 5–8 February

23. Gundlach DJ, Kuo DJ, Nelson CC, Jackson TN (1999) organic thin film transistors with field effect mobility >2 Cm²/V-s, IEEE Electron device Lett., pp 164–165

24. Burns SE et al. (2006) A flexible plastic SVGA e-paper display. SID 06 Digest, p 74

25. Hirao T, Furuta M, Furuta H, Matsuda T, Hiramatsu T, Hokari H, Yoshida M (2006) High mobility top-gate zinc oxide thin film transistors (ZnO-TFTs) for active matrix liquid crystal displays. SID 06 Digest, pp 18–21

26. A. Sazonov A, Stryahilev D, Nathan A, (2002) Low-temperature a-Si:H TFT on plastic films: materials and fabrication aspects, International Conference on Microelectronics, pp 525–528

27. Won SH, Hur JH, Lee CB, Nam HC, Chung JK, Jang J (2004) Hydrogenated amorphous silicon thin film transistors on plastic with organic gate insulator, IEEE Electron device Letters, 25, pp 132–134

28. Sarma KR (2004) a-Si TFT OLED fabricated on low-temperature flexible plastic substrate. Proceeding of the 2004 MRS Spring Conference, San Francisco, CA

29. Hasumi T, Takasugi S, Kanoh K, Kobayashi Y (2006) New OLED pixel circuit and driving method to suppress threshold voltage shift of a-Si:H TFT. SID 06 Digest, p 1547

30. Chaji GR, Nathan A (2006) A stable voltage programmed pixel circuit for a-Si:H AM OLED displays. J Display Technol 2(4):347

31. Khan A, Shiyanovskaya I, Schneider T, Doane JW (2006) Recent progress in flexible drapable reflective cholesteric displays. SID 06 Digest, p 1728

32. McCreary M (2007) Advances in microencapsulated electrophoretic displays. In: Proceeding of USDC Flexible Displays & Microelectronics Conference & Exhibit, 5–8 February

33. Graff G, Burrows PE, Williford RE, Praino RF (2005) Barrier layer technology for flexible displays. In: Crawford GP (ed) Flexible Flat Panel Displays. Wiley, NY

34. Moro M, Chu X, Hirayama H, Krajewski T, Visser RJ (2006) A mass manufacturing process for Barix encapsulation of OLED displays. IMID/IDMC, 23 August

35. Sanford JL, Libsch FR (2003) TFT AM OLED pixel circuits and driving methods. SID 03 Digest, p 10

36. Sarma KR, Roush J, Schmidt J, Chanley C, Dodd S (2006) Flexible active matrix organic light emitting diode (AM OLED) displays. ASID 06 Digest

Chapter 5
Flexible Transition Metal Oxide Electronics and Imprint Lithography

Warren B. Jackson

5.1 Introduction

The previous chapters have discussed inorganic low-deposition temperature materials suitable for flexible applications, such as amorphous and nano-crystalline-silicon (Si) and organic conductors. This chapter presents the results of a recently developed inorganic low-temperature materials system, transition metal oxides (TMOs), that appears to be a very promising, new high-performance flexible electronic materials system. An equally, if not more, important part of this chapter, is the presentation of self-aligned imprint lithography (SAIL) a new fabrication method for flexible substrates that solves the layer-to-layer alignment problem.

The new *materials* system is TMO based consisting of one or more transition metals and oxygen. Some of the more common examples include zinc oxide, zinc tin oxide (ZTO), indium gallium zinc oxide (IGZO) and zinc indium oxide (ZIO). These low-temperature amorphous materials can form the active material for transistors with a performance that significantly exceeds that of amorphous and nanocrystalline Si and approaches that of larger grain poly-Si without the complexities and uncontrolled variability of polycrystalline materials. Thus, this technology can be used to drive organic light-emitting diodes (OLEDs) as well as create electronic circuits, such as shift registers, ring oscillators, and multiplexers on chip. An added feature is the possibility of creating transparent electronics with such materials systems. The entire transistor including the electrodes and the active material can have visible transmittances of 70% or greater. Thus, for applications such as displays, the active electronics would not impact the visual appearance of device and the fill factor of displays with such electronics could approach 100%. Finally, there is the possibility of creating p-type transistors that would permit fabrication of complementary metal oxide semiconductor (CMOS) circuits for flexible applications. Hence, the TMO material system possesses a number of intriguing new possible applications.

W.B. Jackson (✉)
Hewlett-Packard Laboratories, 1501 Page Mill Road, Mail Stop 1198, Palo Alto, CA 94304, USA
e-mail: warren.jackson@hp.com

W.S. Wong, A. Salleo (eds.), *Flexible Electronics: Materials and Applications*, Electronic Materials: Science & Technology, DOI 10.1007/978-0-387-74363-9_5,
© Springer Science+Business Media, LLC 2009

For the purposes of this chapter, the most remarkable feature of the TMO transistors is that these features and performance can be achieved in devices with a maximum processing temperature that is compatible with flexible polymer substrates, such as polyimide (Glass transition temperature $(T_g) = 300°C$) and polyethylene naphthalate (PEN) $(T_g = 200°C)$. Although most metal oxide work published thus far has been accomplished on higher temperature and rigid substrates, such as glass and c-Si, recent work has demonstrated large mobilities for flexible metal oxide transistors using photolithography and standard batch etching processes. However, flexible substrates are not dimensionally fixed during processing, making it virtually impossible to have precise layer-to-layer micron-scale alignment for large dimensionally variable substrates. This important problem of layer-to-layer registration on flexible substrates has been solved using SAIL, a solution that extends the TMO technology as well as amorphous and poly-Si to higher performance short channel devices.

This chapter presents the results in the field of TMO transistors relevant for the emerging field of flexible electronics in the following sections:

(1) A brief discussion of the history of work and background information in the general area of metal oxide transistors relevant for flexible electronics.
(2) The general properties of some of the leading TMO materials.
(3) The device structures used in TMO and flexible electronics along with a description of SAIL
(4) One of the most important sections presents SAIL and amorphous silicon (a-Si) results, indicating that the method can fabricate working devices.
(5) Resulting TMO thin-film transistors (TFTs) on rigid substrates using shadow masks and photolithography on both flexible and rigid substrates along with TMO SAIL results on rigid substrates
(6) Finally, future areas of research for improved large-area flexible electronics.

5.2 Previous Work

TMO materials have been used in transistors for quite some time. One of the first ZnO transistors was made in the 1960s [1]. However, little progress ensued using TMOs as semiconductors although ZnO found uses in surface acoustic devices [2], varistors [3], transparent PN diodes [4], and as a transparent conductor [5]. Interest in ZnO as transistors was revived in about 1997 by Prins et al. [6] and later in Kawasaki et al. and Ohtomo et al. [7, 8], where transistors with mobilities of $100 \text{ cm}^2 \text{ V}^{-1} \text{ s}^{-1}$ were fabricated using spin-on glass as the insulator on glass substrates. Subsequently, there have been many different experiments investigating various aspects of metal oxide transistors.

The current interest in TMO transistors started with Ohya et al. [9] and in the thesis by R. Hoffman 2002 [10], Masuda [11], Nomura [12], Hoffman [13], Nishii

[14], and Carcia [15]. These references led to the current interest in TMO transistors and established convincingly that these materials not only had high Hall mobilities but could actually produce high-performance transistors. Typical mobilities for these initial efforts were on the order of 0.1–1 cm^2 V^{-1} s^{-1} when fabricated using low-temperature dielectrics and 80 cm^2 V^{-1} s^{-1} for high-temperature thermal oxide dielectrics on c-Si. The latter work indicated that transistors with performances approaching that poly-Si were possible using TMO materials as the active layer. Following this initial group of publications, the field has expanded greatly with dozens of papers detailing various aspects of TMO transistors. Various methods of deposition, dielectrics, and TMO combinations were investigated. A list of many of the various results is summarized in Table 5.1 and in [17].

The primary emphasis of most TMO transistor work is to create transparent semiconductors. One important result was that high-performance ($\mu = 20$ cm^2 V^{-1} s^{-1}) transparent transistors could be fabricated at room temperature on glass using silicon oxynitride dielectrics [21] and using atomic layer deposition (ALD) of Al$_2$O$_3$/TiO$_2$ superlattice dielectric [36] (Fig. 5.1). The resulting output curves are very flat and the on–off ratio is very good. This work demonstrated that high-performance TMO transistors on flexible substrates are possible.

Another theme of research in the TMO field is the expansion of candidate TMO material systems used in transistors. The range of viable transistor materials has been enlarged to include materials such as ZnO, ZnSnO, ZnInO, ZnInGaO, and InGaO among others (see Table 5.1). The oxide systems have achieved different levels of performance at a variety of maximum fabrication temperatures and have unique attributes that make them suitable for various applications. The thrust of this work is to decrease the high-temperature annealing step required to reduce the carrier concentration of the TMO. The initial ZnO transistors had limited performance unless the samples were subjected to anneal temperatures exceeding 400°C. The addition of Sn to ZTO has produced stable transistors with reasonable performance using anneal temperatures of 210–250°C [13, 21]. This temperature range is compatible with polyimide but not typical transparent substrates, such as PEN and polyethylene terephthalate (PET). The addition of In into the ZIO devices and IGZO (the addition of Ga) has lowered the temperature needed to produce high-performance devices to below 200°C [21]. Unfortunately, the addition of In and the low temperature of annealing appear to introduce instabilities in the devices that remain poorly understood. A summary of some of the more prevalent material systems and their properties is shown in Table 5.2. A general summary of the trends is that the inclusion of In, Ga, and Sn lowers the maximum processing temperature needed to obtain higher mobilities in part because of the generally higher diffusivities of these metal cations.

Most of the previous TMO work has emphasized the transparency and high mobilities that could be obtained using TMO transistors. For the purposes of this book, the work relating to TMO for flexible electronics is more important. Some of the first work that explicitly addressed issues relevant for the application of TMO to flexible electronics, which reduced the maximum processing temperature of ZnO to 150°C, is Ref. [20]. Later this group produced ZnO transistors on

Table 5.1 Summary of various TMO transistors with properties

Reference No.	Oxide	Substrates	Max. temp (°C)	Deposition method	μ (cm^2 V^{-1} s^{-1})	On–off ratio	V_t (V)	Gate dielectric	Contacts
[11]	ZnO	c-Si	450	PLD	0.03–1	10^5	(−1)–2.5	SiO$_2$/SiN	
[13]	ZnO		700	IBS	0.3–3	10^7	10–15		
[12]	InGaO$_3$(ZnO)$_5$		1,400	PLD	80	10^6	3	HfO$_2$	
[15]	ZnO		25	RMS	0.3–2	10^5–10^6	0		
[16]	ZnO		300	PLD	<4				
[18]	ZnSnO	Glass	600	RMS	14	6×10^6	−4.6	Al$_2$O$_3$–TiO$_2$	
[19]	ZnSnO	Glass	300	RMS	5–15	10^7	0–15	ATO	
[6]	SnO2:Sb							PbZr$_{0:2}$Ti$_{0:8}$O$_3$	
[10]	ZnO	Glass	600	IBS	0.3–2.5	10^7	10–20	ATO	
[14]	ZnO	Glass	150–300	PLD	7	10^7		SiN/CaHfO$_x$	
[9]	ZnO								
[20]	ZnO	Polyester polyimide	RT	RMS	0.4	10^4	7.5	fluoropolymer dielectric	
[20]	ZnO	Glass	RT	RMS	28	3×10^5	19	SiON	GZO
[21]	ZnO	Glass	RT	RMS	20	5×10^5	1.8	SiON	
[22]	ZnO	Polyimide	<120	EB	50	10^5	3.2	Al$_2$O$_3$	
[23]	ZnO		750	PLD	12	10^3	−3	SrTiO$_3$	
[24]	ZnO	Si	250	MOCVD/ALD	0.95	10^6	1.7	SiO$_2$/AlO$_x$ by ALD	
[25]	ZnO	n-c-Si	125, 200, 300, 400, 450	RMS	17	10^5	6	HfO$_2$, Al$_2$O$_3$	
[26]	ZnSnO	Polyimide	250	RMS	14	10^6	−17	SiON, SiO$_2$, SiN	
[27]	ZnSnO	c-Si	300	RMS	20	10^6	0	SiO$_2$	
[28]	ZnSnO	Polyimide	250	RMS	14	10^6	−8.5	SiON, SiO$_2$, SiN	ITO

(continued)

Table 5.1 (continued)

Reference No.	Oxide	Substrates	Max. temp (°C)	Deposition method	μ (cm² V⁻¹ s⁻¹)	On-off ratio	V_t (V)	Gate dielectric	Contacts
[29]	IGZO	PET	RT	RMS	8	10^3		Y_2O_3	
[30]	ZIO	c-Si	RT	RMS	20	10^8	-3.2	SiO_2	
[31]	ZnO	Glass	RT	RMS	20	10^5	21	Al_2O_3 TkO_3	
[32]	Tetracene	Mylar	RT	EV	5×10^{-4}			Mylar	
[33]	Organic	Mylar	RT	EV	1×10^{-4}	10^4		Mylar	
[34]	Organic	Mylar	RT	EV	1×10^{-4}	10^4		Mylar self-aligned	
[35]	IGZO	YSZ	350	R-SPE	80	10^8	5	HfO_4	IZO
[36]	ZnO	glass	RT	RF	27	3×10^5	19	Al_2O_3 TiO_3 e ALD	
[37]	IGZO	PET	RT	PLD	10	10^6	3.2	Y_2O_3	ITO

PLD,Pulsed Laser Deposition; IBS,Ion Beam Sputtering; RMS,Radio frequency Magnetron Sputtering; EB,Electron Beam Sputtering; MOCVD,Metal-Organic Chemical Vapor Deposition; ALD,Atomic Layer Deposition; R-SPE, Reactive solid phase epitaxy; EV,evaporation.

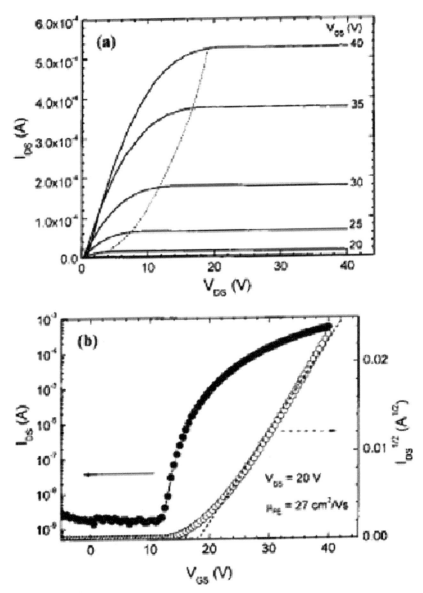

Fig. 5.1 ZnO transistor characteristics on an ITO-coated glass, ALD-deposited Al-Ti-O dielectric, and GZO contacts from [36]. W/L = 1.4

polyester and polyimide substrates [15, 22] using organic fluoropolymer dielectrics and obtained transistors with field effect mobilities of about 0.4 cm^2 V^{-1} s^{-1}, on–off ratios of about 10^4, and threshold voltages (V_T) of 7.5 V. They then replaced the fluoropolymer dielectric with the inorganic dielectric of atomic layer deposited Al$_2$O$_3$ and obtained devices with mobilities of 50 cm^2 V^{-1} s^{-1}, on–off ratios of

Table 5.2 Electronic properties of Zn TMOs

Material	E_g (eV)	μ_H (cm^2 V^{-1} s^{-1})	ρ (Ω cm)	n (carriers cm^{-3})	References
ZnO	3.2–3.3	5–50	10^{-4}	10^{21}	[38–40]
ZnSnO$_3$	3.5	7–12	5×10^{-3}	10^{20}	[40–42]
Zn$_2$SnO$_4$	3.3–3.9	12–26	1–5×10^{-2}	6–30×10^{18}	[43–45]
Zn$_2$In$_2$O$_5$	2.9	12–20	1–4×10^{-3}	3.6×10^{20}	[43, 40, 46]
ZnInGaO	3.0	10		10^{20}	[29]

10^5, and $V_T = 3.2$ V. These devices were deposited using electron beam evaporation through shadow masks. The ALD deposited dielectrics of these transistors are difficult to migrate to a production environment. More recently, ZTO transistors deposited on polyimide using shadow mask patterning have achieved mobilities of 15 cm^2 V^{-1} s^{-1}, on–off ratios of 10^6, on-currents as high as 1.5 mA, and threshold voltages of -17 V [26]. These devices have been obtained using typical production compatible dielectrics such as radio frequency (RF) discharge produced SiN, SiO$_2$, and SiON, and shadow mask patterning, and processing temperatures below 250°C. This work highlighted the need for appropriate contacts such as indium tin oxide (ITO) or Al for high-current applications. Recent work has also been made on applying more flexible substrate-appropriate patterning methods such as SAIL for ZTO [28]. Issues such as channel length scaling and contact resistance have been investigated.

5.3 Properties of Transistor Materials

5.3.1 Semiconductors

The TMO materials space has many possible combinations of transition metals and oxygen each with a unique combination of electronic and materials properties. Many of these combinations in the TMO material space have not been fully investigated particularly the various ternary compounds (those with two or more transition metals). The Zn-based compounds however, are currently generating the most interest and serve as a good representative subclass of the larger TMO group.

The properties of ZnO and related compounds are shown in Table 5.2. The valence band is formed by hybridized Zn $4d$ levels and O $2p$ levels. The conduction band is formed predominately by Zn $4s$ levels. The bandgap is direct and has a value of 3.2–3.3 eV for ZnO and increases to 3.5–3.9 eV with the addition of Sn (Table 5.2). The addition of In tends to lower the bandgap to 2.9 eV. These values are large enough that the material is transparent without much coloration. Hence, they make good candidates for transparent electronics.

The properties of TMOs make them uniquely suited for flexible electronics where low-temperature non-single crystalline materials must be used. The Hall mobility

(μ_H) of single crystalline ZnO is 200 cm^2 V^{-1} s^{-1} and the saturation velocity is large, making high-mobility devices feasible. The measured Hall mobilities in amorphous TMOs are around 5–50 cm^2 V^{-1} cm^{-1}, and the achieved field effect mobilities are roughly of comparable magnitude (Table 5.2). With field effect mobilities of 10–100 cm^2 V^{-1} s^{-1}, milliamp on-currents for flexible devices are obtainable using typical TFT parameters. The high mobility for electrons in the disordered amorphous TMOs arises because of the overlap of the nondirectional 4s orbitals of Zn (or other transition metal cations) creating a conduction band regardless of the angle between nearest neighbors (Fig. 5.2). Hence, the mobility is relatively unaffected by

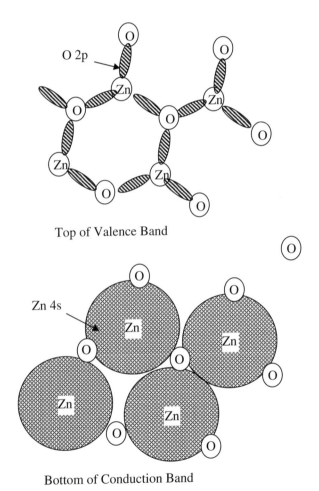

Fig. 5.2 Schematic of electron wavefunction structure for the *top* of the valence band (primarily O 2p) and the *bottom* of the conduction band (Zn 4s). The large overlap with neighboring Zn and spherical symmetry of the Zn 4s levels minimize sensitivity to disorder for the conduction band

Table 5.3 Various miscellaneous properties of ZnO [50]

Density (g cm^{-3})	5.6
Dielectric constant	8.6
Refractive index	2.0
Average thermal expansion Coeff. (/K)	4×10^{-6}

bond angle disorder. Moreover, the ionic nature of the transition metal oxide bonds causes the bonds to be less directional than the typical covalent bonds found in the usual Group IV or III–V semiconductors. Therefore, the large bond angle distortions found at grain boundaries are less apt to result in grain boundary gap state defects. The lack of gap states associated with deformed bonds results in large subthreshold slopes and small turn-on voltages in low-temperature transistors.

A further consequence of the lack of bandgap states and high electron mobility is that the materials often can be easily degenerately doped n type yielding small resistivities, ρ, below 10^{-2} (Ω cm) (Table 5.2). The doping of ZnO occurs by substitutional dopants, Zn interstitials, O vacancies, and/or interstitial H. The typical substitutional dopants include N and P. Oxygen vacancies are used to make the materials highly conductive for use as transparent electrodes for solar cell and photosensor applications. The oxygen vacancies can readily be created by sputtering in an O poor atmosphere. In fact, in TMO transistor fabrication, care must be taken to keep the O-vacancy concentration low so that the channel is sufficiently resistive for enhancement mode transistor operation [15, 22]. There is increasing evidence that the O vacancies may be somewhat mobile resulting in some long-term device instabilities to be discussed later.

There also is evidence that H interstitials act as n-type dopants [47–49]. Exposure of semiconducting ZTO to an H plasma results in a significant enhancement of the conductivity while exposure to an Ar or O plasma does not. Moreover, the H diffusion rate is large even at 200°C. Thus, unintentional H doping during fabrication etching is a concern and must be controlled.

Other useful parameters for ZnO are shown in Table 5.3 [50]. Note that the dielectric constant is somewhat lower than that of a-Si. The values for other TMO are generally similar in magnitude.

5.3.2 Dielectrics

Dielectrics are often one of the most critical materials necessary for a transistor technology. In most electronic applications, shorts are extremely undesirable defects, particularly in array applications. Shorts not only cause the shorted device to fail but often result in failure of any connected device and cause an unacceptable power drain. They also mask failures by other devices. In addition to the highly

Table 5.4 Various dielectrics used in TMO transistor technology [10, 18]

Insulator	E_g (eV)	Relative dielectric constant	Breakdown field (MV/cm)
SiO$_2$	11	3.8	6
SiN	9	7	9
Al$_2$O$_3$	9.5	8	3.5–6
BaTa$_2$O$_5$		22	3.5
HfO$_2$	5.8	16	3
Ta$_2$O$_5$	4.6	24	1.5–3
TiO$_2$	3.7	60	0.2
Y$_2$O$_3$	5.6	12	4
ZrO$_2$	5	19.8	

deleterious effect of shorts, typical large-area applications consist of many devices and crossovers requiring a highly reliable dielectric. Dielectrics that have been successfully used for TMO TFTs are included in Table 5.4 [10, 18]. The dielectrics currently being used in large-area electronics, such as SiO$_2$ and SiN work reasonably well for TMO devices. Typically, if SiN is used, a thin SiO$_2$ buffer layer is used so that the SiN does not directly contact the TMO. One unresolved issue is that the presence of large amounts of H may be one of the problems with SiN in direct contact with the TMO [51]. More recently, evidence has emerged that O vacancies can be created during long-term operation of TMO transistors and result in long-term instability [52–56]. There is possible evidence that the silicon-based oxides are more reactive in creating vacancies. As a result there is interest in less reactive oxides, such as Y$_2$O$_3$, Al$_2$O$_3$, Ta$_2$O$_5$, and HfO$_2$ dielectrics [25, 29, 30, 36, 12]. These oxides have the advantage of having high dielectric constants as well as possibly generating fewer O-vacancy related instabilities. It is not yet clear whether the instability is actually O vacancies or rather H associated with the various dielectrics or due to the motion of metal interstitials. Hence, low H dielectrics may be necessary to solve the instability problem.

5.3.3 Contact Materials

The contact materials needed for TMO have not been completely investigated at this point. For many TMO systems, the metal–TMO contact has been sufficiently conductive to yield high-current devices. However, as shown in the section below, insertion of an intermediate layer between the metal and the TMO can result in significant decrease in the contact resistance. The metals Al, Ti, and TiW make reasonable contacts capable of making devices with more than 1 μA/(W/L) on-current per W/L. The current can be significantly increased, however, if a layer such as ITO is inserted between the TMO and the metal [26, 28]. In the case of ZTO, the ITO layer at the contact increases the on-current to 10 μA [26, 28].

5.4 Device Structures

The structures used for TMO TFTs on rigid substrates are largely the same as those used in other deposited material systems, such as a-Si. The various structures include the staggered bottom gate, collinear bottom gate, and top gate devices. Because fully versatile reproducible etching of the TMO semiconductors is still being developed, most of the early reported work has involved shadow mask structures and transistors on silicon wafers with thermal oxide. It is only recently that more fully photolithography patterned microstructures have been made.

Continuous Bottom Gate. The most prevalent structure for investigating basic material properties and the device potential of the metal oxide system is using a blanket deposited bottom gate structure (Fig. 5.3a,b). Either a layer of metal or a doped silicon wafer serves as the gate electrode. If a crystalline Si wafer is used, the dielectric is usually a thermal oxide, the source and drain electrodes are patterned on the oxidized Si wafer using either shadow masks or photolithography, and finally, the metal oxide material is deposited through a shadow mask to complete the structure. If the back contact is a continuous metal, the dielectric is a deposited dielectric and the other steps are similar to those of a silicon wafer. The shadow mask is used to define the active material to restrict current flow to a region near the contacts. Variations of this structure include the deposition of materials to improve the contact properties such as ITO or Ti. The contact material must be patterned along with the source and drain prior to deposition of metal oxide active material. Also, the contacts can be patterned first followed by deposition of the TMO as in Fig. 5.3b.

The advantages of this structure, particularly for the early phases of work with the new material systems, include the following:

(1) The structure can be made quickly with a minimum number of steps.
(2) The bulk dielectric is thermal Si oxide whose properties are well understood. Therefore any unusual results are solely attributable to the dielectric−metal oxide interface, the metal oxide bulk, or the contacts.
(3) Finally, this structure can be subjected to a full range of temperatures in order to fully investigate the potential of the material system. The continuous gate devices can be deposited using shadow masks to define both the source−drain contacts as well as the TMO channel material.

Bottom gate staggered. In later work, as the ability to etch metal oxides has improved, photolithography has been used to define staggered bottom gate devices (Fig. 5.3c). In this structure, the bottom metal is patterned followed by deposition of the dielectric, the metal oxide active layer, any contact layer, and finally the contact metal. The active region around the device is patterned followed by patterning of the source and drain electrodes and any contact layer. This procedure requires precise alignment of the patterned gate and source–drain electrodes, hence the need for photolithography. The patterning of the source–drain contacts can first be followed by deposition of the semiconductor material (Fig. 5.3d).

(a)

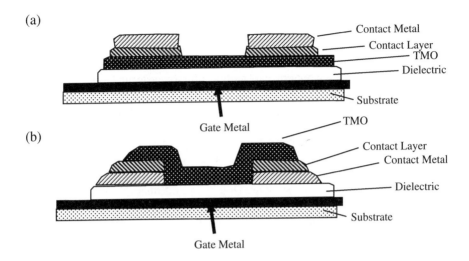

(b)

(c)

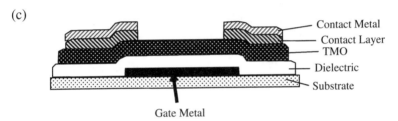

(d)

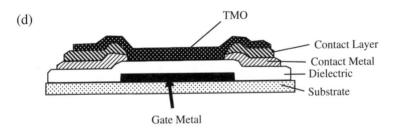

(e)

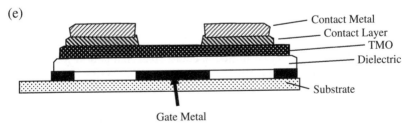

Fig. 5.3 Various transistor structures used in flexible transistors (**a**) continuous bottom gate staggered, (**b**) continuous bottom gate collinear, (**c**) bottom gate staggered, (**d**) bottom gate collinear, and (**e**) constant level transistor. Shadow mask for (**a**) and (**b**), photolithography for (**c**) and (**d**), and imprint lithography for (**a**) and (**e**)

Constant level transistor. This transistor structure is a preferred structure for SAIL defined devices [57, 58]. The material deposition is completed before any etching occurs. The layers are patterned by imprint lithography using SAIL (discussed in detail below). Devices with short channel lengths of 1 μm or less can be made on thin flexible substrates using this structure. There is no step coverage in this structure and the device can be fabricated from a complete metal–dielectric–semiconductor contact stack (Fig. 5.3e). There is also a simplified version of this device which has an unpatterned gate that is useful for transistor development work and slow applications but is unsuitable for higher speed applications (Fig. 5.3a) due to the large source–gate and source–drain overlap capacitances.

5.5 Fabrication on Flexible Substrates

One of the most difficult problems associated with flexible substrates, regardless of the materials used in the technology, is the dimensional variability of the substrate during device fabrication. The combination of large dimensional changes in flexible substrates and the strong dependence of device performance on precise source–drain alignment with respect to the gate require new approaches for fabrication. As mentioned in other chapters, some of these new methods include jetting of masking materials or electronic materials with dimensional adjustment among layers made dynamically during jetting. In this section, a new method for precise layer-to-layer alignment is presented to solve this alignment problem using a roll-to-roll compatible manufacturing technique. This fabrication method is applicable to many other flexible electronic material systems, such as low-temperature poly-Si and a-Si.

The performance of the transistors depends strongly on the source–drain electrode overlap with the gate and the size of the channel. If there is a gap between the source and the gate is 1 μm or larger, the on-current will be significantly reduced by the resistance of the unformed channel. If the pixel capacitance is small compared with the gate capacitance of the transistor, the turn-on time $t_{on} = L^2/(\mu_{FET}V_{ds})$, where L is the channel length, μ_{FET} is the field effect mobility, and V_{ds} is the drain–source voltage. Thus, a short channel greatly decreases the turn-on time of the transistor. Moreover, the on-current depends inversely on L resulting in a much larger on-current as the device size is decreased. On the other hand, a large overlap between the source–drain and the gate creates a significant feed-through capacitance that causes gate transients to be transferred to the pixel. Thus, the ability to align the source–drain contacts with respect to the gate to a level of 1–2 μm is needed to produce devices with speed and voltage transfer fidelity required by present display technologies.

The ability to control the overlap of source–drain layers with the gate to 1 μm accuracies has been quite difficult because the changes in dimensions of large-area flexible substrates in a roll-to-roll environment are quite large. Typically, plastic substrates have thermal expansion coefficients ranging from 10 to 100 ppm/°C and elastic moduli in the range of 0.5–3 GPa compared with 4–9 ppm/°C and 100 GPa for glass. In addition, the flexible plastic substrates are about 50–100 μm thick compared with 3 mm for typical glass displays. For electronics fabricated on a 30 cm wide web, a force as small as 1 N will deform the substrate roughly 300 μm while the thermal-induced dimension changes are 30 μm/°C from one side relative to the other. In addition, as the various overlayers are patterned, the overlayer-induced strain is nonuniform. Hence, the substrate changes size with respect to the mask of the next layer. As a result, it has been virtually impossible to align successive layers to better than about 30 μm, unless the substrate held dimensionally stable by bonding to a more rigid substrate. This later approach has all the expenses and limitations of wafer and glass-based processing, negating the advantage of using low-cost substrates in the first place.

The problem of layer-to-layer alignment on flexible substrates has been solved using SAIL [57, 58]. This fabrication method uses the ability of imprint lithography to fabricate polymer masking structures with precise height control. The information for various layers is encoded within the height of the imprinted structures. Thus, a single imprint step defines all the structures needed to complete the device fabrication. Any change in the size of the substrate affects all layers and, therefore, does not alter the relative alignment of different layers.

5.5.1 Imprint Lithography

For the past 50+ years, integrated circuit fabrication has proceeded through the use of photolithography and photoresist to define the critical features of ever reduced device sizes. As mentioned before, this technology has alignment and cost problems when used with free-standing large-area flexible substrates. The comparatively new fabrication technology, imprint lithography, is a promising new approach for low-cost fabrication of submicron feature electronics on large-area flexible substrates [59–61]. Imprint lithography, or more precisely, soft lithography in the present case creates multilevel patterns that can be used as masks without expensive photolithography. As presented later, the multilevel capability of imprint lithography can solve the layer-to-layer alignment problem inherent in flexible electronics and is compatible with high-throughput batch or roll-to-roll technology. Imprint lithography is useful not only for TMO devices but also for any flexible electronic technology such as a-Si and even has applications for organic electronics if a nonorganic imprint liquid is developed.

The imprint procedure consists of a relatively small number of steps [57, 58]. First, a master is fabricated using traditional photolithography and silicon patterning steps (Fig. 5.4a). Second, a replica stamp is made from the master by coating the master with a release agent and then with polydimethylsiloxane (PDMS) polymer (Fig. 5.4b). The PDMS stamp is compliant and is a remarkably faithful inverse replica of the original master (see Fig. 5.5). This stamp is attached to a flat quartz plate if it is to be used in batch processing or can be attached to quartz roller if roll-to-roll imprint lithography is used (Fig. 5.4c). The substrate plus desired blanket

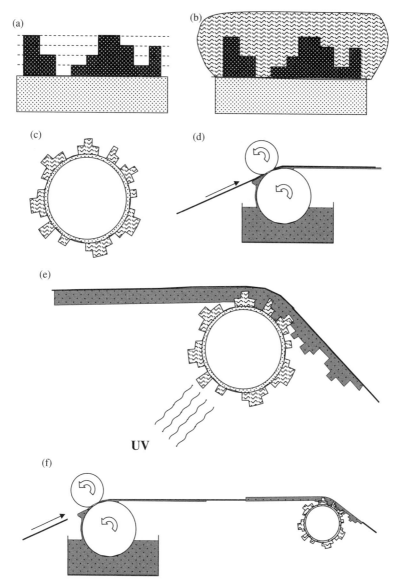

Fig. 5.4 Steps used to create imprint lithography patterns on flexible webs for SAIL fabrication of devices

deposited material is coated with a thin layer of thermal or UV curing epoxy, such as Norland UV curing epoxy (Fig. 5.4d) [62]. The epoxy layer is imprinted by the PDMS stamp and irradiated with UV or heated to polymerize the epoxy within the stamp (Fig. 5.4e,f). The polymerized epoxy replica is released from the stamp and is used as a mask for subsequent processing. An electron micrograph of a multilevel

Fig. 5.5 Multilevel imprint pattern resulting from imprint lithography. The fine scallops are in the original mask arising from the passivation and etch steps of the Bosch process used to create the master

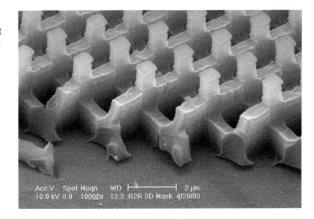

imprinted UV polymer structure is shown in Fig. 5.5. The fidelity of replication from the original master is remarkable. The scallops in the side walls are replicated from the master and arise from the passivation and etch cycles of the Bosch process used to create the high aspect ratio structures in the master. These features indicate that the process is capable of replicating 3-D features smaller than 60 nm.

Imprint lithography has a number of advantages over photolithography. It is capable of producing features as small as 60 nm at rates of centimeters per second. The PDMS stamps last over 3,000 imprints or more and each master can produce perhaps 100 or more stamps. Hence, the cost of the master is amortized over a large number of impressions while the master cost is not that much different than a single silicon wafer. So the imprint stamp costs per imprint are very low. In addition, imprint lithography equipment costs are much smaller than photolithography that requires clean rooms, vibration isolation, large complex optics, and multiple exposures per wafer. Finally, roll-to-roll fabrication facility costs scale roughly as the width of the web rather than as the area of the electronics. For large-area flexible devices, this advantage becomes increasingly important. Finally, the substrate costs are low as well as handling costs. The result of all these cost reductions lead to the expectation that the cost of large-area flexible electronics produced and imprint lithography and roll-to-roll methods can be significantly decreased.

5.5.2 Self-Aligned Imprint Lithography

The ability of imprint lithography to make multilevel layer structures enables the information of multiple mask layers to be encoded in the imprint layer thickness. Because all levels and therefore layer information is imprinted in one imprint process, subsequent changes in substrate size do not result in layer-to-layer misalignment. The SAIL process uses the multilevel imprint mask to create the desired device structures.

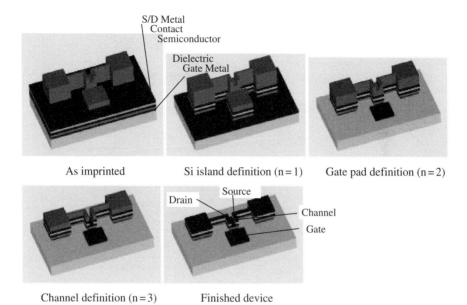

Fig. 5.6 The steps involved in SAIL fabrication of a single transistor. Between each step, the imprint mask polymer thickness is decreased by one level thickness to expose the next layer for patterning. Between $n = 1$ and $n = 2$, the gate is isolated from the bottom gate metal under the source and drain by undercut removal of the bottom metal in the narrow 'fuse' region connecting the source and drain to other pads

The SAIL process begins with a blanket deposition of a complete stack of materials needed for the device as shown in Fig. 5.6. The complete stack consists of a substrate, a gate metal, a dielectric, an active semiconductor (in this case a TMO), an optional contact material, and finally the top metal. A multilevel mask is imprinted on the device stack in order to define the subsequent features. The first etch step removes the entire stack in the region not covered by the imprinted polymer. During etching of the bottom metal, the metal is removed by undercutting the thin bridge portions of the pattern (between $n = 1$ and $n = 2$). This metal undercut isolates the source and drain regions from the gate eliminating possible shorts from the source and drain to the gate while minimizing parasitic source–gate and drain–gate capacitances. Next the imprinted polymer is etched until the next level is removed ($n = 3$). Then the stack layers are removed down to the gate (bottom) metal exposing the gate contact and pad. The polymer mask is then etched until the next polymer layer is eliminated exposing the channel region. The top metal and optional contact material (e.g., ITO) are etched by stopping on the active layer material using timed etching or through the use of selective etches between the contact and the active layer of the transistor. This etch step defines the channel length. Finally, the last level of the imprinted polymer is removed leaving the top metal exposed as shown in Fig. 5.7. The resulting transistor is a constant level device as

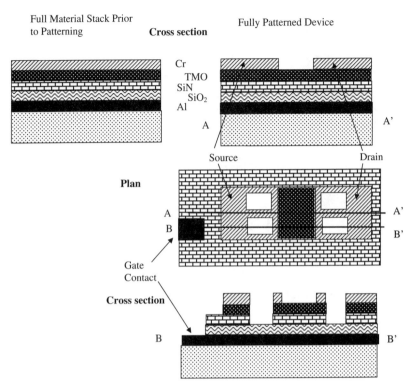

Fig. 5.7 Cross section and plan view of a single layer transistor fabricated by the simplified self-aligned imprint lithography (reefed-SAIL)

described in Fig. 5.3e. The various structures are aligned to within the accuracy of the undercut process.

There is an abbreviated form of the SAIL process, reefed-SAIL, in which the back metal is not patterned but remains as a continuous back contact (Fig. 5.4a). This eliminates one of the mask levels and several etch steps including the fuse step. This process is useful for slow devices in which the source–gate and drain–gate overlap capacitances are not important and for more rapid prototyping investigations. Cross sections and plan views of the devices produced by the abbreviated (reefed-SAIL) and full-SAIL processes are shown in Figs. 5.7 and 5.8 for a ZTO or an a-Si device with a composite dielectric.

An example of a ZTO device fabricated using reefed-SAIL is shown in Fig. 5.9. The pictured reefed-SAIL device is fabricated on a degenerate doped Si wafer as the gate and thermal oxide as the gate dielectric. These devices demonstrate the patterning precision possible with SAIL. Unlike jetting processes where there are problems of drop misplacement, splashing, and uneven drop wetting variations; the structures are sharp and well defined on a submicron scale. For this particular device, the channel region is 2 μm wide. The ability to fabricate very precise aligned and

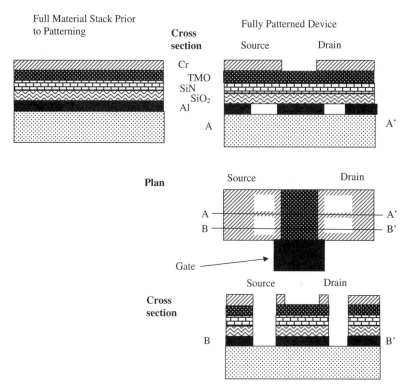

Fig. 5.8 Cross section and plan view of a single layer transistor fabricated by full self-aligned imprint lithography (full-SAIL). Note the isolation of the gate from the bottom layer metal in the source and drain

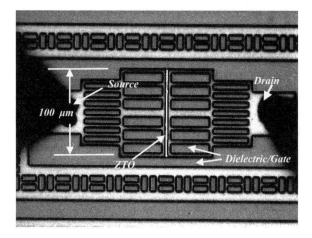

Fig. 5.9 A ZTO transistor fabricated using the reefed-SAIL method on a doped c-Si wafer as the gate electrode and thermal oxide as the gate insulator. The channel region is 2 μm wide

Fig. 5.10 Amorphous silicon transistor fabricated using a full-SAIL process

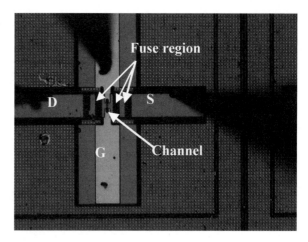

dimensioned devices using SAIL imprint lithography is apparent. The 0.3-μm pattern definition precision greatly exceeds that of shadow masking (roughly 20 μm on flexible substrates and roughly 15 μm for ink-jet-defined device structures). A full-SAIL a-Si device is shown in Fig. 5.10, which includes the fuse regions that are undercut by etching the bottom metal.

5.5.3 SAIL Transistor Results

While the described SAIL fabrication steps appear plausible, the fabrication and characterization of actual devices on flexible substrates indicate that the SAIL process works in practice. Because dry etching recipes for TMOs are still being developed, the reefed-SAIL and SAIL fabrication processes have been used to fabricate a-Si transistors, a mature proven technology that can be used to benchmark the transistors on flexible substrates. Basically, the results presented in this section demonstrate that imprint lithography readily produces flexible short and long channel devices whose performance matches that found for devices on rigid substrates in a-Si. By implication, imprint lithography should be able to generate comparable results for TMO devices.

Amorphous silicon devices were fabricated using SAIL and device material stacks consisting of the following layers. The substrate was 50 μm polyimide with a 100-nm stainless steel layer on the back. A 100-nm Al layer was covered by a 50-nm plasma enhanced chemical vapor deposition (PECVD) SiO$_2$ layer and a 200-nm PECVD SiN layer deposited at 250°C. Then a 100-nm a-Si:H layer was deposited followed by a 100-nm n+ layer and finally 100 nm of Cr comprises the top layer. The devices were imprinted and developed to produce arrays of all combinations of transistors width $W = \{10, 20, 50, 100\}$ μm and length $L = \{1, 2, 5, 10, 20, 50, 100\}$ μm. The normalized transfer characteristics, source–drain current/(W/L),

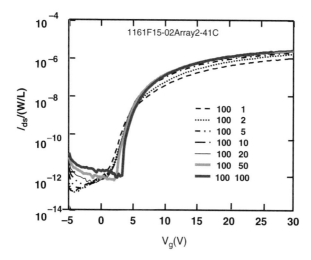

Fig. 5.11 The source–drain current $I_{ds}/(W/L)$ versus V_g for $W = 100$ μm and $L = \{1, 2, 5, 10, 20, 50, 100\}$ μm for SAIL a-Si transistors

for $W = 100$ μm devices with various L are shown in Fig. 5.11. If the current scales with L, the curves should completely overlap. There is some broadening of the exponential slope for the shortest devices but otherwise the L scaling holds. Thus, for the $W/L = 100$ μm /1 μm devices, the on-current is 100 μA which is quite respectable for an a-Si device. Such a device could easily drive OLED pixels. The subthreshold slope is about 0.6 V/decade, the on–off ratio is about 10^5, and V_t is about 3 V. Thus, SAIL can fabricate 1 μm channel length devices in a roll-to-roll process on flexible substrates. The device uniformity is quite decent as well.

SAIL fabrication of arrays has also been undertaken using identical procedures along with the pixel electrode for the backplane of a display. Currently, arrays of 50×50 and 100×100 devices are being fabricated. As with all flexible electronics technology, the yields and susceptibility to flexing remain largely unexplored.

5.5.4 Summary of Imprint Lithography

In summary, imprint lithography can make short channel devices on flexible substrates for large area, flexible, and inexpensive applications. The SAIL process described in this section solves the layer-to-layer alignment problem for micron-scale alignment on dimensionally variable substrates. The abbreviated reefed SAIL process has been used to fabricate isolated TMO transistors, while the full-SAIL process has been used to fabricate amorphous Si arrays and transistors on flexible substrates.

5.6 Flexible TMO Device Results

The previous sections describing flexible transistor fabrication indicated that there are three main ways of fabricating TMO devices: shadow masks, photolithography, and imprint lithography. Section 5.1.2 summarized some important TMO transistor results on rigid substrates; further details can be found in Table 5.1 and the corresponding references. In this section, the performance of various flexible TMO transistors using the above-mentioned fabrication methods is described. While significant TMO work on rigid substrates has been undertaken, flexible TMO work and TMO work using imprint lithography are still in their infancy. It is expected that the TMO results on rigid substrates mentioned above represent appropriate performance benchmarks for flexible electronics because the rigid, higher temperature approaches can utilize any of the flexible technologies. Thus, the results in this section should be compared to rigid substrate results as well as alternative flexible platforms such as a-Si:H.

One of the first reports of devices fabricated on flexible substrates is by Carcia et al. [20], where a mobility of 0.4 cm^2 V^{-1} s^{-1}, an on–off ratio of 10^4, and $V_t = 7.5$ V was achieved. The transistor consisted of a fluoropolymer dielectric, an RF sputter deposited ZnO active layer, and a polyimide substrate and was fabricated at room temperature. The results are shown in Fig. 5.12. This work was one of the first to demonstrate that TMO TFTs could produce flexible transistors with capabilities comparable to a-Si although the off-current is a bit high. Subsequent work by these researchers using a Al$_2$O$_3$ dielectric, ZnO semiconductor, e-beam evaporation on polyimide, and temperatures below 120°C produced devices with mobilities of 50 cm^2 V^{-1} s^{-1}, on–off ratios of 10^5, and $V_t = 3.2$ V [22, 37]. However, the off-current, roughly 10^{-9} A, was too high for a pixel transistor on an LCD. Such a large leakage requires unacceptably large refresh rates and concomitant power dissipation. The mobility and on-current, however, demonstrate that high-performance flexible electronics can be made using TMO as the channel material.

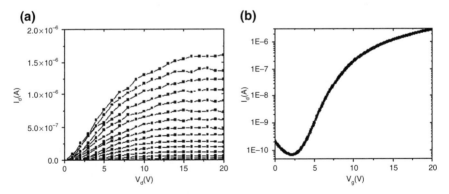

Fig. 5.12 Flexible ZnO TFT curves of (**a**) I_d versus V_d for $V_g = 0, 1, \ldots, 20$ V and (**b**) of I_d versus V_g for $V_{ds} = 20$ V. W/L $= 40\ \mu$m $/400\ \mu$m from [20]

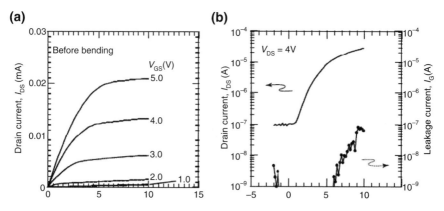

Fig. 5.13 From [29] on PET with a maximum temperature near room temperature

In [29], transistors were fabricated on PET using IGZO as the TMO and Y_2O_3 as the dielectric where the fabrication temperature was kept near room temperature (Fig. 5.13). A mobility of 8 cm^2 V^{-1} s^{-1} was achieved with an on–off ratio of 10^3 and an off-current of 10^{-7} A. Except for the large off-current this is a reasonable transistor. The importance of this work is that PET is clear unlike polyimide, so the devices could be used in principle as part of a backlit liquid crystal display (LCD). Also, this work represents an improvement of the transistor characteristics over the first flexible transistor work. These flexible TMO results are not as good as the results achieved on rigid substrates.

In order to further improve the performance of flexible TMO devices, the contacts must produce sufficient current. In Fig. 5.14, the output characteristics of a ZTO transistor on polyimide using shadow mask definition of ZTO on PECVD deposited SiON, 50-μm polyimide substrate, and a 250°C 10-min anneal [26]. The transistor output curves are shown for contacts consisting of Al on ZTO. The concave current

Fig. 5.14 The output characteristics for a ZTO transistor with Al contacts, SiON dielectric, and a maximum fabrication temperature of 250°C. The concave characteristics are indicative of blocking contacts and suggest that improved contacts can give rise to improved device performance [26]

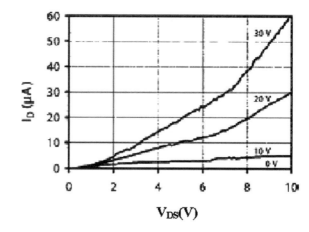

Fig. 5.15 The source–gate capacitance versus dc source-to-gate voltage bias for various frequencies for the transistor with Al source and drain contacts. The measurement frequencies are 0.5, 1, 5, 10, 50, 100, and 500 kHz. The insets indicate the device layout (*upper right*) and the band structure for negative (*left*) and positive (*right*) dc bias conditions

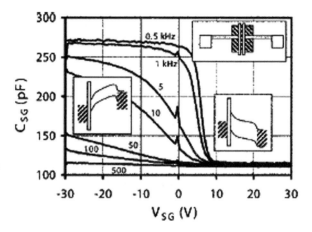

crowding suggests that there is a problem with the contacts. This conclusion about the contacts was further verified by two methods: capacitance and four-probe transistor measurements. In Fig. 5.15, the capacitance of a transistor structure of the source with respect to the continuous gate is shown as a function of V_{sg}. The associated band structure is depicted in the insets and as well as the device geometry. For $V_{sg} > 10$, the device is in depletion and the channel does not conduct. Hence, the capacitance is just the capacitance of the source electrode with respect to the gate regardless of the frequency. For $V_{sg} < 0$ and low frequencies, the channel is turned on so that the capacitance consists of both the source–gate capacitance, the drain–gate capacitance, and the channel–gate capacitance. Moreover, the various capacitances are only the insulator capacitances and do not include the series capacitance of semiconductor. The additional area and lack of the semiconductor series capacitance result in the roughly factor of 3 increase in the capacitance. For higher frequencies, the RC time constant from contact and channel series resistances and the capacitances is large so there is not enough time within one cycle to charge up the channel–gate and drain–gate capacitances. By observing the frequency at which $AR_{contact}C_{ins}2\pi f = 1$, where A is the source–gate contact area, $R_{contact}$ is the contact resistance, C_{ins} is the insulator capacitance, and f is the frequency, the contact resistance $R_{contact}$ can be estimated. Because $C_{ins} = 15$ nF/cm^2 the area of the device is 0.03 cm^2 and $R_{contact} \approx 30$ kΩ. This device is, therefore, not capable of passing much more than about 35 μA without a significant voltage drop across the contact.

A second means of estimating the contact resistance is to measure the voltage drop across the contact using a four-probe method (Fig. 5.16) [26, 28]. The source and drain probes drive a current through the source and drain contacts and the voltage of the two inner probes is measured using a very high impedance instrument, so no current is drawn through the probes. The potential difference between the source and the first contact yields the voltage drop across the source contact, while the voltage difference between the final voltage probe and the drain yields the voltage

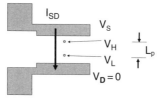

Fig. 5.16 Four-probe configuration for measuring the source and drain contact resistance. Two addition high impedance probes or electrodes are placed between the source and drain to measure the potential drop between the channel and the source and drain electrodes

drop across the drain. The potential drop between the two voltage probes yields the channel voltage drop. In particular, the channel resistance, R_{CH}, the source resistance, R_S, and the drain resistance, R_D are roughly given by

$$R_S = (V_S - V_H)/I_{SD}$$
$$R_D = V_L/I_{SD} \qquad (1)$$
$$R_{CH} = (V_H - V_L)/I_{SD}$$

The expressions for R_S and R_D are approximate in that some channel resistances are included because the probes include at least several microns of the channel. Using these relations, the source, drain, and channel resistances can be estimated [26, 28]. For the Al and ZTO contact, this method yields approximately the same value of the contact. The four-probe method can be used to determine

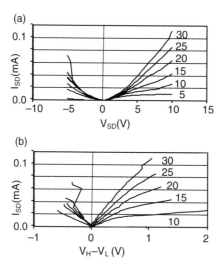

Fig. 5.17 The source–drain current in a ZTO transistor with Al contacts and no ITO layer to improve the contact resistance (**a**). The same current plotted against the voltage drop between the inner voltage sensing probes (**b**). Notice that the current crowding indicative of contact effects has been eliminated

the characteristics without contact impedance effects. In Fig. 5.17, the source–drain current through a ZTO transistor with PECVD SiON dielectric, Al contacts deposited directly on the ZTO, with $W/L = 2000$ μm/200 μm is presented. The field effect mobility is about 0.2 cm^2 V^{-1} s^{-1} for the uncorrected curves. The normal output curves in Fig. 5.17a exhibit current crowding while plotted against the potential difference between the two electrodes eliminates the crowding (Fig. 5.17b). Notice also the difference in horizontal scales is almost a factor of 10. The mobility extracted from the contact corrected curves will, therefore, be about a factor of 10 larger and is about 1.5 cm^2 V^{-1} s^{-1}. Thus, the contact effects dramatically alter the extracted mobility. The four-probe methods enable the true channel mobility to be extracted by eliminating the contact effects. Clearly, however, the most important point is that the contacts must be improved.

By including an ITO layer between the Al and the ZTO, the contact resistance was decreased to about 3 kΩ. In addition, if an oxide buffer layer is included between the SiON and the ZTO, the device performance is greatly improved. For this device the output and transfer characteristics are shown in Figs. 5.18 and 5.19. The incremental mobility is 14 cm^2 V^{-1} s^{-1}(Fig. 5.20) and the on–off ratio is 10^7, and an off-current of 10^{-11} A was produced showing that the high off-current can be reduced while maintaining high on-currents. Moreover, this is a stable current and changes little with stressing. The threshold voltage is somewhat negative but this can be changed by adjusting the dielectric deposition conditions and/or injecting the appropriate charge into the dielectric to adjust the threshold voltage. The maximum incremental saturation mobility is ~14 cm^2 V^{-1} s^{-1} (Fig. 5.20) [26, 28]. The threshold voltage is about −9 V (Fig. 5.21). In summary, the flexible transistor properties are nearly equivalent to a comparable rigid substrate device composed of similar materials. This bodes well for the future of flexible TMO devices.

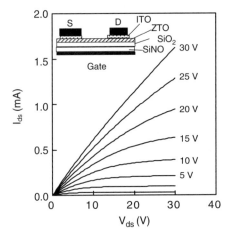

Fig. 5.18 A ZTO transistor deposited on 50 μm polyimide, PECVD SiON, ITO contact layer, Al contacts, and a SiO$_2$ buffer layer ($W/L = 1,000$ μm /100 μm)

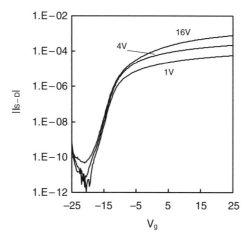

Fig. 5.19 The transfer characteristics of the device in Fig. 5.18

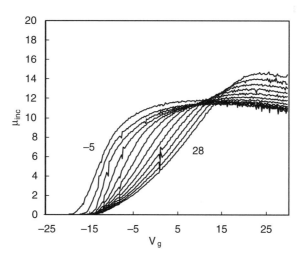

Fig. 5.20 The incremental mobility for the device in Fig. 5.18. The maximum mobility is 15 cm^2 V^{-1} s^{-1}

5.7 Future Problems and Areas of Research

In order to realize the full promise of flexible TMO electronics, there are a number of areas in which further research is desirable. These include the following:

(1) Better control of the carrier density in the TMO channel materials.
(2) Improved low-temperature dielectrics.
(3) Better etching processes for TMO materials.

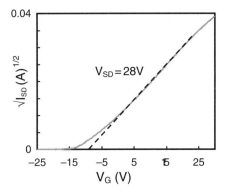

Fig. 5.21 Transfer characteristics of device in Fig. 5.18. $V_T = -9$ V

(4) Improved p-type material.
(5) Device electrical stability.
(6) Flexure and adhesion of the TMO materials on flexible substrates.
(7) Yield validation of flexible substrate fabrication methods.

Progress on these issues could greatly expand the possible areas of application for flexible TMO electronics and improve the control of the TMO material system for cost effective production.

5.7.1 Carrier Density Control

One of the areas that can be difficult to control reliably in TMO materials is the carrier density particularly within the lower temperature constraints imposed by flexible substrates. Because oxygen defects act as n-type dopants, the materials typically must be sputtered in high-oxygen partial pressure conditions. Otherwise the TMO will end up being n-type and unsuitable for an accumulation mode transistor [52–56]. Such n-type materials could in principle be used in depletion mode applications if low off-currents are not required. The high-oxygen partial pressures, however, tend to promote arcing and dielectric breakdown of the underlying gate dielectric in the sputtering environment. More research is needed in understanding the energy levels and doping effects of O vacancies in the various TMO systems particularly as a function of the transition metal cation.

The carrier density is also affected by the presence of H. There is significant evidence that H within many of the TMO materials acts as an n-type dopant [51]. In fact, exposure to an H plasma can change the conductivity of TMO by orders of magnitude [48, 49]. Hence reactive ion etching (RIE) with H-containing materials, such as CH_4 and CF_3H, can be a source of carrier density variation even though these gases are the leading candidates for dry etching TMO materials. Also dielectrics such as PECVD a-SiN dielectrics can also introduce H into the adjacent TMO. Because of the ubiquitous prevalence of H during deposition, etching, and in neighboring materials and the rather high mobility of H in TMOs, postfabrication annealing protocols can be important to control the carrier density. Often annealing for short times above 200°C can significantly reduce the excess carrier density. Presumably this annealing causes H to move to become electrically inactive without releasing H from any dielectric such as SiN that might contain H. Postfabrication H motion could also be a source of instability in the electronic properties of the fabricated devices. Significantly more research is needed in understanding the energy levels introduced by H, the diffusion of H within TMOs, and the interchange of H with dielectrics both during deposition and postdevice fabrication.

5.7.2 Low-Temperature Dielectrics

As with all material systems used on flexible substrates, good low-temperature dielectrics are critical. The dielectrics must have low interface defects with the TMO materials at low deposition temperatures. In bottom gate devices, it is also desirable that the surface roughness of the dielectric be small in order that the field effect mobility approaches that of the underlying TMO. Most importantly, the dielectric must withstand high fields without trapping significant charge. The trapped charge will cause threshold shifts arising from continuous application of gate biases. While charge-trapping-induced threshold voltage shifts are undesirable for all flexible electronic systems, they are particularly undesirable for electronic applications such as driving OLEDs and drive electronics; the potential new application areas for TMO materials. Otherwise one must use more elaborate self-compensating circuits in such applications. Some of the dielectrics that have been used are listed in Tables 5.1 and 5.4, but there has been little investigation into their stability under voltage stress or their tendency to create interface states. Research comparing the stability of the various dielectrics may also shed light on the physical origins of the metastability.

5.7.3 Etching of TMO Materials

For batch fabrication of TMO electronics on flexible substrates, etching continues to be a somewhat difficult area. Both wet and dry etching of TMO materials have undesirable features. Besides the waste disposal issues and expensive process of wet etching, many flexible substrates such as polyimide and PET can absorb water which in turn can cause swelling. Hence, wet etching may require a moisture barrier layer to protect the substrate. The preferable dry RIE etching is also somewhat difficult for TMO materials. The two most promising etching chemistries are CH_4/H_2 and BCl_3 [50]. The BCl_3 chemistry is slow and tends to etch contact metals as well as TMO materials thereby limiting the types of process flows and structures that can be fabricated. The CH_4/H_2 chemistry on the other hand etches the TMO material while leaving metals intact. The chief difficulty with the CH_4/H_2 chemistry is that it tends to leave carbon structures, maybe even carbon nanotubes on surfaces other than the TMO such as the walls of the chambers and source–drain contacts. In normal conditions, one uses oxygen to suppress carbon compound formation but this interferes with the etching of the oxygen in the TMO. The hydrogen concentration is adjusted to help etch surface carbon compounds but as mentioned above, excess atomic H can cause problems for the carrier density in the TMO unless the hydrogen is removed by subsequent annealing steps. The deposition of carbon compounds on the chamber walls often requires periodic mechanically cleaning as the material is remarkable resistant to usual O cleaning procedures. While existing etch protocols are satisfactory, improved etching chemistries and protocols would be quite useful for the wide scale use of TMO electronics.

5.7.4 P-type TMO

Another major area for future research is the deposition of p-type TMO material. P-type material would be highly desirable in order to create CMOS-type circuitry for drive electronics and other electronic applications. Significant effort has been made to develop p-type TMO materials [63–69]. In general, because O vacancies dope n-type TMO, it is difficult to produce p-type material. Whenever a p-type dopant atom is introduced into the growing materials, an O vacancy becomes more energetically favorable. The O vacancies compensate the material. There is promising work using Cu containing materials to help dope the material p-type [63–69]. The Cu cation has an unfilled 4d that lines up with the valence band of most TMO. Cu doping then serves as a p-type dopant. The p-type carrier densities that have been achieved are 3×10^{19} carriers cm^{-3}. Unfortunately, the features of TMOs that make the electron mobility high even for amorphous disorder, results in reduced mobility for holes. The top of the valence band is primarily directional O 2p-type orbitals that are sensitive to disorder. In sufficient quantities, the Cu constituents create a miniband at the top of the valence band that increases the hole conductivity despite the disorder induced broadening of the valence band edge. The best mobilities achieve to date are mobilities of $0.13\,cm^2/V \cdot s$ for nonsingle crystalline material [63–69].

5.7.5 Stability

As in the case of many material systems with disorder and low-temperature dielectrics, TMO devices exhibit metastability. The quantification and understanding of this phenomenon are just beginning. It has been known for a while that ZnO, for example, exhibits a persistent photoconductivity [52–56]. Illumination by bandgap light causes significant changes in conductivity that last for hours or more. This effect has been attributed to O chemisorption initiated by the UV light. The effect is minimized if the transistor has a passivation layer preventing O interaction with the back surface [52–56]. Depending on the composition, annealing history, and exposure to O or H, the electrical characteristics exhibit varying degrees of instability. The instabilities can arise from changes in carrier concentrations, defects in the semiconductor, and/or charge trapping in the insulator. In general, the higher the annealing temperatures, the better the stability and the smaller the initial device-to-device variation in threshold voltages and subthreshold slopes. The presence of In also appears to increase instability. The origin and control of instabilities are of particular interest as one of the many advantages of TMO is their use in edge driver electronics and OLED drivers. The stability requirements of driver electronic devices are typically more demanding because the on-time voltage duty cycle can approach 100%, while in the typical pixel transistor, the on-time duty cycle can be made small. Significant work in this area is required.

5.7.6 Flexure and Adhesion of TMO

Common to all flexible electronic materials and fabrication methods, more work needs to be done characterizing and understanding the effects of flexure on the performance of TMO electronics. The brittle nature of TMO may suggest that flexure maybe a problem. However, the small relative area of the TMO transistor channel region, the small thickness of the TMO material, and the large compliance of the substrate suggest that flexure may not be a significant limitation for TMO electronics. Preliminary studies of devices using brittle a-Si:H as a channel material indicates that metal lines and dielectrics are more of a problem. Virtually no flexing studies on TMOs have been done.

5.7.7 Flexible Fabrication Method Yields

A final area of future study common to most all flexible material systems concerns the yield of various fabrication methods. Because the substrate dimensions are not fixed, the fabrication temperatures are low, and the fabrication methods are relatively nonstandard (not the typical photolithography), there is the all important question of yield. The initial yield of the devices, lines, and crossovers for both imprint lithography and jetting methods of fabrication must be measured and optimized. Work is just beginning to characterize the yield issues of imprint lithography. Dirt-induced defects tend to be eliminated by the imprint process while defects due to imprint stamp wear, propagate, and can result in significant yield loss. Improper stamp filling and bubble defects are less important; they cause at most the failure of one array. Yield issues regarding control and uniformity of etching masks remain unexplored. Jetting methods have problems of splashing, wetting, gaps, and nonuniform coverage issues. The yield issues associated with these flexible substrates transcend the material system and are not specific to the TMO devices.

Issues specific to SAIL that need to be measured include the following:

(1) The yield of the devices and arrays using SAIL is not yet known. There are imprint defects such as voids caused by under filling and excess imprint polymer if there is over filling.

(2) The impact of imprint wears on device yield needs to be measured in order to estimate stamp life time. Defects may form in the stamp due to particles or failure to release from the mold. These defects are then replicated.

(3) The effects of particle densities on yield and whether acceptable yield can be obtained without a clean room.

(4) Finally, the required levels of process uniformities and latitudes that affect device yield should be measured. Because SAIL is so recent, many of these issues require further work. In general, we have been encouraged by the fact that as we have moved toward roll-to-roll fabrication using SAIL, device yields, controllability, and uniformity have markedly improved over batch processing suggesting that SAIL represents a promising fabrication method for large area, inexpensive, and flexible electronics.

5.8 Summary

SAIL is a solution to the layer-to-layer alignment problem for the critical features on flexible substrates. The SAIL process can produce submicron-aligned layers and submicron-sized features, is compatible with both roll-to-roll and high-throughput manufacturing, and is inexpensive. The complex series of steps (deposition, coat with resist, expose resist, develop, etch, and resist removal) required for each layer in the standard photolithography is eliminated. The capability of the SAIL process can be used for any materials system, such as a-Si, low-temperature poly-Si, and laser recrystallized Si as well TMOs. The only materials system for which SAIL does not currently seem straightforward is organics because etch selectivity between the imprint polymer and the organic layer has not been adequately investigated. SAIL should be considered as a possible fabrication methodology in the area of flexible electronics and can enable many of the structures and devices mentioned elsewhere in this book.

The major point of this chapter is the promising possibility of TMOs for flexible electronics. These materials have demonstrated high mobilities, large on–off ratios, and reasonable threshold voltages using fabrication temperatures and procedures compatible with flexible substrates and large-throughput manufacturing methods. Transistors with mobilities as high as 30 cm^2 V^{-1} s^{-1}, on–off ratios of 10^7, $V_t = 3$ V, off-currents of 1 pA, and on-currents in the range of 1 mA have been demonstrated. Moreover, the transistors can be completely transparent so the fill factor can be large. If the stability issue, transition metal etching and appropriate dielectric material selection are solved, the future looks very promising for flexible TMO electronics.

Acknowledgements   I would like to thank G. Herman and R. Hoffman introducing me to the field of transition metal oxide transistors and for making various samples, and C. Taussig, P. Mei, and C. Perlov for many consultations and support during this work.

References

1. Boesen GF, Jacobs, JE (1968) ZnO field-effect transistor. Proc IEEE 56:2094–2095
2. Khuri-Yakub BT, Kino GS (1974) A monolithic zinc-oxide-on-silicon convolver. Appl Phys Lett 25:188–190
3. Mahan, GD (1983) Intrinsic defects in ZnO varistors. J Appl Phys 54:3825–3832
4. Kudo A, Yanagi H, Ueda K, Hosono H, Kawazoe H, Yano Y (1999) Fabrication of transparent p–n heterojunction thin-film diodes based entirely on oxide semiconductors. Appl Phys Lett 75:2851–2853
5. Chopra K, Major S, Pandya D (1983) Transparent conductors – a status review. Thin Solid Films 102:1–46
6. Prins MWJ, Grosse-Holz K, Mller G, Cillessen JM, Giesbers JB (1996) A ferroelectric transparent thin-film transistor. Appl Phys Lett 68:3650–3652
7. Kawasaki M, Tamura K, Saikusa K, Aita T, Tsukazaki A, Ohtomo A, Jin ZG, Matsumoto Y, Fukumura T, Koinuma H, Ohrnaki Y, Kishimoto S, Ohno Y, Matsukura F, Ohno H, Makino T, Tuan NT, Sun PD, Chia CH, Segawa Y, Tang ZK, Wang GKL (2000) Can ZnO eat market in optoelectronic applications. Ext. Abst. 2000 Int Conf. Solid State Deyices ard Materials, Sendai, l28–129
8. Ohtomo A, Kawaski M (2000) Novel semiconductor technologies of ZnO films towards ultra-violet LEDs and invisible FETs. IEEE Trans Electron E83-C:1614–1617
9. Ohya Y, Niwa T, Ban T, Takahashi Y (2001) Thin-film transistor of ZnO fabricated by chemical solution deposition. Jpn J Appl Phys 40:297–299
10. Hoffman R (2002) Development, fabrication, and characterization of transparent electronic devices. Masters Thesis, Oregon State University
11. Masuda S, Kitamura K, Okumura Y, Miyatake S, Tabata H, Kawai T (2003) Transparent thin-film transistors using ZnO as an active channel layer and their electrical properties. J Appl Phys 93:1624–1630
12. Nomura K, Ohta H, Ueda K, Kamiya T, Hirano M, Hosono H (2003) Thin-film transistor fabricated in single-crystalline transparent oxide semiconductor. Science 300:1269–1272
13. Hoffman R, Norris B, Wager J (2003) ZnO-based transparent thin-film transistors. Appl Phys Lett 82:733–735
14. Nishii J, Hossain FM, Takagi S, Aita T, Saik K, Ohmaki Y, Ohkubo I, Kishimoto S, Ohtomo A, Fukumura T, Matsukura F, Ohno Y, Koinuma H, Ohno H, Kawasaki M (2003) High mobility thin film transistors with transparent ZnO channels. Jpn J Appl Phys 42:L 347–349 Part 2, No. 4A
15. Carcia PF, McLean RS, Reilly MH, Malajovich I, Sharp KG, Agrawal S, Nunes G Jr (2003) ZnO thin film transistors for flexible electronics. Mat Res Soc Symp Proc 769:H7.2.1
16. Nishii J, Hossain FM, Takagi S, Aita T, Saikusa K, Ohmaki Y, Kishimoto I, Ohtomo A, Fukumura T, Matsukura F, OhnoY, Koinuma H, Ohno H, Kawasaki M (2003) High mobility thin film transistors with transparent ZnO channels. Jpn J Appl Phys 42:347–349
17. Fortunato E, Hosano H, Granquist C, Wager J (2007) Advances in transparent electronics: From Materials to devices, I, 51(7).
18. Chaing HQ (2003) Development of zinc tin oxide-based transparent thin-film transistors. Master Thesis, Oregon State University

19. Chiang HP, Wager JF, Hoffman, RL, Jeong J, Keszler DA (2005) High mobility transparent thin-film transistors with amorphous zinc tin oxide channel layer. Appl Phys Lett 86: 13503–13505

20. Carcia PF, McLean RS, Reilly MH, Nunes G (2003) Transparent ZnO thinfilm transistor fabricated by rf magnetron sputtering. Appl Phys Lett 82:1117–1119

21. Fortunato E, Pimentel A, Pereira L, Goncalves A, Lavareda G, Aguas H, Ferreira I, Carvalho CN, Martins R (2004) High field-effect mobility zinc oxide thin film transistors produced at room temperature. J Non-Cryst Solids 338–340:806–809

22. Carcia PF, McLean RS, Reilly MH (2005) Oxide engineering of ZnO thin-film transistors for flexible electronics. J Soc Inf Display 13/7:547–550

23. Bellingeria E, Marréa D, Pellegrinoa L, Pallecchia I, Canub G, Vignoloa M, Berninia C, Siria HS (2005) High mobility ZnO thin film deposition on SrTiO3 and transparent field effect transistor fabrication Superlattices and Microstructures 38:446–454

24. Hwang CS, Park SH, Chu HY (2005) ZnO TFT fabricated at low temperature for application active-matrix display. 12th Int. Display Workshops/Asia Display, p1149–1151

25. Carcia PF, McLean RS, Reilly MH (2006) High-performance ZnO thin-film transistors on gate dielectrics grown by atomic layer deposition. Appl Phys Lett 88:123509–123511

26. Jackson WB, Hoffman RL, Herman GS (2005) High-performance flexible zinc tin oxide field-effect transistors. Appl Phys Lett 87:193503–193505

27. Hoffman RL (2006) Effects of channel stoichiometry and processing temperature on the electrical characteristics of zinc tin oxide thin-film transistors. Solid-State Electron 50:784–787

28. Jackson WB, Herman GS, Hoffman RL, Taussig C, Braymen S, Jeffery F, Hauschildt J (2006) Zinc tin oxide transistors on flexible substrates. J Non-Cryst Solids 352:1753–1755

29. Nomura K, Ohta H, Takagi A, Kamiya T, Hirano M, Hosono H (2004) Room-temperature fabrication of transparent flexible thin-film transistors using amorphous oxide semiconductors. Nature 432:488–492

30. Yaglioglu B, Yeom HY, Beresford R, Paine DC (2006) A high-mobility amorphous In_2O_3 – 10 wt% ZnO thin film transistors. Appl Phys Lett 89:062103–062105

31. Fortunato E, Barquinha P, Pimentel A, Gonçalves A, Marques A, Pereira L, Martins R (2005) Fully transparent ZnO thin-film transistor produced at room temperature. Adv Mat 17: 590–594

32. Santato C, Manunza I, Bonfiglio A, Ciroira F, Cosseddu P, Zamboni R, Muccini M (2006) Tetracene light-emitting transistors on flexible plastic substrates. Appl Phys Lett 86: 141106–141109

33. Loi A, Manunza I, Bonfiglio A (2005) Flexible, organic, ion-sensitive field-effect transistor. Appl Phys Lett 86:103512–103514

34. Bonfiglio A, Mameli F, Sanna O (2003) A completely flexible organic transistor obtained by a one-mask photolithographic process. Appl Phys Lett 82(20):3550–3552

35. Ohta H, Nomura K, Hiramatsu H, Ueda K, Kamiya T, Hirano M, Hosono H (2003) Frontier of transparent oxide semiconductors. Solid-State Electron 47:2261–2267

36. Fortunato EC, Barquinha PM, Pimentel AC, Gonçalves AM, Marques AJ, Martins RF, Pereira LM (2004) Wide-bandgap high-mobility ZnO thin-film transistors produced at room temperature. Appl Phys Lett 85:2541–2543

37. Hosono H, Nomura H, Kamiya T (2005) High performance FET using transparent amorphous oxide semiconductors as channel layer on plastic substrate. 12th Int. Display Workshops/Asia Display IDW/AD05 AMD3-1:251–253

38. Ellmer K (2001) Resitivity of polycrystalline zinc oxide films: Current status and physical limit. J Phys D: Appl Phys 34:3097–3108

39. Chopra K, Major S, Pandya D (1983) Transparent conductors – a status review. Thin Solid Films 102:1–46

40. Minami T, Miyata T, Yamamoto T (1998) Work function of transparent conducting multicomponent oxide thin films prepared by magnetron sputtering. Surf Coat Tech 108–109: 583–587

41. Minami T, Takata S, Sato H, Sonhana H (1995) Properties of transparent zinc stannate conducting films prepared by radio frequency magnetron sputtering. J Vac Sci Technol A 13:1095–1099
42. Minami T, Sonohara H, Takata S, Sato H (1994) Highly transparent and conductive zinc-stannate thin films prepared by RF magnetron sputtering. Jpn J Appl Phys 33:1693–1696
43. Wu X, Coutts T, Mulligan W (1997) Properties of transparent conducting oxides formed from CdO and ZnO alloyed with SnO_2 and In_2O_3. J Vac Sci Technol A 15:1057–1062
44. Young DL, Moutinho H, Yan Y, Coutts TJ (2002) Growth and characterization of radio frequecy magnetron sputter-deposited zinc stannate, Zn_2SnO_4, thin films. J Appl Phys 92: 310–319
45. Young DL (2000) Electron transport in zinc stannate (Zn_2SnO_4). Ph. D. thesis, Colorado School of Mines
46. Minami T, Sonohara H, Kakumu T, Takata S (1995) Highly transparent and conductive $Zn_2In_2O_5$ thin films prepared by RF magnetron sputtering. Jpn J Appl Phys 34:971–974
47. Van de Walle CG (2000) Hydrogen as a cause of doping in zinc oxide. Phys Rev Lett 85: 1012–1015
48. Raniero L, Ferreira I, Pimentel A, Goncalves A, Canhola P, Fortunato E, Martins R (2006) Role of hydrogen plasma on electrical and optical properties of ZGO, ITO and IZO transparent and conductive coatings. Thin Solid Films 511–512:295–298
49. Theys B, Sallet V, Jomard F, Lusson A, Rommeluere JF, Teukam Z (2002) Effects of intentionally introduced hydrogen on the electrical properties of ZnO layers grown by metalorganic chemical vapor deposition. J Appl Phys 91:3922–3924
50. Pearton SJ, Norton DP, Ip K, Heo YW, Steiner T (2005) Recent progress in processing and properties of ZnO Progress in Materials. Science 50:293–340
51. Van deWalle C, Nequgebauer J (2003) Universal alignment of hydrogen levels in semiconductors, insulators and solutions. Nature 423:626–628
52. Zhang DH, Brodie DE (1995) Photoresponse of polycrystalline ZnO films deposited by r.f. bias sputtering. Thin Solid Films 261:334–339
53. Zhang DH (1995) Fast photoresponse and the related change of crystallite barriers for ZnO films deposited by RF sputtering. J Phys D: Appl Phys 28:1273–1277
54. Takahashi Y, Kanamori M, Kondoh A, Minoura H, Ohya Y (1994) Photoconductivity of ultra-thin zinc oxide films. Jpn J Appl Phys Part 1, 33:6611–6615
55. Studenikin SA, Golego N, Cocivera M (2000) Carrier mobility and density contributions to photoconductivity transients in polycrystalline ZnO films. J Appl Phys 87:2413–2421
56. Xirouchaki C, Kiriakidis G, Pedersen TF, Fritzsche H (1996) Photoreduction and oxidation of as-deposited microcrystalline indium oxide. J Appl Phys 79:9349–9352
57. Kim H-J, Almanza-Workman M, Chaiken A, Jackson WB, Jeans A, Kwon O, Luo H, Mei P, Perlov C, Taussig C, Jeffrey, F, Braymen S, Hauschildt J (2006) Roll-to-roll fabrication of active-matrix backplanes using self-aligned imprint lithography (SAIL). 6th Int. Meeting Information Display/5th Int. Display Manufacturing Conf. Daegu, Korea, 2006 Digest 1539–1543
58. US Patent 20050176182 (2005)
59. Xia Y, Whitesides GM, (1998) Soft lithography. Annu Rev Mater Sci 28:153–184
60. Quake SR, Scherer A (2000) From micro- to nanofabrication with soft materials. Science 290:1536–1540
61. Rogers JA, Nuzzo RG (2005, February) Recent progress in soft lithography. Mater today 8:50–56
62. Norland Products Inc. www.norlandprod.com
63. Kawazoe H, Yasukawa M, Hyodo H, Kurita M, Yanagi H, Hosono H (1997) P-type electrical conduction in transparent thin films of $CuAlO_2$. Nature 389:939–942
64. Tate J, Jayaraj MK, Draeseke AD, Ulbrich T, Sleight AW, Vanaja KA, Nagarajan R, Wager JF, Hoffman RL (2002) P-type oxides for use in transparent diodes. Thin Solid Films 411: 119–124

65. Park S, Keszler DA, Valencia MM, Hoffman RL, Bender JP, Wager JF (2002) Transparent p-type conducting $BaCu_2S_2$ films. Appl Phys Lett 80:4293–4295
66. Yanagi H, Inoue S, Ueda K, Kawazoe H, Hosono H, Hamada N (2000) Electronic structure and optoelectronic properties of transparent p-type conducting $CuAlO_2$. J Appl Phys 88: 4159–4163
67. Nagarajan R, Draeseke AD, Sleight AW, Tate J (2001) P-type conductivity in CuCr1-x MgxO2 films and powders. J Appl Phys 89:8022–8025
68. Ueda K, Hase T, Yanagi H, Kawazoe H, Hosono H, Ohta H, Orita M, Hirano M (2001) Epitaxial growth of transparent p-type conducting $CuGaO_2$ thin films on sapphire (001) substrates by pulsed laser deposition. J Appl Phys, 89:1790–1793
69. Duan N, Sleigh AW, Jayaraj MK, Tate J (2000) Transparent p-type conducting CuScO2+x films. Appl Phys Lett 77:1325–1326

Chapter 6
Materials and Novel Patterning Methods for Flexible Electronics

William S. Wong, Michael L. Chabinyc, Tse-Nga Ng, and Alberto Salleo

Abstract   The materials considerations and print-processing techniques for fabricating electronic devices on flexible platforms are reviewed. Organic and inorganic semiconductors, dielectrics, and metals for thin-film transistor (TFT) fabrication are presented. Jet-printing techniques for both etch-mask patterning and deposition and patterning of solution-processable polymers will be highlighted. The characterization of low-temperature compatible materials will also be reviewed in regard to conditions that determine device stability and performance in polymeric and silicon-based devices. Finally, an overview of specific applications for organic and inorganic semiconductor devices in backplane, display, and image sensor arrays will be presented.

6.1 Introduction

The scaling of large-area electronics for applications in flat-panel displays, digital X-ray imagers, and flexible electronics is pushing the technological and cost limits of conventional materials and device processing. While conventional processing such as photolithography and vacuum deposition can produce high-performance submicron devices, this capability and the associated cost is not necessary for large-area platforms. An alternative patterning method that can be scaled to very large areas, using noncontact jet printing of electronic image files, may be used to fabricate both silicon-based [1–4] and polymeric-based [5, 6] thin-film transistors (TFTs) having feature sizes limited by the size of the drops (~40 μm). Incorporated with roll-to-roll fabrication [7, 8], jet-printing processes can help to reduce processing and materials cost for flexible electronics fabrication.

W.S. Wong (✉)
Palo Alto Research Center, Palo Alto, CA 94304, USA
e-mail: wsw@parc.com

W.S. Wong, A. Salleo (eds.), *Flexible Electronics: Materials and Applications*, Electronic   143
Materials: Science & Technology, DOI 10.1007/978-0-387-74363-9_6,
© Springer Science+Business Media, LLC 2009

Many new materials systems, based on polymeric semiconductors and flexible substrate materials, are rapidly being developed in order to address the intrinsic difficulties of large-area roll-to-roll processing. Polymeric materials hold promise for low-cost electronics and simplified device processing. Indeed, materials that are solution processable will enable novel deposition and patterning processes borrowed from the graphic arts industry, such as jet printing, for fabrication of TFT devices. Because a large fraction of fabrication costs is due to patterning, an alternative technology based on ink-jet printing and conventional vacuum deposited thin films may provide the required overlap of performance with low-cost fabrication of microelectronic circuits on large-area flexible platforms.

Jet-printing technology can be used to pattern etch masks or active materials to form a desired pattern. The technique has advantages for device processing on flexible substrates: (1) the patterning process is noncontact, (2) layer-to-layer registration is accomplished by digital image processing, and (3) the patterning process originates from an electronic image file that is not fixed to a rigid photomask and can be used for real-time alignment onto a flexible platform. The feature size and spatial control of jet-printed phase-change inks combined with digital imaging and patterning [9–13] in a process referred to as digital lithography onto conventional thin-film electronic materials are optimal for large-area device processing on flexible platforms and are another alternative to direct writing of solution-processable materials in additive patterning processes [5, 14, 15].

The integration of inorganic semiconductors and dielectrics on flexible substrates has the added complexity that high-temperature processing is needed to obtain high-quality materials. No matter what technology is used for flexible displays, if the substrate is an organic polymer then close attention must be paid to the deposition temperatures for all processes used in the fabrication. While plastic substrates have cost advantages and can readily be used in roll-to-roll processes, they have relatively poor characteristics in terms of thermal expansion, mechanical stability, and solvent uptake, when compared to glass and silicon. There are many strategies to address these issues, but in general they require processing to occur below ~170°C. Since this temperature is somewhat lower than that used for conventional processing, it is critical to perform fundamental studies to understand how electronic devices fabricated at these temperatures operate as their defect density is likely to be different than that of devices fabricated at higher temperatures. The replacement of conventional patterning processes with digital lithography is the first step in simplifying the fabrication of TFT devices. In a second step, solution-processable semiconductors and metals can be used to replace conventional vacuum deposited materials enabling fabrication in an ambient environment. While devices fabricated using these materials currently have lower performance compared to that of conventional inorganic thin-film devices, they are being rapidly developed for applications in large-area and flexible electronics. The move to solution-processable materials will enable a transition from a conventional subtractive process for TFT fabrication to an all-additive process where reduced cost can be obtained by combining deposition and patterning in a single process step. Organic materials are more easily processed at the lower temperatures compatible with flexible substrates, but their long-term stability and the origin of defects are poorly understood.

6.2 Materials Considerations for Flexible Electronics

6.2.1 Overview

The semiconductor–dielectric combination in a circuit is one of the most important considerations for flexible electronics. The achievement of stable high-performance TFTs relies on the optimization of the mobility of carriers and the reduction of traps at the semiconductor–dielectric interface. Thus, the semiconductor and dielectric layers cannot be considered completely independent. The desired fabrication technology also dictates which materials are compatible. It is currently possible to fabricate high-quality TFTs from both inorganic and organic materials and we briefly describe these materials here. We also address the possibility of printable conductors that can be used, in principle, with either class of material.

6.2.2 Inorganic Semiconductors and Dielectrics

Hydrogenated amorphous silicon (a-Si:H) is the ubiquitous materials system for large-area electronics and is currently used as a semiconductor in liquid-crystal display (LCD) backplanes and X-ray imagers on glass substrates [16, 17]. The integration of high-performance a-Si:H-based TFT devices onto polymeric substrates requires development of high-quality, low-temperature ($<200°C$) processed TFT devices. The development of such materials has been described in detail in earlier chapters. Particular challenges for flexible substrates are deposition of low mechanical stress layers and the reduction in defect density through optimization of deposition conditions. The most widely used deposition method for both the a-Si:H layer and the gate dielectric layer, silicon nitride (SiN_x), is plasma enhanced chemical vapor deposition (PE-CVD). While there are alternative low-temperature methods, such as sputtering, these techniques have generally not achieved the same quality that is possible with PE-CVD. In addition, since the semiconductor and dielectric layers are generally deposited sequentially in the same reactor, the interface formed between these materials can be highly optimized.

Alternative inorganic materials to amorphous silicon are currently being explored as semiconductors. Poly-silicon is a well-developed materials system that is currently showing promise in OLED display applications. Recently, a form of silicon that can be jet-printed has been described [18]. This material, however, suffered from the need to be processed in a highly inert atmosphere (<1 ppm O_2) and required high-temperature postdeposition annealing. Many semiconducting oxide materials, such as ZnO [19] and ternary compounds, InGaO [20], are being widely studied in test devices. These optically transparent materials have the advantage of being deposited at low temperatures while yielding relatively high field-effect mobilities (1–10 cm^2/Vs). It is, however, difficult to control their doping level during deposition due to electrically active defects formed during deposition. Printable forms of these materials have also been explored. Nanoparticles of ZnO have been used to form TFTs [21], but can require high temperature ($>300°C$) postdeposition anneals for the best performance. TFTs formed with these materials are frequently used with oxide gate dielectric layers, but there are few studies that have explored the best candidates for large-area electronic circuits.

6.2.3 Organic Semiconductors and Dielectrics

Organic semiconductors comprise two broad classes of materials, molecular compounds and polymers (Table 6.1). Both of these classes of materials generally have the feature of either a single conjugated unit or a number of conjugated units that form a planar or nearly planar molecular structure. There are many known semiconducting polymers, but those based on polymers or copolymers of thiophene have generally been the most widely used for TFTs. The design and synthesis of polymeric semiconductors are discussed in chapter 9 in this book. While most of the characterization of these materials has been performed using thermally grown silicon dioxide gate dielectric layers, they are compatible with a variety of dielectrics including organic polymers, PE-CVD SiO_2 and SiN_x, and oxides formed from anodized metals [22]. Both p- and n-type organic materials have been demonstrated with the former being more common than the latter [23]. Improvements in n-type materials suggest that organic materials may have an advantage over a-Si:H in applications where complementary logic is required, for example, simple logic circuits.

The choice of the semiconducting system dictates the deposition method. Molecular materials have generally been deposited by vapor methods and have been used in prototype circuits and backplanes. The best pentacene TFTs have reproducible mobilities around 1 cm^2/Vs [24], but in many cases degradation in performance is observed during subsequent processing, such as formation of an encapsulation layer or patterning by photolithography [25]. Small molecules can be made amenable to solution processing by functionalization. Usually, the solution deposition process causes a reduction in the observed field-effect mobility relative to vapor deposition. The typical reason for this difference is the difficulty in controlling film formation as the solvent evaporates. Recently, a number of materials, such as triisopropylsilyl (TIPS)-pentacene, have been designed that form continuous, highly ordered films from solution and now provide high mobility when deposited from solvent [26]. Semiconducting polymers have recently been demonstrated to have mobilities above 0.5 cm^2/Vs, suggesting that they can have similar performance to molecular materials [27].

Most high-performance organic semiconductors form polycrystalline or semicrystalline films. For many materials, it is found that either thermal or solvent-vapor annealing can lead to improvements in mobility. The orientation of the crystalline domains relative to the flow of current is considered to be important to achieve the highest field-effect mobility. For example, many semiconducting polymers have lamellar structures where a conjugated backbone is separated by alkyl side chains. Transport is poor along the lamellar stacking direction due to the

Table 6.1 Organic semiconducting materials

Semiconductor	Field-effect mobility (cm^2/Vs)	Deposition Method	Reference
Pentacene	0.5-5.0	Vapor	24
α-6T	0.1-1.1	Vapor	54
TIPS-pentacene	0.1 to 1.0	Solution	26
P3HT	0.01 to 0.05	Solution	23
PBTTT	0.1 to 0.5	Solution	27
PQT-12	0.06 to 0.12	Solution	6

insulating alkyl side chains and good through the π-stacking direction and along the backbone of the polymer. The interfacial interaction with the gate dielectric must not prevent the semiconducting layer from adopting this orientation during film formation.

A variety of low-temperature processable gate dielectrics, including inorganic oxides deposited by sputtering, anodization, or PE-CVD and organic dielectric layers deposited by solution coating, are compatible with organic semiconductors [22]. Inorganic dielectrics are generally formed in a bottom-gate configuration whereas organic dielectrics can be either bottom or top gate (Fig. 6.1). Inorganic insulators

Fig. 6.1 Device
architecture for organic
TFTs for (**a**) *bottom* and
(**b**) *top* gated staggered
structures

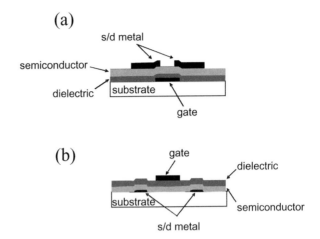

have good electrical stability and low leakage currents (<10 pA/cm^2), are impervious to organic solvents, and are stable in a wide range of environments. Since most inorganic dielectrics must be vapor-deposited or formed using time-consuming chemical bath deposition, research in organic polymeric dielectrics is increasing. The design of an organic dielectric is difficult because solvent interactions with the organic semiconductor must be considered.

For both inorganic and organic dielectrics, as the surface roughness of the gate dielectric increases beyond the molecular size of the semiconductor (>0.3 nm), the field-effect mobility of the TFT generally decreases. For a bottom-gate device, the roughness is caused by contributions from the substrate, gate layer, and dielectric layer. In top-gate devices, the roughness of the semiconducting film itself may dominate. For bottom-gate TFTs formed with pentacene on rough dielectrics, the mobility has been suggested to be affected by changes in grain size, scattering due to the roughness of the surface, and the inability of the carriers to move over the elevated portions of the dielectric due to the applied electric field with the gate [28]. Semiconducting polymers are less crystalline than molecular materials, but similar observations have been reported (Fig. 6.2) [29].

Understanding how molecular interactions at the gate dielectric surface affect the performance of organic TFTs is difficult. For example, deconvolving the effect of chemical interactions from simple changes in dielectric constant is difficult as the two are generally correlated. In addition, the device geometry can also have a strong impact. For example, for bottom-gate TFTs with solution-processed semiconductors, the solvent used for most materials is quite aggressive and can potentially swell or dissolve the gate dielectric. Even if the chemical compatibility can be overcome by other geometries, for example top gate, the roughness of the semiconducting film can still be an issue. Vapor-deposited films of materials, such as pentacene, eliminate these difficulties, but the growth of the films is influenced by the surface as well [30]. Nonetheless there is some understanding of the impact of chemical

Fig. 6.2 Field-effect mobility of PBTTT-C16 coplanar TFTs with a gate dielectric formed from PE-CVD deposited SiO_2/SiN_x at 150°C on glass (*circles*) and polyethylene naphthalate (*triangles*) [29]

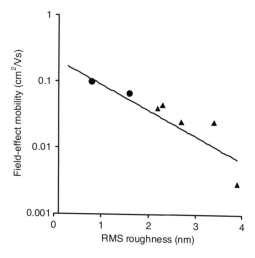

functionalities. Generally, hydrophobic surfaces produce TFTs with higher mobility and lower gate bias stress than hydrophilic ones [22]. Whether this is due to a specific interaction between the chemical moieties of the dielectric and the charge carriers or due to effects such as moisture uptake is unclear at this point. There is evidence that moisture uptake in polymeric dielectrics and semiconducting polymers themselves cause an enhancement in gate bias stress [31]. The interaction with water or hydroxyl groups has been shown to be deleterious for both p- and n-type organic semiconductors [32].

6.2.4 Conductors

In flexible electronics, the metals used for the conductors must meet three criteria: they must have a high enough conductivity to not create large parasitic resistance over the area of the circuit, they must inject charge efficiently into the semiconducting layer, and they must not impact deposition of the gate dielectric. While it is desirable for a single material to meet all requirements, it is not essential. For example, in a-Si:H technology, a doped silicon layer is used for injection and a metal is used for the addressing line and the gate level. In most demonstrations of flexible active-matrix backplanes, the conductors have been deposited from vapor.

Direct deposition of molten metal is usually beyond the thermal tolerance of flexible substrates other than stainless steel. The majority of work on conductors has focused on metallic nanoparticles, and organometallic molecular compounds that can be sintered into continuous metallic films [33]. Metallic nanoparticles of coinage metals such as Au or Ag with sizes of <100 nm have reduced melting points relative to bulk materials and can be sintered at relatively low temperatures (<300°C) [34] or optically with visible lasers [35]. Most nanoparticles are synthesized with an organic surface layer to allow them to be suspended in a solvent without aggregation. While surface layers may be necessary to increase the stability of suspensions of these materials, they are likely to cause injection barriers when used as contacts. For example, contacts formed from silver nanoparticles stabilized with oleic acid appear to show less contact resistance with poly[5,5′-bis(3-dodecyl-2-thienyl)-2,2′-bithiophene] (PQT)-12 than those stabilized with oleylamine [36]. The origin of this difference could be a modification of the work function of the electrode by the layer or by doping of the semiconducting material near the contact.

Semiconducting polymers can form printable conductors when doped and form good contacts to organic semiconductors. Poly(3,4-ethylenedioxythiophene) poly(styrenesulfonate) (PEDOT/PSS) is a widely used printable conductor, but its conductivity is $\sim 10^4$ times lower than most metals [37]. Composite films of polyaniline and single-wall carbon nanotubes have better conductivity than PEDOT:PSS (2 S/cm), but still are significantly less conductive than a bulk metal. While organic materials may be used as local contacts, it is unlikely that they can be used for the addressing lines in large-area circuits that require reasonable switching speeds.

6.3 Print-Processing Options for Device Fabrication

6.3.1 Overview

Printing methods can be broadly classified as noncontact and contact methods. Most high-volume printing, such as packaging, is performed using techniques such as gravure, offset, and flexographic marking where a roller contacts the substrate. The roller may be flat with patterned ink transferred to it from another patterned substrate or the roller itself may have patterned depressions into which the ink is deposited. In electronics, microcontact printing has been demonstrated for patterning of conductors using self-assembled monolayers as resists for chemical etching [38]. Chapter 5 describes the use of imprint methods to fabricate inorganic circuits. Techniques such as electrophotography have been used to print a resist used to fabricate a-Si:H TFTs [39]. We focus here on noncontact printing, in particular ink-jet printing.

Ink-jet printing has been used both to print etch masks for vapor-deposited materials (digital lithography) and for direct deposition of active materials. In digital lithography, the main issues are control of the feature sizes of the printed etch mask and integration of the printing steps into a conventional fabrication process. For printing active materials, attention must be paid to how the material dries or cures into a solid film. In both cases, it is important to understand the fundamental origins of the achievable feature sizes and the ability to control placement of those features.

6.3.2 Control of Feature Sizes of Jet-Printed Liquids

Most conventional ink-jet printers require relatively low-viscosity inks (<100 cP). The dominant factors controlling the sizes of jet-printed features are control of the printhead, mechanical control of the substrate, and the interaction of the printed liquid with the substrate.

Without prepatterning of the substrate, the position and size of the features are first set by the printing hardware. The liquid drop is ejected from the printhead and must travel to the substrate therefore the placement accuracy of the jetted drop can be analyzed by examination of the motion of the printhead (or substrate) and drop during the printing process [40, 41]. Equation (6.1) represents the relative position of a drop with respect to the position of the printhead during ejection of the droplet in the printing direction (x-direction). In this case, u is the printhead translation velocity, v is the drop ejection velocity, s is the distance between the substrate and the printhead, θ is the angle of the drop from the normal direction, and t is the time that the droplet ejects from the head after an applied signal (Fig. 6.3).

$$x \approx \frac{us}{v \cos \theta} + ut + s \tan \theta \qquad (6.1)$$

The deviation from the expected print position in the x-direction is approximated from the differential of Eq. (6.1):

$$\Delta x \approx \frac{us}{v \cos \theta} \left(\frac{\Delta u}{u} + \frac{\Delta s}{s} + \frac{\Delta v}{v} \right) + s \cdot \sec^2 \theta \cdot \Delta \theta - \frac{us}{v} \frac{\Delta \theta}{\sqrt{1-\theta^2}} + u \Delta t + t \Delta u \qquad (6.2)$$

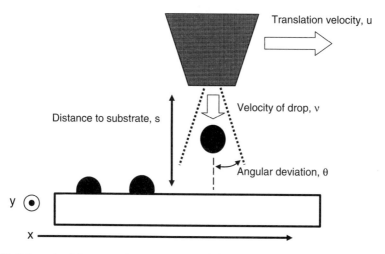

Fig. 6.3 Schematic of the geometric parameters that define the ejection of a droplet onto a substrate surface

The random errors add in quadrature and the values for t and θ are 0s and $0°$, respectively, at the time of ejection. The relationship in Eq. (6.2) then simplifies to

$$\Delta x \approx \frac{us}{v}\left(\frac{\Delta u}{u} + \frac{\Delta s}{s} + \frac{\Delta v}{v}\right) + s \cdot \Delta\theta - \frac{us}{v}\Delta\theta + u \cdot \Delta t \qquad (6.3)$$

The angular directionality of the drop determines in large part the deviation in the y-direction of the printing. Ignoring any temperature variation in the printhead, the deviation orthogonal to the print direction is shown in Eq. (6.4).

$$\Delta y \approx s \cdot \Delta\theta \qquad (6.4)$$

Using typical jet-printing parameters, the expected error in the ejected drop placement is $\sim$5–10 μm, which correlates well to measured droplets printed from a multiejector printhead [42], and is comparable to an analysis of printed conductive polymers [14]. The printed line widths of wax resist using commercially available printheads are $\sim$50 μm, so alignment accuracy of the drop placement is $\sim$20% of the printed line width.

The absolute feature size is set by the printed volume and the interaction of the printed liquid drop and the surface. Typical printed volumes are $\sim$20–100 pL, where gravitational forces acting on the printed drop are negligible compared to surface forces (by $\sim$1,000$\times$). The final geometry of the printed drop can be then estimated from Young's equation which depends on the drop contact angle with the substrate, ϕ, for a given interface energy between the liquid and vapor, γ_{LV}, solid and vapor, γ_{SV}, and solid and liquid, γ_{SL} (Fig. 6.4 inset) and is represented by

$$\gamma_{LV}\cos\phi = \gamma_{SV} - \gamma_{SL} \qquad (6.5)$$

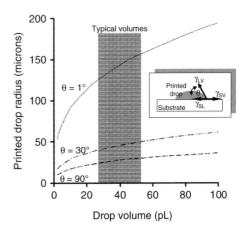

Fig. 6.4 Printed spot radius as a function of drop volume for different liquid–solid contact angles. Contact angle as a function of surface and interface tensions (inset)

The wetting conditions can vary greatly depending on the substrate surface treatment prior to patterning. For conditions in which the contact angle, $\phi = 0$, the printed drop will completely wet the surface while at contact angles greater than 90° nonwetting of the droplet is observed. The latter results in poor adhesion and the former gives large features, both are undesirable conditions. Figure 6.4 shows the calculated variation in spot size as a function of contact angle for different drop volumes with the shaded region showing the typical drop volumes for commercial ink-jet printers.

6.3.3 Jet-Printing for Etch-Mask Patterning

In combination with the spatial resolution of ink-jet printing and digital image processing, digital lithography offers a promising alternative to processing conventional electronic thin-film materials on flexible platforms. The digital lithography process incorporates electronic-digital imaging to register virtual etch masks that are patterned by jet printing onto a surface. The coordinates of the alignment marks are used to reposition an electronic mask layer aligned to the process wafer prior to printing the mask pattern. Phase-change inks are used to achieve the required minimum feature size without the complexity of modulating the surface conditions. In this approach, the substrate temperature is used to "freeze" a drop before it has time to spread on a surface ($0 < \phi < 90°$) allowing both feature size control and good adhesion. Control of printed line widths between 40 and 80 μm by simply varying the substrate temperature by 10°C has been reported [2]. Since the process is noncontact, imperfections such as localized and layer-to-layer distortion on flexible substrates are identified and corrected within a virtual mask before patterning.

A simple demonstration of the digital lithographic patterning techniques is shown in Fig. 6.5. First the pattern is defined by ink-jet printing of the mask (the orthogonal

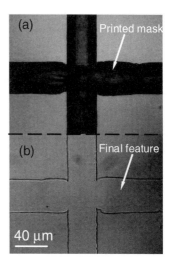

Fig. 6.5 (a) Optical micrograph of a print patterned etch mask. (b) Transfer characteristics of printed pattern onto the thin film after etching and mask stripping (this figure was reproduced from Ref. [40], with permissions by the Society for Information Display)

lines in Fig. 6.5a). The exposed film is etched and then the printed mask is stripped (Fig. 6.5b). The final feature is well aligned with the printed pattern; no substantial under cut of the masked film is seen. The precise image transfer is an important aspect for the wax masking, particularly in the patterning of the source–drain contacts for the TFT and via contact holes in TFT arrays. The use of jet printing for patterning semiconducting materials has the same advantages as digital lithography for etch masks; digital imaging of the substrate prior to deposition allows software adjustment of the printed pattern yielding very good multilayer registration. Jet printing of active layers on electrodes defined by digital lithography in TFTs for both molecular and polymeric semiconductors has also been demonstrated [5].

6.3.4 Methods for Minimizing Feature Size

Fine feature patterning from jet-printed materials is inherently difficult as shown in the range of predicted features sizes in Fig. 6.4. While the drop volume of a printed feature may result in relatively large patterns, the placement of the drops can be controlled to much finer resolution. The spatial control of the drop spacing allows for smaller channel lengths of a TFT than the drop size. For example, 10 μm gaps can be fabricated readily based on considerations from the motion control of the printed drop and from the knowledge of the extent of spreading of the printed drop.

In some cases, small gaps between printable conductors can be created using a "self-aligned" printing method that relies on differences in wettability. By printing onto a previously deposited feature that has a solvophobic coating, the deposited drop dewets from the feature and can form a submicron gap [15, 43]. Such gaps have been created with printable organic and inorganic conductors and used for OTFTs. While it is difficult to control the exact gap, it is likely that the average gap over a wide length scale, ~50 μm, may be controllable from printed device to device.

Although printing small gaps is useful for defining source and drain electrodes with shorter channels, it does not address the size of the gate electrode. Even if the gap between the source and drain contacts is small, the parasitic capacitance caused by a large overlap with the gate electrode degrades the performance of a TFT. In this case, having the source and drain electrode self-aligned to a short-channel gate is preferable to minimize the electrode overlap [44]. Self-aligned coplanar TFTs have been formed using a combination of printed resists and electroplating. Figure 6.6 shows the fabrication process, where the gate electrode can be narrowed by using the printed wax resist as a mask for electroplating. The gate layer itself was then used as a photomask for patterning of the source and drain contacts. After patterning the source–drain contact, the stripping process removed both the photoresist liftoff layer and the printed mask, resulting in the formation of a self-aligned bottom source and drain contacts (Fig. 6.7). The poly(thiophene) device characteristics for conventional and print processed devices (Fig. 6.8) were similar for both types of devices: mobility $= 0.035$ cm^2/Vs from the saturation regime characteristics, $V_T = 0$ V, and an on-to-off current ratio of $\sim 10^7$. Note that the original I–V curves of Fig. 6.8

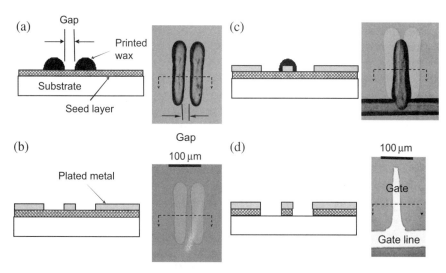

Fig. 6.6 Process flow for digital lithographically defined fine features having a TiW seed layer on a glass substrate as the starting material: (**a**) patterned gaps define the gate electrode channel length and width, (**b**) gate metal deposition accomplished by electroplated Ni metal within the patterned gap followed by mask stripping, (**c**) printed etch-mask patterning to define gate electrode and bus line features, and (**d**) Ni/TiW metal etching and stripping of wax mask for final feature definition. Inset shows optical micrographs of the patterned feature for each step [41]

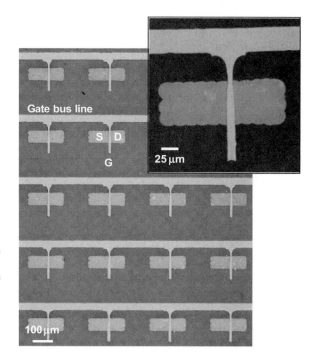

Fig. 6.7 Optical micrographs of the finished TFT structure having self-aligned source (S) and drain (D) contacts to the gate (G) electrode. Inset shows a close up of the TFT structure after coating poly[5,5′-bis(3-dodecyl-2-thienyl)-2,2′-bithiophene] (PQT-12), completing the device fabrication [41]

Fig. 6.8 Current-voltage characteristics for a typical TFT device fabricated using the self-aligned digital lithographic process (solid line). The diamond points shows the I–V characteristics for a conventionally processed TFT measured at $V_{DS} = -5$ V. The inset shows the measured (*circles*) and predicted (*solid line*) output characteristics of the self-aligned TFTs

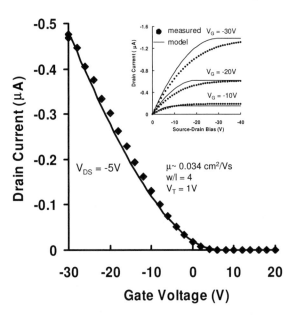

from [41] did not have normalized device dimensions. These results demonstrated that self-alignment of the source–drain contacts to the gate electrode is a useful approach to solving the problems associated with overlap capacitance in OTFTs.

Another approach to better control the feature sizes of printed drops is to prepattern the surface with different regions of differing wettability [45] or topography [46]. For example, the channel length of a TFT has been controlled by defining a bank of polyimide using photolithography and then printing a conducting polymer to form source and drain contacts [45]. The wettability of the substrate can also be patterned without the use of a topographic barrier using techniques such as contact printing or photolithography. While this method is attractive as it allows for better control of features, it requires the use of extra patterning steps in the process with extra processing equipment. If the need for better control of all-additive materials outweighs the additional process complexity, then these approaches may be useful.

6.3.5 Printing Active Materials

The shapes of active materials range from simple island areas to lines over large distances. Printing one drop to form an island is relatively simple as long as the wetting characteristics of the surface are uniform. Printing lines over long distances is more challenging. Since the long lines in a display are generally the address lines, it is important to form them with as few defects as possible because defects would result in opens in the conductors that control the circuit. As a printed line is formed, additional drops add to the previous line that can be thought of as a pump. As more drops are added to the printed liquid line, instabilities, such as necks, may form due to the dynamics of the spreading and drying of the printed liquid line. The mechanism of formation of instabilities in printed liquid lines is complex and has been analyzed in detail [47, 48]. To achieve control of printed lines, one must optimize the printing speed and the wetting characteristics of the printed liquid.

The majority of high-performance organic semiconductors are soluble in polar, low-viscosity, and low-surface energy solvents. Controlling the spreading of these liquids on surfaces when deposited as drops is difficult without physical barriers. Importantly, the surface energy of the gate dielectric in an OTFT cannot be arbitrarily chosen as most solution-processable organic semiconductors perform best on dielectrics with hydrophobic surfaces. While such low-surface energies can in principle decrease the size of printed features, the semiconducting solution is frequently printed on a surface with spatially heterogeneous surfaces. In a coplanar TFT, the contact electrodes will generally be of higher surface energy than the channel causing a printed liquid droplet to dewet from the channel onto the contacts. While it is possible to print contacts on the surface of the semiconductor after it is deposited, control of the feature size of printed metals will be limited since it is currently difficult to arbitrarily change the wetting characteristics of the surface of organic semiconductors.

The drying time of the printed liquid impacts the performance of solution-processable semiconductors. It has been shown that the drying conditions of spin-coated films can lead to changes in performance of factors of 10, even on the same surface [49, 50]. Although the drying conditions of printed drops and spin-coated films are quite different due to the volume of solution, both printed and spin-coated films of semiconducting polymers have been demonstrated to have similar performance when deposited from a high boiling point solvent. While there are no detailed studies of the morphology and microstructure of films of semiconducting polymers deposited by jet printing, the electrical characteristics of the TFTs suggest that the material can self-organize during the drying time of the printed drop.

6.4 Performance and Characterization of Electronic Devices

6.4.1 Overview

Being able to fabricate single devices with the appropriate performance is only a first step toward the use of TFTs in electronic circuits where tens or hundreds of devices are integrated. Circuit designers rely heavily on device stability, reliability, and uniformity across the substrate and correct size scaling to design properly functioning systems. Therefore, all nonideal behavior in TFTs must be well understood in order to realistically design systems comprising multiple devices.

The deposition conditions available for flexible circuits are substantially different than those for conventional devices. In particular, the temperatures that can be used are substantially lower than those for devices on glass substrates. Because of this limitation, the electronic properties of conventional materials as a function of processing temperature must be explored. On the other hand, organic semiconductors represent a relatively unexplored class of materials and the device characteristics of organic TFTs and circuits are still not established.

6.4.2 Bias Stress in Organic Thin-Film Transistors

It is commonly observed that the output current of organic TFTs decreases under operation [51–54]. Such behavior is called bias stress. The main cause of bias stress is a shift of the threshold voltage V_T toward the gate potential. This shift reduces the voltage overdrive of the device and progressively shuts down the transistor during operation. The V_T shift is caused by a sheet of trapped carriers shielding the gate potential [51, 55]. Carriers are trapped in the organic semiconductor material at the dielectric interface, where their concentration is highest during operation. Nevertheless, carrier injection in the dielectric cannot be excluded a priori as a cause of bias stress, especially in devices made with polymeric dielectrics. Another common cause of V_T shift in TFTs is the motion of ions in the dielectric, which on the other hand can be eliminated by careful choice and handling of the materials [56]. A V_T shift toward V_G has been observed as well with gate voltages such as to shut the device off [57, 58]. Studies of the latter type of bias stress are in their initial stages and we will not include them in this chapter. Finally, bias stress effects are heavily influenced by the operating environment. The presence of water at the semiconductor–dielectric interface for instance has been observed to accelerate the V_T shift in polymeric TFTs [59]. Environmental effects are extremely important but are still poorly understood.

Threshold voltage instability makes circuit design extremely challenging. Moreover, the degradation of the electrical output of the TFTs poses a lifetime issue [52, 60]. Understanding, mitigating, and predicting bias stress behavior in organic TFTs is, therefore, of paramount importance in all applications involving these devices. Several charge trapping mechanisms have been proposed and can be found in the literature [61–63]. Here we focus on the modeling of the time dependence of the current decay in organic TFTs in an effort to design optimal biasing conditions that extend the device lifetime in operation.

6.4.2.1 Continuous Biasing

In the simplest bias stress studies, a constant gate bias is applied to turn-on the TFT while the drain current is measured as a function of time. In general, bias stress depends on the charge density in the channel [51]. Typically, as V_T shifts toward V_G, the driving force for trapping – which is related to the gate voltage overdrive – decreases and bias stress slows down. In the linear regime, the charge density in the channel is

$$Q_C = C_0 (V_{GS} - V_T) \tag{6.6}$$

where C_0 is the gate capacitance per unit area V_T is the threshold voltage and V_{GS} is the gate-to-source potential, which in the linear regime is approximately equal to V_{GD}, the gate-to-drain potential. When the transistor is not in the linear regime, V_{GS} is different from V_{GD}, charge density varies along the channel and its average value is [64]:

$$Q_C = \frac{2}{3}C_0\frac{(V_{GS} - V_T)^3 - (V_{GD} - V_T)^3}{(V_{GS} - V_T)^2 - (V_{GD} - V_T)^2} \tag{6.7}$$

In a-Si:H, the shape of the current decay curves does not depend on whether the TFT operates in the linear regime or in the saturation regime. The average charge density, calculated with Eq. (6.6) or Eq. (6.7) is merely a scaling factor. Nevertheless, Eqs. (6.6) and (6.7) indicate that the interpretation of the data is easier when the device operates in the linear regime, where the charge density is approximately constant along the channel. Thus, bias stress data are often measured with $|V_{SD}| << |V_{GS}|$.

Several bias stress mechanisms have been proposed, including bipolaron formation in the semiconductor [61], injection in the dielectric [57], and deep trapping at defects in the semiconductor [65]. More than one of these mechanisms may be simultaneously active. In the absence of a firmly established bias stress mechanism, the V_T shift laws are purely phenomenological. Thus, different time-dependence laws have been proposed for bias stress of organic TFTs under continuous gate bias depending on experimental conditions and materials combinations. For instance, in analogy to bias stress in a-Si:H, a stretched exponential law was recently proposed [66]. In this model, the V_T shift is attributed to the creation of trap states by the gate bias. The threshold voltage has the following time dependence

$$V_T(t) = V_T^0 + \left(V_{GS} - V_T^0\right) \times \left\{1 - \exp\left[-\left(\frac{t}{\tau}\right)^\beta\right]\right\} \tag{6.8}$$

where V_T^0 is the initial threshold voltage, τ is the effective trapping time, and β is a constant related to the activation energy distribution of the localized states created by the gate bias. Bias stress in sexithiophene was found to obey Eq. (6.8), up to a stressing time of 10,000 s [67]. At longer stressing times, deviations from the stretched exponential law were observed and the stress data were better represented by a stretched hyperbola [68]

$$V_T(t) = V_T^0 + \left(V_{GS} - V_T^0\right) \times \left\{1 - \left[\exp\left(\beta\frac{E_{th} - E_A}{kT}\right) + 1\right]^{\frac{1}{1-\alpha}}\right\} \tag{6.9}$$

where E_{th} is the thermalization energy: $E_{th} = kT\ln(vt)$ (v is an attempt-to-escape frequency for trapped charge); E_A is the mean activation energy for trap creation and α is a fitting parameter. The models that give rise to bias stress laws such as those expressed by Eqs. (6.8) and (6.9) assume that the gate bias creates localized electronic states in the semiconductor [69]. In organic TFTs, however, there is no evidence that the V_T shift is due to the creation of new states rather than slow trapping in already existing deep states, thus the application of these equations is somewhat unjustified.

A physical rationalization for the use of a stretched exponential law may be found by adapting a lifetime model successfully applied to OLEDs. A degradation law that closely approximates the experimentally observed stretched exponential degradation in OLEDs is obtained by assuming that the activation energy of the degradation process increases as a function of time [70]. Thus, an analogy with the OLED model can be drawn by postulating that trapping in organic TFTs is an activated process whose activation energy increases as the threshold voltage shifts. As a result, the V_T shift rate decreases as the amount of free charge decreases, in agreement with the observation that the bias stress rate decreases as total stress increases. In OLEDs, these assumptions lead to a differential equation whose numerical solution closely approximates a stretched exponential. It should be noted that while this model may provide a physical justification for the stretched exponential law, it has not been adapted to TFTs yet. Finally, Powell et al. have shown that in certain experimental conditions, the stretched exponential law observed in a-Si:H reduces to a power law [71]:

$$\Delta V_T = A \, |V_G - V_T|^n \, t^\gamma \tag{6.10}$$

V_T shifts that follow a power law have indeed been observed in pentacene under constant biasing with $\gamma \sim 0.05$ and $n \sim 0.75$ [57].

In practical applications, it is very unlikely that organic TFTs will be operated under a continuous gate bias. Yet, all the decay laws proposed in the literature express ΔV_T explicitly as a function of time rather than of the operational parameters of the transistor (e.g., gate voltage overdrive). The bias stress laws derived under continuous bias may provide insight into trapping mechanisms. They cannot, however, be used to predict quantitatively the behavior of devices that are cycled between their on and off states.

6.4.2.2 Pulsed Biasing

Relatively, few studies have been conducted with a pulsed gate bias in order to estimate the lifetime of TFTs under realistic operating conditions. A fundamental difference between bias stress in organic TFTs and bias stress in other disordered materials, such as a-Si:H, is that charge trapped in organic semiconductors is partially released when the gate is turned off [52, 60]. As a result, V_T shifts back toward its original value during the off-time of a pulsed biasing cycle. This effect is extremely important in TFTs used in active-matrix displays as these devices spend most of their duty-cycle in the off-state. The recovery rate depends on trap depth and release activation energy. It was found that the average activation energy was ~ 0.8–1 eV, placing these traps probably deep in the middle of the polymeric bandgap [65]. As a result of the interplay between stress and recovery, and because none of these processes are linear with time, the V_T shift of an organic transistor depends on both the pulse length and the duty-cycle.

Using very general assumptions, one can show that eventually the transistor will stabilize at a unique V_T value that depends only on the operating conditions. Physically, trapping is an equilibrium between free charge and trapped charge. Therefore, the instantaneous trapping rate is a function of the free charge Q_f and the trapped charge Q_t. We define the trapping rate T as:

$$T\left[Q_f, Q_t\right] = \frac{dQ_t}{dt} \tag{6.11}$$

T is therefore positive during gate biasing. Moreover, T is experimentally observed to decrease monotonically as Q_t increases (i.e., bias stress slows down as the device is stressed for a longer time). The release of trapped charge on the other hand depends only on Q_t because there is no free charge when the transistor is turned off. The instantaneous release rate R is negative and its absolute value decreases as Q_t decreases. At steady state, the exact amount of charge trapped during the on-cycle is released during the off-cycle:

$$\int_{On} T\left[Q_f, Q_t\right]dt = -\int_{Off} R\left[Q_t\right]dt \tag{6.12}$$

If there were two possible device states defined by different trapped charge densities $Q_t^{(1)}$ and $Q_t^{(2)}$ that satisfied Eq. (6.12) for the same operating conditions, we would have:

$$\int_{On} \left\{T\left[Q_f^{(1)}, Q_t^{(1)}\right] - T\left[Q_f^{(2)}, Q_t^{(2)}\right]\right\}dt = -\int_{Off} \left\{R\left[Q_t^{(1)}\right] - R\left[Q_t^{(2)}\right]\right\}dt \tag{6.13}$$

If $Q_t^{(1)} > Q_t^{(2)}$, the left-hand side of Eq. (6.13) is negative but the right-hand side is positive. We conclude that $Q_t^{(1)} = Q_t^{(2)}$ and that the steady-state condition is reached at a single value of Q_t. Therefore, an organic TFT will always reach the same steady-state V_T for a given pulsed operating condition regardless of its initial conditions. This prediction was recently confirmed in polythiophene TFTs. Three different polythiophene devices started at different levels of V_T shift at the beginning of the stress cycle but stabilized at the same V_T shift after ~ 1 day of pulsed gate biasing at a 0.5% duty-cycle [60]. The stabilization kinetics depends on the details of the biasing conditions. At low duty-cycles ($<2\%$) the V_T shift in polythiophenes prior to stabilization exhibits a power-law dependence on stress time (i.e., gate on-time), similar to the one expressed in Eq. (6.10). Here the exponent n was found to vary between 2 and 4 and γ varied between 0.35 and 0.40. Because $\gamma < 1$, it takes a long time for the transistor to reach the stabilized V_T, especially at the low duty-cycles encountered in many applications of interest to organic TFTs. For example, in televisions, the duty-cycle is ~0.1%, which may lead to stabilization times as long as 10 days. Therefore, it is important to use the phenomenological knowledge of bias stress kinetics to devise accelerated testing schemes.

In polythiophene TFTs, the stabilized ΔV_T is roughly proportional to the duty-cycle for a given gate bias pulse length [60]. Measurements obtained at higher duty-cycle can, in principle, be used to extrapolate the behavior of the transistors at lower duty-cycle. It is, however, not clear whether such proportionality is robust and occurs with other materials systems. A different approach consists of stressing the device with continuous bias past the stabilized V_T characteristic of the duty-cycle of interest. When the duty-cycle of interest is applied to the prestressed TFT, the device will relax close to the stabilized V_T in a shorter time than it would have taken to reach it from the unstressed state. Qualitatively, the reason for this behavior can be understood as follows. In the unstressed state, the small V_T shift occurring during the short on-time of the cycle is almost entirely compensated by the charge release occurring during the long off-time. Therefore, the incremental ΔV_T per cycle is small. Moreover, as the transistor approaches its stable ΔV_T, the driving force for trapping decreases while the driving force for releasing trapped charge increases because of the accumulated Q_t. Consequently, the ΔV_T per cycle decreases quickly as V_T approaches its stabilized value.

If the device is prestressed with a V_T shift larger than the stabilized V_T on the other hand, it must relax by releasing excess trapped charge. In these conditions, the release rate is large because there is a large Q_t stored in the device and the trapping rate is small. Since the TFT spends most of its time in the off-state, relaxation close to the stabilized V_T is fast. Reaching the exact stabilized V_T may take a long time but this method allows obtaining an accurate estimate of the stabilized V_T very quickly. Indeed, literature data validate this accelerated testing method: it takes less than 1,000 s for a prestressed device to relax to its stabilized V_T while it typically takes tens of hours for an unstressed device to reach the same state (Fig. 6.9) [52, 60].

6.4.2.3 Long-Term Stress Effects

The simple picture of charge trapping and release is slightly changed when extended bias stress experiments (>100 days) are taken into consideration. For instance, the release rate of gate-induced trapped charge depends on the total stress time, which indicates the existence of a distribution of trap states having different release time constants. Traps that are populated after longer stress times release their charge more slowly upon turning the transistor in the off-state. In devices stressed for extended periods of time, the V_T shift may be considered permanent at room temperature but can be partially recovered after thermal annealing.

In addition to the V_T shift, there is recent evidence that extended biasing times (several months) affect carrier mobility in polymeric semiconductors, thus providing a second mechanism of electrical degradation in organic TFTs [60]. The cause of the mobility degradation is currently unknown. It is possible that long-term operation disrupts the ordered microstructure of the semiconducting polymer. Alternatively,

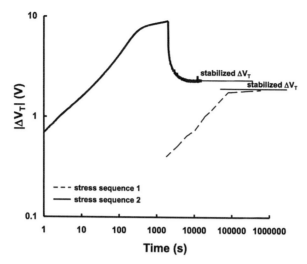

Fig. 6.9 Stabilized ΔV_T and accelerated testing. Stress sequence 1 starts with an unstressed device. The duty-cycle is 0.5%. Stress sequence 2 consists of prestressing the TFT for 2,000 s with a continuous gate bias and letting it relax to its stabilized ΔV_T with a duty-cycle of 0.25%. The stabilized ΔV_T is reached much faster with sequence 2 even though the duty-cycle is lower. Data are taken from [55, 63]. The charge density in the channel is very similar for both devices. In order to account for the different duty-cycles, the stabilized ΔV_T of the device undergoing stress sequence 1 is normalized to the duty-cycle of the device undergoing stress sequence 2 using the duty-cycle scaling of Ref. [63]

the simultaneous presence of charge and impurities in the semiconductor may lead to irreversible damage to the polymer.

All phenomenological theories of bias stress in organic TFTs indicate that the V_T shift increases strongly with charge density in the channel. Thus, there is a great advantage in operating devices at a low gate voltage overdrive. The required TFT current can be reached with a lower charge density by increasing the transistor width or decreasing its length. In most applications, the TFT footprint must be kept small and reducing the device length is the only viable option. On the other hand, non-ideal behavior is, however, often observed in organic transistors with channel length shorter than $\sim$10 μm, which poses challenges to the use of short-channel devices.

6.4.3 Nonideal Scaling of Short-Channel Organic TFTs

Circuit design relies heavily on the correct scaling of devices having different dimensions in order to balance electrical loads throughout a circuit. Contact resistance does not scale with channel length and is, therefore, known to perturb ideal device scaling. The effect of contact resistance on TFT characteristics becomes noticeable when the contact resistance is comparable to the channel resistance. Contact resistance depends on a variety of factors including semiconductor/metal band alignment, contact geometry, and processing conditions. Typically, the effect of contact resistance becomes extremely apparent for devices having a channel shorter than 10 μm [72]. In general, contacts are modeled as a combination of Schottky diodes and ohmic resistances [72, 73]. Although the physics of charge injection from the contact into the semiconductor is still not fundamentally understood, contact resistance models can reproduce device data accurately. Therefore, while the contact resistance may limit device performance, it does not appear to present a fundamental impediment for circuit designers.

In addition to contact resistance, however, devices with short (<10 μm) channels present strong deviations from ideal FET behavior. These deviations are prevalent in the saturation regime and lead to a large overestimate of the mobility in the semiconductor (Fig. 6.10a) [74, 75]. In particular, the transistor does not saturate when $|V_{DS}| > |V_{GS}|$ and the output current is much larger than expected based on classic device scaling (Fig. 6.10b). Finally, short-channel devices display large

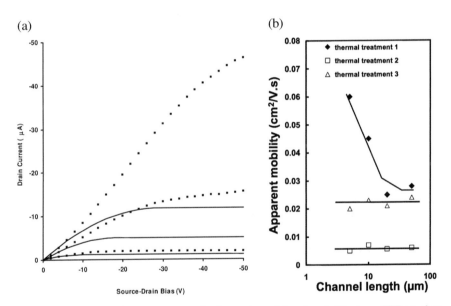

Fig. 6.10 Experimental output curves of a 5-μm channel length polythiophene TFT on glass (*solid squares*) showing extreme nonideal behavior (**a**). The lines are the modeled data according to FET equations using parameters obtained from longer channel devices on the same substrate. The short-channel device yields a higher apparent mobility and does not saturate at $|V_{DS}| > |V_{GS}|$. All experimental output curves cannot be fit with a single value of μ and V_T indicating nonideal behavior. Apparent mobility of polythiophene TFTs as a function of channel length (**b**). Thermal treatment 1 is the standard treatment to obtain high mobility devices (anneal to 140°C and slow cool to room temperature). Nonideal device behavior leads to an apparent mobility increase as the channel length decreases. Thermal treatment 2 is a quench from 140°C. The device behaves according to the FET equations but the mobility is degraded. Thermal treatment 3 is a short (5 min) reanneal of devices having undergone thermal treatment 2. The devices obey standard FET equations at all channel lengths

hysteresis indicative of charge trapping. This nonideality is mostly due to the competition between bulk and channel conductance near the depleted drain electrode. In deep saturation, the bulk current increases with the increasing drain potential leading to a parasitic current path that is not controlled electrostatically by the gate. Thus, the short-channel effect is a body effect due to the presence of bulk semiconductor material above the channel. This effect can be mitigated by fabricating devices with ultrathin semiconductor layers. Alternatively, because the electrical properties of organic semiconductors depend strongly on their microstructure, there is an opportunity to use the materials' microstructure to optimize device performance. For instance, thermal treatments are found to vary field-effect mobility and bulk conductivity independently. A combination of quenching the semiconducting polymer film from above its melting point and reannealing it at a lower temperature for a few minutes greatly reduces the short-channel nonideality in polythiophene TFTs (Fig. 6.11). This behavior may be the result of depressed charge mobility in the bulk because of disorder induced by the thermal treatment. The lowered electrical conductivity in the bulk greatly decreases the contribution from the parasitic current path in saturation.

6.4.4 Low-Temperature a-Si:H TFT Device Stability

While the electrical stability is an important consideration for polymeric TFT devices, it is equally important for low-temperature processed a-Si:H. When

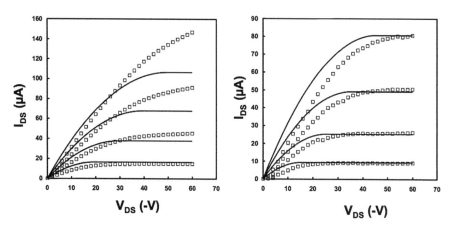

Fig. 6.11 Experimental (*open squares*) and modeled (*lines*) output *curves* of a 10-μm channel length polythiophene TFT on glass. The device on the *left* panel was treated according to thermal treatment 1 of Fig. 6.10b and shows nonideal behavior, as in Fig. 6.10a. The device on the *right* panel was treated according to thermal treatment 3 of Fig. 6.3b and is well modeled by standard FET equations with a single mobility ($\mu = 0.023$ cm^2/V.s) and V_T (-4.5 V), in agreement with long channel TFTs made on the same substrate. Contact resistance is apparent at low V_{DS} due to the bottom contact geometry of the devices

processed at high temperatures ($T_{\text{process}} > 300°C$), a-Si:H TFTs behave in a very predictable manner and have been shown to be very stable in displays and image sensors. As the process temperature is decreased ($T_{\text{process}} < 200°C$), the device performance can be similar to that of devices processed at higher temperatures but the long-term stability may suffer. In this section, we will review the stability performance of a-Si:H-based TFTs processed at various growth temperatures using some of the characterization techniques described earlier.

At process temperatures below 300°C, degradation in the device stability is expected due in part to carrier trapping in the gate dielectric and creation of dangling bond defects in the semiconductor [76, 77]. The use of pulsed and static gate bias on TFTs processed at low temperature ($T < 200°C$) can give insight into the defects in low-temperature materials and help identify the problem areas for development.

In one investigation [78], the examination of the I–V characteristics under pulsed and short-term operation were comparable for TFTs processed at low ($T_{\text{process}} = 120°C$) and high temperatures ($T_{\text{process}} \geq 250°C$). Static bias stress studies are needed to fully characterize the device performance and stability. Figure 6.12 shows a transfer characteristic for a device processed at 170°C measured under pulsed conditions (gate-pulse time = 50 ms). The initial device had a field-effect mobility of ~ 1 cm^2/Vs and a threshold voltage of ~ 2 V. The same device was measured again under pulsed conditions following an applied static gate bias ($V_G - V_T = 20$ V, $V_{DS} = 1$ V) for 60 min. The resulting field-effect mobility and subthreshold slope of the device was unchanged while the V_T showed a ~ 2 V shift (dashed-line curve and inset in Fig. 6.12).

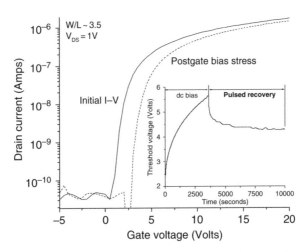

Fig. 6.12 Pulsed transfer characteristics of a low-temperature ($T_{\text{process}} = 170°C$) a-Si:H TFT before and after static gate bias testing for 60 min. The off-current is limited by the sensitivity of the source-measurement unit (~ 50 pA). The inset shows the time dependence of V_T shift and recovery after removing the gate bias. During the pulsed recovery part of the test, $V_G - V_T$ and V_{DS} were 20 and 1 V, respectively, during the on-state and the off-state were 0 and 1 V, respectively

The TFT was then allowed to recover under low duty-cycle pulsed biasing conditions (pulse length = 15 ms, duty-cycle = 0.15%). After removing the static gate bias (time > 3,600 s), the V_T recovers rapidly suggesting the release of trapped charge in the low-temperature dielectric [79, 80], similar to polymeric-based TFTs [60]. The V_T recovery stops after t > 5,000 s, indicating that an irreversible change in the amorphous silicon has occurred due to creation of defect states in the semiconductor, similar to the behavior of high-temperature processed a-Si:H [77].

Figure 6.13 shows the device stability as a function of gate bias (at $V_{DS} = 1$ V) for processing temperatures of 120, 150, 170, and 250°C. While the plot shows an increase in threshold voltage for devices processed at low temperatures, the field-effect mobility and subthreshold slope for these devices remained constant, similar to the result shown in Fig. 6.12. The largest observed V_T shift occurred at the higher gate fields, comparable to what is observed in conventional a-Si:H TFTs processed at high temperature [81]. The stability of the devices fabricated below 200°C is similar to that of devices fabricated on flexible polyimide substrates at temperatures of 250–280°C and glass substrates at temperatures of 300–350°C [82].

6.4.5 Low-temperature a-Si:H p–i–n Devices

The integration of display or sensing media onto the backplane is required to create useful large-area electronics applications. Conventional active-matrix a-Si:H image sensor arrays pose particular challenges because the sensor layer is typically ~2× thicker than the TFT backplane and this layer may experience strain-induced cracking on flexible platforms leading to failure of the sensor array. Stress originating from the thermal expansion coefficient mismatch of the flexible substrate and thin film may be minimized by decreasing the process temperatures ($T \leq 150°C$), but this

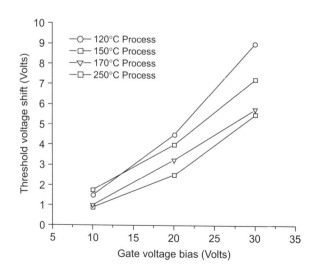

Fig. 6.13 V_T shift in a-Si:H TFTs as a function of gate bias voltage and deposition temperature. Static bias was applied for 60 min at each gate voltage condition

approach typically results in inferior quality materials compared to a conventional high-temperature process [83].

An important criterion for photodiode performance is high sensitivity to light and low dark current under a reverse bias electric field [84–86]. These characteristics allow the image sensor to have a high signal-to-noise ratio and give it a wide range of applications from light to X-ray imaging. As seen for TFTs, the performance of devices processed at low temperatures is not always identical to that of devices processed at high temperatures. Since the p–i–n photodiode is a two carrier device, consideration for charge trapping defects is more complicated than in TFTs.

Figure 6.14 shows the current–voltage (I–V) characteristics of an a-Si:H-based p–i–n structure as a function of processing temperatures. The dark current was found to increase with decreasing deposition temperature at all *i*-layer thicknesses (Table 6.2). As the *i*-layer thickness was further reduced (<500 nm), an increased dark current at high electric field was measured due in part to contact leakage across the junction barrier. The measured ideality factor also increased with decreasing temperature, suggesting an increase in the recombination of the photo generated electron-hole pairs [87]. The ideality factor of conventional a-Si:H is typically between 1 and 2, which was the value found for sensors deposited at $T \geq 180°C$ (see Table 6.2), but increased to values greater than 2 as the process temperature decreased.

At high electric field (≥ 3 V/μm), the photocurrent varied linearly with incident light power over 4 decades for all deposition temperatures (Fig. 6.15a). The external quantum efficiency (η_{EQE}), defined by

$$\eta_{EQ/:E} = (I_{ph}/q)/(P_{inc}/h\nu)$$

where I_{ph} is the photocurrent, P_{inc} is the incident light power, and $h\nu$ is the photon energy, was measured for diodes fabricated in a range of temperatures and

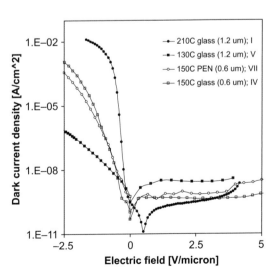

Fig. 6.14 p–i–n diode I–V characteristics for devices made in a range of deposition temperatures and *i*-layer thicknesses on glass and PEN substrates. The Roman numerals indicate the diode characteristics shown in Table 6.2 [86]

Table 6.2 p–i–n sensor configuration and characteristics for devices on glass and PEN substrates [86]

T [°C]	i-layer [μm]	n	η_{EQE} [%]	$\mu\tau$ [10^{-9} cm^2/V]
On glass				
(I) 210	1.20	1.54	71	175
(II) 180	1.33	1.54	60	100
(III) 150	1.10	3.11	43	7
(IV) 150	0.65	2.76	67	7
(V) 130	1.28	14.7	51	2
On PEN				
(VI) 150	1.20	2.76	48	8
(VII) 150	0.60	3.10	70	6

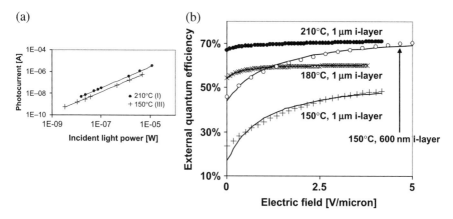

Fig. 6.15 (a) Photocurrent measurements for diodes processed at low and high temperature showing a linear relationship between photocurrent and incident illumination power over 4 decades. (b) External quantum efficiency measurement for p–i–n diodes as a function of applied electric field, for diodes having different deposition temperatures and i-layer thicknesses [86]

with different i-layer thicknesses (with the incident illumination at a wavelength of 528 nm). Fig. 6.15b shows a trade-off between deposition temperature and i-layer thickness. At fixed i-layer thickness, increasing the deposition temperature causes η_{EQE} to increase. In order to obtain a sufficiently high η_{EQE} with low-temperature materials on the other hand, a thinner i-layer or a higher extraction field is required (sensors IV and VII in Table 6.2).

These diode characteristics may be partially explained using the Hecht formula [17]:

$$I_{ph}/I_{sat} = (\mu\tau V_b/d^2)[1 - \exp(-d^2/\mu\tau V_b)] \tag{6.14}$$

where $\mu\tau$ is the effective mobility-lifetime product, d is the i-layer thickness, I_{sat} is the saturation photocurrent, and $V_b = V + V_i$ is the sum of the applied voltage V and the built-in voltage of the photodiode, assumed to be $V_i = 0.5$ V. The fits of the measured data to Eq. (6.14) are within 10% error for $\mu\tau$, which includes contributions from both electrons and holes although the relative contributions cannot be extracted from the data. An increased density of defect trapping will cause low values of $\mu\tau$ since both the electron and hole values are inversely proportional to the defect density. An estimate of the defect density was calculated to be 5×10^{16} cm^{-3}. The low-temperature material was found to have a defect density 20 times higher than that of the high temperature processed material readily explaining the larger ideality factor and the increased leakage current. Furthermore, a simple estimate suggests that the increased defect density would contribute no more than 1 V to the threshold voltage of a TFT made from the same material, which is consistent with the comparable V_T for the low-temperature and the high-temperature TFTs. Reducing the thickness of the photodiode therefore provides reasonable performance and constitutes an adequate compromise between high charge collection, high optical absorption, low leakage current, and low mechanical stress.

6.5 Printed Flexible Electronics

6.5.1 Overview

There have been many demonstrations of flexible circuits using both a-Si:H and organic semiconductors. The majority of effort has been placed on the fabrication of active-matrix backplanes for displays and imagers. Recently, there has been interest in fabrication of radio frequency identification tags on flexible substrates due to the high volumes of tags required if they are widely adopted [88, 89]. We focus here on finished prototypes and circuits fabricated using printing technologies.

6.5.2 Digital Lithography for Flexible Image Sensor Arrays

The fabrication process of active-matrix backplanes using a-Si:H TFTs involves several mask steps that requires precise alignment after each deposition or etching step. The processing on flexible platforms should ideally allow: (1) noncontact patterning, (2) adjustment for local and global distortion of the pattern, (3) electronic imaging for rapid layer-to-layer registration, and (4) additive patterning for rapid throughput for roll-to-roll processing. The digital lithographic process has the potential to reach many of these requirements using a relatively simple tool for patterning while allowing great flexibility on a wide range of substrates and materials.

The fabrication of a-Si:H-based TFT arrays has been successfully demonstrated using the digital lithographic process [1, 2]. Examples of bottom-gate structures have been fabricated using a Kemamide-based wax ejected from a multiejector piezoelectric printhead. Conventional a-Si:H TFT structures having both back-channel and tri-layer stacks have been fabricated using jet printing in place of the photolithographic process. Layer-to-layer registration was accomplished by

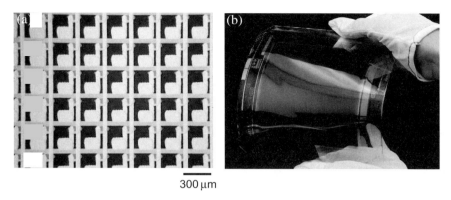

300 µm

Fig. 6.16 (**a**) Optical micrograph of a print-patterned a-Si:H backplane on PEN substrates [86]. (**b**) Photograph of a completed sensor array on PEN

alignment to the gate features whose image and alignment mark location was captured by a camera mounted on a microscope objective.

Figure 6.16 shows a photograph of the completed flexible array. In this example, the substrate was polyethylene naphthalate (PEN) and the TFT fabrication was accomplished using conventional metals (Cr and Al), dielectric (Si_3N_4), and semiconductors (a-Si:H) The I–V transfer and output curves for a typical TFT pixel within the array are shown in Fig. 6.17. The device shows excellent characteristics, comparable to conventional high-temperature devices fabricated on glass substrates.

Once the backplane fabrication has been completed, the integration of the sensor media begins with an encapsulation layer followed by the sensor materials described earlier. Via contacts were etched onto the sensor pixels, followed by patterning and deposition of the pixel contacts. Figure 6.18 shows the successful integration of an a-Si:H TFT array with the low-T a-Si:H p–i–n sensor on PEN along with a schematic of the sensor array cross section. This 3.5 in. diagonal sensor array consisted of 180×180 pixels with 75 dpi resolution. The image is created by light projection through slide film onto the array. The visible line and point defects are mostly due to the process handling of the free standing flexible PEN.

The sensitivity of the sensor array has been measured in terms of its noise equivalent power (*NEP*), expressed as the light flux: $NEP = q_N / (eFl^2 \eta_{EQE})$, where q_N is the minimum detectable charge, F is the fill factor of a sensor pixel, and l is the edge dimension of a square pixel. The image sensor array is operated by external electronics that allows operation up to 20 Hz, with a 14-bit dynamic range and a minimum electronic noise of 1,000–2,000 electrons. For the *NEP* calculation, q_N is defined to be equal to the electronic noise of 2,000 electrons. The sensor fill factor is $F = 0.76$ with $l = 340$ µm. Using the external quantum efficiency of 70% from Fig. 6.15a, the extracted *NEP* is 1.2 pW/cm^2 at 50 ms integration time for a typical sensor array as shown in Fig. 6.19. In addition to a-Si:H, other photosensitive materials such as organic semiconductors have been incorporated with flexible TFT backplanes to demonstrate light-weight, large-area imager sensors [90–94].

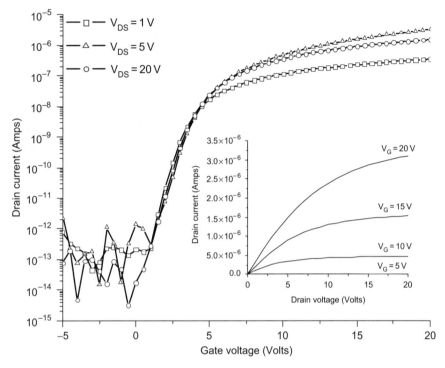

Fig. 6.17 Transfer characteristics of an a-Si:H TFT processed at 150°C on PEN. The W/L was 1, the mobility was ~1 cm^2/V.s, the subthreshold slope was 0.5 V/decade, and the on–off ratio was >10^8. The inset shows the output characteristics

6.5.3 Printed Organic Backplanes

6.5.3.1 Hybrid Fabrication

Many demonstrations of organic electronic devices have used a combination of conventional materials and patterning techniques to form organic active-matrix backplanes [95–97]. The most advanced work has used photolithographically defined electrodes and spin-coated organic semiconductors to fabricate backplanes. There are a number of demonstrations of TFTs that have been fabricated using printable molecular semiconductors and polymers on conventionally patterned electrodes [98]. These demonstrations have shown that jet-printed devices can have identical or similar performance to those fabricated using spin coating once the printing process is optimized.

The combination of digital lithography with an additive printing process for the deposition of polymeric-based semiconductors has been demonstrated as an all-print patterned approach [5, 6]. The printing process is designed to fabricate a bottom-gate TFT with coplanar source and drain contacts (i.e., contacts deposited after the dielectric and before the semiconductor) (Fig. 6.20). The bottom-gate coplanar

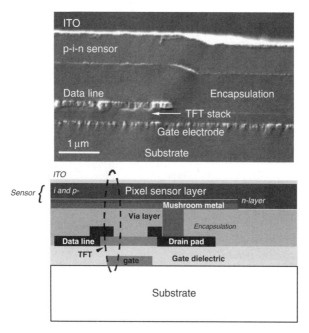

Fig. 6.18 Cross sectional SEM and schematic of an a-Si:H TFT stack and sensor layer on PEN [40]

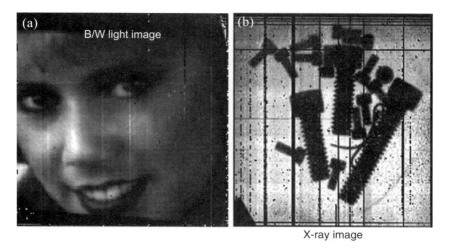

Fig. 6.19 (**a**) Light image [86] and (**b**) X-ray image from an a-Si:H sensor array on PEN fabricated using digital lithography

structure reduces the risk of damage to the polymer from subsequent processing as the semiconducting polymer, PQT-12, is deposited last. Figure 6.20f is an optical micrograph of the pixels within the printed 75 dpi 128 × 128 pixel array. Control of the placement of the printed semiconductor was critical as the semiconducting

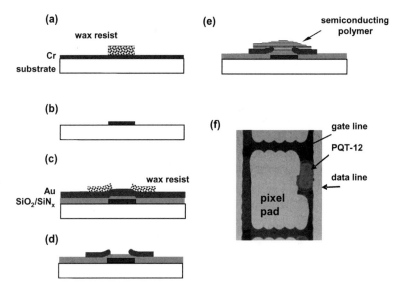

Fig. 6.20 Process flow for fabrication of a TFT array using digital lithography and jet printing of the semiconductor. (**a,b**) The gate electrode is made up of chromium (100 nm) and patterned using printed wax as a resist. (**c,d**) The gate dielectric is SiN$_x$ capped with a thin layer of SiO$_2$ deposited by PE-CVD. The source and drain electrodes are made from gold (100 nm) with a chromium (6 nm) adhesion layer and are patterned using printed wax resist. (**e**) PQT-12 is jet printed from a dispersion in dichlorobenzene ($\sim$0.3% wt). (**f**) Optical micrograph of a finished pixel in a TFT array [100]

island must not extend beyond the gate electrode or it will cause a parasitic leakage current between the electrodes. Also, the well-defined area of jet-printed PQT-12 prevents the formation of a continuous layer of polymer between two TFTs connected by the same gate line, that would form a conduction path, and causes significant cross talk between pixels. The printed TFTs exhibited similar performance to those fabricated by spin coating onto silicon wafers (mobility of 0.05–0.1 cm^2/Vs, and an off-current of $\sim$10^{-12} A, giving an on–off ratio of $\sim$10^6) and had an average mobility of 0.06 cm^2/Vs $\pm$ 30%.

6.5.3.2 All-Printed Electronics

One of the most attractive features of organic materials is the potential for all-additive fabrication of electronics that is a process where all the materials are fabricated by direct deposition without vacuum processing or by photolithography. There are few reported examples of all-additive completed backplane circuits made without the use of photolithography. Large TFT arrays (50 cm × 80 cm) have been fabricated using laser transfer printing to transfer dry films of conductors as the address lines of the array and vacuum deposition of a semiconducting layer of pentacene [99]. The pixels in this demonstration were large ($\sim$0.9 cm × 0.9 cm) and the performance of the TFTs was reported to be comparable to conventionally patterned devices.

Fully printed TFT arrays were fabricated using printed silver nanoparticles as the contacts and address lines, a highly cross-linked organic dielectric and PQT-12 as the semiconducting layer [100, 101]. The pixel size was relatively large, 680 μm × 680 μm, resulting in a backplane with a resolution of 37 dpi (Fig. 6.21a). Individual TFTs were reported to have relatively low mobilities, $\mu = 0.003$ cm^2/Vs, attributed to the interaction between the semiconducting polymer and the dielectric. The backplane pixels fabricated using the method had charging times on the order of 1 ms (Fig. 6.21b). While this response time is slow for applications in video rate displays, it is adequate for large-area applications in signage with media that has relatively slow response times, such as electrophoretic media.

While these examples demonstrate high potential for all-printed device processing as an alternative to conventional vacuum deposition and photolithography, the technology needs to address some deficiencies. Minimum feature size is still relatively large compared to conventional photolithography and development of printing systems with high throughput, small features, accurate drop placement, and low cost could prove challenging. Many of the solution-processable materials currently available do not have the quality and corresponding device performance compared to conventional inorganic semiconductor materials. This situation may change in the coming years since the current level of activity in synthesizing higher quality semiconductors, dielectrics, and metal inks is rapidly rising. Given the high level of interest and the effort being put into printed electronics, the technology has shown rapid development and in the next few years may result in breakthroughs that make it competitive with conventional inorganic devices.

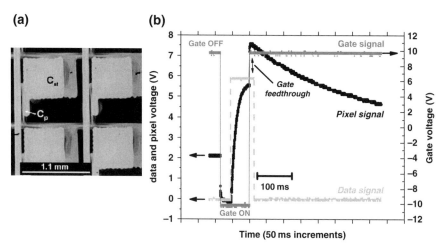

Fig. 6.21 (a) Optical micrograph of a TFT in an all-printed array. The gate lines, data lines, and pixel electrodes were made from sintered silver nanoparticles. The gate dielectric was SU-8 and the semiconductor was PQT-12. (b) Pixel response for an all-printed element [101]. Reproduced by permission of The Electrochemical Society

6.6 Conclusions and Future Prospects

Flexible circuits can now be fabricated using both conventional inorganic and organic materials. Optimization of the electrical performance of materials deposited on plastic is still under way. The results so far suggest that the current level of performance will be adequate for a variety of applications in displays and imagers. Printing technologies have been demonstrated to have great utility for both subtractive and additive patterning. Now that commercial instrumentation is becoming more widely available, adoption of these methods will probably increase. While printing methods present new challenges for control of patterning, they will likely achieve a place where the highest resolution features are not required or where direct deposition of materials is desirable.

The majority of reported work so far on flexible electronics has focused on fabrication on single sheets. One of the potential advantages of flexible substrates is roll-to-roll fabrication, which would dramatically change the method of manufacture of electronics. Development of these techniques is relatively complex as it combines both mechanical design as well as materials processes. Vacuum deposition in a roll-to-roll process has been demonstrated for solar cells and is adaptable for the layers required for a-Si:H TFTs [8]. Solution processes are potentially more easily adaptable as the graphic arts industry already uses liquid inks. Some of the significant challenges for adaptation of the materials demonstrated so far are decreasing annealing times, controlling viscosity of solutions without harming electronic properties of the dried films, and development of high-throughput large-scale printheads. Once theses challenges are overcome then the functionality, performance, and cost should ease the burden of manufacturability and flexible electronics should begin to permeate into mainstream electronic applications.

Acknowledgements   The authors would like to acknowledge the many insights and assistance of experimental data provided by their colleagues and collaborators. The authors would particularly wish to recognize the contributions and efforts of the following colleagues from PARC: Robert A. Street, Rene Lujan, Steve Ready, Beverly Russo, Maryanne Rosenthal, Michael Young, Scott Limb, Sanjiv Sambandan, Jürgen Daniel, Ana-Claudia Arias, Eugene Chow, Vicki Aguilar, and William A. MacDonald of DuPont-Teijin Films. Research performed at PARC was partially supported by the Advanced Technology Program of the National Institute of Standards and Technology (contract #: 70NANB3H3029).

References

1. Wong WS, Ready S, Matusiak R, White SD, Lu JP, Ho J, Street RA (2002) Amorphous silicon thin-film transistors and arrays fabricated by jet printing. Appl Phys Lett 80:610–612
2. Wong WS, Ready SE, Lu JP, Street RA (2003) Hydrogenated amorphous silicon thin-film transistor arrays fabricated by digital lithography. IEEE Electron Dev Lett 24:577–579
3. Creagh LT, McDonald M (2003) Design and performance of ink-jet print heads for non-graphic-arts applications. Mater Res Soc Bull 28:807–811
4. Young R, Tamura Y, Wang CE, Hsieh D (2003) Cost and efficiency comparisons between manufacturing generations and regions. Proceeding of the International Display and Manufacturing Conference, IDMC 2003, pp 285–288

5. Paul KE, Wong WS, Ready SE, Street RA (2003) Additive jet printing of polymer thin-film transistors. Appl Phys Lett 83:2070–2702

6. Arias AC, Ready SE, Lujan R, Wong WS, Paul KE, Salleo A, Chabinyc ML, Apte R, Street RA, Wu Y, Liu P, Ong B (2004) All jet-printed polymer thin-film transistor active-matrix backplanes. Appl Phys Lett 85:3304–3306

7. Jain K, Klosner M, Zemel M, Raghunandan S (2005) Flexible electronics and displays: High-resolution, roll-to-roll, projection lithography and photoablation processing technologies for high-throughput production. Proc IEEE 93:1500–1510

8. Sheats JR (2002) Roll-to-roll manufacturing of thin film electronics. Proc SPIE – Int Soc Opt Eng 4688:240–248

9. Gleskova H, Wagner S, Suo Z (1998) a-Si:H TFTs made on polyimide foil by PE-CVD at 150°C. Mat Res Soc Symp Proc 508:73–78

10. Sarma KR (2004) a-Si TFT OLED fabricated on low-temperature flexible plastic substrate. Mat Res Soc Symp Proc 814:369

11. Zhou L, Jackson T, Brandon E, West W (2004) Flexible substrate a-Si:H TFTs for space applications. Dev Res Conf 1:123–124

12. Gleskova H, Wagner S (2001) DC-gate-bias stressing of a-Si:H TFTs fabricated at 150°C on polyimide foil. IEEE Trans Electron Dev 48:1667–1671

13. Sazonov A, Stryahilev D, Lee CH, Nathan A (2005) Low-temperature materials and thin-film transistors for flexible electronics. Proc IEEE 93:1420–1428

14. Shimoda T, Morii TK, Seki S, Kiguchi H (2003) Ink-jet printing of light emitting polymer displays. Mater Res Soc Bull 28:821–827

15. Sele CW, vonWerne T, Friend RH, Sirringhaus H (2005) Lithography-free, self-aligned ink-jet printing with sub-hundred-nanometer resolution. Adv Mater 17:997–1001

16. Tsukada T (2000) Active-matrix liquid-crystal displays. In: Street RA (ed) Technology and applications of amorphous silicon. Springer-Verlag, Heidelberg

17. Street RA (2000) Large area image sensor arrays. In: Street RA (ed) Technology and applications of amorphous silicon. Springer-Verlag, Heidelberg

18. Shimoda T, Matsuki Y, Furusawa M, Aoki T, Yudasaka I, Tanaka H, Iwasawa H, Wang D, Miyasaka M, Takeuchi Y (2006) Solution-processed silicon films and transistors. Nature 440:783–786

19. Hoffman RL, Norris BJ, Wager JF (2003) ZnO-based transparent thin-film transistors. Appl Phys Lett 82:733–735

20. Nomura K, Ohta H, Takagi A, Kamiya T, Hirano M, Hosono H (2004) Room-temperature fabrication of transparent flexible thin-film transistors using amorphous oxide semiconductors. Nature 432:488–492

21. Noh YY, Cheng X, Sirringhaus H, Sohn JI, Welland ME, Kang DJ (2007) Ink-jet printed ZnO nanowire field effect transistors. Appl Phys Lett 91:043109

22. Veres J, Ogier SD, Leeming SW, Cupertino DC, Khaffaf SM (2003) Low-k Insulators as the choice of dielectrics in organic field-effect transistors. Adv Func Mater 13:199–204

23. Sirringhaus H (2005) Device physics of solution-processed organic field-effect transistors. Adv Mater 17:2411–2425

24. Gundlach DJ, Lin YY, Jackson TN, Nelson SF, Schlom DG (1997) Pentacene organic thin-film transistors – molecular ordering and mobility. IEEE Electron Dev Lett 18:87–89

25. Nomoto K, Hirai N, Yoneya N, Kawashima N, Noda M, Wada M, Kasahara J (2005) A high-performance short-channel bottom-contact OTFT and its application to AM-TN-LCD. IEEE Trans Electron Dev 52:1519–1526

26. Payne MM, Parkin SR, Anthony JE, Kuo CC, Jackson TN (2005) Organic field-effect transistors from solution-deposited functionalized acenes with mobilities as high as 1 cm^2/Vs. J Am Chem Soc 127:4986–4987

27. McCulloch I, Heeney M, Bailey C, Genevicius K, Macdonald I, Shkunov M, Sparrowe D, Tierney S, Wagner R, Zhang W, Chabinyc ML, Kline RJ, McGehee MD, Toney MF (2006)

Liquid-crystalline semiconducting polymers with high charge-carrier mobility. Nat Mater 5:328–333

28. Steudel S, DeVusser S, DeJonge S, Janssen D, Verlaak S, Genoe J, Heremans P (2004) Influence of the dielectric roughness on the performance of pentacene transistors. Appl Phys Lett 85:4400–4402

29. Chabinyc ML, Lujan R, Endicott F, Toney MF, McCulloch I, Heeney M (2006) Effects of the surface roughness of plastic-compatible inorganic dielectrics on polymeric thin-film transistors. Appl Phys Lett 90:233508

30. Nunes G Jr, Zane SG, Meth JS (2005) Styrenic polymers as gate dielectrics for pentacene field-effect transistors. J Appl Phys 98:104053

31. Jung T, Dodabalapur A, Wenz R, Mohapatra S (2005) Moisture induced surface polarization in a poly(4-vinyl phenol) dielectric in an organic thin-film transistor. Appl Phys Lett 87:182109

32. Chua LL, Zaumseil J, Chang JF, Ou ECW, Ho PKH, Sirringhaus H, Friend RH (2005) General observation of n-type field-effect behaviour in organic semiconductors. Nature 343: 194–199

33. Hong CM, Wagner S (2000) Ink-jet printed copper source/drain metallization for amorphous silicon thin-film transistors. IEEE Electron Dev Lett 21:384–386

34. Huang D, Liao F, Molesa S, Redinger D, Subramanian V (2003) Plastic compatible low-resistance printable gold nanoparticle conductors for flexible electronics. J Electrochem Soc 150:412–417

35. Chung J, Ko S, Bieri NR, Grigoropoulos CP, Poulikakos D (2004) Conductor microstructures by laser curing of printed gold nanoparticle ink. Appl Phys Lett 84:801–803

36. Wu Y, Li Y, Ong BS (2006) Printed silver ohmic contacts for high-mobility organic thin-film transistors. J Am Chem Soc 128:4202–4203

37. Groenendaal LB, Jonas F, Freitag D, Pielartzik H, Reynolds JR (2000) Poly(3,4-ethylenedioxythoiphene) and its derivatives: Past, present, and future. Adv Mater 12: 481–494

38. Rogers JA, Bao Z, Baldwin K, Dodabalapur A, Crone B, Raju VR, Kuck V, Katz H, Amundson K, Ewing J, Drzaic P (2001) Paper-like electronic displays: Large-area rubber-stamped plastic sheets of electronics and microencapsulated electrophoretic inks. Proc Nat Acad Sci 98:4835–4840

39. Wagner S, Gleskova H, Sturm JC, Suo Z (2000) Novel processing technology for macro-electronics. In: Street RA (ed) Technology and applications of amorphous silicon, vol 37. Springer-Verlag, Berlin, pp 222–251

40. Wong WS, Chabinyc ML, Limb S, Ready SE, Lujan R, Daniel J, Street RA (2007) Digital lithographic processing for large-area electronics. J Soc Info Display 15:463

41. Wong WS, Chow EC, Geluz-Aguilar V, Lujan R, Chabinyc ML (2006) Fine-feature patterning of self-aligned polymeric thin-film transistors fabricated by digital lithography and electroplating. Appl Phys Lett 89:142118-1–142118-3

42. Street RA, Wong WS, Ready SE, Chabinyc ML, Arias AC, Limb S, Salleo A, Lujan R (2006) Jet printing flexible displays. Mater Today 9:32

43. Zhao N, Chiesa M, Sirringhaus H, Li Y, Wu Y, Ong B (2007) Self-aligned ink-jet printing of highly conducting gold electrodes with submicron resolution. J Appl Phys 101: 064513

44. Thomasson DB, Jackson TN (1998) Fully self-aligned tri-layer a-Si:H thin-film transistors with deposited doped contact layer. IEEE Electron Dev Lett 19:124

45. Sirringhaus H, Kawase T, Friend RH, Shimoda T, Inbasekaran M, Wu W, Woo EP (2000) High-resolution ink-jet printing of all-polymer transistor circuits. Science 290:2123–2126

46. Li SP, Newsome CJ, Kugler T, Ishida M, Inoue S (2007) Polymer thin-film transistors with self-aligned gates fabricated using ink-jet printing. Appl Phys Lett 90:172103

47. Schiaffino S, Sonin AA (1997) Formation and stability of liquid and molten beads on a solid surface. J Fluid Mech 343:95–110

48. Duineveld P (2003) The stability of ink-jet printed lines of liquid with zero receding contact angle on a homogeneous substrate. Phys Fluids 477:175–200
49. Chang JF, Sun B, Breiby DW, Nielsen MM, Solling TI, Giles M, McCulloch I, Sirringhaus H (2004) Enhanced mobility of poly(3-hexylthiophene) transistors by spin-coating from high-boiling-point solvents. Chem Mater 16:4772–4776
50. DeLongchamp DM, Vogel BM, Jung Y, Gurau MC, Richter CA, Kirillov OA, Obrzut J, Fischer DA, Sambasivan S, Richter LJ, Lin EK (2005) Variations in semiconducting polymer microstructure and hole mobility with spin-coating speed. Chem Mater 17:5610–5612
51. Salleo A, Street RA (2003) Light-induced bias stress reversal in polyfluorene thin-film transistors. J Appl Phys 94(1):471
52. Salleo A, Endicott F, Street RA (2005) Reversible and irreversible trapping in poly(thiophene) thin-film transistors. Appl Phys Lett 86:263505
53. Ng TN, Marohn JA, Chabinyc ML (2006) Comparing the kinetics of bias stress in organic field-effect transistors with different dielectric interfaces. J Appl Phys 100:084505
54. Schoonveld WA, Oostinga JB, Vrijmoeth J, Klapwijk TM (1999) Charge trapping instabilities in sexithiophene thin-film transistors. Synth Metals 101:6078
55. Brown AR, Jarrett CP, de Leeuw DM, Matters M (1997) Field-effect transistors made from solution-processed organic semiconductors. Synth Metals 88:37
56. Zilker SJ, Detcheverry C, Cantatore E, de Leeuw DM (2000) Bias stress in organic thin-film transistors and logic gates. Appl Phys Lett 79:1124
57. Knipp D, Street RA, Völkel A, Ho J (2003) Pentacene thin-film transistors on inorganic dielectrics: Morphology, structural properties and electronic transport. J Appl Phys 93:347
58. Gu G, Kane MG, Mau SC (2007) Reversible memory effects and acceptor states in pentacene-based organic thin-film transistors. J Appl Phys 101:014504
59. Chabinyc ML, Endicott F, Vogt BD, de Longchamp DM, Lin EK, Wu YL, Liu P, Ong BS (2006) Effects of humidity on unencapsulated poly(thiophene) thin-film transistors. Appl Phys Lett 88:113514
60. Street RA, Chabinyc ML, Endicott F, Ong B (2006) Extended time bias stress effects in polymer transistors. J Appl Phys 11:114518
61. Street RA, Salleo A, Chabinyc ML (2003) Bipolaron mechanism for bias-stress effects in organic transistors. Phys Rev B 68(8):085316
62. Salleo A, Street RA (2004) Kinetics of bias-stress and bipolaron formation in regio-regular poly(thiophene). Phys Rev B 70(23):235324
63. Lloyd-Hughes J, Richards T, Sirringhaus H, Castro-Camus E, Herz LM, Johnston MB (2006) Charge trapping in polymer transistors probed by terahertz spectroscopy and scanning probe potentiometry. Appl Phys Lett 89:112101
64. Karim KS, Nathan A, Hack M, Milne WI (2004) Drain-bias dependence of threshold voltage stability of amorphous silicon TFTs. IEEE Electron Dev Lett 25:188
65. Salleo A, Chabinyc ML (2006) Electrical and environmental stability of polymer thin-film transistors. In: Klauk H (ed) Organic electronics: Materials, manufacturing and applications. Wiley VCH, Weinheim, Germany
66. Libsch FR, Kanicki J (1993) Bias-stress-induced stretched-exponential time dependence of charge injection and trapping in amorphous thin-film transistors. Appl Phys Lett 62:1286
67. Gomes HL, Stallinga P, Dinelli F, Murgia M, Biscarini F, de Leeuw DM, Muccini M, Müllen K (2005) Electrical characterization of organic based transistors: stability issues. Polym Adv Tech 16:227
68. Gomes HL, Stallinga P, Dinelli F, Murgia M, Biscarini F, de Leeuw DM, Muck T, Geurts J, Molenkamp LW, Wagner V (2004) Bias-induced threshold voltages shifts in thin-film organic transistors. Appl Phys Lett 84:3184
69. Powell MJ, van Berkel C, Franklin AR, Deane SC, Milne WI (1992) Defects pool in amorphous-silicon thin-film transistors. Phys Rev B 45:4160
70. Féry C, Racine B, Vaufrey D, Doyeux H, Ciná S (2005) Physical mechanism responsible for

the stretched exponential decay behavior of aging organic light-emitting diodes. Appl Phys Lett 87:213502

71. Powell MJ, van Berkel C, Hughes JR (1989) Time and temperature dependence of instability mechanisms in amorphous silicon thin-film transistors. Appl Phys Lett 54:1323

72. Street RA, Salleo A (2002) Contact effects in polymer transistors. Appl Phys Lett 81:2887

73. Necliudov PV, Shur MS, Gundlach DJ, Jackson TN (2000) Modeling of organic thin-film transistors of different designs. J Appl Phys 88:6594

74. Chabinyc ML, Lu JP, Street RA, Wu YL, Liu P, Ong BS (2004) Short channel effects in regioregular poly(thiophene) thin-film transistors. J Appl Phys 96:2063

75. Wang GM, Swensen J, Moses D, Heeger AJ (2003) Increased mobility from regioregular poly(3-hexylthiophene) field-effect transistors. J Appl Phys 93:6137

76. Yang CS, Smith LL, Arthur CB, Parsons GN (2000) Stability of low-temperature amorphous silicon thin-film transistors formed on glass and transparent plastic substrates. J Vac Sci Technol B 18:683

77. van Berkel C, Powell MJ (1987) Resolution of amorphous silicon thin-film transistor instability mechanisms using ambipolar transistors. Appl Phys Lett 51:1094

78. Wong WS, Lujan R, Daniel JH, Limb S (2005) Digital lithography for large-area electronics on flexible substrates. J Non-Cryst Solids 352(9–20):1981–1985

79. Choi JB, Yun DC, Park YI, Kim JH (2000) Properties of hydrogenated amorphous silicon thin-film transistors fabricated at 150°C. J Non-Cryst Solids 266–269:1315

80. Ikeda M, Mizutani Y, Ashida S, Yamada K (2000) Characteristics of low-temperature-processed a-Si TFT for plastic substrates. IEICE Trans Electron E83-C:1584

81. Powell MJ, van Berkel C, French ID, Nicholls DH (1987) Bias dependence of instability mechanisms in amorphous silicon thin-film transistors. Appl Phys Lett 51:1242

82. Long K, Kattamis AZ, Cheng IC, Gleskova H, Wagner S, Sturm JC (2006) Stability of amorphous-silicon TFTs deposited on clear plastic substrates at 250°C to 280°C. IEEE Electron Dev Lett 27:111

83. Wang Q, Ward S, Duda A, Hu J, Stradins P, Crandall RS, Branz HM, Perlov C, Jackson W, Mei P, Taussig C (2004) High-current-density thin-film silicon diodes grown at low temperature. Appl Phys Lett 85:2122

84. Vieira M, Louro P, Fernandes M, Schwarz R, Schubert M (2004) Large area p–i–n flexible image sensors. Mater Res Soc Proc 814:I7–11.1

85. Kim KH, Vygranenko Y, Bedzyk M, Chang JH, Chuang TC, Striakhilev D, Nathan A, Heiler G, Tredwell T (2007) High performance hydrogenated amorphous silicon n–i–p photodiodes on glass and plastic substrates by low-temperature fabrication process. Mater Res Soc Proc 989:A19–05

86. Ng TN, Lujan RA, Sambandan S, Street RA, Limb S, Wong WS (2007) Low temperature a-Si:H photodiodes and flexible image sensor arrays patterned by digital lithography. Appl Phys Lett 91:063505

87. Kroon A, van Swaaij RACMM (2001) Spatial effects on ideality factor of amorphous silicon pin diodes. J Appl Phys 90:994

88. Knobloch A, Manuelli A, Bernds A, Clemens W (2004) Fully printed integrated circuits from solution processed polymers. J Appl Phys 96:2286–2291

89. Zielke D, Hubler AC, Hahn U, Brandt N, Bartzsch M, Fugmann U, Fischer T, Veres J, Ogier S (2005) Polymer-based organic field-effect transistor using offset printed source/drain electrodes. Appl Phys Lett 87:123508

90. Yu G, Wang J, McElvain J, Heeger AJ (1998) Large-area, full-color image sensors made with semiconducting polymers. Adv Mater 10:1431

91. Street RA, Mulato M, Lau R, Ho J, Graham J, Popovic Z, Hor J (2001) Image capture array with an organic light sensor. Appl Phys Lett 78:4193

92. Someya T, Kato Y, Iba S, Noguchi Y, Sekitani T, Kawaguchi H, Sakurai T (2005) Integration of organic FETs with organic photodiodes for a large area, flexible, and lightweight sheet image scanners. IEEE Trans Electron Dev 52:2502

93. Peumans P, Yakimov A, Forrest SR (2003) Small molecular weight organic thin-film pho-
    todetectors and solar cells. J Appl Phys 93:3693
94. Schilinsky P, Waldauf C, Hauch J, Brabec CJ (2004) Polymer photodiode detectors: Progress
    and recent developments. Thin Solid Films 451–452:105
95. Huitema HEA, Gelinck GH, van der Putten JBPH, Kuijk KE, Hart CM, Cantatore E, Herwig
    PT, van Breemen AJJM, de Leeuw DM (2001) Polymer electronics: Plastic transistors in
    active-matrix displays. Nature (London, United Kingdom) 414:599
96. Gelinck GH, Huitema HEA, van Veenendaal E, Cantatore E, Schrijnemakers L, van der
    Putten JBPH, Geuns TCT, Beenhakkers M, Giesbers JB, Huisman BH, Meijer EJ, Benito
    EM, Touwslager FJ, Marsman AW, van Rens BJE, de Leeuw DM (2004) Flexible active-
    matrix displays and shift registers based on solution-processed organic transistors. Nat Mater
    3:106–110
97. Burns SE, Cain P, Mills J, Wang J, Sirringhaus H (2003) Inkjet printing of polymer thin-film
    transistor circuits. Mater Res Soc Bull 28:829–834
98. Chang P, Molesa S, Murphy A, Frechet J, Subramanian V (2006) Inkjetted crystalline single-
    monolayer oligothiophene OTFTs. IEEE Trans Electron Dev 53:594–600
99. Blanchet GB, Loo YL, Rogers JA, Gao F, Fincher CR (2003) Large area, high resolution,
    dry printing of conducting polymers for organic electronics. Appl Phys Lett 82:463–465
100. Arias AC, Daniel J, Krusor B, Ready S, Sholin V, Street R (2007) All-additive ink-jet-printed
    display backplanes: materials development and integration. J Soc Info Display 15:485–490
101. Daniel J, Arias AC, Krusor B, Lujan R, Street RA (2006) The road towards large-area elec-
    tronics without vacuum tools. Electrochem Soc Trans 3:229

Chapter 7
Sheet-Type Sensors and Actuators

Takao Someya

7.1 Introduction

Recent intensive research and development of organic field-effect transistors (FETs) [1–6] have been motivated by a new class of applications that cannot be easily realized by conventional electronics based on inorganic semiconductors. Organic transistors are mechanically flexible, thin, lightweight, and shock-resistant, because organic devices are manufactured on plastic films at low (ambient) temperature. Furthermore, manufacturing costs of organic transistor circuits would be inexpensive, even for large areas, when they are fabricated using printing technologies and/or roll-to-roll processes.

There are two major applications for organic transistors. The first one is a flexible display. This new display includes a paper-like display or an electronic paper, where electric inks, electroluminescent (EL) devices, and liquid crystals or other mediums are powered by organic transistor active matrices [4, 7]. The second one is a radio frequency identification (RFID) tag [8, 9]. An organic transistor-based RFID tag may be printed on packages of products, resulting in an inexpensive and robust electronics.

As another application of organic transistors, we demonstrated large-area flexible sensors. The first organic transistor-based large-area sensors are flexible pressure sensor matrices; organic transistor active matrices are used to read out pressure distributions over a large area from a 2-D array of pressure sensor cells. The new pressure sensor can be ideal for electronic artificial skin (e-skin) applications for future generations of robots.

The mobility of pentacene that is known as a high-mobility low-molecular weight semiconductor is typically 1 cm^2/Vs. This value is about two or three orders of magnitude lower than that of poly- or single-crystalline silicon, respectively. Although flexible displays and/or RFID tags require usually high-electronic performance, the

T. Someya (✉)
School of Engineering, Quantum-Phase Electronics Center, The University of Tokyo, 7-3-1 Hongo, Bunkyo-ku, Tokyo 113-8656 Japan
e-mail: someya@ap.t.u-tokyo.ac.jp

W.S. Wong, A. Salleo (eds.), *Flexible Electronics: Materials and Applications*, Electronic Materials: Science & Technology, DOI 10.1007/978-0-387-74363-9_7,
© Springer Science+Business Media, LLC 2009

slow speed is tolerable for most applications of large-area sensors. For e-skin in particular, the integration of pressure sensors and organic peripheral electronics allows one to take advantage of the many benefits of organic transistors, such as mechanical flexibility, large area, low cost, and relative ease of fabrication without suffering many drawbacks. Technical details on e-skins can be seen in journals [10–14] and book chapters [15].

The second example of large-area, flexible sensor based on organic semiconductors is a sheet-type image scanner [16–19]. The device is manufactured on plastic films by integrating organic transistors and organic photodiodes. Organic photodetectors distinguish black and white from the difference of reflectivity between black and white parts on paper. Because the new image-capturing device does not have any optical or mechanical component, it is lightweight, shock-resistant, and physically flexible. It is also human friendly since it could be rolled and carried in a pocket.

Organic transistors are also suitable for large-area actuators. We have manufactured a flexible, lightweight sheet-type Braille display on a plastic film by integrating high-quality organic transistors with soft sheet-type actuators. An array of rectangular plastic actuators is processed from a perfluorinated polymer electrolyte membrane. A small semisphere, which projects upward from the rubber-like surface of the display, is attached to the tip of each rectangular actuator. The present scheme will enable people with visual impairments to carry the Braille sheet display in their pockets and read Braille e-books at any time.

This chapter reports recent progress and remaining issues of organic transistor-based large-area, flexible sensors and actuators. The sheet-type image scanners and sheet-type Braille displays will be described in Sections 7.2 and 7.3, respectively, as examples of large-area, flexible sensors and actuators, respectively.

7.2 Sheet-type Image Scanners

In this section, we report on sheet-type image scanners, particularly, principle of imaging, manufacturing process, and electronic performance. The device is manufactured on plastic films with integrated organic transistors and organic p–n diodes that work as photodetectors. The photodetectors can detect black and white tones by sensing the difference in reflected light from the dark and bright parts of an image. The thin-film pentacene transistors have an 18-μm channel length and a 0.7-cm^2/Vs mobility. The effective sensing area of the integrated device is 5×5 cm^2. The resolution is 36 dpi (dots per inch.), and the total number of sensor cells is 5,184. The total thickness and the weight of the whole device are 0.4 mm and 1 g, respectively. As may be seen in Fig. 7.1, the integrated device formed on a plastic film is mechanically flexible, very thin, and lightweight. Therefore, it is suitable for human-friendly mobile electronics.

Fig. 7.1 An image of a large-area, flexible, and lightweight sheet-type image scanner placed onto the business card for capturing images. The effective scanning size is 50×50 cm^2 and the resolution is 36 dpi. From Fig. 7.1 in Ref. [18] ((c) 2008 IEEE)

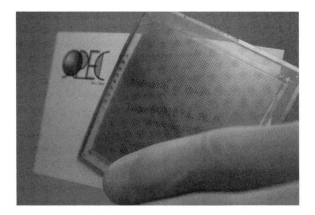

7.2.1 Imaging Methods

The new image scanner does not require any mechanical or optical component. As shown in Fig. 7.2, a linear sensor array is moved from the top to the bottom of a page to capture images in conventional scanners. In the new design, however, a 2-D array of organic photodiodes coupled with organic transistors is used. Instead of a mechanical scanning procedure, the signal of the photodiodes is read out electrically

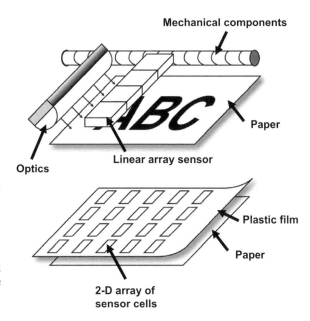

Fig. 7.2 A schematic of a conventional scanner and the present sheet-type scanner. A conventional scanner consists of a linear array sensor and light source. The new scanner consists of a 2-D array of organic photodiodes coupled with organic transistors, which can be read out electrically by the organic transistors. From Fig. 7.5 in Ref. [18] ((c) 2008 IEEE)

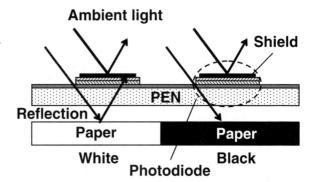

by the organic transistors, avoiding the need to use any movable part. As a result, the device is thin, lightweight, and mechanically flexible.

How can the new scanner distinguish between black and white? If all incident light reaches directly to the active layers, photodetectors cannot distinguish between black and white, as may be seen in Fig. 7.3. Thus, we prepared light-shielding layers to prevent photodetectors from being exposed to direct incident light. Now direct light cannot reaches to the active layers. The incident light passing though transparent regions is reflected on white part of paper and reaches to the active layers, while that on black does not go to the active layers. In this way, the present device can distinguish between black and white.

7.2.2 Device Structure and Manufacturing Process

The chip picture and a circuit diagram are shown in Fig. 7.4a,b, respectively. The cross-sectional device structure is schematically illustrated in Fig. 7.5 along with chemical structure of each layer. Organic FET matrix and photodiode matrix have been manufactured separately on different plastic films and then laminated with each other with silver pastes or anisotropic conductive films to electrically couple the sensor array to the backplane.

A 72×72 matrix of organic transistors with top contact geometry is manufactured using a fine shadow mask technique. The base film is a 125-μm-thick transparent heat-resistant poly(ethylene naphthalate) (PEN) film. The base film is heated at 190°C for 1 h prior to the deposition of thin-film layers. This prebake process is very important to suppress many issues associated with the shrinkage of the PEN film during annealing processes. Note that the temperature of prebaking (190°C) is slightly higher than the maximum process temperature of 180°C, which is needed for a cross-linking process of polyimide gate dielectric layers, as mentioned later. A 150-nm-thick gold layer with a 5-nm-thick chromium adhesion layer is deposited in the vacuum evaporator with shadow masks to form a gate electrode. Then, polyimide precursors (Kemitite CT4112, Kyocera Chemical Co. Ltd., Japan) are spin coated and cured at 180°C to form 630-nm-thick gate dielectric layers [20].

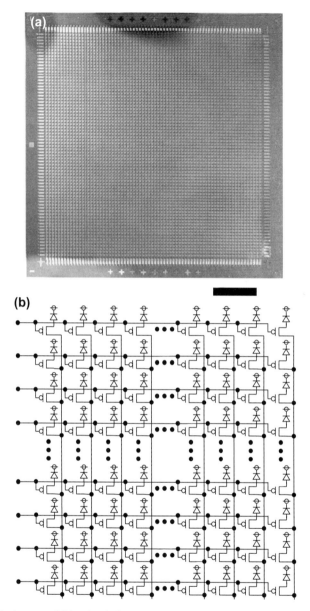

Fig. 7.4 (**a**) An image and (**b**) a circuit diagram of the present sheet-type image scanner consisting of organic transistors integrated with organic photodiodes. Scale bar is 1 cm. From Fig. 7.2 in Ref. [18] ((c) 2008 IEEE)

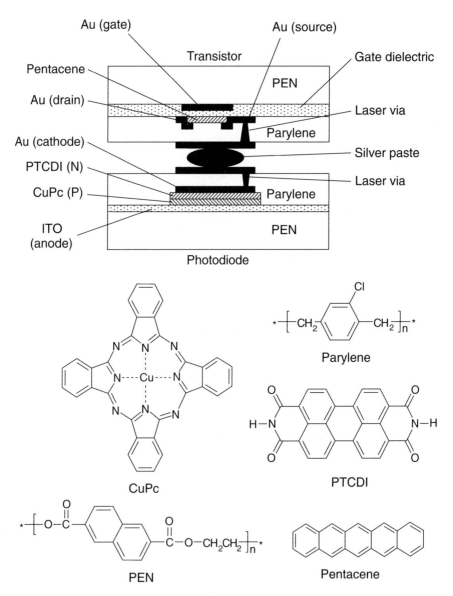

Fig. 7.5 A cross-sectional illustration of the present device consisting of organic transistor and organic pn-diode. The chemical structure of each organic layer is also shown. From Fig. 7.3 in Ref. [18] ((c) 2008 IEEE)

A 50-nm-thick pentacene layer is deposited to form the channel region. A 60-nm-thick gold layer is evaporated through shadow masks to form the source and drain electrodes of the transistors. Figure 7.6a shows the magnified image of four transistors before integrating with organic diodes. The periodicity is 700 μm, which

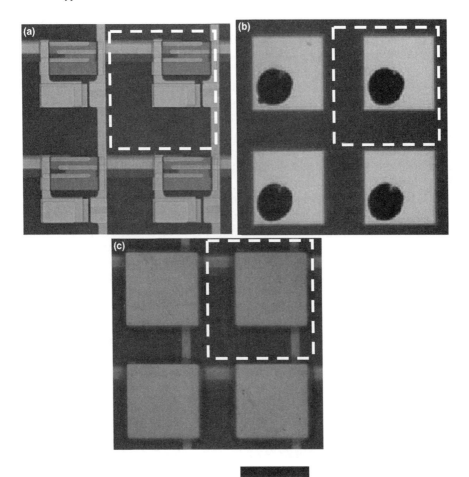

Fig. 7.6 (a) A magnified image of four transistors before integrating with organic diodes. (b) A magnified image of four contact pads with silver paste islands before laminating organic transistor films and organic diode films. (c) A magnified image of four sensor cells integrating organic transistors and organic photodiodes. The whole transistor regions are covered by the photodiodes. The channel length L and width W of the transistors are 18 and 400 μm, respectively. The periodicity as indicated by *dashed line* is 700 μm. From Fig. 7.4 in Ref. [18] ((c) 2008 IEEE)

corresponds to resolution of 36 dpi. The channel length L and width W of the transistors are 18 and 400 μm, respectively.

The photodiodes are separately manufactured on the different films. The base film of photodiodes is a PEN film coated with indium tin oxide (ITO). The surface of the ITO-coated films is cleaned with organic solvent and subsequently a UV ozone cleaner. The resistivity and surface smoothness of the ITO layers are very important. The resistivity of the present ITO layers is 95 Ω/sq. In the devices with ITO having higher resistivity, the current is limited by the ITO layer rather than that of the organic layers. Thus, the change of current induced by light is limited by the resistivity of the ITO. A 30-nm-thick p-type semiconductor of copper phthalocyanine (CuPc) and a 50-nm-thick n-type semiconductor of 3,4,9,10-perylene-tetracarboxylic-diimide (PTCDI) are deposited in a vacuum system. A 150-nm-thick gold layer is deposited in the cathode electrodes. Gold is chosen as the cathode electrode for the final structures because Au allows us to obtain reliable interconnection when using a laser drilling via process. The size of the cathode electrodes and periodicity of the photodiodes used to integrate with organic transistors are 450×450 and 700×700 μm^2, respectively. Some test structures containing the smaller photodiodes are manufactured under the same process conditions for comparison.

After using the fabrication process described above, both films with organic FETs and photodiodes are transferred to the vacuum chamber without exposure to air and uniformly coated with a 2-μm-thick poly-monochloro-para-xylylene (parylene) passivation layer. Spots of parylene on the electrodes are removed by a numerically controlled (NC) CO_2 laser drilling machine to form a via to create the electrical interconnections. Then, the two films are laminated with each other. For vertical interconnections, we used silver paste islands patterned by a microdispenser or anisotropic conductive films (Anisolm, Hitachi Chemical Co. Ltd.). The magnified image of four contact pads with silver paste islands before laminating the organic transistor films and organic diode films is shown in Fig. 7.6b, while the magnified image of four sensor cells integrated with the organic transistors and organic photodiodes is shown in Fig. 7.6c. The entire transistor region is covered by the photodiode within each pixel.

7.2.3 Electronic Performance of Organic Photodiodes

We report on electronic performance of stand-alone photodiodes, which are characterized under ambient environment with adequate device sealing or under nitrogen environment without sealing.

Figure 7.7 shows typical current–voltage characteristic of the manufactured organic photodiodes with 450×450 μm^2 gold cathode electrodes under illumination of light with different light intensities. The light source is a halogen lamp with a cold filter. The light intensity is changed from 0 to 175 mW/cm². In the case without light illumination, the threshold voltage of the forward-biased photodiodes is 4 V, while the breakdown voltage is -17 V. We plot the density of photocurrent with reverse bias of -4 V as a function of the light intensity in Fig. 7.7b. The current is linearly proportional to the light intensity up to 100 mW/cm².

One of the organic photodetectors is positioned on a sheet of white paper that has a black region printed by a laser printer. The device is illuminated uniformly from the top surface with light intensity of 40 mW/cm². Figure 7.8 shows that I–V curves measured on white and black parts. The photocurrent ratio of 8:1 is obtained at a voltage bias of -4 V.

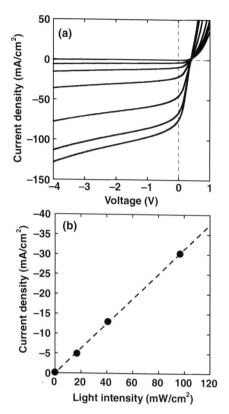

Fig. 7.7 (**a**) A typical current–voltage characteristic of the manufactured organic photodiodes with $450 \times 450\ \mu m^2$ gold cathode electrodes under illumination of light with different intensities. The light intensities are 0, 20, 40, 100, 140, 170, and 180 mW/cm². (**b**) Photocurrent density with reverse bias of –4 V is plotted as a function of the light intensity. From Fig. 7.7 in Ref. [18] ((c) 2008 IEEE)

The reduction of device dimensions is crucial to increase spatial resolution of these image scanners. We have prepared photodiodes with various sizes of gold cathode electrodes from $1 \times 1\ mm^2$ to $50 \times 50\ \mu m^2$. The photocurrent density is measured under illumination of light (70 mW/cm²) as shown in Fig. 7.9. When the device dimensions is reduced down to $50 \times 50\ \mu m^2$, photocurrent density decreases by only 25%, which is sufficient to achieve spatial resolution of 250 dpi.

7.2.4 Organic Transistors

Figure 7.10 shows typical characteristic of the manufactured pentacene transistor. The measured mobility is 0.7 cm²/Vs at an operating voltage of −60 V. The device failure is due to gate leakage. The initial yield strongly depends on the

Fig. 7.8 I–V characteristics of a 36 dpi-photodiode array without transistors are measured without light, with light on *black* and *white* regions. From Fig. 7.10 in Ref. [18] ((c) 2008 IEEE)

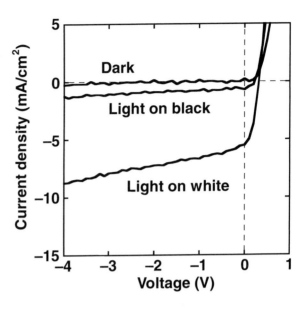

Fig. 7.9 Photocurrent density of the pn-photodiode with various sizes of gold cathode electrodes from 1×1 mm^2 to 50×50 μm^2. Measurements are performed under illumination of light of 70 mW/cm^2. From Fig. 7.12 in Ref. [18] ((c) 2008 IEEE)

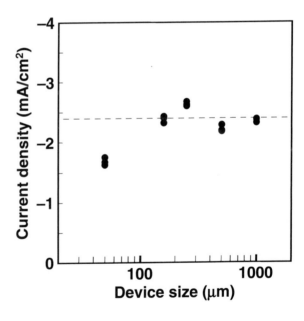

thickness of the polyimide gate dielectric layers. Lowering the operation voltage is not very difficult by introducing thinner gate dielectric layers or the use of high-k materials. The polyimide precursors are spun coat at a revolution rate of 4,500 rpm without dilution, the polyimide layers are 630 nm after the curing

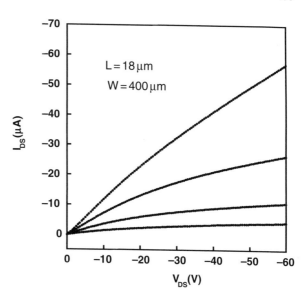

Fig. 7.10 Typical characteristic of the manufactured p-type organic transistor. A source–drain current (I_{DS}) is shown as a function of source–drain voltage (V_{DS}) swept from 0 to –60 V. A gate voltage (V_{GS}) is changed from 0 to –60 V with a step of –10 V

process. For the devices with 630-nm-thick polyimide gate dielectric layers, the initial yield exceeds 99%. We can easily reduce the thickness of gate dielectric layers if we dilute the source materials by n-methyl-2-pyrrolidone (NMP) and/or increase the revolution rate during spin coating. Transistors with a thinner gate dielectric (240 nm) have also been fabricated. These transistors are used for sheet-type Braille displays and will be described in the following section. The devices exhibit mobility above 0.1 cm^2/Vs and can be operated at the voltage less than 10 V. Since there is a trade-off between the operating voltage and the device yield, the thickness of the gate dielectric layers has to be designed appropriately.

7.2.5 Photosensor Cells

We measured one of the sensor cells consisting of one transistor and one photodetector under illumination of various intensities of light up to 70 mW/cm^2. Figure 7.11 shows a typical I–V characteristic of the sensor cell. When the gate-to-source voltage, $V_{GS} = -80$ V, is applied, the drain-to-source current, I_{DS}, is proportional to the light intensity up to 40 mW/cm^2 and then saturates at a light intensity of 60 mW/cm^2. Therefore, in order to obtain a high-contrast ratio between black and white, a maximum light intensity for white imaging of less than 40 mW/cm^2 is required.

We next prepared a 10×10 organic photodiode matrix without the organic transistors. The effective sensing area of each sensor cell is 50×50 μm^2, with a periodicity of 100 μm. The photocurrent dispersion for the photodiodes, with light illumination on the black and white areas at 80 mW/cm^2, is shown in Fig. 7.12.

Fig. 7.11 The electric performance of one sensor cell of the 36-dpi integrated device. The power supply V_{DD} is –2 V. I_{DS} versus light intensity with different V_{GS} bias. From Fig. 7.11 in Ref. [18] ((c) 2008 IEEE)

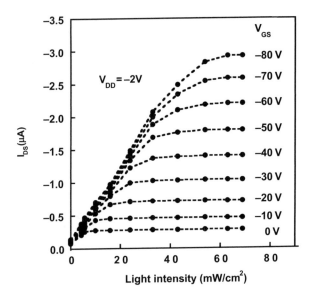

Fig. 7.12 The dispersion of hundred devices of the 250-dpi organic photodiode matrix without transistors. The effective size of each photodiode is $50 \times 50~\mu m^2$. From Fig. 7.16 in Ref. [18] ((c) 2008 IEEE)

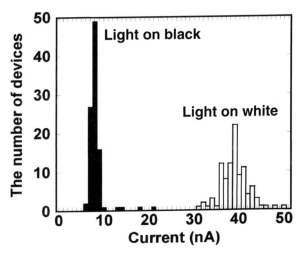

Although a fairly large distribution of photocurrent is observed, the performance distribution can be compensated and corrected by calibration of each sensor cell. Such a situation is very different from display applications which requires very high uniformity of each pixel.

We have positioned a sheet of paper with white capital letters of "U" and "O" prepared by a laser printer underneath the photodiode matrix and measured photocurrent of each detector with light illumination (80 mW/cm²). The photocurrent array mapping is shown in Fig. 7.13.

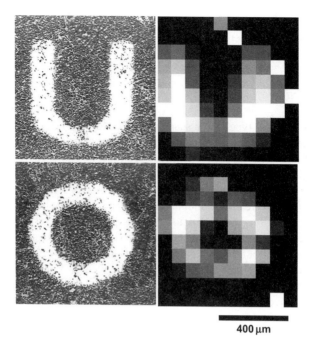

400 μm

Fig. 7.13 White capital letters of "U" and "O" prepared by a laser printer are placed onto the 250-dpi organic photodetector matrix without organic transistors. Photocurrent of each detector is measured under light (80 mW/cm^2). The mapping of normalized photocurrents is compared with an image taken by a commercial scanner with 250 dpi. The size of each image is 0.8 × 0.8 mm^2. From Fig. 7.17 in Ref. [18] ((c) 2008 IEEE)

7.2.6 Issues Related to Device Processes: Pixel Stability and Resolution

In addition to the intrinsic materials quality and performance, there are also issues related to device processing. The first issue is the reliability and stability of organic transistors and diodes. The performance of these devices without encapsulation layers changes over time (typically over a period of hours). This instability is the most stringent problem related to organic transistors, but it can be overcome by introducing adequate encapsulation layers similar to organic electroluminescent devices.

The second issue is the spatial resolution. Although the present resolution is 36 dpi with organic transistors, almost all the practical applications require at least 250 dpi resolutions. We have confirmed that an organic photodiode array can distinguish between black and white in reflection geometry up to 250 dpi, but the bottleneck for integrating organic transistor backplanes to the sensor media is the via process that allows the electrical coupling between the laminated films. The diameter of the via holes processed by a typical CO$_2$ laser drilling machine ranges between 50 and100 μm. However, we expect to reduce the size of the via holes and improve the pixel resolution with the use of shorter wavelength lasers such as pulsed-excimer lasers or yttrium aluminum garnet (YAG) lasers.

7.2.7 A Hierarchal Approach for Slow Organic Circuits

The drawback of sheet-type image scanners is the slow operating frequency due to the low-carrier mobility in organic transistors compared to inorganic TFTs. Thus, it is crucial to improve the device switching speed for practical use. We verified that a hierarchal approach, using a double-wordline and double-bitline structure, reduces the addressing delay by a factor of 5 and reduces the power consumption by a factor of 7. One approach to realize a double-wordline and double-bitline structure is by implementing a 3-D integrated stack to fabricate the organic scanner (Fig. 7.14).

A prototype array has been fabricated having 64×64 pixels that occupy an effective sensing area of 80×80 mm^2. The pixel size, in this case, was 1.27×1.27 mm^2, which corresponds to 20 dpi resolution. The supply voltage is 40 V, however, a voltage imposed between an anode and cathode of an organic photodiode should be limited to 5 V to avoid Zener avalanche breakdown of the photodiode.

7.2.8 The Double-Wordline and Double-Bitline Structure

The double-wordline and double-bitline structure is introduced in an organic transistor-based sheet-type scanner to reduce the line delays and the power consumption. Figure 7.15 shows the schematic of the array design. The 64×64 pixels are divided into 8×8 blocks so that each block has 8×8 pixels. Every pixel comprises an organic photodiode and pixel selector made of an organic transistor.

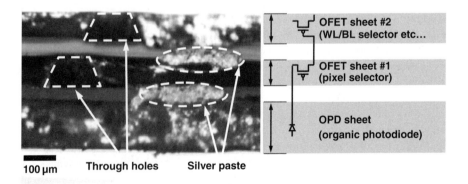

Fig. 7.14 A cross-sectional picture of stacked three organic sheets and corresponding circuit diagram. From Fig. 7.11 in Ref. [19] ((c) 2008 IEEE)

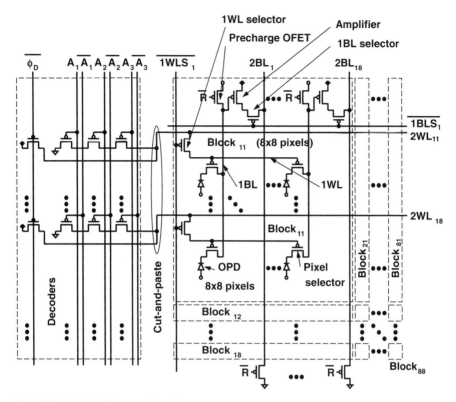

Fig. 7.15 A circuit diagram of the sheet-type scanner with the double-wordline and double-bitline structure. From Fig. 7.2 in Ref. [19] ((c) 2008 IEEE)

A first wordline (1WL) connects to a second wordline (2WL) through a first-wordline selector (1WL selector). A 1WL activates the gates of the pixel selectors to specify a local row address. A 1WL selector selects a 1WL with a first-wordline select signal ($1WLS_X$). A second-wordline decoder (2WL decoder) drives a 2WL, which will be in the following sections. A similar notation is used for the bitlines. A first bitline (1BL) is a local bitline in a block. A precharge gate precharges a 1BL with the precharge signal (R), and this signal predischarges a second bitline (2BL) before the readout operation. An amplifier amplifies a first-bitline voltage. A first-bitline selector (1BL selector) selectively transfers the amplified voltage to a 2BL with a first-bitline select signal ($1BLS_X$).

The decoder used in the scanner does not draw an active leakage current due to the dynamic operation. The switching OFETs are connected in series. This decoder is a ratioless circuit without the precharge OFET sized, and thus more tolerant of process, threshold-voltage, and supply-voltage variation/fluctuation than the conventional ratio-type one.

The double-wordline structure reduces wordline delay by a factor of 6. The double wordline structure can potentially reduce the dynamic power by the same factor

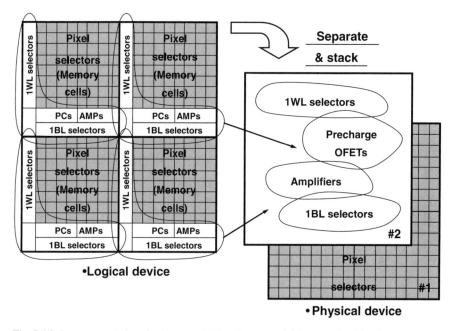

Fig. 7.16 In memory design, for instance, 1WL selectors are laid out on the side of memory cells and amplifiers are positioned at the bottom of the memory cells due to a logical device, however in a scanner, such kind of placement is not allowed to keep uniform distribution of pixels because it is a physical device and position is meaningful. From Fig. 7.8 in Ref. [19] ((c) 2008 IEEE)

as well as the wordline delay since the circuits operate on a block-by-block basis, where the capacitance associated with the operation is lower than the single word-line scheme. This becomes especially important when random access is employed for intelligent image capturing.

Although the hierarchical structure like the double-wordline and double-bitline structure is well known for memories, a situation is different for sensor applications. In memory applications, the first-wordline and sensor amplifiers are laid out by shifting the memory cells to the sides. This approach is because memory is a logical device, but for sensors, pixels cannot be shifted because the pixel density changes and uniform sensing becomes impossible (Fig. 7.16). We cannot rearrange the distribution of the pixel uniformly. Moreover, since the organic FET is large, only a single organic transistor is allowed per pixel and there is no room left for the peripheral circuits in the pixel region. Thus, the 1WL selector and sense amplifiers, the 1WL selectors, and the precharge organic transistors are placed on a separate sheet, #2, and stacked onto a pixel selector sheet, #1. The entire structure is thus laminated to form a 3-D stack, integrating all the components of the circuit. This design is why the development of a 3-D stacked organic transistor sheet is essential.

By introducing the double-bitline structure, a bitline can be divided into small segments and the bitline capacitance is reduced. In addition, due to the 3-D stack integration, an amplifier can be put near the segmented bitline. This arrangement means that a photodiode just draws charge out of a relatively small capacitance. In the future, we could improve the sensitivity of the photodiode in order to generate more photocurrent, making it possible to use ambient light as a light source.

7.2.9 A New Dynamic Second-Wordline Decoder

As a 2WL decoder, a serial-connected decoder is introduced for lower power and higher speed operation than the parallel-connected static one used in the electronic-skin (e-skin) [10–15]. In the e-skin application, switching organic FETs are connected in parallel, and the load transistor must be small because of the normally on load and sizing requirements. Thus, the falling time is slow, and in addition, bias voltage adjustment is needed to cancel the variability of the threshold voltage, which results in a leakage current in the order of micro-Amps. On the other hand, the new decoder does not draw active leakage due to its dynamic operation. The new decoder is a ratio-less circuit, and thus it has a wider margin of operation.

The layout style is novel and takes advantage of the "cut-and-paste" customization [12]. For example, five switching organic transistors connected in series are prepared in advance. Next, if only 1 out of 8 decoders having only 3 switching organic transistors are needed, then the three switching organic transistors can be removed from the prefabricated decoder and laminated to a 2WL pad as shown in Fig. 7.15. This "cut-and-paste" customization reduces the mask cost as well as a nonrecurring engineering cost because we do not need to design or make new masks for various sizes.

7.2.10 Higher Speed Operation with Lower Power Consumption

The measured operation waveforms are shown in Fig. 7.17 together with a sketch of stimulus signals. Successful operation is observed. Both the conventional single-wordline and single-bitline device and proposed double-wordline and double-bitline device are manufactured for comparison. Falling time of the 1BL from V_{DD} to 90% of V_{DD} in the proposed structure is 3 ms but in the conventional scheme is 17 ms. The wordline delay can be reduced by a factor of >5.

Now, we would like to explain the photocurrent-sensing scheme. After a 1BL is precharged and wordline is asserted, a photocurrent according to the light intensity discharges a first-bitline capacitance, and the first-bitline voltage starts decreasing. The falling time depends on the photocurrent. An amplifier amplifies the first-bitline voltage, which starts to pull-up the second-bitline voltage. Thus, the rising time of the 2BL is a function of the photocurrent, by which black and white pixels are discriminated. This scheme is a kind of current-to-time conversion.

When the sense voltage is set to 30 V, the readout time in the conventional single-bitline scheme is 18 ms while that in the proposed double-bitline structure is 3 ms, achieving a factor of 6 improvement in the wordline delay. In the

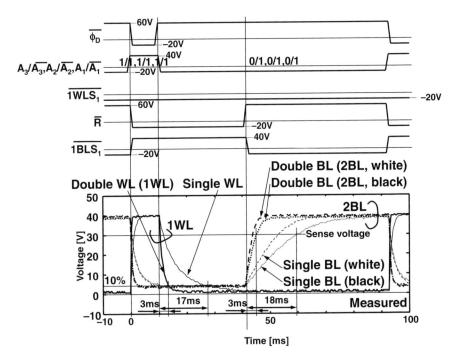

Fig. 7.17 Comparison of measured waveforms between "single-wordline and single-bitline scheme" and "double-wordline and double-bitline structure". Falling times of the word lines are 17 and 3 ms, respectively, realizing the wordline activation time by a factor of 5. From Fig. 7.15 in Ref. [19] ((c) 2008 IEEE)

conventional single-wordline and single-bitline scheme, a cycle time of 39 ms is typical while in the proposed double-wordline and double-bitline structure the cycle time is reduced to 7 ms, a cycle time reduction by a factor of 5. Measured power for the proposed structure is 900 μW for a 7 ms cycle and 350 μW for a 39 ms cycle. The conventional scheme has a measured power of 2.5 mW for a 39 ms cycle, which is seven times larger than the proposed structure. Although the improvement factors in delay and power consumption are not exactly the same as 8 (8/64), this can be ascribed to the effects of capacitance of the additional transistors needed for the hierarchal structures.

7.2.11 New Applications and Future Prospects

The new scanner is thin, lightweight, and flexible. The present scanner is suitable for mobile electronics and could be easily carried in a pocket. Beyond the portability feature, there would be unique applications of the new scanner. For example, it can be bent to conform over a printed page in book-scanning applications. It would be also suitable for the recording of fragile and historically valuable documents. A label affixed on a curved surface, such as a bottle of wine, could also be accurately and conveniently scanned.

In the future, it is predicted that the number of pixels per wordline and bitline will be increased to more than 2,048, and pixel size reduced to less than 1/16 of its current size. In such a large-area and high-resolution scanner, the double-wordline and double-bitline structure further reduces delay and power as well. The scan-out time for the conventional single-wordline and single-bitline scheme will be an order of 10^3 s while the proposed double wordline and bitline scheme is estimated to reduce this time by a factor of about 40, taking the scan time down to 10 s. The power consumption can also be decreased by the same factor. The proposed approach can be applicable to other types of large-area organic transistors sensors including e-skin and solves fundamental issues in large-area sensor electronics.

7.3 Sheet-Type Braille Displays

In this section, we describe a sheet-type Braille display on a plastic film by integrating organic FETs with soft actuators. An array of rectangular plastic actuators is processed from a perfluorinated polymer electrolyte membrane. A small semisphere, which projects upward from the rubber-like surface of the display, is attached to the tip of each rectangular actuator. The effective display size is 4×4 cm^2. Each letter consists of 3×2 Braille dots and the total number of dots is 144; thus, 24 Braille letters can be displayed. The total thickness and weight of the entire device are 1 mm and 5.3 g, respectively. The present scheme will enable people with visual impairments to carry the Braille sheet display in their pockets and read Braille e-books at any time.

7.3.1 Manufacturing Process

The sheet-type Braille displays are manufactured by laminating three layers: (1) the organic transistor sheet, (2) the polymeric actuator sheet, and (3) the cover layer. Since all the materials except the metal electrodes are made of soft materials, the entire system is thin, lightweight, and mechanically flexible (Fig. 7.18). A picture of a sheet-type Braille display is shown in Fig. 7.19a. Some sections of the layers have been intentionally removed in order to reveal the internal three-layer structure, although the sizes of all three layers are identical. Figure 7.19b shows a circuit diagram of the Braille display.

Figure 7.20 shows the cross-sectional structure of a single Braille cell comprising one transistor and one actuator. When a voltage is applied to the polymeric actuators, the sheet-type actuators bend, as shown in Fig. 7.20. The hemisphere placed on the actuator rises with the voltage supply and pushes up a rubber-like surface. An organic transistor active matrix is used to address the pop-up dots.

Fig. 7.18 An image of the Braille sheet display which is manufactured on a plastic film integrating the active matrix of organic transistors with polymer actuator array based on a perfluorinated polymer electrolyte membrane. The device is a thin, lightweight, and flexile sheet Braille display. From Fig. 7.1 in Ref. [23] ((c) 2008 IEEE)

The manufacturing processes of organic transistors are similar to the methods described in Section 7.2. Polyimide or PEN films are used as substrates for the organic transistors. Figure 7.21 shows images of the organic transistor sheet of different magnifications: (a) the entire chip and (b) one transistor. The channel length L and the width W of the transistors are 20 μm and 49 mm, respectively, which corresponds to a W/L ratio of 2,450. This large W/L ratio is required to ensure a good time response of the actuators. The size of the entire active matrix is 4×4 cm^2, while that of a single transistor is 1.5×1.5 mm^2.

The sheet-type actuators are made of an ionic polymer metal composite (IPMC) [21]. By employing electroless plating, electrodes are formed on both surfaces of a 300-μm-thick perfluorinated ion-exchange membrane, Nafion (NE-1110, DuPont) [22]. The actuator sheet is immersed in lithium chloride solution to exchange the protons inside the membrane with lithium ions. This process facilitates a large displacement, high speed, and a large generating force. As shown in Fig. 7.22, the actuator sheet is mechanically processed using a NC cutting machine to form an array of 12×12 rectangular actuators whose size is 1×4 mm^2. Each rectangular actuator remains connected to the parent sheet while being isolated electrically. For the isolation, insulating grooves are prepared on one side of the actuator sheet by using an NC drilling machine and the other side is not processed and is used as a common electrode. Plastic hemispheres with a radius of 0.9 mm are attached on the top of the actuators.

Finally, the transistor and actuator sheets are laminated together. The via interconnections between the electrode pads of the transistors and the surface electrodes of the actuators are realized by anisotropic conductive tapes or silver pastes. The laminated sheets are covered by a plastic frame made of 650-μm-thick

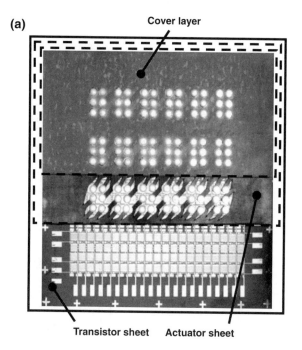

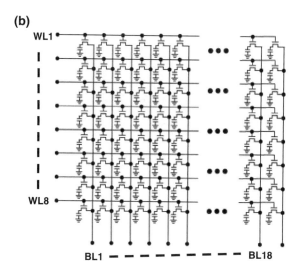

Fig. 7.19 (**a**) A picture of the device assembly. Some parts of the device are removed intentionally to show inner structures of this device. This device is composed of three layers, an organic transistor sheet, a polymeric actuator sheet, and a cover layer. (**b**) Circuit diagram of the Braille sheet display. Each polymeric actuator is connected to one organic transistor. The *vertical* and *horizontal* lines represent bit and word lines, respectively. From Fig. 7.2 in Ref. [23] ((c) 2008 IEEE)

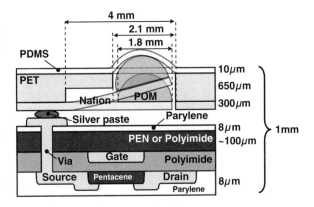

Fig. 7.20 A cross-sectional illustration of a single Braille dot of this device. An organic transistor is connected to a polymeric actuator with silver paste patterned by a microdispenser. A semisphere is attached to the tip of each actuator. From Fig. 7.3 in Ref. [23] ((c) 2008 IEEE)

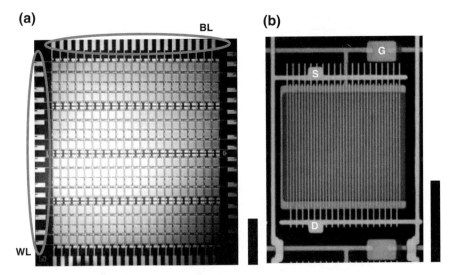

Fig. 7.21 (a) Pictures of a whole of the organic transistor active matrix sheet, which includes 144 transistors. The scale is 1 cm. (b) A further magnified view of a single transistor. G, D, and S indicate gate, drain, and source electrodes, respectively. The scale is 1 mm. From Fig. 7.4 in Ref. [23] ((c) 2008 IEEE)

poly(ethylenentereapthalate) (PET) whose surface is coated by a polydimethylsiloxane (PDMS) film. The PDMS layer is fluorinated to obtain a smooth surface.

7.3.2 Electronic Performance of Braille Cells

The characteristics of discrete transistors are investigated prior to integration. All electrical measurements were performed in air by using a semiconductor parameter analyzer. The typical IV characteristics of the organic transistors are shown

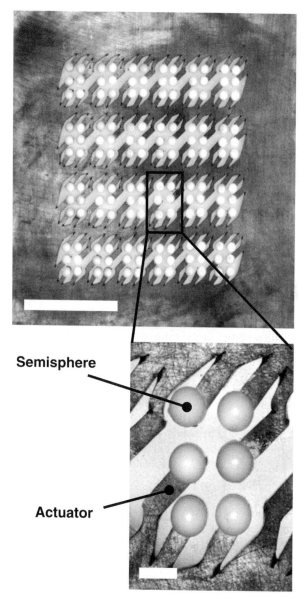

Fig. 7.22 Picture of whole sheet of polymeric actuators (12 × 12 array) and magnified image of actuators for one letter (3 × 2 dots). The scale for the picture of whole sheet and magnified image are 2 cm and 2 mm, respectively. From Fig. 7.5 in Ref. [23] ((c) 2008 IEEE)

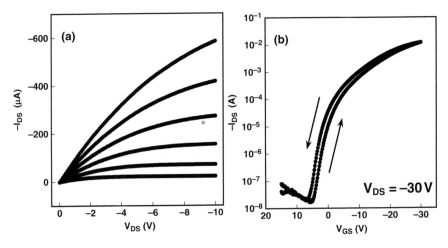

Fig. 7.23 (a) V_{DS}–I_{DS} characteristics of the organic transistors measured by gate voltage bias V_{GS} from 0 to –10 V with a step of –2 V. (b) V_{GS}–I_{DS} characteristic measured by $V_{DS} = -30$ V and gate voltage bias V_{GS} from 15 to –30 V. From Fig. 7.6 in Ref. [23] ((c) 2008 IEEE)

in Fig. 7.23a,b. The mobility in the saturation regime is 1 cm^2/Vs and the on–off ratio is 10^6. In order to reduce the operation voltage to 10 V, the thickness of the gate dielectric layer is set to a value as small as 240 nm. A large current (-600 μA) is obtained at low voltages ($V_{GS} = V_{DS} = -10$ V) for the organic transistors.

One of the stand-alone actuators (1 × 4 mm^2) is characterized before its integration with the organic transistors. The time response of the actuators is measured. The rectangular voltage of ±3 V is applied to the actuator at a repetition rate of 2 Hz. Figure 7.24a,b shows the voltage between the two actuator electrodes and the displacement of the actuators, respectively, as a function of time. The displacement and the generating force of the actuators are measured as a function of voltage for three actuators with identical sizes of 1 × 4 mm^2 and plotted in Fig. 7.25a,b, respectively. The force was measured using a load cell. The displacement and the generating force increase with the applied voltage and peak at 0.4 mm and 1.5 gf, respectively, at 2.5 V. Although the performance of the actuators depends on their structural parameters such as width, length, and thickness, the present design using actuators with a size of 1 × 4 mm^2 exhibits a good performance that is suitable for application in Braille sheet displays.

Characterization of the Braille cells is described next. The time response of the actuators is measured at different gate voltages – $V_{GS} = 0, -10, -20$, and -30 V. The power supply voltage (V_{DD}) is -10 V for the up states and 10 V for the down states. Figure 7.26a,b shows the voltage between two actuator electrodes and the displacement of the actuators, respectively, as a function of time. The time required to obtain a displacement of 0.2 mm decreases with the increase in V_{GS} and reduces to 0.9 s at $V_{GS} = -30$ V, thus indicating that a frame rate of 1 Hz would be feasible. Figure 7.27a shows one of the Braille dots moving upward and downward. Four

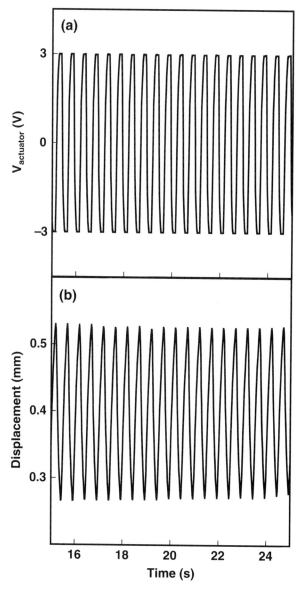

Fig. 7.24 A stand-alone polymeric actuator of 4 mm in length and 1 mm in width is measured. When a series of rectangular voltage of ±3 V is input, the frequency response of the actuators extends up to 2 Hz. The voltage between two electrodes of actuator and displacement of actuator is shown in (**a**) and (**b**) in the same time scale, respectively. From Fig. 7.7 in Ref. [23] ((c) 2008 IEEE)

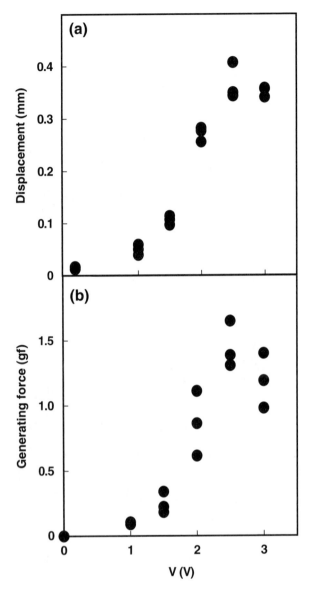

Fig. 7.25 (a) The displacement and (b) the generating force of an actuator are plotted as a function of the input voltage. From Fig. 7.7 in Ref. [23] ((c) 2008 IEEE)

Fig. 7.26 The displacement of a Braille cell when gate voltage bias $V_{GS} = 0, -10, -20,$ and -30 V is measured as a function of time. The power supply voltage (V_{DD}) is rectangular voltage of ± 10 V. The voltage between two electrodes of actuator and displacement of actuator is shown in (**a**) and (**b**), respectively. From Fig. 7.8 in Ref. [24] ((c) 2008 IEEE)

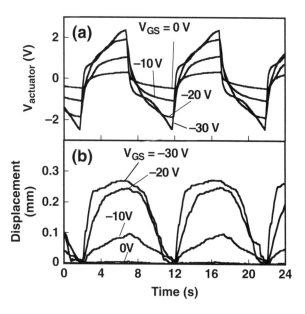

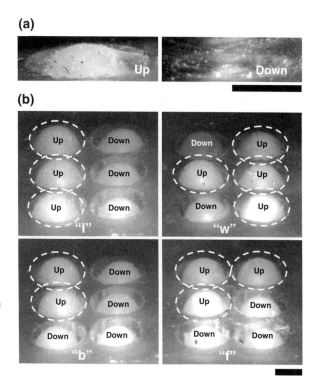

Fig. 7.27 (**a**) Magnified pictures of one Braille dot moving *upward* and *downward*. The scale is 1 mm. (**b**) Pictures of Braille sheet display showing the characters "l," "w," "b," and "f" in the American Braille style. From Fig. 7.9 in Ref. [24] ((c) 2008 IEEE)

Braille letters displayed by the present device are shown in Fig. 7.27b. We will describe the readability of the present device later.

7.3.3 Organic Transistor-based SRAM

Each Braille dot has an organic transistor static random access memory (SRAM) to compensate for the slow transition of the actuator. The transition time of the soft actuator is about 1 s. When an actuator array is sequentially driven, it takes more than 1 min to change the present Braille displays. Organic FET SRAM techniques are employed to increase the speed of the actuator.

Figure 7.28 shows the circuit of the SRAM and the driver for one actuator. In the SRAM cell, a foremost concern is a slow write-time of DATAb, because DATAb has no access transistors. Our design target for the write-time of the whole SRAM (= 144 cells) is within 2 s. Figure 7.29 shows the measured waveforms of DATA and DATAb during a write-operation. When BL is low, the transition time of DATAb is 2 ms. In contrast, when BL is high, the transition time of DATAb is 40 ms, because the drive current of M1 in Fig. 7.28 is small. This slow transition time can be hidden in the SRAM system level by pipelining the write-operation. After the input data are written to all SRAMs within 2 s, all the actuators are driven all at once using the

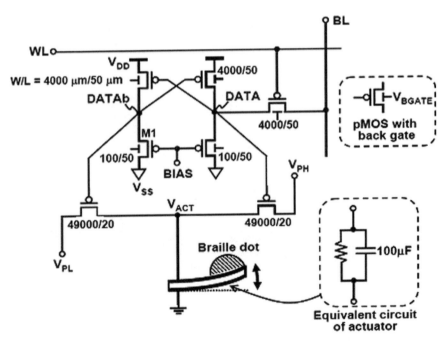

Fig. 7.28 Circuit of organic transistor SRAMs and the driver for one actuator. From Fig. 7.4 in Ref. [24] ((c) 2008 IEEE)

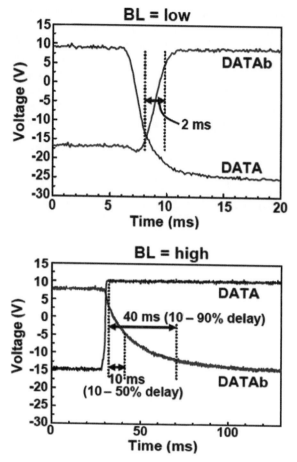

Fig. 7.29 The measured waveform of the Braille cell in the write-operation mode. From Fig. 7.6 in Ref. [24] ((c) 2008 IEEE)

drivers depending on the data. In this way, the time required to change the whole Braille cell is reduced from 144 to 3 s, which satisfies our design target.

7.3.4 Reading Tests

The readability of the present Braille display is examined by four visually impaired individuals. When the device exhibits "Na" and "Wa" in the Japanese Braille format, all four individuals are able to recognize the letters correctly. This result demonstrates the feasibility of this new design that integrates organic transistors and polymer actuators for realizing Braille sheet displays.

However, it is known that the ability of reading Braille varies across the visually impaired and, in general, a larger displacement and a larger force should help them in reading Braille more easily. In particular, the variations in the displacement and the generating force of the Braille dots must be minimized for better readability, although the present device shows fairly large performance variations. These variations may be ascribed to the unevenness in the electroless plating process and/or the fluctuations in the size of the actuators that arise due to the inaccuracy of the mechanical process. These imperfections would be removed by optimizing the process conditions.

7.3.5 Future Prospects

The new design in this study offers an attractive scheme for reducing the thickness and weight of Braille displays and yet maintains a reasonable performance. Commercial Braille displays that utilize piezoelectric or solenoid actuators are available. Such actuators can control displacement with a high accuracy and also obtain large forces; however, their miniaturization is complicated. The proposed lightweight, thin, and flexible sheet-type Braille displays can be easily carried in pockets; therefore, they are suitable for mobile applications such as Braille e-books. It is expected that the sheet-type Braille can be easily implemented in many digital and information appliances including cell phones without necessitating major changes in the design of the parent appliances. Thus, Braille will be more conveniently used by the visually impaired in various situations in the future.

7.4 Summary

This chapter reported on large-area, flexible sensors and actuators based on organic transistors. As an example of large-area sensors, a sheet image scanner integrating organic transistors and organic photodetectors was described. The new sheet-type scanner with light shielding layers can distinguish between black and white in the reflection geometry. Then, a sheet-type Braille display was reported as an example of large-area, flexible actuators. The new Braille display was fabricated on a plastic film by integrating high-quality organic transistors and soft sheet-type actuators. In those two types of devices, all the components are manufactured on plastic films. Therefore, it is lightweight, shock-resistant, and flexible and is suitable for human-friendly mobile application.

Acknowledgements The research work in this chapter was carried in collaboration with Prof. Takayasu Sakurai, Prof. Makoto Takamiya, Dr. Tsuyoshi Sekitani, Dr. Hiroshi Kawaguchi, Yusaku Kato, and Yoshiaki Noguchi.

References

1. Drury CJ, Mutsaers CMJ, Hart CM, Matters M, de Leeuw DM (1998) Low-cost all-polymer integrated circuits. Appl Phys Lett 73:108–110
2. Crone B, Dodabalapur A, Lin YY, Filas RW, Bao Z, LaDuca A, Sarpeshkar R, Katz HE, Li W (2000) Large-scale complementary integrated circuits based on organic transistors. Nature 403:521–523
3. Gelinck GH, Geuns TCT, de Leeuw DM (2000) High-performance all-polymer integrated circuits. Appl Phys Lett 77:1487–1489
4. Gelinck GH, Huitema HEA, van Veenendaal E, Cantatore E, Schrijnemakers L, van der Putten JBPH, Geuns TCT, Beenhakkers M, Giesbers JB, Huisman BH, Meijer EJ, Benito EM, Touwslager FJ, Marsman AW, van Rens BJE, de Leeuw DM (2004) Flexible active-matrix displays and shift registers based on solution-processed organic transistors. Nat Mater 3: 106–110
5. Huitema E, Gelinck G, van der Putten B, Cantatore E, van Veenendaal E, Schrijnemakers L, Huisman BH, Leeuw DM (2003) Plastic transistors in active-matrix displays. ISSCC Dig. Tech. Papers, pp 380–381
6. Brederlow R, Briole S, Klauk H, Halik M, Zschieschang U, Schmid G, Gorriz-Saez JM, Pacha C, Thewes R, Weber W (2003) Evaluation of the performance potential of organic TFT circuits. ISSCC Dig. Tech. Papers, pp 378–379.
7. Rogers JA, Bao Z, Baldwin K, Dodabalapur A, Crone B, Raju VR, Kuck V, Katz H, Amundson K, Ewing J, Drzaic P (2001) Paper-like electronic displays: Large-area rubber-stamped plastic sheets of electronics and microencapsulated electrophoretic inks. Proc Natl Acad Sci USA 98:4835–4840
8. Baude PF, Ender DA, Haase MA, Kelley TW, Muyres DV, Theiss SD (2003) Pentacene-based radio-frequency identification circuitry. Appl Phys Lett 82:3964–3966
9. Baude PF, Ender DA, Kelley TW, Haase MA, Muyres DV, Theiss SD (2003) Organic semiconductor RFID transponders. IEDM Tech. Dig., pp 191–194.
10. Someya T, Sekitani T, Iba S, Kato Y, Kawaguchi H, Sakurai T (2004) A large-area, flexible pressure sensor matrix with organic field-effect transistors for artificial skin applications. Proc Natl Acad Sci USA 101:9966–9970
11. Someya T, Kato Y, Sekitani T, Iba S, Noguchi Y, Murase Y, Kawaguchi H, Sakurai T (2005) Conformable, flexible, large-area networks of pressure and thermal sensors with organic transistor active matrixes. Proc Natl Acad Sci USA 102(35):12321–12325
12. Kawaguchi H, Someya T, Sekitani T, Sakurai T (2005) Cut-and-paste customization of organic FET integrated circuit and its application to electronic artificial skin. IEEE J Solid-State Circ 40:177–185
13. Someya T, Kawaguchi H, Sakurai T (2004) Cut-and-paste organic FET customized ICs for application to artificial skin. ISSCC Dig. Tech. Papers, pp 288–289
14. Someya T, Sakurai T (2003) Integration of organic field-effect transistors and rubbery pressure sensors for artificial skin applications. IEDM Tech. Dig., pp 203–206
15. Someya T, Sakurai T (2006) Large-area detectors and sensors. In: Klauk H (ed) Organic electronics. Wiley-VCH, Weinheim, pp 395–410
16. Someya T, Iba S, Kato Y, Sekitani T, Noguchi Y, Murase Y, Kawaguchi H, Sakurai T (2004) A large-area, flexible, and lightweight sheet image scanner. IEDM Tech. Dig., pp 580–581
17. Kawaguchi H, Iba S, Kato Y, Sekitani T, Someya T, Sakurai T (2005) A sheet-type scanner based on a 3D-stacked organic-transistor circuit using double wordline and bitline structure. ISSCC Tech. Dig., pp 365–368
18. Someya T, KatoY, Iba S, Kawaguchi H, Sakurai T (2005) Integration of organic field-effect transistors with organic photodiodes for a large-area, flexible, and lightweight sheet image scanner. IEEE Trans Electron Dev 52(11):2502–2511

19. Kawaguchi H, Iba S, KatoY, SekitaniT, Someya T, Sakurai T (2006) A 3D-stack organic sheet-type scanner with double-wordline and double-bitline structure. IEEE Sensors J 6(5): 1209–1217
20. KatoY, Iba S, Teramoto R, Sekitani T, Someya T, Kawaguchi H, Sakurai T (2004) High mobility of pentacene field-effect transistors with polyimide gate dielectric layers. Appl Phys Lett 84:3789–3791
21. Shahinpoor M, Bar-Cohen Y, Simpson JO, Smith J (1998) Ionic polymer-metal composites (IPMCs) as biomimetic sensors, actuators and artificial muscles – a review. Smart Mater Struct 7:R15–R30
22. Fujiwara N, Asaka K, Nishimura Y, Oguro K, Torikai E (2000) Preparation of gold-solid polymer electrolyte composites as electric stimuli-responsive materials. Chem Mater 12: 1750–1754
23. Kato Y, Sekitani T, Takamiya M, Doi M, Asaka K, Sakurai T, Someya T (2007) Sheet-type Braille displays by integrating organic field-effect transistors and polymeric actuators. IEEE Trans Electron Dev 54(2):202–209
24. Takamiya M, Sekitani T, Kato Y, Kawaguchi H, Someya T, Sakurai T (2007) An organic FET SRAM with back gate to increase static noise margin and its application to Braille sheet display. IEEE J Solid-State Circ 42(1):93–100

Chapter 8
Organic and Polymeric TFTs for Flexible Displays and Circuits

Michael G. Kane

8.1 Introduction

Organic and polymeric thin-film transistors are a natural complement to flexible substrates. This is because organic thin-film transistors (OTFTs) can be made using a very low temperature process, not much above room temperature, allowing electronic circuits and systems to be made on plastic films.[1] The organic and polymeric materials that can be used as semiconductors, dielectrics, and conductors are themselves flexible like the substrate, so that a complete flexible electronic system is possible.

There are several benefits of building an electronic system on a flexible substrate. In some applications, there may be advantages to mechanical flexibility in actual use, as with a rugged, unbreakable, rollable display. In other cases, flexibility may not be needed in the application, but benefits may derive from the lower manufacturing costs of continuous roll-to-roll fabrication. For example, low-cost displays fabricated in a web-based line might end up laminated to a hard, rigid material, which may be curved or flat, glass or plastic. There may also simply be cost advantages associated with the use of the organic or polymeric materials. The fact that OTFTs can be fabricated at low temperatures leads to lower manufacturing costs. Manufacturing cost depends on process temperatures, because in general higher-temperature processing entails higher capital costs, more expensive substrate materials, and lower throughput because of the time required for temperature ramping. Furthermore, if additive, printing-like processes are used for some or all of the layers, the costs associated with materials and photolithography, two of the most expensive components of thin-film transistor (TFT) manufacturing, are reduced or eliminated.

M.G. Kane (✉)
Sarnoff Corporation, CN5300, Princeton, NJ 08543, USA
e-mail: mkane@sarnoff.com

[1]For simplicity we use the term *organic thin-film transistor* and the acronym OTFT to represent both small-molecule organic and polymer thin-film transistors.

W.S. Wong, A. Salleo (eds.), *Flexible Electronics: Materials and Applications*, Electronic Materials: Science & Technology, DOI 10.1007/978-0-387-74363-9_8,
© Springer Science+Business Media, LLC 2009

The most compelling near-term application for OTFTs on flexible substrates is active matrix displays. Nevertheless, a case can also be made for using the technology in nondisplay applications. Various types of imaging and sensing arrays can benefit from mechanical flexibility. In addition, types of electronic systems other than pixel arrays may have advantages when fabricated on a flexible substrate. There is a very large potential market for very low-cost radio frequency identification (RFID) tags for item-level tracking. Significant efforts are underway to develop an RFID technology that can be manufactured in a continuous roll-to-roll line to reduce tag cost.

In this chapter, we begin by discussing several transistor parameters that are important for using OTFTs in electronic systems. Then we examine a number of applications for OTFTs on flexible substrates. The greatest amount of attention is given to active matrix displays, the nearest-term application. We consider active matrix liquid crystal displays (AMLCDs), active matrix electrophoretic displays, and active matrix organic light-emitting diode (AMOLED) displays. In addition, we look at using OTFTs for more general circuit applications, such as integrated display drivers and RFID tags.

8.2 Important Organic TFT Parameters for Electronic Systems

The field-effect mobility, μ_{FE}, is often cited as a performance metric for comparing different OTFT materials and fabrication methods. But the field-effect mobility is only one of several parameters that are important for using OTFTs in electronic systems. In this section, we discuss the field-effect mobility as well as other important parameters.

8.2.1 Field-Effect Mobility

In the standard metal–oxide–semiconductor field-effect transistor (MOSFET) drain-current equations, μ_{FE} is a proportionality factor that relates the drain current I_D to the gate and drain voltages V_{GS} and V_{DS}, the threshold voltage V_t, the channel width W and length L, and the gate dielectric capacitance per unit area C_{ox}. The standard MOSFET drain-current equation for an n-channel device in the linear region of operation ($V_{DS} < V_{GS}-V_t$) is:[2]

$$I_D = \frac{W}{L}\mu_{FE}C_{ox}\left[(V_{GS} - V_t)V_{DS} - V_{DS}^2/2\right], \qquad (8.1a)$$

---

[2]For p-channel MOSFETs, all applied voltage polarities and current directions are reversed.

which reduces to

$$I_D = \frac{W}{L}\mu_{FE}C_{ox}(V_{GS} - V_t)\,V_{DS} \tag{8.1b}$$

when V_{DS} is small ($V_{DS} << V_{GS}-V_t$). In the saturation region of operation ($V_{DS} \geq V_{GS}-V_t$), the standard equation is:

$$I_D = \frac{W}{L}\frac{\mu_{FE}C_{ox}}{2}(V_{GS} - V_t)^2. \tag{8.1c}$$

Experimentally, one finds different values for μ_{FE} in the linear and saturation regions. Therefore, a distinction is often made between the linear and saturation field-effect mobilities. In this chapter, we do not distinguish between the two. However, if the two have significantly different values, a careful circuit designer will do so.

The field-effect mobility is distinct from the more physically fundamental *carrier mobility* μ, which has the same units but relates average carrier velocity v to applied electric field E through $v = \pm\ \mu E$, where the positive sign is used for holes and the negative sign for electrons. Carrier mobility is only one factor in the field-effect mobility. In this chapter, we use the term *mobility* to refer to the field-effect mobility.

Experimentally, the mobility is found to depend on V_{GS} and V_{DS}. Therefore, it is more straightforward to view it as a small-signal bias-dependent quantity, analogous to the small-signal gain of an amplifier rather than the large-signal quantity of Eq. (8.1a, b, c). Then at each bias point mobility is a proportionality factor that relates how small changes in gate voltage V_{GS} produce changes in drain current I_D. In this case, in the linear region, for small drain voltages ($V_{DS} << V_{GS}-V_t$), mobility is defined as

$$\mu_{FE} = \frac{L}{W C_{ox} V_{DS}}\frac{\partial I_D}{\partial V_{GS}}, \tag{8.2a}$$

and in the saturation region mobility is defined as

$$\mu_{FE} = \frac{2L}{W C_{ox}}\left[\frac{\partial\sqrt{I_D}}{\partial V_{GS}}\right]^2. \tag{8.2b}$$

The resulting values for mobility cannot be indiscriminately plugged into Eq. (8.1a, b, c). Figure 8.1 shows the bias-dependent mobility in the saturation region for a p-channel pentacene OTFT, derived from measurements using Eq. (8.2b).

There have been efforts to model the bias dependence of the mobility. The best-known model is based on the variable-range hopping (VRH) model of transports in which carriers move by hopping in an exponential density of localized states [1]. In this model, the physical basis for the increase in mobility with gate voltage is that, as the gate voltage is increased, lower-lying states in the semiconductor's density of

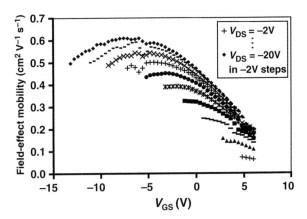

Fig. 8.1 Measured voltage-dependent mobility in the saturation region for a pentacene OTFT with SiO_2 gate dielectric. The mobility is plotted only for data points in the saturation region

states are filled, so that additional charges occupy higher-energy states and require less energy to hop to nearby states. This model leads to a dependence of mobility on gate voltage of the form:

$$\mu_{FE} = \mu_0 \left(\frac{V_{GS} - V_t}{V_{AA}} \right)^{\gamma} \tag{8.3}$$

where γ depends on the energy distribution of localized states, and typically has a value of about 0.5. The parameter V_{AA} serves to render the factor in parentheses dimensionless so that μ_0 has standard units for mobility. V_{AA} can be arbitrarily set to 1 V if μ_0 is defined as the mobility at $V_{GS}-V_t = 1$ V. The gate-voltage dependence of the mobility in Eq. (8.3) results in the linear dependence of drain current on gate voltage in Eq. (8.1b) being replaced by a superlinear dependence having an exponent $1+\gamma$. Similarly, the quadratic dependence in the saturation region in Eq. (8.1c) is replaced by a dependence with an exponent $2+\gamma$.

Strictly speaking, the VRH transport model applies only to amorphous materials and is not applicable to more ordered organic semiconductors, such as vacuum-sublimed pentacene. Nevertheless Eq. (8.3) is often used as an empirical, curve-fitting model for the mobility of TFTs made with nonamorphous organic semiconductors. For example, it can be used to fit the data in Fig. 8.1 over a limited gate-voltage range, but not over the entire range, since the mobility decreases at higher gate voltages. Furthermore, the dependence of mobility on drain voltage seen in Fig. 8.1 is not treated in this model. A similar model has been used for the gate-voltage-dependent mobility in amorphous silicon thin-film transistors (a-Si TFTs) and is used in the Level 15 a-Si TFT model in the circuit simulation program AIM-SPICE. Methods have been developed to extract γ from measured TFT data [2, 3, 4]

Usually, a single value for mobility is cited rather than a set of curves or mobility model parameters. In this case, the convention is to report the maximum mobility over a range of gate voltages for a specified drain voltage, either in the linear region or in the saturation region. A similar method of specifying mobility is used for the bias-dependent mobility of silicon MOSFETs. For creating an initial circuit design prior to computer-aided simulations that use more accurate models, it is best to estimate an average mobility over the expected range of voltages from curves like those in Fig. 8.1, rather than using the maximum mobility value, which represents best-case device behavior at a single bias point.

8.2.2 Threshold Voltage

Like the mobility, the starting point for defining the threshold voltage V_t is the standard MOSFET equations in Eq. (8.1a, b, c). Experimentally, one finds different values for the threshold voltage in the linear and saturation regions of operation. In addition, there may be a drain-voltage dependence. Here, we will neglect these effects and refer simply to a single threshold voltage V_t.

The threshold voltage can be extracted from measurements in the linear region by plotting I_D versus V_{GS} for small drain voltages ($V_{DS} \ll V_{GS}-V_t$), and extrapolating the line to $I_D = 0$. Similarly, the threshold voltage can be extracted from measurements in the saturation region by plotting $\sqrt{I_D}$ versus V_{GS} and extrapolating to $I_D = 0$. In either case, it is found experimentally that the drain current turns off more gradually than the standard MOSFET equations predict. One reason for this is that the gate-voltage-dependent mobility decreases as V_{GS} approaches V_t. Another reason is that there is a subthreshold region for $V_{GS} < V_t$ in which a small drain current still flows. As a result, typically the extrapolation to $I_D = 0$ is performed using a tangent to the curve at the point of maximum slope, or using two points on the curve that bracket the operating region. Figure 8.2 shows measured data for the same pentacene OTFT as in Fig. 8.1, but in the linear region ($V_{DS} = -0.1$ V). The straight line in Fig. 8.2 is tangent to the drain-current curve at the point of maximum slope, allowing the linear-region threshold voltage to be calculated as +6.5 V.

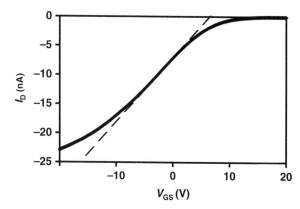

Fig. 8.2 Measured linear-region drain-current data (solid line, $V_{DS} = -0.1$ V) and the tangent at the point of maximum slope (*dashed line*) for the same pentacene OTFT as in Fig. 8.1. The tangent allows the threshold voltage to be calculated as +6.5 V

Threshold voltages in OTFTs are typically large and uncontrolled, and values as high as tens of volts are not uncommon. This is not well understood at present, though it may result from a high (and uncontrolled) density of trapped charge in the highly defected organic semiconductor. For electronic applications, low threshold voltages are required for keeping power supply and drive voltages low. The importance of low threshold voltages for the usefulness of OTFTs is often neglected. The application of self-assembled monolayers such as octadecyltrichlorosilane to the gate dielectric before semiconductor deposition helps to reduce the magnitude of V_t and render it more controllable. Using a gate dielectric with higher capacitance also reduces the threshold voltage, although often at the expense of higher gate leakage and lower yield.

8.2.3 Subthreshold Swing

The simplified physical models underlying Eq. (8.1a, b, c) predict that the drain current is zero for all $V_{GS} \leq V_t$. In reality, as Fig. 8.2 indicates, these equations do not correctly represent the drain current in the subthreshold region, where the power-law dependence of the drain current on gate voltage makes a transition to an exponential dependence. The standard drain-current equation in the subthreshold region has the form:

$$I_D = \frac{W}{L} K \mu_{FE} C_{ox} \left(1 - e^{-qV_{DS}/kT}\right) e^{qV_{GS}/nkT} \tag{8.4}$$

where K is a constant that depends on materials and device structure, n is the ideality factor, k is Boltzmann's constant, and T is the absolute temperature. Thermodynamics demands $n \geq 1$.

The subthreshold region is shown more easily using a logarithmic scale for the drain current, since plotting Eq. (8.4) on a log scale yields a straight line, as illustrated in Fig. 8.3. The subthreshold behavior can be specified by means of n or the subthreshold slope $\partial \log I_D / \partial V_{GS}$, but it is more common to cite the subthreshold swing $S = \left[\partial \log I_D / \partial V_{GS}\right]^{-1}$, the inverse of the subthreshold slope. Typically, the units used for subthreshold swing are volts/decade, so that the value represents the increment in gate voltage needed to change the drain current by a factor of 10. Experimentally, it is found that S varies through the subthreshold region, as Fig. 8.3 illustrates. At the low end the subthreshold characteristic smoothly merges into a leakage current characteristic, and at the upper end it turns into the saturation characteristic. Therefore, the value that is typically cited is the maximum slope of the log I_D–V_{GS} characteristic, analogous to what is done with mobility in the region above threshold.

The relationship between S and the ideality factor n is

$$S = n kT \ln(10) \text{ V/decade} \tag{8.5a}$$

$$= 60 n \text{ mV/decade at } 300 \text{ K} \tag{8.5b}$$

Fig. 8.3 The transfer characteristic of a MOSFET exhibits an exponential dependence of drain current on gate voltage below the threshold voltage. As the gate voltage is further reduced, the exponential characteristic merges with a leakage current characteristic. Often the value that is cited for the off-current I_{off} is the minimum current

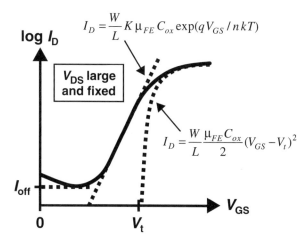

A subthreshold region is found in all MOSFETs, but long-channel single-crystal silicon MOSFETs with light substrate doping have a nearly ideal subthreshold region in which the ideality factor is very close to unity, so that the change in gate voltage needed to change the drain current by a factor of 10 is only 60 mV at 300 K. Nonideal subthreshold characteristics ($n > 1$) occur when a change in gate voltage does not produce a corresponding equal change in the surface potential of the semiconductor. Increasing the gate dielectric capacitance reduces the subthreshold swing by improving the coupling between the gate and the semiconductor surface. Typical values of subthreshold swing at 300 K for some representative TFT technologies are:

$$S = 200 - 500 \text{ mV/decade (polysilicon TFTs)}$$
$$= 500 \text{ mV} - 1 \text{ V/decade (a-Si TFTs)}$$
$$= 500 \text{ mV} - 5 \text{ V/decade (OTFTs)}$$

From an electronic system point of view, small subthreshold swing is desirable because less gate voltage excursion ΔV_{GS} is needed to turn the transistor from fully off to fully on. Indeed, one may view the total ΔV_{GS} required to take the transistor from fully off to fully on as having two parts. The subthreshold slope determines the voltage excursion that must take place below the threshold voltage, and the mobility determines the required excursion above threshold. OTFTs often have subthreshold swings that are large compared to silicon devices, due to localized states in the energy gap near the highest occupied molecular orbital (HOMO) or lowest unoccupied molecular orbital (LUMO) levels. For OTFTs the voltage excursion required below threshold often has about the same magnitude as that required above thresholds and the dielectric breakdown requirements, power supply requirements, and power dissipation are adversely affected.

8.2.4 Leakage Currents

Two types of leakage current have to be considered:

- Drain leakage current I_D when the transistor is off. That is, the gate voltage V_{GS} biases the device below the subthreshold region. This current is often called the off-current.
- Gate leakage current I_G under all transistor conditions, both on and off currents.

The off-current is seen in the log I_D–V_{GS} curve of Fig. 8.3 at gate voltages below those producing the exponential subthreshold characteristic. In general, the off-current is not independent of the gate and drain voltages. In some cases, it increases as the gate voltage is reduced further, as Fig. 8.3 illustrates. A similar effect is seen in polysilicon (poly-Si) TFTs, where it is attributed to field-assisted tunneling through mid-gap states present at grain boundaries; reducing the gate voltage increases the high fields at the drain edge of the gate, increasing the tunneling rate. The value that is typically cited for the off-current in TFTs is the minimum drain current at a specified drain voltage. For initial design work prior to more accurate computer-aided simulations that take into account the entire drain-current characteristic, a better figure to use for off-current is the worst-case leakage over the expected range of gate and drain voltages, rather than the minimum leakage found at a single voltage point.

For the polymer dielectrics and low-temperature inorganic dielectrics often used in OTFTs, the gate current can be significant. Indeed, sometimes the off-current is dominated not by drain-to-source leakage but by drain-to-gate leakage. Determining whether the off-current arises from leakage from drain to source or drain to gate is the first step in identifying the physical source of the off-current.

8.2.5 Contact Resistance

Making good electrical contact to organic semiconductors is sometimes difficult. As a result there are often large parasitic resistances in series with the source and drain. In an OTFT, these contact resistances typically depend on the gate voltage, decreasing with gate voltage, just like the channel resistance. Nevertheless, the contact resistances are distinct from the channel resistance because they do not scale with channel length, but rather are channel-length independent. Indeed, sometimes they can be modeled simply as a fixed added channel length. For example, the contact resistances may behave like an extra 2 μm of channel length, so that an OTFT with channel length L behaves as if it had a channel length $L + 2$ μm. The contact resistance can be determined by fabricating OTFTs of various channel lengths, plotting channel resistance versus channel length, and extrapolating channel resistance to a channel length of zero. An alternative method for determining channel resistance that avoids uncertainties due to device-to-device variations is the four-probe technique. The OTFT has four electrodes, with the outer two used as source and

Fig. 8.4 Contact resistance in OTFTs can lead to "hooked" output characteristics, as in this pentacene OTFT

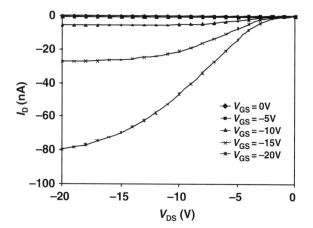

drain electrodes and the inner two used as voltage-sensing leads that penetrate into the channel region. Recent studies of OTFT contact resistance are found in Refs. [5, 6].

In some OTFTs, the contact resistance is nonlinear, depending on drain voltage, and for small V_{DS} a large fraction of V_{DS} may be dropped across the contact resistances rather than the intrinsic FET, while at larger V_{DS} the fraction of V_{DS} dropped across the contacts may decrease, and the contact resistances may be less significant. This leads to "hooked" output characteristics (Fig. 8.4). Since small values of V_{DS} are used to extract linear mobility, the measured linear mobility may be significantly smaller than the measured saturation mobility.

The total resistance of the FET is the sum of the channel resistance and the contact resistances. Because the channel resistance scales with channel length but the contact resistances are independent of channel length, the performance of short channel devices can be strongly degraded by contact resistance, and it is common for OTFTs to exhibit lower mobility as channel length is reduced because contact resistance comes to dominate the total resistance. The term *ohmic contact* is often used phenomenologically to describe a contact that has low enough resistance that it can be neglected compared to the channel resistance. Thus, ohmic contacts might be obtained at longer channel lengths but not at shorter channel lengths.

8.2.6 Capacitances and Frequency Response

The significant capacitances between the terminals of an OTFT are the gate–source and gate–drain capacitances. The drain–source capacitance is much smaller and can usually be ignored. The gate–source and gate–drain capacitances are each made of two capacitances in parallel, one due to the parasitic overlap capacitance between gate and source/drain and the other due to the capacitance

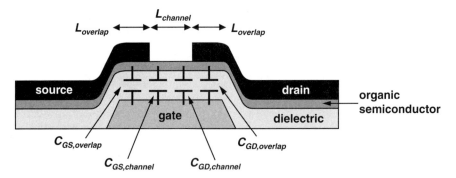

Fig. 8.5 The gate–source capacitance C_{GS} and the gate–drain capacitance C_{GD} are each composed of two components, one due to the overlap between gate and source/drain and the other due to the capacitance between gate and channel, which is partitioned between source and drain

between gate and channel (Fig. 8.5). The channel is not a terminal of the device, so that the capacitance between gate and channel must be partitioned between source and drain to obtain device terminal capacitances. Thus, $C_{GS} = C_{GS,overlap} + C_{GS,channel}$ and $C_{GD} = C_{GD,overlap} + C_{GD,channel}$. The overlap capacitances are voltage independent and can be approximated as fixed parallel plate capacitances, with $C_{GS,overlap} = C_{GD,overlap} = W\, L_{overlap}\, C_{ox}$, where $L_{overlap}$ is the overlap length.

In a MOSFET the channel capacitance is a complicated bias-dependent quantity. The standard model can be found in [7].[3] The overlap length $L_{overlap}$ responsible for the overlap capacitance can be large in OTFTs, sometimes 10 μm or more. It may be larger than the channel length L itself. Thus, the overlap contributes a relatively large capacitance in OTFTs. The reason for the overlap is that the fabrication processes used to make OTFTs, like those for a-Si TFTs, do not produce source and drain electrodes that are self-aligned to the gate. In order to ensure that the channel induced by the gate reaches the source and drain, a misalignment tolerance must be designed into the layout of these regions. This misalignment tolerance is the overlap length. For OTFTs on plastic substrates, the dimensional instability of the plastic film requires additional allowance for misalignment, so that $L_{overlap}$ may have to be even larger.

The effect of the overlap capacitance is evident from its effect on the unity-gain frequency f_t, the frequency at which the transistor's current gain falls to unity. In a simple model for the frequency-dependent current-gain, this frequency is:

$$f_t = \frac{\mu_{FE}(V_{GS} - V_t)}{2\pi\, L(L + 2L_{overlap})}. \tag{8.6}$$

---

[3]The standard MOSFET model also includes capacitances to a fourth terminal, the body, which does not exist for a TFT, which has only three terminals. An accurate TFT capacitance model is therefore expected to deviate from an accurate four-terminal MOSFET model.

A model that includes the effect of contact resistance is found in Ref. [8]. When $L_{overlap} \ll L$, as is the case for conventional self-aligned silicon CMOS processes, one obtains the well-known silicon scaling law in which the maximum operating frequency scales as L^{-2}. Large performance gains come from reducing the channel length, motivating amazingly successful efforts to reduce this transistor dimension in silicon ICs. However, if the amount of overlap is fixed by large-area registration capabilities and cannot be reduced as channel length is reduced, the silicon scaling law is replaced by one in which the maximum operating frequency ultimately scales as L^{-1}, altering significantly the economics of integration.

The overlap capacitance could be reduced dramatically if a self-aligned OTFT process could be developed. A self-aligned silicon process uses the gate as an *in situ* ion-implantation mask to form source and drain regions that are well aligned to the gate. An equivalent process does not yet exist for OTFTs. It would require a top-gate TFT geometry, together with a liquid- or vapor-phase doping process that renders the organic semiconductor highly conductive in a stable manner.

8.2.7 TFT Nonuniformity

Although the performance of OTFTs is lower than that of silicon MOSFETs, it is adequate for many applications, as long as the transistor parameters are uniform and predictable. However, nonuniform or unpredictable parameters are difficult to tolerate in electronic systems. For example, in displays, transistor nonuniformities can lead to visible variations in color or gray level, which is aesthetically displeasing and commercially nonviable. Random brightness variations across the display, known as *mura*, are more of a problem than smooth variations, because the low spatial frequencies in smooth variations are less visible to the human visual system. This has been an issue for poly-Si TFTs, which exhibit random variations in threshold voltage and off-current due to the random grain structure of the poly-Si.

In our laboratory, we have developed routines for the analysis of OTFT nonuniformity using automated probing and characterization of transistor arrays, followed by statistical analysis. We analyzed spatial parameter variations, including parameter correlations between pairs of closely spaced OTFTs, in order to predict the performance of matched pairs such as might be used in current mirror circuits. The distributions show smooth variations as well as random variations that cause nonuniformity at short distance scales.

8.2.8 Bias-Stress Instability and Hysteresis

Bias-stress instability and hysteresis are memory effects in which the DC characteristics of a transistor at a given time depend on voltages applied to the device in the past. These memory effects are undesirable in electronic systems, although there are design methods that can improve the tolerance to memory effects.

Generally, *bias-stress instability* refers to long-term changes in transistor characteristics that do not saturate but continue without limit until the device is rendered useless. *Hysteresis* refers to short-term reversible shifts that lead to looping in the measured characteristics, depending on which direction the bias voltages are swept in. There is no sharp distinction between the two, and they may arise from the same or similar physical causes.

These phenomena deserve careful attention because of their relevance to OTFT device physics and to the performance of OTFT electronic systems. Several studies have been performed [9, 10, 11, 12, 13], and the field for polymer TFTs is reviewed in [14]. In general, the bias-stress instability of OTFTs fabricated on inorganic gate dielectrics behaves as follows: the primary effect of positive gate bias is to shift the threshold voltage to more positive voltages, and the primary effect of negative bias is to shift the threshold voltage to more negative voltages. A similar effect is observed in a-Si TFTs [15] and to a lesser extent in poly-Si TFTs [16]. In our laboratory, we have found that the on-state bias-stress instability in pentacene OTFTs with SiO_2 gate dielectric is somewhat worse than in a-Si TFTs, and the off-state instability is much worse. The details depend strongly on the type of organic semiconductor used, and on the presence of oxygen and water vapor, just as with organic light-emitting diodes (OLEDs), which may lead to similar encapsulation requirements.

Hysteresis can be seen in the looping pentacene OTFT characteristics in Fig. 8.6. The same phenomenon also manifests itself as an overshoot or undershoot in the drain current when a gate-voltage step is applied. Similar effects have been observed in a-Si TFTs [17]. The details of this drain-current transient have been used to determine whether the responsible trapping states in pentacene OTFTs are electron traps or hole traps [18] .

When a polymer gate dielectric is used, there is an additional complicating factor. Slow polarization of the dielectric can cause an instability in a direction opposite to the bias-stress instability and hysteresis in organic semiconductors, so that there are two competing mechanisms, with a possible crossover between them after a certain stress period [19, 20]. Slow polarization in a polymer dielectric is often due

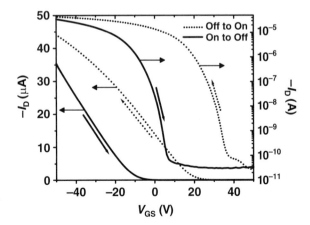

Fig. 8.6 Hysteresis leads to looping transistor characteristics, as seen in the linear-region transfer characteristics of this OTFT made using pentacene on thermal SiO_2. The drain current is plotted on both a linear scale (*left-hand vertical axis*) and a logarithmic scale (*right-hand vertical axis*)

Fig. 8.7 A comparison of the frequency-dependent capacitance $C(f)$ of two polymer dielectrics with slow polarizability and a dielectric without slow polarizability

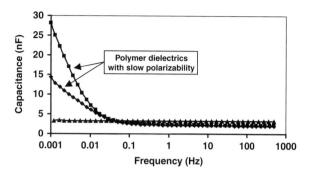

to residual polar solvent in the dielectric or water absorption from the air. It is natural to characterize this type of dielectric behavior by analyzing the frequency-dependent capacitance $C(f)$ at sufficiently low frequencies. Figure 8.7 compares $C(f)$ of two polymer dielectrics that exhibit slow polarizability with a dielectric that does not. At very low frequencies, below 100 mHz, the capacitances of the two dielectrics increase significantly. In the time domain this leads to memory effects in which voltages applied in the past affect how the OTFT responds to signals applied in the present.

As a result of hysteresis, measurements of μ_{FE} and V_t can depend strongly on the gate sweep direction. This can be seen in the way the transfer characteristics in Fig. 8.6 depend on sweep direction. Extracted OTFT parameters have sometimes been contaminated by hysteresis effects. An IEEE standard has been written in an effort to prevent these and other errors in OTFT characterization [21]. Attempts to derive μ_{FE} and V_t in the presence of hysteresis, perhaps by averaging the on-to-off and the off-to-on characteristics, are probably doomed to failure. It is preferable to extract parameters only when hysteresis is small.

8.3 Active Matrix Displays

8.3.1 Introduction

For low-resolution displays, such as those in digital watches, directly driven segmented elements are used, where each element has its own external connection. This is not possible for higher-resolution displays, since too many external leads would be required. As a result, high-resolution flat-panel displays use a matrix architecture in which the display is an array of pixels arranged in rows and columns. By definition, an active matrix display has one or more electronic switching elements in each pixel, while a passive matrix has no switching elements. The substrate containing the switching elements of an active matrix is often called the *backplane*. The origin and evolution of the active matrix as a method of display addressing are reviewed in Ref. [22].

It is more costly to include switching elements than to leave them out, so active matrix displays cost more to produce than passive matrix displays. However, it is more difficult to scale passive matrix displays to large sizes than active matrix displays. The mathematical expression of the scaling limitation on passive matrix liquid crystal displays (LCDs) is known as the *iron law of multiplexing*. It relates the number of rows in the display to the maximum possible contrast ratio [23]. Increasing the number of rows, and by implication the resolution of the display, reduces the contrast ratio achievable with a given liquid crystal material. Analogous limitations exist for other types of passive matrix displays. The basic issue is that in the absence of a switch at each pixel, it may be difficult to address an individual pixel at the intersection of a row and column without to some extent addressing the other pixels in the same row and/or column.

As a result of the limitations of passive matrix displays, today's very competitive LCD market is dominated by AMLCDs, sometimes called TFT LCDs. Passive matrix LCDs are found only in small, low-cost applications such as mobile phones and handheld games. The plasma display panels (PDPs) used in large flat-screen televisions are passive matrix because conventional switching elements cannot tolerate the required high voltages. Indeed, scaling to higher resolutions has been difficult for PDP technology.

In this chapter, we consider the use of OTFTs in active matrix flat-panel displays. Nowadays, commercial AMLCDs use a-Si or poly-Si TFTs as the switching elements. a-Si TFTs are the dominant technology, but manufacturers of small to medium-sized AMLCDs, intended for use, for example, in personal digital assistants (PDAs) and digital cameras, are making increasing use of poly-Si TFTs because their higher performance permits row and column driver circuits to be integrated directly on the display glass, reducing display module cost and shortening product development times.

The flat-panel display technologies that can benefit from an OTFT backplane are AMLCDs, electrophoretic displays, and AMOLED displays. There are other flat-panel technologies that are not compatible with TFT backplanes, either because the voltages are too high, for example plasma displays and inorganic electroluminescent displays, or because the area is too large for an active matrix, for example LED displays such as those used for video signage. In this section, we discuss the use of OTFTs for AMLCDs and electrophoretic displays together, since these display technologies are similar, and then we discuss the use of OTFTs for AMOLED displays. Other treatments of the application of OTFTs to active matrix displays can be found in [24, 25].

8.3.2 Liquid Crystal and Electrophoretic Displays

8.3.2.1 Introduction

The first reported OTFT display was a 16×16 pixel electrophoretic display demonstrated by researchers at Bell Laboratories and E Ink Corporation in 2001 [26]. Figure 8.8 shows operation of the $5''$ square display, which used pentacene TFTs on a flexible film of polyethylene terephthalate (PET).

Fig. 8.8 The first reported OTFT display, an electrophoretic display using pentacene TFTs on a flexible PET film. From Ref. [26]. Used with permission. Copyright 2001, National Academy of Sciences, USA

Later in 2001, OTFT AMLCDs using polymer-dispersed liquid crystal (PDLC) material were reported by two independent groups at about the same time. A team from Sarnoff Corporation, Penn State University, and Kent State University reported 16×16 PDLC displays using pentacene TFTs on polyethylene naphthalate (PEN) substrates [27]. The high mobility of the pentacene TFTs permitted a refresh rate of 60 Hz, with line times deliberately shortened to the 69 μs appropriate for a quarter VGA (QVGA, 320×240) display so that the measured display performance would apply to a larger display. This was the first group reported video-capable AMLCD on a plastic substrate. The second group, at Bell Laboratories, reported 2×3 PDLC displays on PET substrates, but used only static rather than dynamic driving signals [28]. Still later in 2001, Philips demonstrated a video-rate 64×64 PDLC display on a glass substrate using a solution-deposited polythienylenevinylene precursor film that was converted into a semiconductor [29]. Recent significant results include a 10″ diagonal SVGA (800×600) electrophoretic display using polyfluorene polymer OTFTs from Plastic Logic (Fig. 8.9) [30] and a 15″ diagonal XGA (1024×768) color twisted nematic (TN) AMLCD using pentacene TFTs on glass from Samsung (Fig. 8.10) [31], which is the largest and highest resolution OTFT displays reported to date. Table 8.1 lists these and other reported AMLCDs and electrophoretic displays using OTFT backplanes.

Fig. 8.9 A 10″ diagonal SVGA (800×600) electrophoretic display using polyfluorene polymer OTFTs on PET. From Ref. [30]. Permission for Reprint courtesy Society for Information Display

Fig. 8.10 A 15″ diagonal XGA (1024×768) color LCD using pentacene TFTs on glass. From Ref. [31]. Permission for Reprint courtesy Society for Information Display

Table 8.1 Active matrix LCDs and electrophoretic displays using organic TFTs

Organization	Display type	Semiconductor	Substrate	Pixel count	References
Bell Labs	Electrophoretic	Pentacene	PET	16×16	[26]
Sarnoff, Penn State, Kent State	PDLC	Pentacene	PEN	16×16	[27]
Bell Labs	PDLC	Pentacene	PET	2×3	[28]
Philips	PDLC	Polythienylenevinylene precursor	Glass	64×64	[29]
Plastic Logic	PDLC	Polyfluorene polymer	Glass	80×60	[94]
Polymer Vision	Electrophoretic	Pentacene precursor	PEN	320×240	[95]
ERSO/ITRI	Color TN LC	Pentacene	Glass	64×128	[96]
Samsung	Color TN LC	Pentacene	Glass	1024×768	[31]
Sony	Monochrome TN LC	Pentacene	Glass	160×120	[97]
Hitachi	Color TN LC	Pentacene	Glass	80×80	[98]
Plastic Logic	Electrophoretic	Polyfluorene polymer	PET	800×600	[30]

8.3.2.2 Electro-Optic Response of Liquid Crystal Materials

Conventional AMLCDs use a TN cell configuration, in which a nematic liquid crystal lies with its molecules in a twisted orientation between the two glass plates, with optical polarizers attached to the outer surfaces of the glass. One of the glass plates is the TFT backplane. The thickness of the liquid crystal material between the two glass plates is called the *cell gap*. Display operation relies on the response of polarized light to the electrically controlled orientation of the birefringent liquid

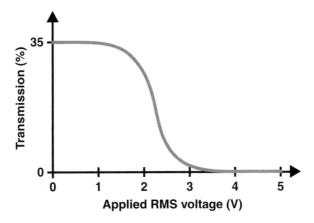

Fig. 8.11 Electro-optic curve for a typical twisted-nematic liquid crystal cell

crystal molecules.[4] From the perspective of electronic design, the significant fact is that a nematic liquid crystal responds to the root mean square (RMS) average of the applied field. Figure 8.11 shows a typical transmission versus RMS voltage characteristic of a TN cell. When 5 V is applied, the normally clear cell turns black.

Some OTFT AMLCDs have used PDLC material. In a PDLC cell, the nematic liquid crystal is dispersed as droplets in a solid polymer matrix sandwiched between the two glass plates. Unlike a TN LCD, a PDLC display does not use polarized light, but relies on electrically controlled scattering of unpolarized light. However, the electro-optic properties of the two are similar. A PDLC cell switches from a scattering state to a clear state in response to the applied RMS voltage. By placing a black background behind the cell, its appearance can be switched from milky white in the scattering state to black in the clear state, creating the appearance of black print on paper, although the reflectivity of the white scattering state is not as high as with paper.

Although TN LCDs are the mainstream display technology and provide a better appearance than PDLC displays, PDLC material is sometimes chosen for OTFT AMLCDs because it is more compatible with plastic substrates. In particular:

(1) The birefringence of typical plastics makes it difficult to maintain the polarization control required in a TN cell, but PDLC cells do not use polarized light.
(2) A TN cell is sensitive to the cell-gap changes that occur when a plastic display is flexed, but a PDLC cell is unaffected by flexing.

---

[4]We consider only LCDs that use nematic liquid crystals as the electro-optic material. There are other, less commonly used liquid crystal materials, such as cholesteric and ferroelectric liquid crystals.

(3) Unlike a TN cell, a PDLC cell is insensitive to gas permeation through the plastic substrates causing air bubbles to form inside the cell.

With both types of LCD, it is important that the applied voltage averaged over a period of seconds not have a significant DC component. A DC voltage component leads to the transport of ionic impurities within the cell, allowing an ion-induced compensating DC potential to build up. In itself this is not harmful, but the effect of a built-in DC potential is that the liquid crystal no longer responds to the applied signal alone, but rather to the sum of the applied voltage and the built-in potential. In addition, DC components larger than a few hundred millivolts can lead to irreversible electrolysis of the liquid crystal material.

It is fortuitous that the RMS-responding property allows the requirement that DC voltage components be avoided to be easily satisfied. The display driver circuits that provide data voltages to the display alternate the polarity from one frame to the next. Although it is the responsibility of the display driver circuits to produce the appropriate alternating-polarity data signals, the display designer must consider whether the pixel circuit contributes a DC artifact, due, for example, to charge injection when the pixel switch is shut off.

8.3.2.3 Electro-Optic Response of Electrophoretic Materials

Electrophoretic materials contain a charged pigment dispersed in an optically contrasting material, or a mixture of oppositely charged contrasting pigments dispersed in a neutral material. By applying an electric field, the pigment particles are moved to the front or back surface of the display, producing an electro-optic effect. A display made with electrophoretic material operates in a reflective mode using ambient light. Changing the optical state requires moving the pigment, so electrophoretic displays typically do not operate at video rates, but have response times of 100–300 ms. Because the only force on the pigment particles is the applied field, the movement of the pigment, and therefore its optical state, is only a function of the time-integrated electric field $\int_0^t E(\tau)\,d\tau$.[5] When no electric field is applied, there is no applied force, so that the optical state does not change, at least not over short periods. Thus, an electrophoretic material is a bistable material whose reflectivity depends on this time integral. Bistability can provide significant display power savings in applications that leave an image on the display for a period, such as electronic books and maps, because the display module can be powered down at these times. Typical data voltages are in the range of -15 V to $+15$ V, where negative voltage drives the display toward one optical state and positive voltage drives it toward the opposite state. Therefore, an electrophoretic display is not an RMS-responding display like an LCD, but rather responds to opposite polarities by producing opposite optical responses.

---

[5]This analysis is approximate, because it ignores forces between the particles and the effect of initial pigment conditions.

8.3.2.4 Liquid Crystal and Electrophoretic Display Architecture

The architecture of OTFT liquid crystal and electrophoretic displays is not unlike that of their a-Si TFT and poly-Si TFT counterparts. The basic design is shown in the schematic of four pixels in Fig. 8.12 .

The display consists of an array of TFT switches arranged in rows and columns, one TFT per pixel. For color displays each pixel is divided into three subpixels, each containing a red, green, or blue color filter, and there is one TFT per subpixel. All TFTs in the same row have their gates connected to a common row line (or *select line*), and all TFTs in the same column have a source–drain terminal connected to a common column line (or *data line*). The other source–drain terminal of each TFT connects to a storage capacitor C_{storage} internal to the pixel, and to a pixel electrode that faces the electro-optic material, that is, the liquid crystal or electrophoretic material. The other electrode of the storage capacitor is connected to a capacitor return line tied to an external DC potential. The previous select line is often used instead of a separate return line, since it provides a convenient nearly DC potential. On the other side of the cell gap is an unpatterned common electrode shared by all pixels and connected to an external DC voltage V_{com}. The electrodes on both sides of the cell gap are made from a transparent conductor, such as indium tin oxide (ITO); for reflective displays only the common electrode needs to be transparent, but the pixel electrode can be opaque. Figure 8.13 shows a portion of a transmissive OTFT backplane using pentacene devices on a PEN substrate [27]. The large electrode in the middle of each pixel is the ITO pixel electrode. The storage capacitor is formed between this electrode and a separate capacitor return line that runs over it along the row direction, with the gate dielectric lying in between.

In the fabrication of OTFT backplanes there are complications related to the order in which the organic semiconductor and source–drain contacts are deposited, and how they are patterned. The organic semiconductor is very sensitive to processing after it is deposited. This leads to two problems:

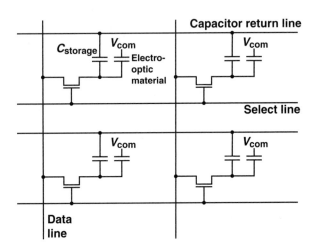

Fig. 8.12 Schematic of an array of four active matrix LCD or electrophoretic display pixels

Fig. 8.13 A portion of an OTFT backplane for an LCD display using pentacene devices on a PEN substrate

Active island patterning. The semiconductor must be patterned in order to avoid large leakage currents through ungated areas. But it is difficult to pattern the organic semiconductor using standard thin-film photolithographic methods, because these methods require the semiconductor to withstand organic solvents. Attempts to use these methods have failed. Patterning using shadow masking is not a practical option because the pattern is too fine and the alignments are too tight. Penn State University developed a technique for use with pentacene in which a photosensitized polyvinyl alcohol (PVA) film is used as a photoresist [32]. We used this method to pattern the pentacene in the display shown in Fig. 8.13. Another method developed by IBM uses parylene as a hard mask to protect pentacene from photolithographic chemicals [33, 34]. Ink-jetting a soluble semiconductor solves the patterning problem because deposition and patterning are performed at the same time.

Top versus bottom contacts. The source and drain electrodes can contact the organic semiconductor from above or below. The best OTFT performance is usually obtained with top contacts, that is, when the source–drain electrodes lie on top of the semiconductor, probably because with bottom contacts the semiconductor does not form a good film at the abrupt electrode edges, and because the metallization undergoes postdeposition processing that degrades the contact formed subsequently with the semiconductor [6]. However, top contacts require source–drain deposition and patterning after the semiconductor is deposited, which is problematic. Again, shadow masking is possible but impractical. Sarnoff developed a process for photolithographically patterning top contacts on pentacene [35], but it has not been widely adopted. Most OTFT backplanes have used photolithographically patterned bottom contacts and suffered the performance reduction.

The active matrix operates as follows. By applying a voltage pulse to one of the select lines, the switches in that row are turned on, and analog voltage levels applied to the data lines by the column driver circuits pass through the switches, charging each pixel's internal capacitance. The select lines are pulsed sequentially, row by row, and thus all the pixels in the display are written with analog voltages. Then the process starts again for the next display frame. The duration of each pulse is the line time T_{line}, and the time required to write all the pixels in the display is the frame time $T_{frame} = N_{row} T_{line}$, where N_{row} is the number of rows in the display. The display is rewritten at the refresh rate $R_{refresh} = 1/T_{frame}$.

The basic considerations in display design are (1) the pixel capacitance must be charged through the switch to a voltage accuracy consistent with the required gray-scale resolution (or color depth) of the display, and (2) the leakage of the switch must not permit the pixel voltage to decay excessively during the time when the switch is off. The first consideration favors using a wide TFT for the switch, while the second favors using a narrow TFT. There is an optimum width for the switch that allows about the same amount of voltage misconvergence during charging as voltage decay during the off period, minimizing the RMS voltage error averaged over the frame time. An additional consideration in a transmissive display is that the area occupied by the switch uses up pixel aperture, and this is another factor favoring narrow TFTs.

Some OTFTs have contact resistance effects like those shown in Fig. 8.4. A hooked output characteristic like this prevents good pixel convergence in a line time. The pixel voltage converges to a level that differs from the data voltage by approximately the turn-on voltage of the hooked characteristic. If this offset error was uniform across the display it could be compensated for all pixels together by adjusting the common electrode voltage V_{com}. However, the error is very likely to be nonuniform, and OTFTs with poor contacts are unlikely to be suitable for displays.

The leakage requirements for the pixel switch must be met over the full range of possible pixel voltages and data line voltages, even in the presence of data voltage inversion, in which the data line polarity may be the opposite of the pixel voltage polarity, producing a large drain–source voltage across the TFT. Furthermore, leakage requirements must be met not just on average, but by all (or nearly all) of the switches in the array, since leaky switches lead to visible pixel defects. a-Si TFTs are able to satisfy these stringent off-current requirements. However, the off-currents of OTFTs are typically much higher than those of a-Si TFTs. Allowance can be made for large statistical scatter in TFT leakage by using double or triple series-gated devices, provided that the leakage through each gated region is independent of the others. But in general displays with high gray-scale resolution and color depth will be difficult to make using OTFTs until off-currents are reduced significantly, while maintaining high mobilities for pixel charging.

8.4 Active Matrix OLED Displays

8.4.1 Introduction

Most OLED displays currently used in consumer products are passive matrix OLED (PMOLED) displays. Only a few products, such as the Kodak EasyShare LS633 digital still camera, the Sony Clie VZ-90 PDA, and the iRiver Clix2 media player have used AMOLED displays. But the advantages of AMOLED displays over PMOLED displays are compelling. They arise from the fact that light emission from the passive matrix occurs one line at a time, whereas in the active matrix emission is continuous. Continuous operation leads to higher power-efficiency, longer operational lifetime, and less demands on the current-handling capacity of the driver circuits.

The first AMOLED display was a 16×16 array using a poly-Si TFT back-plane, demonstrated in 1998 by a team from Sarnoff Corporation, Planar America, Kodak, and Princeton University [36]. Later that year Seiko Epson demonstrated an 800×236 AMOLED display, also using poly-Si TFTs [37]. Most AMOLED backplanes since then have used poly-Si TFTs because of their high mobility. The prospects were considered to be poor for making AMOLED backplanes using lower mobility transistors such as a-Si TFTs [38].

However, in 1995 our laboratory began investigating the use of a-Si TFTs for AMOLED displays as a low-cost alternative to poly-Si TFTs. We concluded that a-Si TFT AMOLED backplanes can be competitive, but are sensitive to the bias-stress instability exhibited by a-Si TFTs. In 1997, Princeton University demonstrated the integration of a single a-Si TFT with an OLED [39]. In 2003, Chi Mei Optoelectronics and IBM demonstrated a 20″ WXGA (1280×768) color AMOLED display based on a-Si technology, proving that a-Si backplanes can be used for large-area AMOLED displays [40]. There have been other, more recent demonstrations, most notably a 40″ 1280×800 a-Si color AMOLED display from Samsung in 2006 [41]. A review of a-Si technology for AMOLED displays is found in [42]. The mobility of OTFTs is similar to that of a-Si TFTs, so these demonstrations imply the feasibility of large-area OTFT AMOLED displays.

The first demonstrations of using OTFTs to drive OLEDs came nearly simultaneously in 1998 from independent groups at Bell Laboratories and Cambridge University [43, 44]. These were single "smart pixels," each incorporating one polymer-based OTFT and one OLED. Since AMOLED displays require at least two TFTs per pixel, these demonstrations were not examples of complete AMOLED pixels.

In 2004, Pioneer reported the first OTFT AMOLED display, containing an 8×8 array of two-TFT pixels using pentacene OTFTs on a glass substrate [45]. In 2005, Penn State University demonstrated a 48×48 OTFT AMOLED display on a PET substrate [46], and in 2006 Samsung demonstrated a 64×3×120 OTFT AMOLED display on a glass substrate [47]. Both displays used two-TFT pixels. Table 8.2 lists these and other significant OTFT AMOLED display results.

8.4.1.1 Electro-Optic Response of Organic Light-Emitting Diodes

The electro-optic behavior of OLEDs is similar to that of inorganic LEDs. The current-voltage (I–V) characteristic is diode-like. The current that flows under forward bias causes light emission. Under reverse bias little current flows, and there is no light emission. As with inorganic LEDs, the light output of OLEDs is proportional to the forward current over a wide current range. That is, the quantum efficiency is nearly constant. However, at very high current densities the quantum efficiency drops because of high exciton densities in the organic materials. Among

Table 8.2 Active matrix OLED displays using organic TFTs

Organization	Semiconductor	Substrate	Specifications	References
Cambridge University	Polythiophene	Glass	1×1, single TFT	[43]
Bell Laboratories	Polythiophene	Silicon wafer	1×1, single TFT	[44]
University of Tokyo	Pentacene	Glass	1×1, single TFT	[99]
Pioneer	Pentacene	Glass	8×8, two TFTs	[45]
Dong-A University	Pentacene	PET	64×64, single TFT	[100]
Penn State University	Pentacene	PET	48×48, two TFTs	[46]
Victor and NHK	Pentacene	PEN	16×16, two TFTs	[101]
Samsung	Pentacene	Glass	$64 \times 3 \times 120$, two TFTs	[47]

the reasons for PMOLED displays having lower efficiencies than AMOLED displays are the high voltage and the low quantum efficiency at high current levels.

The operational lifetimes of OLEDs vary widely and depend on the materials used and the degree of encapsulation. In severe cases, OLEDs may last for less than 100 h of operation. OLEDs using more advanced materials that are well-encapsulated last more than 10,000 h. Lifetime issues have slowed the commercialization of the technology, and a great deal of effort has been devoted to improving OLED materials and developing encapsulation methods to serve as a barrier to oxygen and water vapor. From the perspective of display design, the primary lifetime effects are:

(1) *Luminance decay.* The quantum efficiency of an OLED decreases over its operational lifetime, so that at a given current its light output decreases.
(2) *Voltage increase.* The voltage required for a given forward current to flow through an OLED increases over its operational lifetime.

Figure 8.14 illustrates these effects. Both also occur in inorganic LEDs, but the timescales involved are so long that they are ignored in all but the most demanding LED display applications.

8.4.1.2 OLED Display Architectures

Figure 8.15 illustrates PMOLED and AMOLED display architectures, with only one pixel of the AMOLED display shown for simplicity. The pixel in Fig. 8.15b is the simplest AMOLED pixel, a two-TFT pixel. It was proposed by Brody in 1975 for use with any electro-optic material that requires the flow of current [48]. More complex pixels using more than two TFTs offer performance advantages at the cost of additional complexity.

A PMOLED display consists of a set of row electrodes and column electrodes, with an OLED formed at each intersection (Fig. 8.15a). A row is selected for light emission by applying a select voltage pulse, forward biasing the diodes in that row.

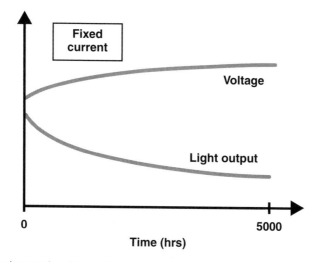

Fig. 8.14 An increase in voltage and a decrease in light output at a fixed current level are the two effects limiting the operational lifetime of OLEDs

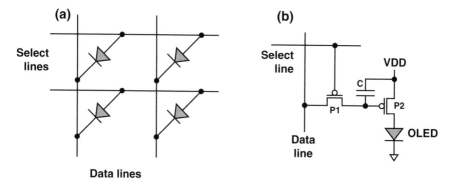

Fig. 8.15 (**a**) Passive matrix and (**b**) active matrix OLED display pixels

Currents proportional to the image data for the selected row are applied to the column lines by the driver circuits. All the unselected rows are reverse biased.

In an active matrix, data are written into a selected row by applying a select pulse to the TFT switches in the row. Each pixel's capacitance is charged to the data voltage, just as in an AMLCD. Thus, TFT P1 is used in the same way as the switch in an AMLCD, and our discussion of the switch in an AMLCD in Section 8.3.2.4 applies here as well, with similar considerations for charging and leakage. Unlike an AMLCD pixel, the AMOLED pixel contains a drive transistor P2 for converting the stored voltage to a current that drives the OLED continuously. The reason the pixel needs more than one TFT is that, unlike a liquid crystal, an OLED is not a capacitor that can hold the stored data voltage, but is a current-drawing element that would quickly dissipate stored charge if the OLED were connected directly to the switch.

Therefore, one or more additional TFTs are needed in an AMOLED pixel to allow the data voltage to be held as it controls the OLED current.

The advantage of AMOLED over PMOLED displays arises because emission in a passive matrix occurs one line at a time, so that each OLED element operates at high peak currents and low duty-cycle. The duty-cycle in a PMOLED display is approximately equal to the inverse of the number of rows. For example, in an SXGA (1280×1024) display the duty cycle is $\sim 0.1\%$. The peak current of an OLED pixel may be 1 mA or more. High OLED currents lead to reduced power efficiency and operational lifetime, and also place greater demands on the current capacity of the row driver circuits, which may have to handle currents of hundreds of milliamperes on each output (although not simultaneously). In contrast, in AMOLED displays each OLED element operates at nearly 100% duty cycle, independently of the number of rows in the display. Because of this, it is generally agreed that high-resolution OLED displays will require an active matrix for efficiency and long lifetime.

The demands that an AMOLED display places on TFT performance are more stringent than those placed by liquid crystal and electrophoretic displays. Table 8.3 compares the TFT requirements these types of displays. The different TFT requirements arise from the fact that liquid crystal and electrophoretic active matrix display pixels have a single TFT that is used as a switch, while an AMOLED pixel has one or more TFTs used as a switch, plus one or more used to drive current. A transistor used as a switch typically has stringent requirements for off-current, because it must not leak significant charge when it is off, but the requirements for mobility, uniformity, and stability are modest. A transistor that is used to drive current has stringent requirements for mobility, uniformity, and stability, but off-current requirements are lenient. Because the AMOLED pixel has both types of transistor, the TFT technology must meet more stringent requirements than required by liquid crystal and electrophoretic pixels.

As with color AMLCDs, in a color AMOLED display each pixel is divided into three subpixels, one each for red, green, and blue. The different colors can be produced by using three sets of OLED materials with different emission spectra. Alternatively, one white-emitting OLED material can be used with three color filters, or one blue-emitting material with down-converting phosphors.

In the simplest and most common AMOLED process, the drive transistor is connected to the anode of the OLED, and there is a large common cathode shared by all the OLEDs. ITO works well as an OLED anode, and this is the material commonly used. The cathode is usually an opaque low-work function metal. Thus, light

Table 8.3 TFT requirements of AMLCD, active matrix electrophoretic, and AMOLED displays

TFT parameter	LC and electrophoretic display	AMOLED display
Mobility	≥ 0.1 cm^2 V^{-1} s^{-1}	≥ 1 cm^2 V^{-1} s^{-1}
Off-current	≤ 1 pA	≤ 1 pA
Uniformity	Moderately important	Very important
Stability	Moderately important	Very important

emission in an AMOLED display is usually through the backplane. This is called bottom emission. As a result, the TFTs get in the way of the light emission, and the use of large TFTs or many TFTs in the pixel reduces the emission fill-factor of the display. The OLEDs can always be driven at high current densities to obtain adequate brightness in spite of low fill-factor, but at the expense of efficiency and lifetime, for the same reasons as in PMOLED displays. This is a particular problem with a-Si TFT and OTFT AMOLED displays, since the low mobilities of these technologies require the use of wide drive transistors. One way out of this problem is to use top-emitting OLEDs, which have a transparent cathode, so that the TFTs in the backplane do not obstruct light emission.

In a-Si TFT technology only n-channel devices are available, while in OTFT technology p-channel devices have had the best performance and the greatest ease of processing. For the switch used in AMLCD and AMOLED pixels, it makes no difference whether an n-channel or a p-channel TFT is used, since they are equivalent if one inverts the polarity of the select pulse. However, for the drive transistor a p-channel TFT is preferred over an n-channel TFT for the following reason. When OLEDs are fabricated on a TFT backplane using a conventional OLED process, the TFT is connected to the OLED anode, not its cathode. Because the anode is the terminal through which current enters a diode, connecting this terminal to the TFT ties it to the drain of a p-channel TFT but the source of an n-channel TFT. As a result, for a p-channel drive transistor the V_{GS} of the drive transistor is directly determined by the data voltage V_{data} stored in the pixel, since $V_{GS} = V_{data} - V_{DD}$. If the supply voltage V_{DD} is high enough to keep the TFT in the saturation region, the TFT acts as a current source with the current set by the data voltage, which works well.

However, with an n-channel drive transistor the OLED is in the TFT's source, so that the TFT does not function as a current source. Thus, unlike poly-Si TFT and OTFT technologies that provide p-channel TFTs, a-Si technology cannot implement the two-TFT pixels well. More complex a-Si pixels can overcome the difficulties from having the OLED in the source of the TFT, but the issue remains as a complication. In this respect the availability of p-channel devices gives OTFTs an advantage over a-Si TFTs for AMOLED displays.

8.4.1.3 Nonideal Behavior in AMOLED Pixels

Ideally the drive TFT would be a perfect voltage-controlled current source in which the current delivered to the OLED is a function of the data voltage alone, and this function is uniform across the display. Even if the current is a complicated nonlinear function of the data voltage, this function can be determined in advance and the data can be warped to take the nonlinearity into account. Unfortunately, the current is not a function of the data voltage alone. For one thing, a transistor is not an ideal current source. In reality, the current I_D through the drive transistor depends on the voltage V_{DS} across the transistor, even in the saturation region. Thus, the current varies with the OLED voltage V_{diode}.

There are other important ways in which the drive TFT fails to be a perfect, uniform voltage-controlled current source. The voltage-to-current conversion performed by the drive transistor depends on the TFT's threshold voltage and mobility, and these parameters are nonuniform across the display. This has been a particular problem for poly-Si AMOLED backplanes because poly-Si TFTs exhibit large nonuniformities due to the random nature of the poly-Si grain structure. But even a-Si TFTs must cope with this problem. While a-Si TFTs are typically uniform at the beginning of operational life, they quickly become nonuniform as they undergo image-dependent threshold voltage shifts from bias-stress instability. Hysteresis is expected to be a source of image-dependent nonuniformity over shorter timescales, an effect that has been observed in poly-Si TFT AMOLED displays [49]. In the present state of the technology OTFTs are expected to exhibit initial nonuniformity as well as image-induced nonuniformity.

One way of dealing with TFT nonuniformities is to demand better initial uniformity and less bias-stress instability and hysteresis from the TFT technology. There has been progress in this direction using new OTFT materials and processes. However, an alternative is to develop pixel designs that are more tolerant of TFT nonuniformity. AMOLED pixels that perform electronic self-compensation for TFT nonuniformities were first demonstrated by a team from Sarnoff Corporation, Planar America, Kodak, and Princeton University. In 1998, this team demonstrated 16×16 poly-Si AMOLED arrays using four-TFT self-compensating pixels that exhibited better brightness uniformity than arrays using conventional two-TFT pixels [36]. In 1999, a full self-compensating QVGA (320×240) AMOLED display using the four-TFT pixels and integrated row and column drivers to control the compensation circuitry was demonstrated [50]. Reviews of electronic self-compensation methods can be found in [51–55]. Optical self-compensation methods have also been demonstrated with a photodetector in each pixel to sense OLED light output, and apply feedback to correct for nonuniformity [56]. It should be noted that this is the *only* approach that compensates for OLED luminance decay over operational life, since none of the others detects light output. However, the method introduces new sources of nonuniformity arising from the photodetector and its associated circuitry, and it is unclear whether the net effect is an improvement in display uniformity.

Additional discussion and analysis of the use of OTFTs for AMOLED displays can be found in [57, 58]. To date only the conventional two-TFT pixels have been implemented with OTFTs, although a paper design for a self-compensating four-OTFT AMOLED pixel has been presented [58]. It is only a matter of time before the more complex pixels that have been investigated using poly-Si and a-Si TFTs are evaluated for providing compensation for OTFT nonuniformities.

8.5 Using Organic TFTs for Electronic Circuits

8.5.1 Thin-Film Transistor Circuits

8.5.1.1 Comparison with Silicon CMOS

The use of TFTs for electronic circuits has been led by manufacturers of small- and medium-sized AMLCDs, who are making increasing use of poly-Si TFTs for the integration of row and column display drivers onto the display glass. The integration of driver circuits on the display substrate can reduce display module cost, shorten product development times, and reduce the complexity and improve the reliability of the display interconnects. The cost advantage is most dramatic for small displays, because the cost of the driver circuits and the interconnects is a larger fraction of the total display module cost for smaller displays than for larger ones. Similar arguments could be made for integrated driver circuits for TFT-based imagers such as digital X-ray imagers, although integrated drivers have not yet made inroads in this application space.

The argument is less compelling for general purpose electronic circuits. Conventional CMOS silicon IC technology is able to produce MOSFETs for the astonishingly low cost of a few microcents to a few hundred microcents per transistor, depending on the type of IC and the level of integration. Absent any special considerations, such as the interconnect arguments that apply to displays and imagers, other types of TFT circuits must compete with silicon ICs on the basis of cost per transistor. Although the cost of processed TFT substrates per unit area will be lower than the areal cost of silicon ICs, the dimensions of MOSFETs used in silicon ICs are so much smaller than typical TFT dimensions that it will be difficult to reduce the cost of a TFT to be lower than the cost of a silicon MOSFET. Fabrication methods that promise to reduce cost per unit area, such as printing processes for OTFTs, but at the expense of increased transistor size because of poorer dimensional tolerances, may actually increase the cost per transistor, reducing competitiveness for these types of circuits.

Of course, TFT technology must compete with silicon technology not only on the basis of cost but also performance. It is well known that the performance of OTFT circuits is limited by the low mobility of the transistors, but there are other limiting factors. One significant factor is the absence of a good organic CMOS technology. The major advantages of CMOS over single-channel circuits and systems are (1) power dissipation is lower, especially static dissipation in digital systems; (2) signal voltages can easily swing all the way from one power supply voltage to the other; (3) the performance of the system is more tolerant of variations in transistor parameters, so that fabrication yields are higher; (4) circuit gain is higher, leading to larger noise margins in digital systems; and (5) design methods are simpler.

There are two obstacles to organic CMOS: the lack of good n-channel devices, and the process difficulty of integrating two types of organic TFTs on one substrate. The advantages of CMOS exist even if one transistor type is inferior to the other. But to derive the full benefit of CMOS, high performance is required from both device types, because otherwise system performance will be dominated by the lower performing device. For example, a low-mobility n-channel OTFT in a CMOS inverter must be scaled to large widths to provide pull-down capability that matches the pull-up capability of the p-channel OTFT, but this means that gate input capacitances are dominated by the large n-channel devices. Rise and fall times will be balanced, but slow. A good case can be made that the benefits of a complementary technology outweigh the gains achieved from achieving modest mobility improvements in a single-channel process, and that more effort to develop organic CMOS is warranted.

Various analog and digital OTFT circuits have been demonstrated. Ring oscillators are commonly used to demonstrate dynamic circuit operation, but the condition for a ring oscillator to oscillate is modest. It is only necessary that at the DC bias point an overall gain greater than unity is obtained through the inverter chain. Small-amplitude oscillations will occur, even if the gain is greater than unity over only a small voltage range. An IEEE standard has been written in an effort to render OTFT ring oscillator measurements more meaningful [59]. Clocked digital circuits provide a more realistic assessment of speed than ring oscillators, since proper functionality requires large internal voltage swings that approach saturated binary logic levels. Two of the most complex digital circuits have been a 4-bit parallel-to-serial converter containing 171 p-channel OTFTs [60] and a 15-bit code generator containing 326 p-channel OTFTs [61]. The highest level of OTFT integration reported to date was a 120-stage shift register from Philips [62]. The circuit was fabricated on a plastic substrate using a soluble pentacene precursor. It contained 2,130 OTFTs and operated at a maximum clock frequency of 2 kHz. The circuit is suitable for use as an integrated display row driver, and could drive a QVGA display at a frame rate of 8 Hz. There have been fewer demonstrations of analog circuits. Single-ended and differential OTFT amplifiers were studied in [63, 64].

8.5.1.2 Digital OTFT Design

In the absence of an organic CMOS technology, single-channel design methods must be used. Older texts on MOS circuits are good sources of information on single-channel design [e.g., 65]. In general, the situation is as follows for digital logic. There are three basic types of single-channel inverters. We show them in Fig. 8.16, together with a conventional CMOS inverter. For simplicity, we show the single-channel inverter circuits using n-channel transistors, but these circuits also work with p-channel transistors if the polarities of all power supplies are inverted.

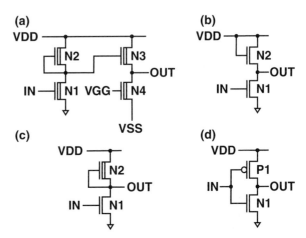

Fig. 8.16 Four logic families: (**a**) single-channel depletion-mode logic, (**b**) single-channel enhancement-mode logic, (**c**) single-channel enhancement/depletion-mode logic, and (**d**) CMOS logic

The first type of inverter (Fig. 8.16a) uses only depletion-mode transistors, that is, transistors that are on when $V_{GS} = 0$. High inverter gain, and therefore large noise margin, is obtained when driver transistor N1 has a larger W/L ratio than load transistor N2, which places the input switching point below zero volts. This necessitates a level-shifting output stage using a source follower N3 biased by a current source N4 to shift the output-low level below zero volts. The amount of level shifting is determined by V_{GS} of N4 (which is equal to $V_{GG}-V_{SS}$) and the relative sizing of N3 and N4. Source follower N3 not only provides level shifting but also serves as a good output driver because of its low output impedance. We employed this configuration in our early work on clocked OTFT circuits, which used depletion-mode pentacene devices [63].

The second type of inverter (Fig. 8.16b) uses only enhancement-mode transistors, that is, transistors that are off when $V_{GS} = 0$. The input switching point is above zero volts, so no level shifter is needed. However, diode-connected load transistor N2 is unable to pull the output higher than a threshold voltage below the V_{DD} supply, limiting output voltage swing. In addition, the impedance of a diode-connected device depends strongly on the voltage across it, so that the pull-up capability of the load transistor diminishes as the output approaches the high logic level. Thus, the delay times of the low-to-high and high-to-low transitions are typically mismatched.

The third type of inverter (Fig. 8.16c) uses both enhancement- and depletion-mode transistors. An enhancement-mode device is used for driver N1 and a depletion-mode device is used for load N2. The input switching point is above zero volts, so no level shifter is needed. Load transistor N2 is able to pull the output to the positive supply rail, and the output transition times can be matched. This inverter configuration typically provides the best performance with the lowest complexity of the three types of single-channel inverters. However, it requires transistors with two different threshold voltages, one positive and one negative, which may be just as high a technological hurdle as providing the n-channel and p-channel devices that CMOS requires.

However, recent work on OTFTs with a single semiconductor layer combined with two gates, one on the top and one on the bottom, has shown that two different threshold voltages can be obtained in a relatively straightforward way [66–68]. The two gates can be used in various ways. The top gate and bottom gate can be used directly to provide the different threshold voltages; a DC bias can be applied to the top gate to adjust the threshold voltage of the bottom-gate transistor; or the two gates can be tied together to provide higher on-currents and steeper subthreshold slope [67]. The double-gate approach was recently used to fabricate inverters [69] and was also used in an OTFT static RAM cell, where a DC bias was applied to the top gate to compensate for poor threshold voltage control in the fabrication process [70]. It is likely that digital circuits using enhancement-mode and depletion-mode OTFTs produced by means of double gates will be demonstrated soon.

We have discussed inverter configurations, but logic gates to perform NAND and NOR functions can be constructed from the single-channel inverter by adding one additional driver transistor per input. It is preferable to place driver transistors in parallel rather than in series to minimize the resistance of the driver transistor network and maximize speed. With n-channel transistors, this implies favoring NOR gates over NAND gates; with p-channel transistors NAND gates are favored.

8.5.2 Frequency Limitations of OTFTs

A rough idea of the frequency limitations of an OTFT is obtained by again considering the unity-gain frequency f_t, given in a simple model by Eq. (8.6),

$$f_t = \frac{\mu_{FE}(V_{GS} - V_t)}{2\pi\, L(L + 2L_{\text{overlap}})}.$$

We can estimate the unity-gain frequency of a pentacene OTFT made with current technology and driven at reasonable voltage levels. The field-effect mobility μ_{FE} of a good device is about 0.5 cm^2 V^{-1} s^{-1} and the DC overdrive bias $(V_{GS}-V_t)$ is $\sim$10 V. For a process with good dimensional tolerances, the channel length L and the overlap length L_{overlap} are both about 5 μm. This yields a unity-gain frequency of 1.1 MHz.

Digital and analog circuits can be designed to operate up to about 20–30% of f_t. In the case of digital circuits one way of viewing this is that the waveforms produced by these circuits should saturate at the power supply rails, requiring the circuits to reproduce harmonics well above the fundamental frequency. Typical analog circuits use closed-loop operation with negative feedback, and therefore must operate no higher than a fraction of f_t in order for the open loop gain to be much larger than one. Thus with current technology a pentacene-based OTFT circuit can operate up to a few hundred kilohertz.

It was noted in Section 8.2.6 that an OTFT's unity-gain frequency could be improved significantly by developing a self-aligned process that reduces L_{overlap} to a very small value, eliminating the overlap capacitances. In the above example this would triple the unity-gain frequency. An alternative is to neutralize these capacitances using circuit methods. Unlike the channel capacitance, which is distributed along the length of the device, the overlap capacitances are accessible at the terminals of the device and can therefore be resonated out over a band of frequencies by well-known methods used in high-frequency amplifiers.[6] These tuned methods are usually not applicable to general digital and analog circuits, which operate in a broadband mode, possibly down to DC. But for narrowband RF circuits these methods may allow the effects of L_{overlap} to be eliminated, so that one must cope only

---

[6]It might be objected that parasitics can only be resonated out if their values are predictable. Since the overlap capacitances depend on alignment between different metallization levels, they are expected to be quite variable. However, layout techniques can render parasitics well-defined and predictable. For example, by laying out a TFT with one source electrode in the middle of the device, and two drain electrodes connected in parallel, one on each side of the source, the overlap capacitances can be made constant, because as the overlap on one side gets larger the overlap on the other side gets smaller.

with the distributed channel capacitance due to L, which corresponds to the intrinsic propagation delay of the device and sets a limit on the highest operating frequency.

8.5.3 Integrated Display Drivers

A matrix-addressed display requires driver circuits to generate the signals that are applied to the row and column lines. An AMLCD row driver consists of a shift register that applies a select pulse to the row lines sequentially. The shift register typically operates at clock frequencies below 100 kHz, within the capability of OTFTs. An AMLCD column driver generates an analog voltage on each column that depends on the image data for the pixels in that column. External driver chips have a digital-to-analog converter (DAC) for each column of the display. Integrated column drivers typically do not put all this functionality onto the display. With current poly-Si integrated drivers, a few high-speed DACs are located on an external silicon chip, and the integrated column driver circuitry has an analog demultiplexer with a shift register and switches to sample the high-speed analog data sequentially onto the columns at a rate of a few megasamples to tens of megasamples per second.

The performance of poly-Si TFTs is very good for integrated display drivers. Although poly-Si TFTs often have high off-currents, this is generally not important for display driver circuits. Furthermore, complementary n-channel and p-channel devices are available with poly-Si TFT technology, so that CMOS design methods can be used.

It is considerably more difficult to make integrated drivers with OTFTs because of low mobility and bias-stress instability, and because organic CMOS technology is not yet available. The situation is similar with a-Si TFTs, which likewise have low mobility and bias-stress instability, and provide only n-channel TFTs. Nevertheless, a-Si is a lower-cost technology than poly-Si technology, and this has spurred efforts to develop a-Si integrated drivers. As long ago as 1987, the David Sarnoff Research Center and Thomson LCD developed the self-scanned amorphous silicon integrated display (SASID) technology, which permitted the integration of a-Si TFT row and column drivers [71]. SASID consisted primarily of a set of circuit design methods that allowed driver circuits to tolerate the performance limitations of a-Si TFTs. For example, a SASID row driver maintains the TFTs in each stage in the off state except when that stage is generating a select pulse, minimizing bias-stress induced threshold shift. SASID column drivers implement only the demultiplexing switches on the display, leaving the DACs and the shift register on an external silicon chip. Thales Avionics manufactures a 480×480 cockpit display using a-Si row drivers based on the SASID technology, and work on a-Si drivers continues to the present time [72].

Integrated drivers have not yet been implemented with OTFTs. However, the application of OTFTs to integrated drivers is expected to be similar to the situation with a-Si TFTs, with similar design approaches used. In spite of the difficulties, novel circuit design methods will allow low-cost OTFT drivers to be integrated on the display.

8.5.4 Radio Frequency Identification Tags

8.5.4.1 Introduction

RFID tags have been proposed as an application for OTFTs on flexible substrates because very low cost tags would open up the possibility of item-level tracking, supplementing or replacing printed bar codes. Item-level RFID tags have several advantages over bar codes. Manual scanning is not required, permitting faster and more automated item inventory and tracking; stored information can be updated; and loss, theft, and counterfeiting can be more easily deterred. General treatments of RFID technology can be found in [73, 74].

The use of RFID tags on shipping containers and pallets has demonstrated its value. For these applications, a tag cost of several dollars or more can be justified, especially when tag reuse is possible. But tracking of individual consumer items requires such a low tag cost to be economically viable that RFID has not made inroads yet, except for high-priced items and controlled pharmaceuticals. It has been estimated that simply to break even on the cost versus benefits trade-off, an RFID label for a $40 item must cost between 5 and 10 cents. Prices for the lowest-cost silicon-based RFID tags are currently about 7 cents in large volumes for tags that are unconverted (not yet in label form), and 10–12 cents for tags in the form of self-adhesive labels.[7] In contrast, printed bar codes add virtually no packaging cost to an item, although there are infrastructure and labor costs associated with reading and tracking.

This is the reason OTFTs are considered for this application, with a goal of producing a very low cost "penny tag." However, the cautions of Section 8.4.1.1 apply to RFID tags. Unless the cost per transistor of OTFT technology is lower than that of silicon IC technology, OTFTs will not be cost-competitive. On the other hand, it has been pointed out that, even if the cost per transistor of OTFTs is higher than that of silicon ICs, OTFTs could still be cost-competitive. By fabricating the OTFT RFID circuit on the tag itself, the cost of mounting a silicon RFID chip on the tag is avoided, so that OTFT technology has a cost advantage due to lower assembly costs. While this is true, there is a compensating disadvantage to fabricating the OTFT circuit directly on the tag: in order to accommodate the antenna, the circuit must be fabricated on a substrate significantly larger than the area required by the transistors alone. As a result, the transistor throughput of the OTFT fabrication facility is reduced by this large area factor, and the cost per transistor of the RFID circuit is scaled up by approximately the same factor.

---

[7]Here we are not considering so-called one-bit tags, which simply signal their presence to the reader. This type of tag is used in stores as an anti-theft device, and is deactivated during purchase. It does not contain electronic circuitry, but uses a simple physical effect such as the magnetization properties of a metal. One-bit tags can be made for only a fraction of a cent.

RFID tags can be active or passive. Active tags contain a battery for power, while passive tags draw their power from the RF signal broadcast by the tag reader.[8] Active tags can be used at a greater distance from the reader because of the on-tag power source, but are more expensive. The advantage of OTFTs is their potentially low cost, so the technology is only considered for passive tags.

Four carrier frequency bands are currently used for RFID tags:

Low-frequency <135 kHz band. The near field of an electromagnetic wave occurs at distances less than $\lambda/2\pi$ than from the antenna, where λ is the free-space wavelength. The far field occurs at larger distances. At 135 kHz the near field is found at distances closer than 354 m, so that tags in this band operate in the near field of the reader. Therefore, the antenna for an RFID tag operating in this band does not function as a conventional radio antenna, communicating with the reader via a propagating electromagnetic wave. Instead the antenna is a coil that provides magnetic coupling to a coil in the reader, like the coupling between primary and secondary windings of a transformer. Wound wire coils with many turns are typically used to achieve high inductance with low resistance, sometimes with a ferrite core to improve the magnetic coupling. These coils are expensive, so that this type of tag is not considered for low-cost item-level tags, and the cost advantage of OTFT technology is not likely to be compelling. This is unfortunate because OTFTs can operate without much difficulty in this frequency range.

There is one possible way tags operating in the low-frequency range might be produced at low cost. Reader and tag can be coupled using capacitive rather than inductive coupling. A capacitive coupling antenna can be printed at low cost. Motorola considered commercializing this approach at 13.56 MHz with their BiStatix technology, which used carbon ink electrodes printed on a paper label, with a silicon IC attached to the electrodes. However, there are significant disadvantages to capacitive coupling. Nearby conductive objects disrupt the electric field. Furthermore, RFID tags typically use an inductor and capacitor in parallel at the input to form an *LC* tank circuit with resonance at or near the carrier frequency, in order to boost the voltage generated by the carrier to a level that can drive the tag electronics. But since the goal of the capacitive coupling approach is to eliminate the expensive coil, an on-tag resonant circuit is not practical, and the tag's read range is limited by low induced voltages.

High-frequency 13.56 MHz band. Tags operating in this band also operate in the near field of the reader, since the near field is found at distances closer than $\lambda/2\pi$ or 3.5 m. The higher frequency permits a planar spiral inductor rather than a wound wire coil to be used for magnetic coupling to the reader, significantly reducing cost. Typically an etched metal inductor is used. This is an ideal frequency band for low-cost RFID tags and is generally considered to be the best opportunity for OTFT RFID tags. However, at present it is challenging for electronic components fabricated using OTFT technology to operate reliably at frequencies this high. The

---

[8]The reader in an RFID system is sometimes referred to as the *interrogator*, and the tag is often referred to as the *transponder*.

challenges associated with using OTFTs at high frequencies for RFID tags are discussed in Section 8.4.4.2, together with possible methods of meeting them.

Ultra high frequency 915 MHz and microwave 2.45 GHz bands. Tags operating in these bands operate in the far field of the reader, typically using a dipole antenna to receive the signal transmitted by the reader, and to reflect back a modulated wave. Because of signal interference issues near metallic and liquid items, these bands are not likely to be useful for item-level tagging, although there is ongoing work to improve the interference characteristics in these bands. These carrier frequencies will be out of the range of components fabricated using OTFT technology in the foreseeable future.

There have been a few demonstrations of very simple OTFT RFID tags. 3M demonstrated the first OTFT RFID circuits: pentacene-based circuits for an 8.8 MHz 1-bit tag in 2003 [75] and for a 1.2 MHz 8-bit tag in 2004 [76]. In both circuits the carrier powered a free-running ring oscillator, with one circuit producing a detectable square wave at the reader and the other producing an 8-bit code. In 2006, PolyIC demonstrated a similar tag architecture operating at 13.56 MHz using poly-hexylthiophene TFTs and vertical organic diodes on a polyester substrate [77]. Also in 2006, Philips demonstrated pentacene-based 13.56 MHz tags with a capacitively coupled antenna on a plastic substrate that allowed simple 2-bit and 6-bit codes to be detected by the reader, and the most complex OTFT tag demonstrated to date, a 125 kHz tag containing 1938 transistors that allowed a 64-bit code programmed into the tag to be read [78].

8.5.4.2 Using OTFT Technology for RFID Tags

A block diagram of a simple passive near-field RFID system is shown in Fig. 8.17. The reader's coil L_1 generates an RF magnetic field that induces a current in the tag's coil L_2. Coil L_2 and capacitor C_1 form an *LC* tank circuit resonant at or near the carrier frequency, which boosts the AC voltage generated by the carrier, and

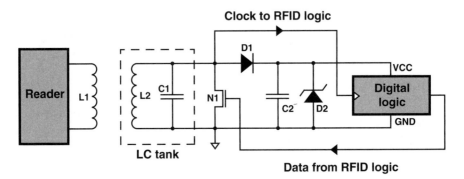

Fig. 8.17 Schematic of a simple passive near-field RFID system. The reader at *left* communicates with the tag at *right* via the magnetic coupling of coils L_1 and L_2

also makes the tag more detectable by the reader. The AC signal is rectified by diode D_1 and filtered by supply capacitor C_2 to generate the DC supply voltage V_{CC} that powers the tag's digital logic. The single diode D_1 provides only half-wave rectification and can be replaced by a full-wave rectifier for higher efficiency. A regulator diode D_2 is used to clamp the DC supply voltage seen by the logic, or a voltage regulator circuit is used, since changes in the distance to the reader produce large variations in field strength. Without some type of voltage regulation, several hundred volts can be induced close to the reader, which is likely to damage the tag electronics.

In a conventional silicon-based tag the RF signal is used not only to power the tag but also to generate a clock for the tag's digital logic, as shown in Fig. 8.17. The carrier signal is divided down by a digital frequency divider, so that only the first-stage frequency divider operates at the carrier frequency, while the rest of the logic operates at lower frequencies.

After a start-up period during which the supply capacitor is charged, the tag begins to output stored data, either in response to the clock only or in response to commands sent from the reader by modulating the carrier. Data are sent from the tag to the reader by switching on and off shunt transistor $N1$, modulating the loading of the tank with the output data. The load modulation is detected at the reader by monitoring the AC voltage across coil L_1. When the tag's tank circuit is within the range of the reader, the reader's coil is loaded in the same way that the primary winding of a transformer is loaded by the secondary. When $N1$ is on, the quality factor Q of the tank is reduced and the loading of the reader's coil is altered. An alternative is to place $N1$ in series with a capacitor, modulating the tuning of the tank circuit rather than its Q. With proper design the switching of $N1$ does not interfere with power or data from the reader.

Thus, the RF carrier incident on the tag from the reader performs three functions:

(1) Powering the tag
(2) Signaling the tag with a clock or code
(3) Detecting the tag

These three functions require some components in the tag to operate at the carrier frequency. Below we consider the challenges of implementing these functions in an OTFT technology. Then, we examine an additional RFID function that may be difficult to implement in OTFT technology: memory.

Powering the Tag

To supply power to the tag, AC current at the carrier frequency must pass through the coil, capacitors C_1 and C_2, and rectification diode D_1. There is no serious difficulty producing a coil and capacitors that function at RF frequencies in an OTFT process, although it is challenging to produce a low-resistance, high-Q coil and a capacitor with a well-controlled capacitance using low-cost printing methods [79]. However,

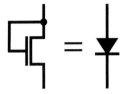

Fig. 8.18 A diode-connected enhancement-mode transistor is functionally equivalent to a diode

fabricating a diode in an OTFT process that operates at RF frequencies with low voltage-drop is challenging, particularly at the preferred 13.56 MHz frequency. There are two approaches. One is to use an enhancement-mode OTFT in a diode-connected configuration, with its gate and drain connected together (Fig. 8.18). Diode-connected transistors are sometimes used in silicon RFID ICs. They have the advantage of requiring no additional processing steps. The other approach is to use a vertical organic diode, which requires additional processing.

A diode-connected transistor has the same frequency limitations as the underlying transistor. In Section 8.4.2 the unity-gain frequency of a typical pentacene OTFT was calculated to be about 1 MHz. It has been shown that ripple requirements for the RFID supply voltage set an even lower limit on the carrier frequency than is set by the unity-gain frequency [80]. Apart from frequency response considerations, the forward voltage-drop across the diode should be as small as possible, since the tag's supply voltage is reduced by this voltage-drop or, in the case of a full-wave rectifier, two voltage-drops. But the turn-on voltage of a diode-connected MOS transistor is equal to its threshold voltage, which can be large and uncontrolled in OTFTs. In addition, it is advantageous for the diode to have as small a parasitic capacitance as possible to minimize carrier feedthrough, which produces ripple on the DC supply, but a diode-connected OTFT has a large parasitic overlap capacitance. In spite of its disadvantages, this approach to rectification in OTFT technology is advocated because of its simplicity. Using a circuit developed for silicon RFID tags that permits full-wave rectification with only a single diode voltage-drop [81], workers at OrganicID fabricated and tested a rectifier with 3 μm channel-length pentacene OTFTs that allowed a 10 V RMS signal induced in the antenna coil to produce a DC voltage of nearly 8 V at 1 MHz, 5.5 V at 5 MHz, and 1 V at 20 MHz [82].

A vertical organic diode can operate at higher frequencies than a diode-connected OTFT because of the smaller distance between the metal contacts. However, the performance advantage comes at the expense of additional processing to form the vertical structure. In a simple transit-time model, the upper frequency limit of a semiconductor device is proportional to L^{-2}, where L is the spacing between the contacts. In a vertical diode this spacing is the thickness of the organic stack, whereas in a transistor it is the channel length. In this simple model, a vertical diode with a 100-nm organic film has a frequency response that is 10^4 times higher than a diode-connected transistor using the same material with a 10-μm long channel. However, this model overestimates the advantage of the vertical diode. Because of the field effect the carrier density is higher in a diode-connected transistor than in a vertical diode, which results in higher carrier mobility. The VRH transport model described in Section 8.2.1 explains this phenomenon.

Researchers at UCLA fabricated vertical $Cu/C_{60}/Al$ diodes using a 100-nm layer of C_{60}, and observed rectification in response to an applied 1 MHz signal [83]. A UC Berkeley group showed that iodine doping of a vertical Al/pentacene/Au diode improved its frequency response by reducing the series resistance [84]. Workers at IMEC and the University of Leuven built a vertical Au/pentacene/Al diode and studied its usefulness as an RFID supply rectifier [80]. The 300 nm pentacene layer was grown at a high deposition rate to obtain small dense grains, thereby preventing shorting between the metal electrodes. The authors showed that an 18 V amplitude signal from the antenna coil gave a filtered DC voltage of 11 V at 14 MHz and 8 V at 50 MHz. Calculations suggested that the diode should be usable as a supply rectifier up to hundreds of megahertz. Indeed, organic photodiodes with a carefully design vertical structure have been demonstrated with a measured bandwidth of 430 MHz [85], demonstrating that vertical organic diodes can have a frequency response adequate for 13.56 MHz RFID tags.

Signaling the Tag

Conventional silicon IC technology yields transistors that operate at frequencies in the microwave band. As a result, in a silicon-based tag the RF signal at the carrier frequency is typically used to generate a clock for the tag's digital logic. Thus, the entire tag operates synchronously with the reader. However, if the transistors do not have sufficiently high frequency response to operate at the carrier frequency, the RF signal cannot directly provide a clock. Based on our calculation of unity-gain frequency in Section 8.4.2, OTFTs are expected to have this limitation in the preferred 13.56 MHz band. One way to circumvent the limitation is to use a local free-running low-frequency oscillator to provide a clock, rather than deriving a clock from the carrier. That is, use the carrier for power but not for timing. Thus, the tag operates asynchronously rather than synchronously with the reader. The disadvantage of this approach is that typically the characteristics of OTFTs are not well controlled, so that the oscillator frequency will vary widely from tag to tag and is likely to depend sensitively on supply voltage, environment, etc., making detection of the tag's asynchronous output data difficult.

An approach that maintains synchronous operation without requiring OTFTs to operate at the carrier frequency is shown in Fig. 8.19 . The RF signal from the reader is amplitude-modulated with a low-frequency subcarrier. For example, a 13.56 MHz carrier might be modulated at 10 kHz. The modulation of the carrier need not affect the ability of the carrier to power the tag. The RF signal must be rectified and filtered to detect only the low-frequency component. This is done with a high-frequency diode and filter that are separate from the power supply rectifier and filter. In Fig. 8.19, the filter is shown as a parallel RC circuit. The roll-off frequency $f_{-3db} = 1/2\pi RC$ is set to a value between the carrier and subcarrier frequencies, in

Fig. 8.19 Deriving the RFID clock from a low-frequency subcarrier. Diode D_1 is the only organic component that operates at the carrier frequency

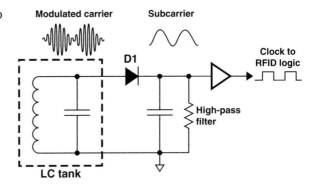

order to filter out the rectified carrier and leave the low-frequency modulation. The filter may be unnecessary, since the OTFT logic is not able to follow the carrier frequency anyway and will simply follow the low-frequency envelope. However, some sort of current sink like the resistor shown in Fig. 8.19 is needed to allow a path for low-frequency current through the detection diode to return to ground.

Government regulations on the power spectrum of the signal transmitted by the reader may restrict the subcarrier to frequencies lower than the desired clock frequency. This restriction can be overcome by using a phase-locked loop with a frequency divider to lock the frequency of a local oscillator on the tag to a multiple of the subcarrier frequency, thus providing a clock for the tag at the desired frequency.

Detecting the Tag

The tag's output circuitry presents special difficulties for OTFT technology. Although the load modulation transistor is switched at the low frequency of the tag's output data, nevertheless when the switch is on it must be able to pass the carrier frequency in order to alter the Q or the tuning of the tank circuit. Therefore, the unity-gain frequency of the switch must be at least as high as the carrier frequency. If the unity-gain frequency of the switch is lower than the carrier frequency, the RF signal will be attenuated as it passes through the switch. Under the right conditions, at sufficiently high power levels and short read ranges, load modulation may be detectable at the reader. However, good tag performance at a reasonable power level and read range is not expected of an approach that uses the load switch at a frequency higher than its unity-gain frequency.

An alternative to using a transistor as a load switch is to use an organic diode for load modulation. An example of this approach is shown in Fig. 8.20. By applying the output data to diode D_1, the tank circuit is load-modulated at low frequencies, which can be detected by the reader in the conventional way. Blocking capacitor C_1 prevents low-frequency currents from flowing through the coil. Load modulation by the diode can be capacitive, using the diode as a varactor whose capacitance changes with voltage under reverse bias, or ohmic, altering the resistance of the diode by

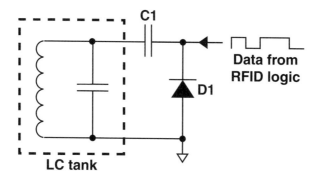

Fig. 8.20 Load modulation by modulating the impedance of diode D_1. In reverse bias the diode's capacitance can be modulated, and in forward bias its resistance can be modulated. Capacitor C_1 prevents large low-frequency currents from flowing through the coil. The diode is the only organic component that operates at the carrier frequency

switching it between forward and reverse bias. The capacitive approach will dissipate less power, since the diode does not draw DC current under reverse bias. Heterojunctions between organic layers can behave like conventional p–n junctions in inorganic semiconductors, exhibiting similar capacitance-voltage characteristics and allowing an organic varactor to be made. Indeed, a capacitance variation of greater than 2:1 has been measured as the applied reverse bias is varied from 0 V to 10 V on organic photodiodes with 430 MHz bandwidth [86], more than adequate for RFID applications at 13.56 MHz.

Memory

An additional RFID circuit function that may be difficult to implement in a low-cost OTFT technology is nonvolatile memory. For some RFID applications, a factory-programmed read-only memory (ROM) is sufficient. A ROM can easily be implemented in OTFT technology using laser cutting of a metallization pattern. For example, the Electronic Product Code (EPC) uses a 96-bit identifier that is unique for each EPC tag, and thus each tag can carry a ROM that is personalized in the manufacturing process. But in many cases field-programmability is required to allow the user to write information into the tag in service. Examples of this are expiration dates for perishable goods and flight information for checked baggage. It is also useful during production testing. Field-programmability is available in current silicon-based tags, either as programmable ROM (PROM), which is write-once, or as electrically erasable PROM (EEPROM), which is rewritable.

At the present time OTFT technology has not been used to implement PROM or EEPROM functions. However, some components of a programmable memory technology have been demonstrated. Researchers at UC Berkeley recently demonstrated a low-power organic antifuse array using a polyvinylphenol insulator and pentacene addressing diodes on a plastic substrate [87]. In an antifuse memory, programming takes place by applying a sufficiently high voltage across an insulator to physically damage it, creating a short circuit. Diodes are required for robust matrix addressing. Provided that the peripheral components for the antifuse array could be fabricated in an OTFT process, this technology could be used as PROM in low-cost RFID tags. There has also been work by many researchers on bistable memory effects that could form the basis for EEPROM in organic RFID tags. Bistability has been observed in vertical organic devices [88–91] as well as lateral OTFT-like devices with functionalized gate dielectrics [92, 93].

8.6 Conclusion

Organic TFT technology on flexible substrates holds unique promise for providing a low-cost route to fabricating electronics. Furthermore, the use of flexible substrates opens up the possibility of mechanically flexible electronic systems. However, the dominant technology, silicon CMOS, has achieved astonishing levels of performance improvement and cost reduction, and displacing it will be a formidable challenge, even in a limited application space. Applications are needed that permit OTFTs to be gracefully and incrementally introduced as a low-cost alternative to existing technologies. Demanding applications such as RFID tags are likely to require new circuit and system designs in order to coax high enough performance from OTFT technology to take advantage of its low cost. For less demanding applications such as active matrix displays, OTFT fabrication could be introduced into standard TFT display manufacturing lines incrementally and with little fanfare. Initially, PECVD and photolithographic processes on glass substrates could be replaced by lower-cost additive, printing-like processes, with roll-to-roll manufacturing on flexible substrates introduced later as OTFT backplane technology becomes mainstream.

Acknowledgements The author would like to acknowledge G. Gu for many helpful discussions about organic thin-film transistors.

References

1. Vissenberg M, Matters M (1998) Theory of the field-effect mobility in amorphous organic transistors. Phys Rev B 57:12964–12967
2. Slade HC (1997) Device and material characterization and analytic modeling of amorphous silicon thin film transistors. PhD dissertation, University of Virginia
3. Cerdeira A et al. (2001) New procedure for the extraction of basic a-Si:H TFT model parameters in the linear and saturation regions. Solid-State Electron 45:1077–1080
4. Fadlallah M et al. (2006) Modeling and characterization of organic thin film transistors for circuit design. J Appl Phys 99:104504-1–104504-7
5. Pesavento PV et al. (2006) Film and contact resistance in pentacene thin-film transistors: Dependence on film thickness, electrode geometry, and correlation with hole mobility. J Appl Phys 99:094504-1–094504-10

6. Gundlach DJ et al. (2006) An experimental study of contact effects in organic thin-film transistors. J Appl Phys 100:024509-1–024509-13
7. Tsividis YP (1999) Operation and modeling of the MOS transistor, 2nd ed. Oxford University Press, New York
8. Wagner V et al. (2006) Megahertz operation of organic field-effect transistors based on poly(3-hexylthiophene). Appl Phys Lett 89:243515-1–243515-3
9. Katz HE et al. (2002) Organic field-effect transistors with polarizable gate insulators. J Appl Phys 91:1572–1576
10. Knipp D et al. (2003) Pentacene thin-film transistors on inorganic dielectrics: Morphology, structural properties, and electronic transport. J Appl Phys 93:347–355
11. Salleo A et al. (2005) Reversible and irreversible trapping at room temperature in poly(thiophene) thin-film transistors. Appl Phys Lett 86:263505-1–263505-3
12. Han SH, Jang J (2006) Chemical and electrical stabilities of organic thin-film transistors for display applications. J Soc Inf Display 14:1097–1101
13. Gu G et al. (2007) Reversible memory effects and acceptor states in pentacene-based organic thin-film transistors. J Appl Phys 101:014504-1-1-9
14. Salleo A, Chabinyc M (2006) Electrical and environmental stability of polymer thin-film transistors. In: Klauk H (ed) Organic electronics: Materials, manufacturing, and applications. Wiley-VCH, Weinheim, pp 108–131
15. Powell MJ et al. (1992) Defect pool in amorphous-silicon thin-film transistors. Phys Rev B 45:4160–4170
16. Hack M et al. (1993) Physical models for degradation effects in polysilicon thin-film transistors. IEEE Trans Electron Dev 40:890–897
17. Dresner J (1991) Dynamic changes in characteristics of a-Si transistors during fast pulsed operation. IEEE Trans Electron Dev 38:2673–2676
18. Gu G et al. (2005) Electron traps and hysteresis in pentacene-based organic thin-film transistors. Appl Phys Lett 87:243512-1–243512-3
19. Zilker SJ et al. (2001) Bias stress in organic thin-film transistors and logic gates. Appl Phys Lett 79:1124–1126
20. Jung T et al. (2005) Moisture induced surface polarization in a poly(4-vinyl phenol) dielectric in an organic thin-film transistor. Appl Phys Lett 87:182109-1–182109-3
21. IEEE Std 1620-2004, IEEE standard for test methods for the characterization of organic transistors and materials. IEEE, New York
22. Brody TP (1996) The birth and early childhood of active matrix - a personal memoir. J Soc Inf Display 4:113–127
23. Alt PM, Pleshko P (1974) Scanning limitations of liquid-crystal displays. IEEE Trans Electron Dev ED-21:146–155
24. Huitema E et al. (2002) Polymer-based transistors used as pixel switches in active-matrix displays. J Soc Inf Display 10:195–202
25. Martine S et al. (2003) Organic-polymer thin-film transistors for active-matrix flat-panel displays? J Soc Inf Display 11:543–549
26. Rogers JA et al. (2001) Paper-like electronic displays: Large-area rubber stamped plastic sheets of electronics and microencapsulated electrophoretic inks. Proc Nat Acad Sci 98:4835–4840
27. Kane MG et al. (2001) AMLCDs using organic thin-film transistors on polyester substrates. SID Int Symp Dig Tech Papers, vol 32, pp 57–59
28. Mach P et al. (2001) Monolithically integrated flexible display of polymer-dispersed liquid crystal driven by rubber-stamped organic thin-film transistors. Appl Phys Lett 78:3592–3594
29. Huitema HEA et al. (2001) Plastic transistors in active-matrix displays. Nature 414:599
30. Burns SE et al. (2006) A flexible plastic SVGA e-paper display. SID Int Symp Dig Tech Papers, vol 37, pp 74–76
31. Hong MP et al. (2005) Recent progress in large sized and high performance organic TFT array. SID Int Symp Dig Tech Papers, vol 36, pp 23–25

32. Sheraw CD et al. (2002) Organic thin-film transistor-driven polymer-dispersed liquid crystal displays on flexible polymeric substrates. Appl Phys Lett 80:1088–1090

33. Kymissis I et al. (2002) Patterning pentacene organic thin-film transistors. J Vac Sci Technol B 20:956

34. DeFranco JA et al. (2006) Photolithographic patterning of organic electronic materials. Org Electron 7:22

35. Gu G et al. (2004) Organic thin-film transistor with photolithographically patterned top contacts and active layer. Device Res Conf Dig, pp 83-84

36. Dawson RMA et al. (1998) Design of an improved pixel for a polysilicon active matrix organic light emitting diode display. SID Int Symp Dig Tech Papers, vol 29, pp 11-14

37. Shimoda T et al. (1998) High resolution light emitting polymer display driven by low-temperature polysilicon thin-film transistor with integrated driver. Int Display Res Conf, pp 217–220

38. Kanzaki K, Sakamoto M (2001) Direction of low-temperature p-Si technology. SID Int Symp Dig Tech Papers, vol 32, pp 242-245

39. Wu CC et al. (1997) Integration of organic LEDs and amorphous Si TFTs onto flexible and lightweight metal foil substrates. IEEE Electron Dev Lett 18:609–612

40. Tsujimura T et al. (2003) A 20-inch OLED display driven by super-amorphous-silicon technology. SID Int Symp Dig Tech Papers, vol 34, pp 6–9

41. Chung K et al. (2006) Large-sized full-color AMOLED TV: Advancements and issues. SID Int Symp Dig Tech Papers, vol 37, pp 1958–1963

42. Nathan A et al. (2004) Amorphous silicon backplane electronics for OLED displays. IEEE J Select Top Quant Electron 10:58–69

43. Sirringhaus H et al. (1998) Integrated optoelectronic devices based on conjugated polymers. Science 280:1741–1744

44. Dodabalapur A et al. (1998) Organic smart pixels. Appl Phys Lett 73:142–144

45. Chuman T et al. (2004) Active matrix organic light emitting diode panel using organic thin-film transistors. SID Int Symp Dig Tech Papers, vol 35, pp 45–47

46. Zhou L et al. (2005) Pentacene TFT driven AMOLED displays. IEEE Electron Dev Lett 26:640–642

47. Suh M et al. (2006) 4.0 inch organic thin-film transistor (OTFT) based active matrix organic light emitting diode (AMOLED) display. SID Int Symp Dig Tech Papers, vol 37, pp 116–118

48. Brody TP et al. (1975) A 6×6-in 20-lpi electroluminescent display panel. IEEE Trans Electron Dev ED-22:739–748

49. Kim B-K et al. (2004) Recoverable residual image induced by hysteresis of thin-film transistors in active matrix organic light-emitting diode displays. Jpn J App Phys 43:L482–L485

50. Dawson RMA et al. (1999) A poly-Si active matrix OLED display with integrated drivers. SID Int Symp Dig Tech Papers, p 30

51. Dawson RMA et al. (1998) The impact of the transient response of organic light-emitting diodes on the design of active matrix OLED displays. Int Electron Devices Meeting, pp 875–878

52. Dawson RMA, Kane MG (2001) Pursuit of active matrix organic light-emitting diode displays. SID Int Symp Dig Tech Papers, vol 32, pp 372–375

53. Fish D et al. (2002) A comparison of pixel circuits for active matrix polymer/organic LED displays. SID Int Symp Dig Tech Papers, vol 33, pp 968–971

54. Sanford JL, Libsch FR (2003) TFT AMOLED pixel circuits and driving methods. SID Int Symp Dig Tech Papers, vol 34, pp 10–13

55. Nathan A et al. (2005) Driving schemes for a-Si and LTPS AMOLED displays. J Display Tech 1:267–277

56. Fish D et al. (2005) Optical feedback for AMOLED display compensation using LTPS and a-Si:H technologies. SID Int Symp Dig Tech Papers, vol 36, pp 1340–1343

57. Jackson TN et al. (1998) Organic thin-film transistors for organic light-emitting flat-panel display backplanes. IEEE J Select Top Quant Electron 4:100–104

58. Aerts WF et al. (2002) Design of an organic pixel addressing circuit for an active-matrix OLED display. IEEE Trans Electron Dev 49:2124–2130
59. IEEE Std 1620.1-2006, IEEE standard for test methods for the characterization of organic transistor-based ring oscillators. IEEE, New York
60. Krumm J et al. (2004) A polymer transistor circuit using PDHHT. IEEE Electron Dev Lett 25:399–401
61. Drury CJ et al. (1998) Lost-cost all-polymer integrated circuits. Appl Phys Lett 73:108–110
62. van Lieshout PJG et al. (2004) System-on-plastic with organic electronics: A flexible QVGA display and integrated drivers. SID Int Symp Dig Tech Papers, vol 35, pp 1290–1293
63. Kane MG et al. (2000) Analog and digital circuits using organic thin-film transistors on polyester substrates. IEEE Electron Dev Lett 21:534–536
64. Gay N et al. (2006) Analog signal processing with organic FETs. IEEE Int Solid-State Circuits Conf, pp 278–279
65. Carr WN, Mize JP (1972) MOS/LSI design and application. McGraw-Hill, New York
66. Iba S et al. (2005) Control of threshold voltage of organic field-effect transistors with double-gate structures. Appl Phys Lett 87:023509-1–023509-3
67. Gelinck GH et al. (2005) Dual-gate organic thin-film transistors. Appl Phys Lett 87:073508-1–073508-3
68. Morana M et al. (2005) Double-gate organic field-effect transistor. Appl Phys Lett 87:153511-1–153511-3
69. Koo JB et al. (2006) Pentacene thin-film transistors and inverters with dual-gate structure. Electrochem Solid-State Lett 9:G320–G322
70. Takamiya M et al. (2006) An organic FET SRAM for braille sheet display with back gate to increase static noise margin. IEEE Int Solid-State Circuits Conf, pp 276-277
71. Stewart RG et al. (1995) Circuit design for a-Si AMLCDs with integrated drivers. SID Int Symp Dig Tech Papers, vol 26, pp 89–92
72. Lebrun H et al. (2005) Design of integrated drivers with amorphous silicon TFTs for small displays: Basic concepts. SID Int Symp Dig Tech Papers, vol 36, pp 950–953
73. Finkenzeller K (2003) RFID handbook, 2nd ed. Wiley, West Sussex, England
74. Paret D (2005) RFID and contactless smart card applications. Wiley, West Sussex, England
75. Baude PF et al. (1993) Organic semiconductor RFID transponders. Int Electron Devices Meeting Tech Dig, pp 191–194
76. Baude PF et al. (2004) Pentacene based RFID transponder circuitry. Device Research Conf Dig, pp 227–228
77. Böhm M et al. (2006) Printable electronics for polymer RFID applications. IEEE Int Solid-State Circuits Conf, pp 270–271
78. Cantatore E et al. (2006) A 13.56 RFID system based on organic transponders. IEEE Int Solid-State Circuits Conf, pp 272–273
79. Subramanian V et al. (2005) Progress toward development of all-printed RFID tags: Materials, processes, and devices. Proc IEEE 93:1330–1338
80. Steudel S et al. (2005) 50 MHz rectifier based on an organic diode. Nature Mater 4:597–600
81. Masui S et al. (1999) A 13.56 MHz CMOS RF identification transponder integrated circuit with a dedicated CPU. IEEE Int Solid-State Circuits Conf, pp 162-163
82. Rotzoll R et al. (2006) Radio frequency rectifiers based on organic thin-film transistors. Appl Phys Lett 88:123502-1–123502-3
83. Ma L et al. (2004) High-speed and high current density C60 diodes. Appl Phys Lett 84:4786–4788
84. Huang D, Subramanian V (2006) Iodine-doped pentacene schottky diodes for high-frequency RFID rectification. Device Research Conf Dig, pp 219–220
85. Peumans P et al. (2000) Efficient, high-bandwidth organic multilayer photodetectors. Appl Phys Lett 76:3855–3857
86. Peumans P et al. (2003) Small molecular weight organic thin-film photodetectors and solar cells. J Appl Phys 93:3693–3723

87. Mattis B, Subramanian V (2006) A field-programmable antifuse memory for RFID on plastic. Device Research Conf Dig, pp 215–216
88. Ma LP et al. (2002) Organic electrical bistable devices and rewritable memory cells. Appl Phys Lett 80:2997–2999
89. Bozano LD et al. (2004) Mechanism for bistability in organic memory devices. Appl Phys Lett 84:607–609
90. Tondelier D et al. (2004) Metal/organic/metal bistable memory devices. Appl Phys Lett 85:5763–5765
91. Leong WL et al. (2007) Charging phenomena in pentacene-gold nanoparticle memory device. Appl Phys Lett 90:042906-1–042906-3
92. Singh TB et al. (2004) Nonvolatile organic field-effect transistor memory element with a polymeric gate electret. Appl Phys Lett 85:5409–5411
93. Naber RCG et al. (2005) Low-voltage polymer field-effect transistors for nonvolatile memories. Appl Phys Lett 87:203509-1–203509-3
94. Sirringhaus H et al. (2003) Active matrix displays made with printed polymer thin-film transistors. SID Int Symp Dig Tech Papers, vol 34, pp 1084–1087
95. Huitema HEA et al. (2003) A flexible QVGA display with organic transistors. Proc. Int. Display Workshop. Fukuoka, Japan, pp 1663–1664
96. Ho J-C et al. (2004) Pentacene organic thin-film transistor integrated with color twisted nematic liquid crystals display (CTNLCD). SID Int Symp Dig Tech Papers, vol 35, pp 1298–1301
97. Nomoto K et al. (2005) A high-performance short-channel bottom-contact OTFT and its application to AM-TN-LCD. IEEE Trans Electron Dev 52:1519–1526
98. Kawasaki M et al. (2006) High-resolution full-color LCD driven by OTFTs using novel passivation film. IEEE Trans Electron Dev 53:435–441
99. Kitamura M et al. (2003) Organic light-emitting diodes driven by pentacene-based thin-film transistors. Appl Phys Lett 83:3410–3412
100. Seong R-G et al. (2005) Flexible AMOLED backplane technology using pentacene TFTs. Int Symp Super-Functionality Organic Devices, pp 146–149
101. Mizukami M et al. (2006) Flexible AMOLED panel driven by bottom-contact OTFTs. IEEE Electron Dev Lett 27:249–251

Chapter 9
Semiconducting Polythiophenes for Field-Effect Transistor Devices in Flexible Electronics: Synthesis and Structure Property Relationships

Martin Heeney and Iain McCulloch

9.1 Introduction

Interest in the field of organic electronics has burgeoned over the last 10 years, as the continuing improvement in performance has transitioned the technology from an academic curiosity to the focus of intense industrial and academic research. Much of this interest is driven by the belief that organic materials will be readily amenable to low-cost, large-area deposition techniques, enabling both significant cost savings and the ability to pattern flexible substrates with active electronics. Potential applications include thin-film transistor (TFT) backplanes for a variety of display modes including active matrix liquid crystal displays (AMLCDs), flexible displays such as e-paper, disposable item level radio frequency identity (RFID) tags, flexible solar cells, and cheap and disposable sensors.

The primary figure of merit for charge transport in organic semiconducting materials is the charge carrier mobility. A current target for an organic TFT is to achieve a mobility of around 1 cm^2/Vs, which would equal the approximate performance of the amorphous silicon currently used as semiconductor in AMLCD TFT backplanes. The opportunity to replace amorphous silicon with a printable organic semiconductor of similar electrical performance in an additive process on flexible substrates is a potentially attractive cost proposition to device manufacturers.

In principle each of the components of the organic field-effect transistor (OFET) can be an organic material and several fully organic device examples have been reported [1–4]. In practice, however, one or more of the components is usually nonorganic. For the purpose of comparative evaluation of materials, a particularly common device set-up comprises of n-doped silicon as the gate electrode, with a layer of thermally grown silicon dioxide on top as the dielectric layer, and patterned source and drain electrodes. The dielectric provides a very smooth and homogeneous surface onto which to deposit the organic semiconductor, and this device

M. Heeney (✉)
Department of Materials, Queen Mary, University of London, Mile End Road, London, E1 4NS, UK
e-mail: m.heeney@qmul.ac.uk

W.S. Wong, A. Salleo (eds.), *Flexible Electronics: Materials and Applications*, Electronic Materials: Science & Technology, DOI 10.1007/978-0-387-74363-9_9,
© Springer Science+Business Media, LLC 2009

architecture provides a useful tool for the rapid screening of semiconductor material properties. Although each component has an important role in the performance of the overall device, only the role of the semiconducting material is considered in this chapter.

Organic semiconductors can be classified into two broad classes: small molecule semiconductors which are discreet, chemically distinct compounds, and semiconducting polymers. The exemplary small molecule semiconductor is pentacene, which is a fused oligoacene consisting of five linearly fused benzene rings. Charge carrier mobilities of around 1 cm^2/Vs for thin films were first reported in 1997 by Jackson and coworkers [5]. Since then, the performance has steadily improved, as a better understanding and control of thin-film morphology and device optimization was achieved, with recent reports quoting mobilities as high as 3.5 cm^2/Vs for thin-film devices [6]. Devices fabricated on single crystals of pentacene have demonstrated impressive mobilities of 35 cm^2/Vs [7], giving some indication of the possible upper limits for thin-film devices. However, pentacene has some drawbacks as a semiconducting material. In the presence of oxygen and light, pentacene readily undergoes photooxidation, and unencapsulated devices operated in ambient air rapidly deteriorate [8]. In addition, pentacene is very poorly soluble in most organic solvents, and is most readily processed by vacuum-deposition techniques. This excludes the possibility of solution patterning of the semiconductor by inexpensive and widely available techniques such as inkjet printing. Although notable progress has been made in the development of vapor deposition techniques which are able to pattern the growth of the semiconductor, either by use of prepatterned surface templates to direct the crystal growth [9–11] or by organic vapor-phase deposition through an appropriate nozzle [12, 13], solution deposition remains an attractive prospect, especially for high-throughput and/or large-area applications. The introduction of bulky alkynyl silyl groups in the central 6,13-positions of the pentacene ring has been shown to significantly improve both solubility and solution stability by Anthony and coworkers [14–17]. In addition, the substituents dramatically alter the crystalline packing of material, resulting in the formation of highly ordered 2-D slipped stack arrays in certain derivatives of substituted pentacenes or related dithienoanthracenes. These materials display highly impressive mobilities in excess of 1 cm^2/Vs for drop-cast films [16]. However, potential problems still remain, in particular issues relating to interlayer mixing during the solution deposition of subsequent dielectric or encapsulation layers, control of the crystallization process over large substrate areas, and the anisotropic in-plane transport that causes device-to-device nonuniformity.

Polymeric semiconductors have many potential advantages over small molecules. Higher solution viscosities can be achieved, opening up the possibility to formulate solutions compatible with printing techniques such as gravure and flexography [18]. The rheological properties of polymer formulations are also advantageous in the fabrication of cohesive, conformal thin films from solution casting processes. Thin-film reticulation, which often occurs on drying from solution on a low-energy substrate, can be controlled through optimization of polymer molecular weight and polydispersity. Fabrication of multilayer device stacks from solution deposition processes requires that each layer is impervious to the solvents and temperatures that are subsequently used during manufacture of the device. Polymers, with their narrow solubility parameter window, negligible vapor pressure, and high bulk viscosity, typically can be exposed to a wider range of solvents while remaining inert, thus expanding the choice of materials that can be used in devices. For example, top-gate transistor devices incorporating small molecule semiconductors require deposition of the dielectric layer from a very narrow range of solvents on the extremes of high and low polarity. The dielectrics that can therefore be employed need to be soluble in this narrow solvent range, which limits choice. The mechanical properties of polymer semiconductors are also superior to small molecules, making thin polymer films more compatible with flexible processing or flexible operation. Isotropic in-plane transport can be achieved from solution deposition of even highly crystalline polymer semiconductors, due to their small thin-film crystalline domain sizes (typically tens of nanometers) relative to the transistor channel length. This characteristic results in low device-to-device performance variability, which is particularly important in TFT devices for organic light-emitting diodes (OLED) applications.

Both amorphous and crystalline polymer semiconductors have been widely studied. Amorphous polymers such as polyarylamines have demonstrated excellent ambient stability with mobilities of around 0.005 cm^2/Vs and isotropic in-plane transport achieved in transistors with high work function electrodes and low-k dielectrics, with minimal thermal annealing of the semiconductor required [19]. Most high performing crystalline semiconducting polymers are at least in part comprised of either a fluorene or a thiophene unit in the backbone. Polyfluorenes are rigid rod polymers which can be rendered soluble in organic solvents by appropriate substitution at the bridging C9-position. Alkyl substituted polyfluorenes can exhibit high-temperature liquid crystalline phases which can be exploited to achieve the optimal microstructure in transistor devices [20]. This class of polymer has very low lying highest occupied molecular orbital (HOMO) energy levels, leading to poor charge injection in OFET devices and poor charge transport, most likely due to the poor backbone packing attributed to the nonplanar projection of the alkyl groups at the bridging position of the fluorene unit. Incorporation of a bithiophene unit to form the alternating copolymer poly(9,9-dioctylfluorene-co-bithiophene) (F8T2) resulted in an increase in both HOMO energy level (from −5.8 to −5.4 eV) and improved charge transport [21]. Orientation of the polymer backbones can be achieved by a thermal annealing step at the mesophase temperature utilizing a rubbed polyimide alignment layer as substrate [22]. Mobilities of up to 0.02 cm^2/Vs have been reported for this polymer, with good ambient stability. The highest mobilities measured in solution processed all-polymer FET devices have been exhibited by thiophene polymers. Thiophene is an electron-rich, planar aromatic heterocycle, which can form a variety of conjugated polymers when coupled appropriately. The crystalline nature of many functionalized thiophene derivatives contributes to their excellent charge transport properties. The most widely studied semiconducting polymer for charge transport applications is poly(3-hexylthiophene) (P3HT), which will be highlighted in this article.

9.2 Polymerization of Thiophene Monomers

9.2.1 General Considerations

The choice of synthetic route for the preparation of polythiophene derivatives is an important factor in determining the electrical performance of the resulting polymer. Molecular weight, polydispersity, defects in the polymer backbone and impurities levels are all influenced by the choice of synthetic route, and these can have a significant influence on electrical performance. Several groups have investigated the influence of molecular weight upon transistor performance for various classes of thiophene polymer [23–28]. All studies have shown a beneficial effect of increasing molecular weight, and although a plateau region is typically reached, average molecular weights above 20,000 KDa are usually desirable. At very high-molecular weights (Mn >150 KDa), the high polymer viscosity can hinder crystallization during annealing, resulting in a nonoptimal mobility. A systematic study of the influence of polydispersity upon charge carrier mobility has not been reported, but high polydispersity would be expected to reduce the crystallinity of the polymer and therefore be detrimental to device performance. Impurities resulting from the synthesis such as catalyst residues can have a deleterious impact on device performance [29], but they can generally be removed by appropriate purification techniques such as sequestration, washing, or reprecipitation [30]. Rather harder to remove are defects which are chemically bound into the polymer backbone, and the best strategy is to minimize defects by the choice and optimization of synthetic route.

9.2.2 Synthetic Routes for the Preparation of Thiophene Polymers

Many synthetic routes have been described for the preparation of thiophene containing polymers [31–33], but they can usually be categorized by one of three synthetic methods: chemical oxidation, electrochemical oxidation, or transition metal catalyzed cross-coupling chemistries. These will be described in more detail in the following sections. Unsubstituted polythiophene is very insoluble in most organic solvents, making it difficult to process once prepared and is of limited use for transistor devices. The introduction of solubilizing substituents, typically straight chain alkyl groups, into the 3 and/or 4 positions of the thiophene ring greatly enhances the solubility of the resulting polymers. The introduction of groups into the 3-position of the thiophene ring results in a noncentrosymmetric monomer, polymerization of which can afford three possible side chain regiochemistries in the polymer backbone, head-to-tail (HT), head-to-head (HH), or tail-to-tail (TT) (Fig. 9.1). Head-to-head linkages result in stronger steric interactions that cause twisting of the adjacent monomers out of the backbone plane, interrupting the conjugation length. In comparison, steric interactions are reduced in HT or TT couplings and the planarity of the backbone is maintained. A number of studies have shown that absorption

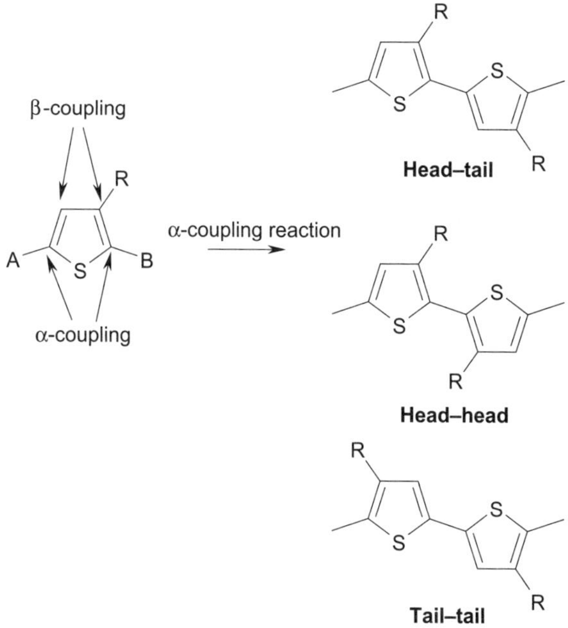

Fig. 9.1 Possible
regiochemical isomers
formed by the polymerization
of 3-alkylthiophene

wavelength [34], conductivity [35], field-effect charge carrier mobility [36], and photovoltaic performance [37] are all critically influenced by the percentage of HT couplings in the polymer backbone. The percentage of HT couplings present is commonly referred to as the regioregularity (RR) of the polymer, and is strongly dependent on the synthetic method used to make the polymer.

9.2.2.1 Chemical Oxidation Routes

The synthesis of polythiophene by chemical oxidation was first reported in 1984 by Sugimoto and coworkers, who described the polymerization of thiophene by chemical oxidation with iron (III) chloride in chloroform (Scheme 9.1). Other oxidants such as copper (II) perchlorate [38] or vanadium acetyl acetate [39] have also been used. The initial step of the mechanism is the oxidation of the monomer to a radical cation. The subsequent step is the subject of some debate, but occurs either by the dimerization of two radical cations [40, 41] or by the electrophilic addition of a neutral thiophene monomer to a radical cation [42]. For both mechanisms, the couplings occur at the positions of highest electron density on the monomer, which for thiophenes tends to be the alpha (α) positions rather than the beta (β) positions, with a difference in reactivity of about 95:5 [43]. Thus, polymerization occurs mainly at the $\alpha\alpha'$- positions, shown in Fig. 9.1. However undesired $\alpha\beta$ miscouplings can occur as a defect in the main chain [44], having a detrimental impact on the conjugation and crystallinity of the polymer. The use of trialkylsilyl substituents as sacrificial

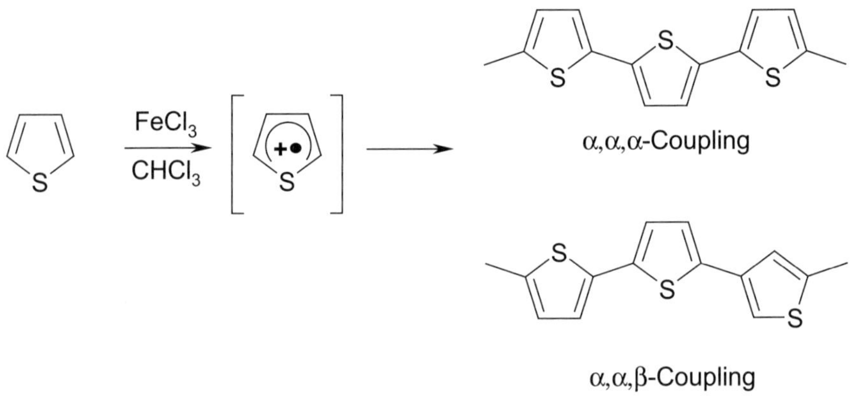

Scheme 9.1 Oxidative coupling of thiophene with ferric chloride results in predominately α-coupled polymer, but defects can be introduced by the presence of undesirable αβ-couplings

leaving groups to direct the oxidative coupling have been investigated as a possible solution [45]. In this case, higher molecular weights and lower polydispersities in comparison to standard chemical oxidation are also reported. Although polymerization preferentially occurs at the α-positions, there is little differentiation between the two α-positions of a 3-alkylthiophene because of the small difference in electron density at the two positions. Some regioselectivity can be achieved thanks to steric effects, but generally the oxidative polymerization of 3-alkylthiophene's tends to give polymers with low RR, typically around 70–80% [46]. Detailed studies by Amou have shown that careful control over the temperature and reagent concentration of the oxidative coupling reaction can afford P3HT with RR up to 90% [47]. Nevertheless, highly RR P3AT tends to be synthesized by alternative routes, in part because of problems with reproducibility of the synthesis [48, 49].

The oxidative coupling reaction is particularly attractive for the polymerization of centrosymmetric monomers, since the problems with controlling RR are not a consideration. The reaction has been used by several groups to prepare substituted bithiophene [50, 51] and terthiophene derivatives [52–54]. Molecular weights vary widely according to the exact monomer and conditions used, from several thousand to values above 20,000 KDa.

The attractiveness of the chemical oxidation method is partly due to the simplicity of the reaction. There is no need to synthesize monomers containing reactive halogen or organometallic reactive groups. On the other hand, the potential presence of miscouplings in the backbone is a serious drawback. In addition, the polymers are synthesized in an oxidized or doped form, and must be reduced to their neutral form during work-up. The large excesses of iron salts (typically up to 4 equivalents) in the reaction can result in significant quantities (up to 0.5 wt%) of metal salts contaminating the final polymer [55]. The presence of iron impurities has been shown to have a significant detrimental effect on the photostability of poly(3-alkyl)thiophene thin films [56, 57].

9.2.2.2 Electrochemical Oxidation Routes

Electrochemical oxidation involves the in-situ polymerization of solution of monomer onto a working electrode (commonly ITO or glassy carbon disk) in the presence of a suitable electrolyte. The polymer forms directly onto the working electrode, where it can be studied by a variety of spectroscopic techniques. The method is most useful for the study of insoluble polymers, which would otherwise be difficult to isolate, purify, and analyze. In addition, the technique is best suited to regiosymmetrical monomers, since there is little control over the coupling of positions of similar electronic charge density in the monomer (for example, in the 2,5-positions of 3-alkylthiophene). Although soluble polymers can also be prepared, problems can exist with dissolution of the growing polymer back into the electrolyte solution, limiting molecular weight. In addition, the technique is not readily amenable to scale-up and only small amounts of polymer can usually be prepared. The polymers are also deposited in a doped state, where a counter-anion from the electrolyte solution has been incorporated into the polymer matrix. The resulting materials can be difficult to de-dope fully, and therefore one might expect high off-currents during transistor operation. The technique has been reviewed in detail by Roncali [43].

9.2.2.3 Transition Metal Catalyzed Cross-coupling Methodologies

The use of cross-coupling methodologies has been well explored in the synthesis of polythiophenes, with the synthesis of RR poly(3-alkyl)thiophene attracting particular interest. These methods have the advantage of using low catalyst loadings (typically around 1 mol%), so catalyst impurities can be more readily removed [58]. In addition, the regiochemistry of polymer is usually fixed by appropriate substitution of the monomer. There are two general approaches to the polymerization (Scheme 9.2). In the first, a difunctional monomer is utilized containing both an organometallic group and a halogen leaving group (typically bromine or iodine), the so-called AB polymerization. This has the advantage of maintaining the 1:1 stoichiometry that is required to achieve high-molecular weights according to the Carother's equation. In addition, the regiochemistry defined in the monomer is largely preserved in the polymer backbone, so this approach is particularly suited to the polymerization of nonregiosymmetrically substituted monomers. The drawback to the AB approach is that the monomers can be difficult to synthesize and purify. An alternative approach uses two different monomers each functionalized with either two halogen groups or two organometallic leaving groups, a so-called AA plus BB approach (Scheme 9.2). This affords alternating copolymers. In this case, it is highly desirable to have centrosymmetrically substituted monomers, since there is usually little differentiation in reactivity between the two groups in each monomer. For example, in Scheme 9.2, if R and R' are different substituents then polymerization via the AA plus BB approach will result in a regiorandom arrangement of R and R' along the polymer backbone. In order to maintain the 1:1 stoichiometry and afford high-molecular weight polymers, it is necessary to have high purity monomers, so crystalline monomers are highly desirable.

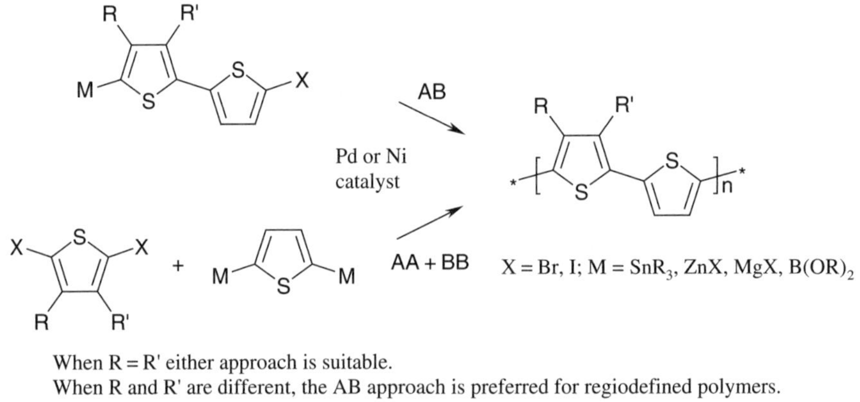
R
R'
M
S
S
X
AB
Pd or Ni
catalyst
R
R'
S
S
n
X
S
X
+
M
S
M
AA + BB
X = Br, I; M = SnR3, ZnX, MgX, B(OR)2
R
R'
When R = R' either approach is suitable.
When R and R' are different, the AB approach is preferred for regiodefined polymers.

Scheme 9.2 General cross-coupling strategies to conjugated polymers

Various organometallic groups such as organotin (Stille reaction), organoboron (Suzuki reaction), organozinc (Negishi reaction), or organomagnesium (Grignard reaction) have been explored. The latter two are reactive intermediates that cannot be readily isolated and stored, so they tend to be formed in situ during the polymerization process. Organo-tin and boron reagents tend to be stable, readily isolated intermediates that can be prepared and purified before use. Although the precise details vary according to the organometallic group used, the mechanism of the cross-coupling reactions can be generalized as follows. In the initial step, oxidative insertion of the catalyst into the carbon halogen bonds take place, followed by a transmetallation step between the resulting halogen transition metal complex and the organometallic group to afford a transition metal complex. In the final step, reductive elimination results in the formation of a new aryl–aryl bond and regeneration of the active catalyst. The final desired reductive elimination step can be in competition with a ligand rearrangement step, which results in either chain termination or an undesirable incorporation of ligand into the conjugated backbone. The rearrangement can be prevented by the use of bulky ligands [59]. In addition, homo-coupling of either the aryl halide or the organometallic groups can reduce control of the regiochemistry in the polymer backbone, as well as altering the 1:1 stoichiometry of the reaction mixture. These side reactions can be reduced by utilizing a transition metal catalyst in a zero oxidation state, rather than the commonly used +2 state [60]. Thus, the exact conditions of polymerization (transition metal, ligand, solvent, temperature, and optional base) often need optimization on a case-by-case basis to ensure high purity and high-molecular-weight material.

Organomagnesium Based Cross-coupling Polymerizations

One of the first chemical syntheses of unsubstituted polythiophene utilized the nickel catalyzed cross-coupling of 2-bromo-5-magnesiobromothiophene, prepared by insertion of magnesium metal into 2,5-dibromothiophene, to afford low-molecular-weight polythiophene [61]. Similar methods were used to prepare poly(3-alkyl)thiophene from the insertion of magnesium metal into 2,5-diiodo-3-alkylthiophenes [62]. However, the polymers prepared in these cases tended to have low regioregularities. Holdcroft demonstrated that this was due to the formation of mixtures of mono- and di-Grignard reagents during the magnesium insertion reaction, which tended to cross-couple to give large amounts of HH linkages [63]. The first synthesis of RR PAHT was reported by McCullough and coworkers, who used a bulky lithium amide base to selectively deprotonate the 5-position of 2-bromo-3-alkylthiophene (Scheme 9.3) [64]. No halogen scrambling was seen at low temperatures ($-78°C$), and the resulting organolithium intermediate could be transmetallated into a Grignard reagent by treatment with $MgBr_2$. Addition of a nickel (II) catalyst subsequently initiated the polymerization. Highly RR polymers were produced, with HT couplings over 98% and molecular weights (Mn) in the range of 20–40 K. Similar results were obtained from the polymerization of mono-alkylated bithiophenes [65].

McCullough and coworkers later reported an improved preparation of RR P3HT which did not require low temperatures, the so-called Grignard Metathesis reaction or GRIM reaction (Scheme 9.3) [66]. In this method, a solution 2,5-dibromo-3-alkylthiophene, which is more readily prepared and purified than 2-bromo-3-alkylthiophene, was treated with one equivalent of an alkyl Grignard reagent. A metathesis reaction occurred to afford an alkylbromide and a mixture of 2-bromo-5-magnesiobromo-3-alkylthiophene and 5-bromo-2-magnesiobromo-3-alkylthiophene in about an 85:15 ratio, with no di-Grignard reagent formed. Interestingly, treatment of this mixture with a $Ni(dppp)Cl_2$ catalyst resulted in the formation of highly RR polymers, in good molecular weight, low polydispersities, and

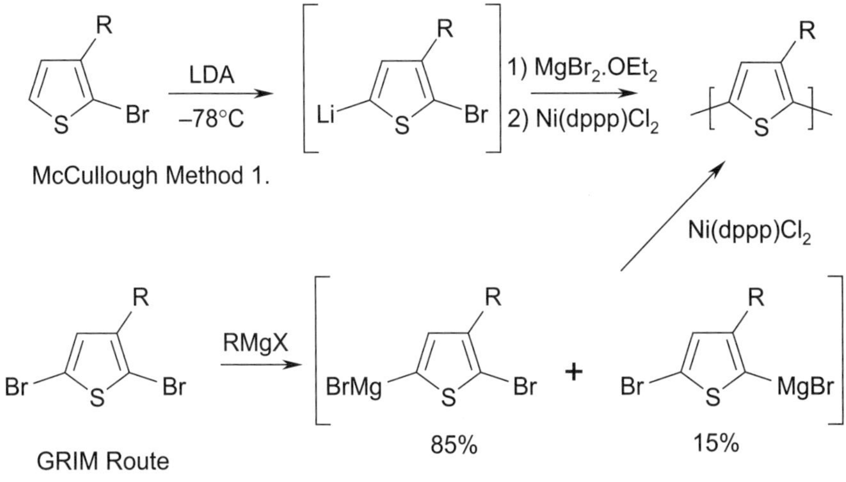

Scheme 9.3 Two methods of preparation of regioregular poly(3-alkyl)thiophene by the McCullough group

high yield. Bolognesi reported a similar study utilizing 2,5-diiodo-3-alkylthiophenes around the same time [67]. Detailed mechanistic studies [68] have shown that the formation of highly RR polymers from the mixture of Grignard reagents can be rationalized due to the slow rate of head-to-head coupling reactions in combination with the tendency for only one TT miscoupling to occur per polymer chain. Subsequent studies have found that the reaction proceeds via a chain-growth mechanism [69–71] whereupon the catalyst remains associated with the growing polymer chain, either as an "associated pair" [70] or via an intramolecular transfer [69] to the chain end after reductive elimination. The ability to prepare P3ATs with narrow poly-dispersities and defined endgroups has been exploited by several groups to form interesting block copolymer structures [72, 73], where blocks of highly RR P3HT can be combined with amorphous or crystalline nonconjugated polymers. These approaches offer the attractive combination of the processing properties of noncon-jugated polymers with the functional properties of conjugated materials, as well as the ability to drive self-organization through phase separation. For example, McCul-lough and coworkers reported the formation of well-defined nanowires in thin films of P3HT-polystyrene or poly(methyl acrylate) copolymers [72, 73].

Whilst organomagnesium chemistry is probably the method of choice for the preparation of RR P3HT and its analogues [74, 75], it has not been widely adopted for the preparation of other thiophene polymers. This is most likely due to prob-lems in selectively preparing monofunctional Grignard reagents in more complex systems [76].

Organozinc Based Cross-coupling Polymerizations

An alternative approach to RR P3HT was reported in 1992 by Rieke and cowork-ers [34, 77]. They treated a solution of finely dispersed, highly reactive "Rieke" zinc (Zn*) with 2,5-dibromo-3-alkylthiophene at low temperature. Zinc insertion occurred mainly at the 5-position, although small amounts of the two isomers were always formed, with the regioselectivity being dependent on the temperature of addition. Addition of a transition metal catalyst resulted in high-molecular-weight polymers. The authors were able to control the degree of RR by choice of cata-lyst and phosphine ligand; large transition metals such as palladium led to lower degrees of RR than smaller metals like nickel, and bidentate phosphine ligands like diphenylphosphinoethane (DPPE) led to higher degrees of RR than more labile ligands like triphenylphosphine. Regioregularities greater than 98% were reported for the best combination {Ni(dppe)Cl$_2$} [34, 77]. This route has also been used to prepare RR poly(3-alkylthio)thiophene [78] and mono-alkylated poly(bithiophenes) [79]. The Rieke route offers the advantage of high-functional group tolerance. Organozinc reagents are much less nucleophilic than Grignard reagents, so sensi-tive functionalities such as ester groups can be incorporated onto the side chains. The organozinc intermediate can also be generated in situ by transmetallation of an organolithium intermediate with zinc chloride, in a similar approach to the first McCullough method [65, 80]. If the organolithium intermediate is generated at low temperature using nonnucleophilic bases such as LDA, base-sensitive functionalities such as cyano groups can be included on the backbone [65].

Organoboron Based Cross-coupling Polymerizations

The cross-coupling of an organoboron reagent with an aryl halide in the presence of a base and transition metal catalyst is called the Suzuki reaction. It has been widely and successfully utilized for the preparation of many conjugated polymers, especially those containing fluorene or phenyl backbones [81]. However, there have been considerably fewer examples of the successful synthesis of thiophene containing polymers by the Suzuki reaction. One reported problem is that the electron-rich thiophene boronates can be prone to deboronation during the polymerization, leading to low-molecular weights. This can be improved by using more stable thiophene boronate esters [82, 83]. Another problem is early termination of polymerization by the transfer of aryl groups from the triphenylphosphine ligands used in the polymerization [82]. By utilizing ligand free cross-coupling conditions, Bidan reported the successful synthesis of poly(3-octylthiophene) in reasonable molecular weights and RR (Scheme 9.4) [84]. Janssen reported a similar increase in molecular weight with ligand free conditions for thiophene copolymers, although the degree of polymerization was still very low [85]. Higgins and coworkers later explored the use of bulky, electron-rich phosphine ligands to inhibit ligand transfer and facilitate the oxidative addition of palladium catalyst in the electron-rich thiophene halide (Scheme 9.4) [83].

Stille Reaction

The Stille reaction has been widely used in the preparation of thiophene polymers generally affording good results [86, 87]. The polymerization occurs under neutral conditions, allowing the introduction of acid or base sensitive functionality onto the polymer backbone. Bao and coworkers reported one of the first detailed studies of the Stille polymerization, utilizing a variety of different monomers [60]. They observed that stannyl thiophenes exhibited much higher reactivity than aryl or vinyl stannanes, affording higher molecular weight polymers. They attributed this to the electron donating effect of the sulfur atom in thiophene, which may accelerate the rate-limiting transmetallation step of the catalytic cycle. Iraqi et al. reported the synthesis of RR P3HT by the polymerization of 2-iodo-3-hexyl-5-tri-

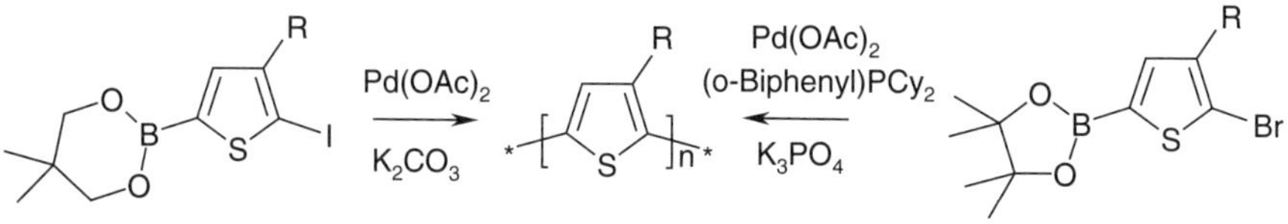

Scheme 9.4   Preparation of P3HT by Suzuki coupling

n-butylstannylthiophene in an AB type polymerization [88]. The RR was greater than 96% with molecular weight around 20,000 KDa. Interestingly, the polymer could be isolated with the tributylstannyl endgroups still intact, which could then be utilized in additional cross-coupling chemistry. Poly(bithiophenes) containing one solubilizing side chain were similarly prepared in high-molecular weight and good RR by the polymerization of a bifunctional monomer [89]. Similar polymerizations of AB type thiophene monomer have been used to prepare polythiophenes with functionalized side chains [90–93]. McCullough reported the use of the Stille polymerization to make vinyl substituted polythiophenes in 90% RR from the reaction of 2,5-dibromo-3-hexylthiophene and 1,2-bis(tributylstannyl)ethylene [94]. It is unusual to have such high RR from the reaction of an AA monomer with a BB monomer. In this case, it may originate from preferential oxidative addition at the less sterically hindered 5-position to generate a vinyl stannane AB-type dimer, which then couples with itself to generate the polymer. Yamamoto et al. also reported the preparation of a thiazole–thiophene copolymer in great than 90% RR from the reaction of 2,5-dibromo-3-alkylthiazole with a bis(trimethylstannyl)thiophene [95]. Such high RR may also stem from preferential oxidative addition at the 2-position of the thiazole. Copolymers of thiophene with fluorene [96], naphthalene [97], benzimidazole [98], siloles [87], and bithiophenes [86] have all been reported by an AA plus BB type Stille polymerization. Tierney and coworkers reported that AA plus BB type Stille polymerizations in all thiophene systems could be accelerated under microwave heating, reducing reaction times from days to minutes and affording excellent yields of polymer with high-molecular weights [99]. These conditions were utilized to prepare a range of copolymers containing fused thieno(2,3-b)thiophene [100] or thieno(3,2-b)thiophene [101] in the backbone (Scheme 9.5).

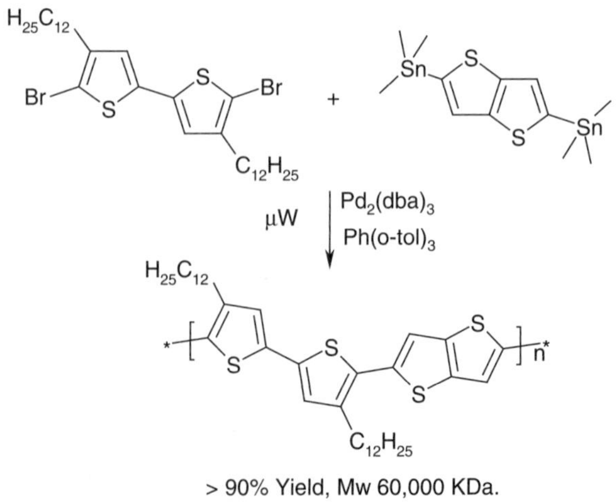

> 90% Yield, Mw 60,000 KDa.

Scheme 9.5 Preparation of thieno(3,2-b)thiophene copolymers by the microwave accelerated Stille reaction

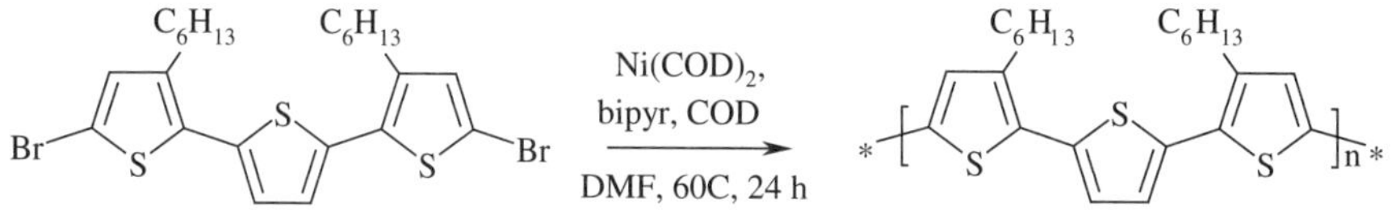

Scheme 9.6 Synthesis of Poly(3,3″-dihexylterthiophene) as a representative dehalogenative coupling procedure [86]

9.2.2.4 Dehalogenative Polymerization

Yamamoto first reported the polymerization of 2,5-dibromothiophene by heating with an excess of zero valent nickel catalyst bis(1,5-cyclooctadiene nickel (0) $Ni(COD)_2$) in the presence of triphenylphosphine (Scheme 9.6) [102]. Polymerization occurs exclusively at the 2,5-positions, and the yield is usually high. The polymerization of 2,5-dibromo-3-alkylthiophene was also reported [102]. In this case, the polymerization mainly afforded head-to-head and tail-to-tail couplings due to preferential oxidative addition at the less sterically hindered 5-position of the monomer. Polymerization also occurs with the more reactive di-iodo monomer.

The high yield, functional group compatibility and simplicity of the reaction have made it widely utilized in the synthesis of many conjugated polymers [103]. The reaction is best suited to the polymerization of centrosymmetric monomers where RR is not an issue. For more complex polymers, the main issue is the synthesis of the required di-halogenated starting monomer in sufficient purity. One drawback to the reaction is the use of stoichiometric amounts of expensive and highly toxic nickel catalyst. Catalytic versions of the reaction have been utilized where the nickel (II) produced by the coupling of two aryl monomers is reduced in situ by a reductant, such a metallic zinc, to afford the reaction Ni (0) catalyst [103].

9.3 Poly(3-Alkylthiophenes)

The observation that the alkyl side chains of polyalkylthiophenes when regioarranged in a head-to-tail configuration can give rise to enhanced charge carrier mobility [36] has provoked widespread interest in this class of polymer. RR P3HT has emerged as a benchmark semiconducting polymer due to its ready availability, ease of processing from solution, and its promising electrical properties. When the number average molecular weight (Mn) is greater than 20 KDa with a corresponding polydispersity of less than 2, and the RR is greater than 96%, P3HT is a highly crystalline polymer with a sharp melting temperature of 240°C, as can be seen from the differential scan calorimetry (DSC) shown in Fig. 9.2a. A glass transition temperature (Tg) of about 150°C can also be observed in this scan, although other physical analytical techniques such as dynamic mechanical analysis (DMA) have suggested that the Tg is lower than this, at around 50°C. The electron-rich, π-conjugated and highly planar backbone contribute to a high HOMO energy level of about –4.6 eV, as measured by ultraviolet photoelectron spectroscopy (UPS) shown in Fig. 9.2c, and

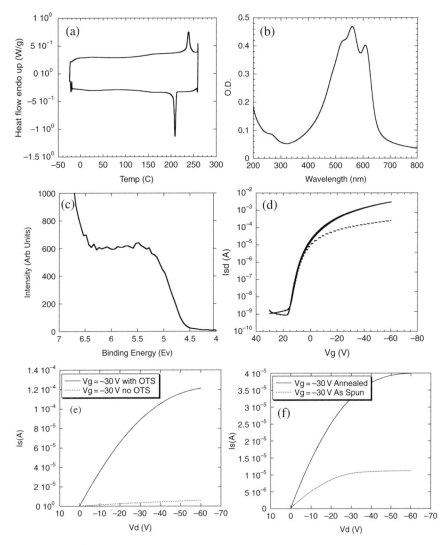

Fig. 9.2 Characterization of regioregular poly(3-hexylthiophene) comprising (**a**) differential scan calorimetry (DSC) graph (second scan), (**b**) thin-film UV-vis absorption spectrum, (**c**) Ultraviolet photoelectron spectroscopy (UPS) spectrum, and (**d**) OFET transfer characteristics in both linear (*dashed line*) and saturation (*solid line*) regime from a doped silicon bottom-gate device with thermally grown 230 nm thick SiO₂ dielectric, treated with an octyltrichlorosilane (OTS) SAM, with lithographically patterned gold electrodes where L = 10 μm and W = 1 cm and spin coated (from o-dichlorobenzene) P3HT annealed at 100°C for 10 min under nitrogen. The saturation mobility reached 0.1 cm²/Vs; (**e**) output characteristics of a P3HT transistor (architecture same as (**d**), with 20 μm channel length) with and without treatment of the dielectric interface with the hydrophobic OTS SAM, and (**f**) output characteristics of a P3HT transistor (architecture same as (**e**)) before and after annealing at 100°C for 10 min

a thin-film electronic absorption maximum of about 560 nm, as shown in Fig. 9.2b. The additional fine structures that can be observed at wavelengths greater than the absorption maximum are attributed to solid-state aggregation [104].

9.3.1 Electrical Properties

When the head to tail regioregularity of P3HT exceeds about 96%, charge carrier mobilities of up to 0.1 cm^2/Vs have been exhibited in an inert atmosphere [105]. The transfer characteristics shown in Fig. 9.2d are from P3HT device with a RR of 98%. The transistor architecture was a bottom contact, bottom-gate device with SiO_2 dielectric, gold source, and drain electrodes, with a channel length of 10 μm. Both forward and reverse curves are shown, illustrating negligible hysteresis, at source drain voltages of 5 V (dashed line) and 60 V (solid line). Both curves show a positive threshold voltage of about +15 V, which is believed to be due to both slight intrinsic doping of the semiconductor as well as dipole effects from the dielectric. The in-plane electrical properties of these crystalline films are dramatically influenced by many device and fabrication properties. For example, treatment of a bare SiO_2 dielectric surface with a hydrophobic self-assembled monolayer (SAM), such as octyltrichlorosilane (OTS) [106], results in improved P3HT mobility and higher on-currents, as demonstrated in the output characteristics shown Fig. 9.2e. It is believed that a contributing factor to reduce the mobility on bare silicon oxide surface, in addition to the presence of surface trap sites [107], is that the more polar surface can "anchor" polymer chains, inhibiting large-scale self-assembly at the critical dielectric–semiconductor interface. This leads to faster nucleation and a less organized semiconductor microstructure. Particularly on the treated, low-energy surface, an annealing step also improves charge carrier mobility, even at temperatures that appear to be lower than the thin-film Tg. This is illustrated in the output characteristics shown in Fig. 9.2f for both as cast and annealed devices, where the annealed device has a higher output current by a factor of about 4. It has also been observed that P3HT devices often show an increase in saturation mobility with decreasing channel lengths, up to about a channel length of 2–3 μm, at which point short-channel injection effects then dominate to lower the mobility.

Several groups have examined the role of the side chain in influencing electrical performance. Early reports suggested that hexyl and octyl side chains gave similar mobilities, whereas dodecyl side chains exhibited much lower mobilities of about 10^{-6} cm^2/Vs [108]. Subsequent studies [109, 110] demonstrated that hexyl was the optimum chain length, with a significant decrease in the charge carrier mobility as the chain length increased to octyl, followed by further decreases as the chain length increased further. The ππ distance is similar across the series [108], so the decrease may be due to an increase in the fraction of insulating side chains in the polymer. If the polymer lamellae are not well aligned in the plane of the substrate and direction of charge flow, hopping or tunnelling between the insulating chains may be required. Since the rate of hopping is dependent on distance, longer chains would be expected to show a detrimental effect.

The effect of the bulkiness of the side chain on field-effect mobility has also been examined by Bao and Lovinger [111]. They prepared a series of RR polythiophenes containing bulky or highly polar substituted endgroups. These showed low degrees of crystallinity and ordering, and poor field effect mobilities of around 10^{-6} cm^2/Vs. Introduction of a chiral alkyl side chain was shown to maintain crystallinity, but to increase the $\pi\pi$ stacking distance to 4.3 Å (vs. 3.8 Å for P3HT). A reduction in FET charge carrier mobility of about one order of magnitude was observed.

9.3.2 Thin-film Device Processing and Morphology

The microstructure of thin-film P3HT is influenced by both intrinsic molecular properties and fabrication conditions, and has a critical influence on the electrical properties. Adjacent polymer backbones have been shown to stack together face to face (π-stacked) and the sheet-like structures that are formed from extended inter chain π stacking are vertically separated by the alkyl chains that extend from the backbone, forming a crystalline lamella structure (see inset in Fig. 9.3a). The polymer RR is one key molecular factor that can influence the degree of crystallinity and has been shown to affect the orientation of the π conjugated thiophene backbone ring planes as they assemble from solution on a substrate. Highly RR P3HT was observed to assemble such that the thiophene backbone exhibited a planar conformation, with the π conjugated plane oriented orthogonal, or "edge-on" to the substrate plane. This was elucidated by the 2-D wide-angle XRD data reported by Sirringhaus and coworkers [105] with the optimized orientation shown in Fig. 9.3a, where the out

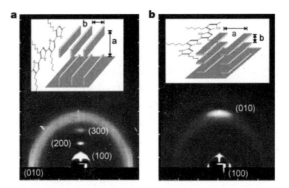

Fig. 9.3 Wide-angle X-ray scattering of thin P3HT films of (**a**) 96% and (**b**) 81% regioregularity [105] (source: Nature, London). Only the highly regioregular sample shows evidence of the out of plane (labeled in the *vertical* direction as "a" on the 2-D image) scattering from (100) lamella planes oriented in the plane of the substrate. The (010) π stacking scattering from the backbone ring planes can be seen in the *horizontal* axis of the image in (a), indicating that the backbone plane is oriented "edge-on" with respect to the substrate, corresponding to the molecular orientation illustrated on the inset. In contrast, the scattering image in (**b**) shows the (010) scattering peak as an arc, centered on the *vertical* axis b

of plane scattering of highly RR P3HT was obtained from the lamella formed from edge-on $\pi\pi$-stacked planar backbones and oriented in the plane of the substrate. A lower RR P3HT obtained a "face-on" orientation of the conjugated backbone to the substrate, as shown in Fig. 9.3b, resulting in low mobilities.

In a transistor device, this π stacking is in the plane of the accumulation layer formed between the source and drain electrodes, which is the optimal orientation for charge transport and devices therefore typically exhibit the highest charge-carrier mobility. Another molecular factor that has a strong influence on the microstructure is the polymer molecular weight. Low-molecular weight, high-RR P3HT ($\sim$5 KDa) is highly crystalline, with a rod-like microstructure observed by atomic force microscopy (AFM) measurements, in which the rod width does not exceed the length of the polymer chains. Higher molecular weight, high-RR P3HT ($>$30 KDa) is less crystalline with small nodule like crystallites. The low-molecular-weight films, although more crystalline, exhibit lower mobilities. Many reasons for this have been proposed. It appears that the higher molecular weight P3HT has better defined and more connected grains, whereas the low-molecular-weight P3HT has more defined grain boundaries [23]. An enhanced out of plane twisting in low-molecular-weight polymer backbone conformation has also been proposed as an explanation for the difference in mobility [112]. The deviation from planarity decreases the effective conjugation length and reduces the efficiency of charge hopping. A study in the high-mobility regime has correlated increasing molecular weight with increasing crystalline quality within domains, with fewer chain ends per domain or "nanoribbon" as well as the possibility for individual polymer chains to bridge between domains at high-molecular weight [25]. However, at high-molecular weights ($>\sim$50 KDa) there is an increase in crystalline disorder, possibly due to slower crystallization kinetics. It was also observed that charged polaron delocalization is significantly larger as the molecular weight increases. The sensitivity of higher molecular weight P3HT morphology development on the processing solvent and conditions has also been recognized. Higher boiling solvents allow the self-assembly of crystalline domains to occur in a more organized and controlled time frame [113], leading to improved electrical performance. The solvent evaporation rate, controlled by spin casting from solution at different spinning rates [114], was also shown to influence the orientation of the backbone plane with respect to the substrate. A low-evaporation rate promoted a more edge-on orientation, which is more optimal for charge transport. Similarly, dip-coating techniques have been utilized to fabricate highly ordered thin-film layers that show mobilities up to 0.2 cm^2/Vs [115].

9.3.3 Doping and Oxidative Stability

The fundamental electrochemistry that governs the reactivity of neutral p-type organic semiconductors in the presence of both oxygen and water has been previously described by DeLeeuw et al. [116]. Although there are many other factors that can contribute to the instability of π conjugated aromatics, it is necessary to ensure that the electrochemical oxidation process is not thermodynamically favorable. One oxidative process involving the reaction of both oxygen and water with the semiconductor will occur at a potential of <0.5 V (vs. SCE) corresponding to an ionization potential of the semiconductor of <4.9 eV from vacuum. Conjugated thiophene polymers have electron-rich π electron systems with relatively high energy HOMO levels rendering them susceptible to this process. High electron affinity acceptors [117] capable of extracting electrons from the HOMO of the polymer, should therefore be avoided to ensure the stability of these π conjugated systems. There are many reports in the literature that have observed instabilities in OFET performance in ambient air [118, 119] and attributed this to an interaction with molecular oxygen [56]. Charge transfer complexes between thiophene and oxygen have been proposed, which can generate reversible charged states and a doping effect on transistor performance. In a p-type transistor device, this manifests over continued ambient exposure time to an increase in acceptor states in the band tail, giving both a rise in the transistor Off-current, and a shift to more positive turn on voltage, as illustrated in Fig. 9.4.

On photoexcitation, the charge transfer complex may promote the formation of singlet oxygen which can then react with the polymer leading to irreversible chemical degradation. Very recently, Chabinyc [120] has proposed that in the absence of light, oxygen is in fact not a strong dopant for thiophene polymers, but rather that ozone, with an electron affinity of −2.1 eV, is more likely to be responsible for doping. While the ozone molecule remains intact, the doping process is reversible. However, on dissociation, an exothermic reaction between ozone and the polymer backbone can occur resulting in carbon–carbon bond cleavage and corresponding

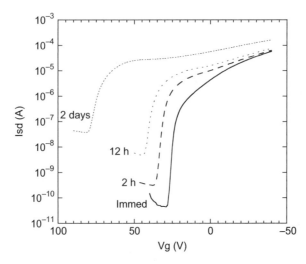

Fig. 9.4 Transfer characteristics of a bottom gate, bottom contact OFET device with a poly(3-hexyl)thiophene semiconductor on continual exposure to ambient environment

reduction of the π conjugation. This proposed mechanism for instability fits well with the evidence that top-gate devices typically exhibit enhanced stability. In this architecture, the semiconductor is protected from the environment by the dielectric and gate layers which may act as sacrificial surface for reaction with ozone, particularly as they will not be an effective barrier to the diffusion of ambient oxygen or water.

In the presence of both light and oxygen, significant photooxidation of the conjugated thiophene chromophore can occur [121, 122]. Singlet oxygen can be formed from photosensitization by the excited state of the polymer, which can then undergo a 1,4-Diels-Alder addition to the thienyl double bonds, breaking the π electron system. Further chemistry leading to free radical assisted chain scission was also proposed [121].

9.4 Polythiophene Structural Analogues

A chemical design strategy to reduce the HOMO energy level of thiophene polymers, in an attempt to improve ambient stability has been reported [119]. Chemical modification of the thiophene backbone structure was used to generate controlled changes in both the backbone conformation and the microstructure, as well as modification of the electronic energy levels of the molecular orbitals. This is illustrated in Fig. 9.5, where a simple rearrangement of the regiopositioning of the alkyl side chain on a thiophene backbone dramatically changes the polymer backbone conformation. The tail-to-tail regiopositioning of the alkyl groups shown in the polyterthiophene (A, Table 9.1, **2**) ensure that there are no steric interactions between neigh-

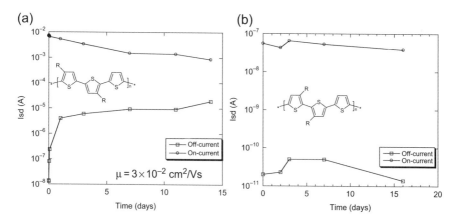

Fig. 9.5 Effect of continuous exposure to ambient air on the storage stability of bottom gate bottom contact transistor devices comprised of (**a**) poly(3,3″-dioctylterthiophene) semiconductor with an ionization potential of 5.0 eV and (**b**) poly(4,4″-dioctylterthiophene) semiconductor with an ionization potential of 5.6 eV. On and Off currents of the devices are plotted versus exposure time

Table 9.1 Charge carrier mobilities of polythiophene semiconductors

Polymer	Monomer 1	Monomer 2	Saturated Charge Carrier Mobility	Ref.
1		–	2×10^{-2}	[119]
2			3×10^{-2}	[119]
3			1×10^{-5}	[119]
4			1×10^{-5}	[145]
5			6×10^{-4}	[119]

Table 9.1 (continued)

6			0.003 [123]
7			4 x 10^{-4} [124]
8			0.02 [22]
9			0.06 [87]
10			7×10^{-4} [128]
11			3.4 x 10^{-4} [129]
12			0.0025 [130]

Table 9.1 (continued)

13	C$_4$H$_9$(C$_2$H$_5$)CHCH$_2$		0.02 [131]
14		CO$_2$C$_{12}$H$_{25}$ / H$_{25}$C$_{12}$O$_2$C	0.06 (TC) [132]
15		CO$_2$C$_{12}$H$_{25}$	0.004 (TC) [132]
16		CO$_2$C$_{12}$H$_{25}$ / H$_{25}$C$_{12}$O$_2$C	[133]
17		C$_{12}$H$_{25}$ / C$_{12}$H$_{25}$	0.14 [55]
18		H$_{21}$C$_{10}$ C$_{10}$H$_{21}$	0.01 [54]

Table 9.1 (continued)

19			0.03	[100]
20		–	3×10^{-5}	[138]
21			0.63	[101]
22		–	1×10^{-6}	[144]
23			0.30	[142]
24			0.20	[142]

Table 9.1 (continued)

25	$C_{16}H_{33}$ ... $H_{33}C_{16}$		0.02	[142]
26	$C_{16}H_{33}$ ... $H_{33}C_{16}$		0.007	[144]

boring alkyl chains, allowing a highly planar backbone conformation and optimal π orbital delocalization along the backbone. The head-to-head regiopositioning of the alkyl groups in polyterthiophene (A, Table 9.1, **3**) leads to significant steric interactions and a resultant twist between the planes of adjacent thiophene rings. This twist not only reduces the π orbital overlap but prevents a closely packed π-stacked lamella microstructure, thus inhibiting crystallization. The reduced π orbital overlap lowers the HOMO energy level by 0.6 eV compared to the planar terthiophene (**2**), and as illustrated in Fig. 9.5b, transistor devices formed from the lower lying HOMO level semiconductor exhibit stable On and Off currents on continual exposure to ambient conditions, whereas transistors from the higher lying HOMO energy polymer show a sharp initial rise in Off-current, most likely due to oxidative doping. A negative consequence of the backbone conformation twist, and subsequent amorphous morphology, is that the charge carrier mobility of terthiophene (**3**) is reduced by over three orders of magnitude when compared to the crystalline analogue (**2**).

Alternative approaches to increase the ionization potential of semiconducting polymers have utilized less electron-rich comonomers in the polymer backbone (Table 9.1). For example, the introduction of both benzene (**4**) and naphthalene (**5**) into the backbone results in an increase in ionization potential of 0.4 eV (benzene) (Merck unpublished results) and 0.45 eV (naphthalene) [119] in comparison to P3HT, and an increase in the ambient stability of the unencapsulated devices. However, in these cases mobility was reduced, possibly due to steric interactions between the solubilizing alkyl chains on the thiophenes and the ortho hydrogen atoms on the 6- and 6,6-membered benzene and naphthalene rings, resulting in reduced intrachain conjugation and reduced interchain ππ overlap. Such interactions could also result in the large increase in ionization potential. Other electron-deficient 6-membered rings, such as 1,2-pyrazine [123] (**6**) or perfluorobenzene [124] (**7**), have also been introduced into the backbone. These are able to planarize the backbone through the formation of intrachain S–N or S–F bonding interactions. Increases in ionization potential can therefore be attributed solely to an electron withdrawing effect. Ambient stabilities were not reported, but charge carrier mobilities were on the order of 10^{-3} cm^2/Vs. The less electron-rich monomer can also include the solubilizing side chains that are required for processable polymers, as is the case in the extensively studied poly(9,9-dioctylfluorene-co-bithiophene) (F8T2) (**8**) [21, 22, 125]. This polymer exhibits a high-temperature nematic liquid crystal phase, and can be preferentially orientated in the transistor channel by the use of a polyimide alignment layer. In this case, charge carrier mobilities up to 0.02 cm^2/Vs, with good ambient stability were reported [22]. Attempts to increase mobility by increasing the thiophene content in the backbone were unsuccessful [126]. Very recently a series of copolymers containing fused electron-poor siloles and thiophene have been reported (**9**) [22]. The mobilities were as high as 0.06 cm^2/Vs with good ambient stability reported.

Donor–acceptor (D–A) type polymers utilizing a range of acceptor monomers have been reported by several groups [127], mainly for use as photovoltaic materials. For suitable acceptors, strong intramolecular charge transfer interactions can occur between D and A units allowing the polymers to exhibit small bandgaps. In this case, the polymer tends to adopt the HOMO energy level of the donor unit, and the lowest occupied molecular orbital (LUMO) of the acceptor [127]. Thus, rather small ionization potentials and small bandgaps can be observed, and the materials have a tendency to dope in ambient air. This may not be significant for photovoltaic applications, but is undesirable for transistor materials. Nevertheless, the FET mobility has been reported for several materials. Polymers incorporating thieno(3,4-*b*)pyrazine (**10**) have been reported with moderate charge carrier mobilities [128]. However, high off-currents in the devices suggested unintentional doping by the ambient atmosphere. Copolymers of unsubstituted thiophene and thiadiazole have also been reported (**11**), with moderate charge carrier mobilities [129]. Due to the lack of solubilizing substituents, transistor devices were prepared from strongly acidic trifluoroacetic acid solutions, into which the basic thiadiazole groups could be dissolved. Soluble copolymers of alkylated thiazole and thiophene have also been reported (**12**) [130]. X-ray scattering data on thin film indicated the polymers ordered into closely packed lamellae, with a π–π distance of 3.65 Å. Transistor devices exhibited charge carrier mobilities up to 0.0025 cm^2/Vs, although ambient stability was not reported. Very recently a copolymer of benzothiadiazole and cyclopenta(2,1-*b*;3,4-*b*')dithiophene (**13**) was reported that exhibited FET mobilities of 0.02 cm^2/Vs [131].

An alternative approach reported by Frechet and coworkers [132] was to introduce the electron withdrawing ester group as a substituent on a thiophene polymer backbone (**14, 15**). This served to both solubilize the polymer and increase the ionization potential in comparison to the simple alkyl chain substituent analogues. The polymer still exhibited crystalline morphology with charge carrier mobilities up to 0.06 cm^2/Vs in top contact mode, slightly lower than the analogous alkyl analogue [55], but the polymers exhibited much improved ambient operation. An analogous polymer incorporating a fused thieno(3,2-b)thiophene (16) has recently been reported with slightly lower mobility in bottom contact transistors [133].

Bauerle and coworkers first reported a regiosymmetric thiophene oligomer with a 3,3″-dialkyl-quaterthiophene repeat unit which exhibited a strong tendency for self-organization through interdigitation of the solubilizing alkyl chains [134, 135]. Polymers based upon the same unit, poly(3,3″-dialkyl-quaterthiophenes) (PQT), were later reported (17) that could also form highly ordered thin-film structures with FET charge carrier mobilities up to 0.14 cm^2/Vs [55]. Annealing of the polymer within its liquid crystalline mesophase was beneficial to the formation of ordered films [136]. Furthermore, the additional conformational freedom of the unalkylated bithiophene monomer was attributed as the reason for reduced conjugation along the thiophene backbone, leading to a modest increase of 0.1 eV in ionization potential over P3HT. Nevertheless, this was reportedly sufficient to impart a significant improvement in ambient stability, with transistors operating for over 1 month in ambient air in the absence of light. The same group reported that a terthiophene derivative (18) also exhibited good air stability, with a mobility of 0.01 cm^2/Vs [137].

9.5 Thienothiophene Polymers

9.5.1 Poly(Thieno(2,3-b)Thiophenes)

Increasing the ionization potential of thiophene polymers has been shown to be effective in improving ambient stability. Reducing the π electron delocalization below the effective conjugative length of the backbone will result in a lowering of the HOMO energy level, thus improving stability. This has been typically achieved through torsional manipulation of the conjugated backbone planarity. An undesirable consequence of this strategy is that the backbone twisting typically suppresses the close packed, π-stacked microstructure that is optimal for charge transport and thus mobility is often compromised. Thieno(2,3-b)thiophene is a planar, electron-rich heterocycle, which when coupled in the 2,5-position is unable to form a fully conjugated pathway between coupled units due to the central cross-conjugated double bond. This is illustrated in Fig. 9.6a, where delocalization between aromatic units at the 2- and 5- positions cannot be achieved. Copolymers of thieno(2,3-b)thiophene and 4,4′-dialkyl-2,2′-bithiophene (referred to as pBTCT poly(bithiophene-cross-conjugated thiophene)) were prepared by Stille coupling (19), with alkyl chain lengths from C8 to C12 [100]. It is expected that for steric reasons, where rotationally feasible, the sulfur atoms along the backbone will prefer to maximize their spacial separation from each other across the short axis of the polymer, as illustrated in Fig. 9.6d, where the sulfurs in adjacent thiophene units arrange in an "anti" configuration across the backbone, as do the sulfurs in coupled thiophene–thienothiophene units. A consequence of this conformation is that the polymer long axis has a "crank-shaft" like shape, with two distinct side chain

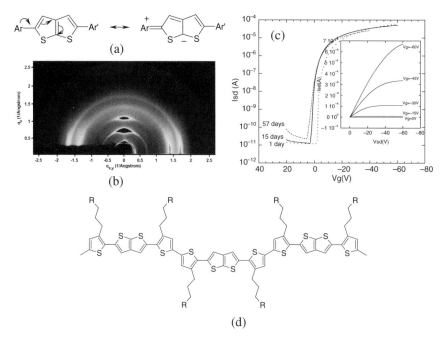

Fig. 9.6 (**a**) Resonance structures of thieno(2,3-b)thiophene illustrating the inability to delocalize electrons across the 2–5 positions, (**b**) 2-D X-ray scattering at grazing incidence of an annealed film of pBTCT-C12 on SiO$_2$/OTS (source M. Chabinyc, PARC), (**c**) transfer characteristics of a bottom gate, bottom contact OFET device with pBTCT-C10 semiconductor (inset: output characteristics at increasing gate voltages), and (**d**) extended pBTCT polymer structure illustrating the preferred sulfur atom and side-chain conformation

separation distances. The pBTCT polymer series were reported to exhibit a 0.4 eV lowering of the HOMO energy level in comparison to P3HT [100]. This lowering of the HOMO energy level contributed to the enhanced ambient stability achieved by these copolymers. Transistor devices fabricated from a pBTCT semiconducting polymer showed only very minor changes in transfer characteristics when measured over a storage period of up to 2 months in air, as shown in Fig. 9.6c.

Charge carrier mobilities of about 0.04 cm^2/Vs can be achieved by this polymer class, with corresponding On–Off ratios around 10^6 in air. The linearity of the output characteristics at low source–drain voltages (inset in Fig. 9.6c) suggest that even with the reduction in HOMO energy level, good charge injection from gold electrodes is still possible. Two thermal transitions can be observed by DSC for all polymers, attributed to both side chain melt at lower temperature and main chain melt at higher temperature. On annealing, highly ordered and crystalline polymer films can be obtained. The 2-D X-ray scattering image, shown in Fig. 9.6b of a thin film of pBTCT (M. Chabinyc – unpublished results) at grazing incidence, exhibits peaks corresponding to the lamellar spacing along the q$_z$ direction, with up to three orders of reflection observed. High resolution grazing X-ray scattering measurements showed a (010) peak assigned to a π-stacking distance of 3.67 Å. An RR polymer homo polymer of 3-alkylthieno(2,3-b)thiophene was also prepared (**21**). In this case, the reduced delocalization along the polymer backbone contributes to a much reduced charge carrier mobility of 10^{-5} cm^2/Vs [138].

9.5.2 Poly(Thieno(3,2-b)Thiophenes)

Recently, poly(2,5-bis(3-alkylthiophen-2-yl)thieno(3,2-b)thiophene) copolymers (pBTTT, **21**), shown in Fig. 9.7a, were reported [101] with high charge carrier mobilities. This has been attributed to the large crystalline domains formed by assembly of the planar conjugated backbones into closely π-stacked sheet-like lamellae, hence giving rise to excellent 2-D transport. pBTTT is an alternating A-B-A-B copolymer of thieno(3,2-b)thiophene and 4,4-dialkyl 2,2-bithiophene. Both monomers are centrosymmetric, and on polymerization, the repeat unit has a rotational symmetry along the polymer long axis. In addition, the optimal conformation of the backbone has an all-trans sulfur orientation across the backbone short axis with a resultant regularity in spacing between alkyl side chains, unlike the pBTCT polymer described in the previous section. This helps facilitate optimal backbone and side chain packing, leading to a highly ordered microstructure. The delocalization of electrons from the thienothiophene aromatic unit into the backbone is less favorable than from a single thiophene ring, due to the larger resonance stabilization energy of the fused ring over the single thiophene ring. This reduced delocalization along the backbone, as well as the reduced inductive electron donation from the fewer alkyl chains per repeat unit, causes a lowering of the polymer HOMO level compared to P3HT, and therefore improved ambient stability.

   The polymer can be oriented with the backbone long axis in the plane of the substrate, and the thiophene ring plane oriented orthogonal to the substrate plane with the appropriate surface treatment. It has been shown [139] from both spectral elipsometry and near edge X-ray absorption fine structure spectroscopy (NEXAFS), that on annealing, the lamella lies almost exactly in the plane of the substrate, and that the backbone conjugated plane is tilted with respect to the lamella plane, as shown in Fig. 9.7b. The alkyl side chains also appear, from polarized IR measurements, to be well ordered in a trans configuration, and are oriented tilted to the backbone. Significant side chain interdigitation has also been identified [140], a process that, due to the symmetry of the repeat units, can be facilitated by simple rotation round the backbone axis, unlike P3HT. The combination of the spacing density between adjacent alkyl chains on the polymer backbone, and the tilt angle of the side chains facilitates this interdigitation, which "registers" the lamella layers. In thin-film form, large lateral domains of dimensions in the micron length scale, comprised of many polymer chains in length can be developed, as can be seen in Fig. 9.7c. Very thin films (20–30 nm) were also observed to have terrace like topography, as illustrated in Fig. 9.7d in which the height of each step correlates well to the cross-sectional width of the polymer backbone and tilted side chains. The presence of both side

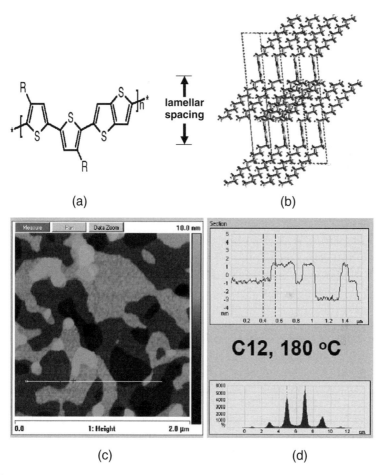

Fig. 9.7 (**a**) Poly(2,5-bis(3-alkylthiophen-2-yl)thieno(3,2-b)thiophene) (pBTTT) repeat unit, (**b**) representation of the molecular orientation of the polymer, depicting the projection of the side chains orthogonal to the substrate plane, extending from the backbone in an all trans configuration, and vertically interdigitating between backbones from the nearest lamella above and below; the side view of the polymer illustrates that conjugated aromatic rings of the backbone adopt a planar conformation with no twist, that the lamella that are formed from the closely stacked backbone packing are coplanar to the substrate with a tilt from horizontal in the backbone plane within the lamella (source: P Brocorens, U. Mons), (**c**) AFM image of annealed polymer film, revealing large domains of micron dimensions (source: J. Kline, NIST), and (**d**) topology (above) in nanometer scale, and histogram (below) of height distribution across the film following the line shown in (**c**), illustrating the terrace like topography, the height of which corresponds to the width of the polymer (backbone and side chains) (source: J. Kline, NIST)

chain ordering and interdigitation is in contrast to that of RR P3HT, in which the side chains were concluded [141] to be "liquid like" and noninterdigitated as observed by polarized FTIR. This improved crystallinity and ordering manifests as improved charge carrier mobility in FET devices.

OFET devices were fabricated from pBTTT polymer solutions and hole mobility values of up to 0.8 cm^2/Vs were reported in a nitrogen atmosphere. These values approach that of high performing evaporated small molecule devices and are comparable to amorphous silicon. In contrast to P3HT, there is less than a factor of 2 change in charge carrier mobility on varying the alkyl chain length from C10 to C18, with a maximum mobility observed at a chain length of C14. In bottom gate, bottom contact devices, in which the active semiconductor layer is the exposed top surface, the effect of different ambient conditions has been evaluated. Exposure to unpurified, ambient air in which the humidity is ∼50%, results in an initial increase in the Off-current of the device, and therefore a drop in the On–Off ratio as can be seen in Fig. 9.8b. In filtered, low humidity air, transistor devices remain very stable over time. Transfer characteristics recorded over a period of up to 72 days, as shown in Fig. 9.8a, show very little change in Off-current or threshold voltage, with a small decrease in On-current corresponding to a drop in mobility by about a factor of 3.

Analogues of the pBTTT polymer series were designed in order to further improve the ambient stability, while preserving the excellent electrical properties [142]. Alkyl side chains were introduced at the 3 and 6 positions of thieno(3,2-b)thiophene by a Negishi coupling of an alkyl zinc halide with 3,6-dibromothieno(3,2-b)thiophene. The HOMO polymer of this monomer has previously been reported by Matzger [143]. In this case, the steric strain between alkyl chains of adjacent monomers thieno(3,2-b)thiophene causes severe twisting of the backbone. Consequently, charge carrier mobility was low according to our measurements [144]. Matzger also reported a mono-alkylated homo polymer of

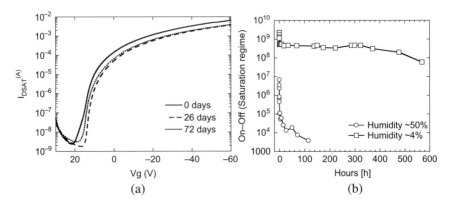

Fig. 9.8 (a) Transfer characteristics of a bottom gate, bottom contact OFET device with pBTTT semiconductor on continuous exposure to filtered, low-humidity (∼4% RH) air and (b) On–Off ratio of pBTTT bottom gate, bottom contact transistors continuously exposed to filtered, low-humidity air or ambient air

thieno(3,2-b)thiophene, but solubility was too low to fully characterize the materials [143]. We prepared a series of copolymers of 3,6-dialkylthieno(3,2-b)thiophene with a range of comonomers, as detailed in Table 9.1. Copolymers with both unsubstituted thiophene (**23**) and bithiophene (**24**) comonomers gave rise to liquid crystalline polymers, with high charge carrier mobility. The bithiophene copolymer, for example, had a highly crystalline thin-film microstructure, although AFM images reveal crystalline domain sizes of less than tens of nanometers on OTS functionalized substrates, dimensions that do not approach the micron scale domains of pBTTT fabricated under identical conditions. Specular x-ray scattering plot shown in Fig. 9.9a show that even as-cast films are highly ordered, with 4 orders of reflection observed from the crystalline lamella, and an interlamella spacing observed to be about 21.6 Å. The intralamella π stacking distance was measured by 2-D XRD at about 3.9 Å (M. Chabinyc, unpublished results), which is larger than that measured for pBTTT (3.7 Å). This more loosely packed backbone density may indicate a less planar backbone conformation, which is also supported by the hypsochromic λ max shift in the solid-state UV spectra of about 15 nm compared to pBTTT. This reduced planarity may also contribute to the slightly increased ionization potential of 0.05 eV as measured by UPS. Bottom-gate transistor devices were fabricated in a nitrogen environment, with charge carrier mobilities of up to 0.2 cm^2/Vs recorded. Significantly, these devices were identified to be remarkably stable in fully ambient conditions in the dark, as shown in Fig. 9.9b, with only a small threshold voltage shift observed on storage in air over 12 days, and a charge carrier mobility of up to 0.1 cm^2/Vs observed in air. The monothiophene copolymer (**23**) also exhibited high charge carrier mobility, in this case up to 0.3 cm^2/Vs in nitrogen. It appears, however, that this analogue is not as stable in ambient as the bithiophene polymer even though the HOMO energy levels, measured by ambient UPS (Riken AC-2), were quite similar.

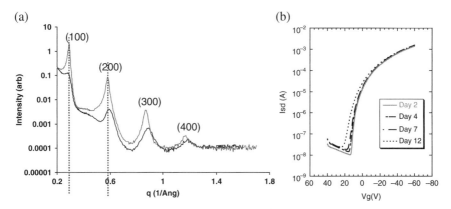

Fig. 9.9 (**a**) High-resolution specular X-ray scattering from a film of DATTT-16 on OTS/SiO2 as spun (*lower line*) and annealed at 180°C (*upper line*), corresponding to a lamellar spacing of 20.8 Å as spun and 21.6 Å after annealing (source: M. Chabinyc, PARC). (**b**) Transfer characteristics of bottom gate, bottom contact transistors with DATTT-16 continuously exposed to ambient air

The angle formed from the projection of the two bonds that link a thiophene monomer at the 2 and 5 positions is less than 180°. This makes rotation of the thiophene ring around the long axis of the polymer a cooperative motion requiring rotation of neighboring thiophene units. Replacement of these nonrotationally invariant 2,5 thiophene links, with a linear 2,5 thienothiophene link gives rise to polymers with a fully linear backbone (**25, 26**) as described in Table 9.1. These linear, all thienothiophene polymers had a correspondingly lower lying HOMO than either of the thiophene analogues, and also exhibited two thermal transitions, most likely a side chain and a lamella melt. Initial measurements reported mobilities of 0.02 cm^2/Vs for the thieno(3,2-b)thiophene analogue and 0.007 cm^2/Vs for the thieno(2,3-b)thiophene analogue.

9.6 Summary

Semiconducting polymers containing conjugated thiophene derivatives are attractive candidates for organic electronic transistor devices, due to their combination of high charge carrier mobilities and the potential for low cost processing. Molecular design principles have been employed in the identification of high performing thiophene copolymers and several illustrations of this were highlighted in the chapter. Highly crystalline, well ordered and oriented polymer thin-film microstructures have been shown to be necessary to achieve optimal semiconducting electrical properties. In particular, liquid crystalline thienothiophene polymers with their propensity to efficiently orient and organize within the mesophase temperature range have shown remarkably high mobilities, combined with reasonable ambient stabilities. Optimization of synthetic routes and purification techniques, as well as the appropriate polymer molecular weights, processing conditions, choice of solvent, surface treatments, and dielectric interfaces are all important to achieve incremental device performance improvements. Challenges remain, however, to further improve stability when exposed to the combination of oxygen, humidity and light, and to tailor the solubility of these aromatic materials to processing friendly solvents. Manipulation of the highest occupied molecular orbital energy level of the conjugated backbone was shown to be an effective strategy to improve stability. With the promising advances in understanding of the morphological and molecular origins of optimal electrical performance made by many groups in recent years, there is exciting potential for further improvements.

Acknowledgements The authors would like to thank the following colleagues and collaborators for their valuable contributions to the chapter: Maxim Shkunov, David Sparrowe, Weimin Zhang, Steve Tierney, Clare Bailey, Warren Duffy, Kristijonas Genevicius, Joe Kline, Michael Chabinyc, Dean DeLongchamp, Eric Lin, Lee Richter, Alberto Salleo, Mike Toney, Patrick Brocorens, Wojciech Osikowicz, Jui-Fen Chang, Henning Sirringhaus, and Masayoshi Suzuki.

References

1. Sandberg HGO, Backlund TG, Osterbacka R, Stubb H (2004) Adv Mater 16:1112
2. Drury CJ, Mutsaers CMJ, Hart MMCM, de Leeuw DM (1998) Appl Phys Lett 73:108
3. Garnier F, Hajlaoui F, Yassar A, Srivastava P (1994) Science 265:1684
4. Sirringhaus H, Friend RH, Shimoda T, Inbasekaran M, Wu W, Woo EP (2000) Science 298:2123
5. Gundlach DJ, Lin YY, Jackson TN, Nelson SF, Schlom DG (1997) IEEE Electron Dev Lett 18:87
6. Kelley T, Boardman L, Dunbar T, Muyres D, Pellerite M, Smith T (2003) J Phys Chem B 107:5877
7. Jurchescu OD, Baas J, Palstra TTM (2004) Appl Phys Lett 84:3061
8. Maliakal A, Raghavachari K, Katz H, Chandross E, Siegrist T (2004) Chem Mater 16:4980–4986
9. Briseno AL, Tseng RJ, Ling MM, Falcao EHL, Yang Y, Wudl F, Bao Z, (2006) Adv Mater 18:2320
10. Briseno AL, Mannsfeld SCB, Ling MM, Liu S, Tseng RJ, Reese C, Roberts ME, Yang Y, Wudl F, Bao Z (2006) Nature 444:913
11. Briseno AL, Aizenberg J, Han YJ, Penkala RA, Moon H, Lovinger AJ, Kloc C, Bao Z (2005) J Am Chem Soc 127:12164
12. Shtein M, Peumans P, Benziger JB, Forrest SR (2003) J Appl Phys 93:4005
13. Burrows PE, Forrest SR, Sapochak LS, Schwartz J, Fenter P, Buma T, Ban VS, Forrest JL (1995) J Cryst Growth 156:91
14. Sheraw CD, Jackson TN, Eaton DL, Anthony JE (2003) Adv Mater 15:2009
15. Anthony JE, Eaton DL, Parkin SR (2002) Org. Lett 4:15
16. Payne MM, Parkin SR, Anthony JE, Kuo CC, Jackson TN (2005) J Am Chem Soc 127:4986
17. Anthony JE, Brooks JS, Eaton DL, Parkin SR (2001) J Am Chem Soc 123:9482
18. Lawrence D, Kohler J, Brollier B, Claypole T, Burgin T (2004) Organic & Molecular Electronics (Ed. Gamota DR, Brazis P, Kalyanasundaram K, Zhang J) Kluwer Academic Publishing 2004 161
19. Veres J, Ogier SD, Leeming SW, Cupertino DC, Khaffaf SM (2006) Adv Funct Mater 13:199
20. Knaapila M, Stepanyan R, Lyons BP, Torkkeli M, Monkman AP (2006) Adv Funct Mater 16:599
21. Sirringhaus H, Kawase T, Friend RH, Shimoda T, Inbasekaran M, Wu W, Woo EP (2000) Science 290:2123
22. Sirringhaus H, Wilson RJ, Friend RH, Inbasekaran M, Wu W, Woo EP, Grell M, Bradley DDC (2000) Appl Phys Lett 77:406
23. Kline RJ, McGehee MD, Kadnikova EN, Liu J, Frechet JMJ, Toney MF (2005) Macromolecules 38:3312
24. Zen A, Pflaum J, Hirschmann S, Zhuang W, Jaiser F, Asawapirom U, Rabe JP, Scherf U, Neher D (2004) Adv Funct Mater 14:757
25. Chang JF, Clark J, Zhao N, Sirringhaus H, Breiby DW, Andreasen JW, Nielsen MM, Giles M, Heeney M, McCulloch I (2006) Phys Rev B 74:115318/1
26. Hamilton R, Bailey C, Duffy W, Heeney M, Shkunov M, Sparrowe D, Tierney S, McCulloch I, Kline RJ, DeLongchamp DM, Chabinyc M (2006) Proc SPIE 6336:633611/1
27. Pokrop R, Verilhac JM, Gasior A, Wielgus I, Zagorska M, Travers JP, Pron A (2006) J Mater Chem 16:3099
28. Verilhac JM, Pokrop R, LeBlevennec G, Kulszewicz-Bajer I, Buga K, Zagorska M, Sadki S, Pron A (2006) J Phys Chem B 110:13305
29. Krebs FC, Nyberg RB, Jorgensen M (2004) Chem Mater ASAP 18:5684
30. Garrett CE, Prasad K (2004) Adv Synth Catal 346:889
31. McCullough R (1998) Adv Mater 10:93
32. Pron PRA (2002) Prog Polym Sci 27:135

33. Chan HSO, Ng SC (1998) Prog Polym Sci 23:1167
34. Chen TA, Wu X, Rieke RD (1995) J Am Chem Soc 117:233
35. McCullough RD, Tristram-Nagle S, Williams SP, Lowe RD, Jayaraman M (1993) J Am Chem Soc 115:4910
36. Bao Z, Dobabalapur A, Lovinger AJ (1996) Appl Phys Lett 69:4108
37. Kim Y, Cook S, Tuladhar SM, Choulis SA, Nelson J, Durrant JR, Bradley DDC, Giles M, McCulloch I, Ha CS, Ree M (2006) Nat Mater 5:197
38. Inoue NB, Velazquez BF, Inoue M (1988) Synth Met 24:223
39. Hayakawa T, Fukukawa KI, Morishima M, Takeuchi K, Asai M, Ando S, Ueda M (2001) J Polym Sci A: Polym Chem 39:2287
40. Waltman RJ, Bargon J, Diaz AF (1983) J Phys Chem 87:1459
41. Barbarella G, Zambianchi M, Di Toro R, Colonna M Jr, Iarossi D, Goldoni F, Bongini A (1996) J Org Chem 61:8285
42. Wei Y, Chan CC, Tian J, Jang GW, Hsueh KF (1991) Chem Mater 3:888
43. Roncali J (1992) Chem Rev 92:711
44. Engelmann G, Jugelt W, Kossmehl G, Welzel HP, Tschuncky P, Heinze J (1996) Macromolecules 29:3370
45. Lere-Porte JP, Moreau JJE, Sauvajol JL (1996) J Organometallic Chem 521:11
46. Leclerc M, Diaz FM, Wegner G (1989) Makromol Chem 190:3105
47. Amou S, Haba O, Shirato K, Hayakaw T, Ueda M, Takeuchi K, Asai M (1999) J Polym Sci A: Polym Chem 37:1943
48. McCullough RD (1998) Adv Mater 10:92
49. Abdou MSA, Lu X, Xie ZW, Orfino F, Deen MJ, Holdcroft S (1995) Chem Mater 7:631
50. Chan HSO, Ng SC (1998) Prog Poly Sci 23:1167
51. Andreani F, Salatelli E, Lanzi M, Bertinelli F, Fichera AM, Gazzano M (2000) Polymer 41:3147
52. Gallazzi MC, Castellani L, Marin RA, Zerbi G (1993) J Polym Sci A: Polym Chem 31:3339
53. Ng SC, Ong TT, Chan HSO (1998) J Mater Chem 8:2663
54. Ong BS, Wu Y, Jiang L, Liu P, Murti K (2003) Synth Met 19:1384
55. Ong BS, Wu Y, Liu P, Gardner S (2004) J Am Chem Soc 126:3378
56. Abdou MSA, Orfino FP, Son Y, Holdcroft S (1997) J Am Chem Soc 119:4518
57. Abdou MSA, Lu X, Xie ZW, Orfino F, Deen MJ, Holdcroft S (1995) Chem Mater 7:631
58. Krebs F, Nyberg CRB, Jorgensen M (2004) Chem Mater 16:1313
59. Goodson FE, Wallow TI, Novak BM (1998) Macromolecules 31:2047
60. Bao Z, Chan WK, Yu L (1995) J Am Chem Soc 117:12426
61. Yamamoto T, Sanechika K, Yamamoto A (1980) J Polym Sci, Polym Lett Ed 18:9
62. Sato M, Tanaka S, Kaeriyama K (1986) Chem Commun 873
63. Mao H, Holdcroft S (1992) Macromolecules 25:554
64. McCullough RD, Lowe RD (1992) Chem Commun 70
65. Greve DR, Apperloo JJ, Janssen RAJ (2001) Eur J Org Chem 18:3437
66. Loewe RS, Khersonsky SM, McCullough RD (1999) Adv Mater 11:250
67. Bolognesi WPA, Bajo G, Zannoni G, Fannig L (1999) Acta Polym 50:151
68. Loewe RS, Ewbank PC, Liu J, Zhai L, McCullough RD (2001) Macromolecules A 34:4324
69. Miyakoshi R, Yokoyama A, Yokozawa T (2005) J Am Chem Soc 127:17542
70. Sheina EE, Liu J, Iovu MC, Laird DW, McCullough RD (2004) Macromolecules 37:3526
71. Yokoyama A, Miyakoshi R, Yokozawa T (2004) Macromolecules 37:1169
72. Jinsong Liu ES, Kowalewski T, McCullough RD (2002) Angew Chem 114:339
73. Radano CP, Scherman OA, Stingelin-Stutzmann N, Mueller C, Breiby DW, Smith P, Janssen RAJ, Meijer EW (2005) J Am Chem Soc 127:12502
74. Koeckelberghs G, Vangheluwe M, Samyn C, Persoons A, Verbiest T (2005) Macromolecules ASAP 38:5554
75. Babudri F, Colangiuli D, Farinola G, Naso F (2002) Eur J Org Chem 16:2785
76. Coppo P, Cupertino DC, Yeates SG, Turner ML (2003) Macromolecules 36:2705

77. Chen TA, Rieke RD (1992) J Am Chem Soc 114:10087
78. Wu X, Chen TA, Rieke RD (1995) Macromolecules, 28:2101
79. Kowalik J, Tolbert LM, Narayan S, Abhiraman AS (2001) Macromolecules 34:5471
80. Zhai L, McCullough RD (2002) Adv Mater 14:901
81. Schluter AD (2001) J Polym Sci Part A: Polym Chem 39:1533
82. Jayakannan M, van Dongen JLJ, Janssen RAJ (2001) Macromolecules 34:5386
83. Liversedge IA, Higgins SJ, Giles M, Heeney M, McCulloch I (2006) Tett Lett 47:5143
84. Guillerez S, Bidan G (1998) Synth Met 93:123
85. Jayakannan M, van Hal PA, Janssen RAJ (2002) J Polym Sci A: Polym Chem 40:2360
86. Kokubo H, Sato T, Yamamoto T (2006) Macromolecules 39:3959
87. Usta H, Lu G, Facchetti A, Marks TJ (2006) J Am Chem Soc 128:9034
88. Iraqi A, Barker GW (1998) J Mater Chem 8:25
89. Lere-Porte JP, Moreau JJE, Torreilles C (2001) Eur J Org Chem 7:1249
90. McCullough RD, Ewbank PC, Loewe RS (1997) J Am Chem Soc 119:633
91. Bjørnholm T, Greve DR, Reitzel N, Hassenkam T, Kjaer K, Howes PB, Larsen NB, Bøgelund J, Jayaraman M, Ewbank PC, McCullough RD (1998) J Am Chem Soc 120:7643
92. Malenfant PRL, Frechet JMJ (2000) Macromolecules 33 :3634
93. de Boer B, van Hutten PF, Ouali L, Grayer V, Hadziioannou G (2002) Macromolecules 35:6883–6892
94. Loewe RS, McCullough RD (2000) Chem Mater 12:3214
95. Yamamoto T, Arai M, Kokubo H, Sasaki S (2003) Macromolecules 36:7986
96. Tsuie B, Reddinger JL, Stzing GA, Soloducho J, Katritzkyb AR, Reynolds JR (1999) J Mater Chem 9:2189
97. Nehls BS, Asawapirom U, Fuldner S, Preis E, Farrell T, Scherf U (2004) Adv Funct Mater 14:352
98. Nurulla ATI, Shiraishi K, Sasaki S, Yamamoto T (2002) Polymer 43:1287
99. Tierney S, Heeney M, McCulloch I (2005) Synth Met 148:195
100. Heeney M, Bailey C, Genevicius K, Shkunov M, Sparrowe D, Tierney S, McCulloch I (2005) J Am Chem Soc 127:1078
101. McCulloch I, Heeney M, Bailey C, Genevicius K, MacDonald I, Shkunov M, Sparrowe D, Tierney S, Wagner R, Zhang W, Chabinyc ML, Kline RJ, McGehee MD, Toney MF (2006) Nat Mater 5:328
102. Yamamoto T, Morita A, Miyazaki, Maruyama T, Wakayama H, Zhou Z, Nakamura Y, Kanbara T, Sasaki S, Kubota K (1992) Macromolecules 25:1214
103. Yamamoto T (2002) J Organomet Chem 653:195
104. Brown PJ, Thomas DS, Kohler A, Wilson JS, Kim JS, Ramsdale CM, Sirringhaus H, Friend RH (2003) Phys Rev B 67:064203
105. Sirringhaus H, Brown PJ, Friend RH, Nielsen MM, Bechgaard K, Langeveld-Voss BMW, Spiering AJH, Janssen RAJ, Meijer EW, Herwig P, de Leeuw DM (1999) Nature 401:685
106. Salleo A, Chabinyc ML, Yang M, Street R (2002) Appl Phys Lett 81:4383
107. Chua LL, Zaumseil J, Chang JF, Ou ECW, Ho PKH, Sirringhaus H, Friend RH (2005) Nature 434:194
108. Bao Z, Feng Y, Dodabalapur A, Raju VR, Lovinger AJ (1997) Chem Mater 9:1299
109. Babel A, Jenekhe SA (2005) Synth Met148:169
110. Zen A, Saphiannikova M, Neher D, Asawapirom U, Scherf U (2005) Chem Mater 17:781
111. Bao Z, Lovinger AJ (1999) Chem Mater 11:2607
112. Zen A, Hirschmann PJS, Zhuang W, Jaiser F, Asawapirom U, Rabe JP, Scherf U, Neher D (2004) Adv Funct Mater 14:757
113. Chang JF, Sun B, Breiby DW, Nielsen MM, Sölling TI, Giles M, McCulloch I, Sirringhaus H (2004) Chem Mater 16:4772
114. DeLongchamp DM, Vogel BM, Jung Y, Gurau MC, Richter CA, Kirillov OA, Obrzut J, Fischer DA, Sambasivan S, Richter LJ, Lin EK (2005) Chem Mater 17:5610

115. Wang G, Swensen J, Moses D, Heeger AJ (2003) J Appl Phys 93:6137
116. de Leeuw DM, Simenon MMJ, Brown AR, Einerhand REF (1997) Synth Met 87:53
117. Foster R (1969) Organic charge-transfer complexes (organic chemistry: A series of monographs), vol 15. Academic Press, New York
118. Meijer EJ, Detcheverry C, Baesjou PJ, van Veenendaal E, de Leeuw DM, Klapwijk TM (2003) J Appl Phys 93:4831
119. McCulloch I, Bailey C, Giles M, Heeney M, Love I, Shkunov M, Sparrowe D, Tierney S (2005) Chem Mater 17:1381
120. Chabinyc ML, Street RA, Northrup JE (2007) Appl Phys Lett 90:123508.
121. Abdou MSA, Holdcroft S (1993) Macromolecules 26:2954
122. Rost H, Ficker J, Alonso JS, Leenders L, McCulloch I (2004) Synth Met 145:83
123. Yasuda T, Sakai Y, Aramaki S, Yamamoto T (2005) Chem Mater 17:6060
124. Crouch DJ, Skabara PJ, Lohr JE, McDouall JJW, Heeney M, McCulloch I, Sparrowe D, Shkunov M, Coles SJ, Horton PN, Hursthouse MB (2005) Chem Mater 17:6567
125. Salleo A, Street RA (2003) J Appl Phys 94:471
126. Brennan DJ, Chen Y, Feng S, Godschalx JP, Spilman GE, Townsend PH, Kisting SR, Dibbs MG, Shaw JM, Welsh DM, Miklovich JL, Stutts D (2004) Mat Res Soc Symp Proc 814:1
127. Roncali J (1997) Chem Rev 97:173
128. Zhu Y, Champion RD, Jenekhe SA (2006) Macromolecules 39:8712
129. Yamamoto T, Yasuda T, Sakai Y, Aramaki S (2005) Macromol Rapid Commun 26:1214
130. Yamamoto T, Kokubo H, Kobashi M, Sakai Y (2004) Chem Mater 16:4616
131. Muehlbacher D, Scharber M, Morana M, Zhu Z, Waller D, Guadiana R, Brabec C (2006) Adv Mater 18:2884
132. Murphy AR, Liu J, Luscombe C, Kavulak D, Frechet JMJ, Kline RJ, McGehee MD (2005) Chem Mater 17:4892
133. Tierney S, Bailey C, Duffy W, Hamilton R, Heeney M, MacDonald I, Shkunov M, Sparrowe D, Zhang W, McCulloch I (2007) PMSE Preprints 96:180
134. Baeuerle P, Fischer T, Bidlingmeier B, Stabel A, Rabe JP (1995) Angew Chem Int Ed Engl 34:303
135. Azumi R, Gotz G, Debaerdemaeker T, Bauerle P (2000) Chem Eur J 6:735
136. Zhao N, Botton GA, Zhu S, Duft A, Ong BS, Wu Y, Liu P (2004) Macromolecules 37:8307
137. Ong BS, Wu Y, Jiang L, Murti K (2004) Synth Met 142:49
138. Heeney M, Bailey C, Tierney S, McCulloch I (2006), Pat Appl WO2006021277
139. DeLongchamp DM, Kline RJ, Lin EK, Fischer DA, Richter LJ, Lucas LA, Heeney M, McCulloch I, Northrup JE (2007) Adv Mater 19:833
140. Kline RJ, DeLongchamp DM, Fischer DA, Lin EK, Richter LJ, Chabinyc ML, Toney MF, Heeney M, McCulloch I (2007) Macromolecules 40:7960–7965
141. Gurau MC, Delongchamp DM, Vogel BM, Lin EK, Fischer DA, Sambasivan S, Richter LJ (2007) Langmuir 23:834
142. Heeney M, Bailey C, Duffy W, Shkunov M, Sparrowe D, Tierney S, Zhang W, McCulloch I (2006) PMSE Preprints 95:101
143. Zhang X, Koehler M, Matzger AJ (2004) Macromolecules 37:6306
144. Heeney M, Wagner R, McCulloch I, Tierney S, Pat Appl WO2005111045
145. Merck, unpublsihed results

Chapter 10
Solution Cast Films of Carbon Nanotubes for Transparent Conductors and Thin Film Transistors

David Hecht and George Grüner

10.1 Introduction: Nanoscale Carbon for Electronics, the Value Proposition

In recent years, we have seen the emergence of novel electronic materials ranging from conducting plastics to advanced composites. These materials have been developed with the aim of replacing inorganic materials that have been improved to near perfection decades ago. The attractiveness of using these materials in electronic applications lies not necessarily in increased performance, but in inexpensive, room-temperature fabrication, and attributes such as mechanical flexibility.

Various new forms of nanoscale carbon have emerged with novel attributes. The discovery of C_{60} [1], a "new form of carbon" was quickly followed by the discovery of carbon nanotubes [2], and recently graphene flakes [3]. In all cases, the interesting and useful properties derive from the fact that carbon materials based on sp^2 bonds, such as are found in graphite, lie between metals, and semiconductors (graphite is a semimetal). This, together with boundary conditions imposed at the nanoscale, leads to remarkable electrical properties.

Currently, carbon nanotubes hold the most promise for applications. The tubes have excellent electrical properties, and these properties are tunable using chemical means. Thus, they can potentially replace electrical conductors and even semiconductors in a variety of applications that require electronic materials.

As is the case for all areas of nanotechnology, manufacturability and system integration are required for the successful exploitation of the attributes of carbon nanotubes. This is an objective difficult to achieve if, for example, devices incorporating single tube elements are fabricated. For this reason, an alternative avenue that exploits large-scale statistical averaging over many tubes is more promising. A random network of carbon nanotubes deposited on a substrate is an obvious – and perhaps most straightforward – realization of this concept. Such a network is also

D. Hecht (✉)
Department of Physics and Astronomy, Los Angeles, CA, USA
e-mail: Dhecht@GMAIL.com

W.S. Wong, A. Salleo (eds.), *Flexible Electronics: Materials and Applications*, Electronic   297
Materials: Science & Technology, DOI 10.1007/978-0-387-74363-9_10,
© Springer Science+Business Media, LLC 2009

referred to as a "thin film," although this characterization may be misleading for networks with significantly less than full coverage of a substrate.

Various technical parameters of the films (low sheet resistance, high optical transparency in the visible, and IR spectral ranges) have been established and compared with that of other materials. Such comparisons, together with the demonstrated performance of devices that have been fabricated to date, establish carbon nanotubes as a competitive material in the area known as plastic, printed, flexible, or macroelectronics.

This chapter is intended as a summary of the current status of the field, a status that is characterized by being at the stage where promising early feasibility studies and prototyping are being closely followed by the exploration of issues such as manufacturability and product competitiveness.

10.2 Carbon NT Film Properties

10.2.1 Carbon Nanotubes: The Building Blocks

Carbon nanotubes, discovered at Hyperion in the 1970s and also by S. Ijima in 1991, can be thought of as a sheet of graphene bent into a cylindrical shape. They exist both as a single rolled tube (single-walled nanotube or SWNT), or as a series of nested tubes of ever-increasing diameter, similar to the Russian Matrioshka dolls (multiwalled nanotube or MWNT) [4]. Depending on the roll-up direction (or chirality) along the graphene sheet, either semiconducting or metallic electronic states are created. Due to the strong carbon–carbon bonding between the atoms in the tube, and the near perfection of the lattice, SWNTs possess mobilities on the order of $1,00,000 \, \text{cm}^2\text{V}^{-1}\text{s}^{-1}$ and conductivities up to $4,00,000 \, \text{Scm}^{-1}$ [5]. The semiconducting tubes have a bandgap that is inversely proportional to the tube diameter; the gap energy is on the order of 0.7 eV for tubes that are 1 nm in diameter [6]. Theoretical and experimental studies have also established the work function of SWNT networks to be in the range 4.7–5.2 eV [7, 8], high enough to meet the requirement for anodes in several types of photonic devices.

10.2.2 Carbon Nanotube Network as an Electronic Material

While devices based on individual tubes may have an impact on the silicon roadmap, manufacturability and the variation of electronic attributes remain a significant challenge. Devices that incorporate a large number of tubes, thus leading to statistical averaging of their electronic properties, may solve the issue of manufacturability and reproducibility. The simplest architecture that achieves this goal is a network of randomly oriented nanotubes that form a "film" on a surface; although these films can in principle consist of either SWNT or MWNT, only films of SWNTs will be

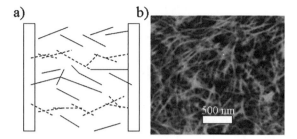

Fig. 10.1 Nanowire network, NT FILM architecture. (**a**) Schematic of a nanotube network above the percolation threshold, with the dashed lines indicating conducting pathways. (**b**) AFM image of a typical nanotube network

discussed in this chapter. The notion of a "film" is somewhat misleading, in particular when the average thickness of a "film" is less than a monolayer; we will however use this notation throughout this review. The materials architecture is illustrated in Fig. 10.1a, where the dashed line indicates a conducting pathway; a typical atomic force microscope (AFM) image of such film is shown in Fig. 10.1b. The concept of a nanotube network attempts to solve the major issues limiting single tube devices. A network can be made to arbitrarily large areas, limited only by the deposition process. Statistical averaging over the many tubes in the network limits the device-to-device variation. Lack of a preferred direction in a random array makes network devices much simpler to manufacture than single tube devices. Some of the other highlights of a nanotube network are listed below.

Electrical conductance: The high conductivity of the individual components of the films leads to a highly conducting network. It is analogous to the way in which a freeway system allows one to travel more quickly than the slower, but more direct, surface roads. The charge carrier transport through this network is thought to be limited not by the conductivity along the nanotubes themselves, but by the large intertube resistance associated with barriers to charge propagation that arise at the tube–tube junctions as will be discussed below. The highest reported conductivity of a nanotube film reported to date is 6,000 Scm^{-1} [9], about three orders of magnitude lower than that of a single nanotube. However, the conductivity is sufficient for a variety of applications.

Optical transparency: A network of highly 1-D wires, with large aspect ratios has high transparency – approaching 100% for truly 1-D wires with aspect ratio approaching infinity. This is in contrast to networks formed of low aspect ratio components such as discs (illustrated in Fig. 10.2d), where substantial coverage of the surface – and thus small optical transparency – is needed for electrical conduction.

Flexibility: Just like a spider web, a NT Film is a highly flexible material. A random network of wires has, as a rule, significantly higher mechanical

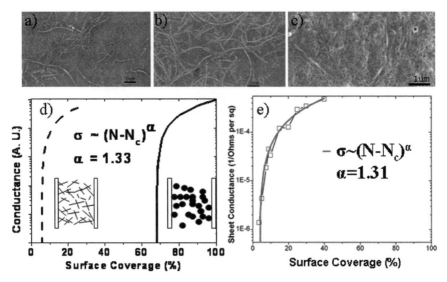

Fig. 10.2 SEM images of nanotube networks of varying densities below (**a**), near (**b**), and well above (**c**) the percolation density. (**d**) Theoretical plot of the conductance versus area coverage for a 2-D surface covered with sticks (*dashed, left*) and discs (*solid, right*). Notice that the critical density for sticks (aspect ratio of 100) is ~5%, while that of discs is ~67%. (**e**) Experimentally measured sheet conductance versus surface coverage for NT networks seen in "a–c". It follows the expected power law with a critical exponent of 1.31

>   flexibility than a solid film, making the architecture eminently suited in par-
>   ticular for applications requiring flexibility.
>   *Fault tolerance:* Breaking a conducting path leaves many others open, and the
>   pathways for current flow will be rearranged. This concept, called fault toler-
>   ance, is used in many areas, from telephone networks to networks of power
>   lines. The same concept applies here.

All these attributes of nanotube films are of significance for printed flexible electronics, where mechanical flexibility is essential, and optical transparency is often
required.

10.2.3 Electrical and Optical Properties of NT Films

NT films are characterized by two types of disorder. The first is topological: above
a certain critical density (tubes per unit area) there is a large number of conducting
pathways, connecting, say, a source and drain electrode, and the number of pathways is a strong function of the tube density [10]. At the same time, the individual
SWNTs are separated by barriers for electron intertube transfer [11]. Due to differences in tube chiralities, and variations in the contact force between the individual
tubes, there is a distribution of barrier heights; for charge propagation from tube

to tube, this leads to a second source of randomness. As expected for a material with substantial randomness, charge transport is fundamentally different from band transport in materials like a good metal or a single crystal silicon. This leads to a conductivity that is concentration, temperature, frequency, and voltage dependent. Such dependencies and comparison with relevant theories can be used to assess the underlying transport mechanism.

10.2.3.1 Concentration Dependent Conductivity

One of the more important notions with respect to applications of NT films is the concept of percolation. Percolation is a phenomenon that describes how paths form across a surface as the density (i.e., objects/area, referred to below as N) of the network increases; theoretical work has established how the number of paths should scale with density for networks consisting of a variety of objects, such as discs and sticks. Percolation, in the case of NT films, refers to the propagation of electric charge across the random network of tubes, separated by barriers, and it is assumed that the network conductance is proportional to the number of formed pathways. There are three distinct regimes where the electronic properties of the network are substantially different; these three regions are highlighted in Fig. 10.2a–c. At low densities, a continuous path across the surface cannot be formed; the conductance (G) here is zero. As the density increases, the NT film reaches a critical density (N_c, called the percolation threshold), where paths are first beginning to form. At this critical density, theory predicts that the conductance of a percolating network varies as:

$$G \sim (N - N_c)^{\alpha} \tag{10.1}$$

The critical density N_c depends on the geometry of the object in the network; theoretical models have established that for a network comprised of identical sticks of length L_T:

$$N_c = \frac{5.7}{L_T^2} \tag{10.2} \text{[12]}$$

Figure 10.2d illustrates the advantage of using longer aspect ratio materials such as nanotubes (sticks) as opposed to nanoparticles (discs): the higher aspect ratio material forms percolating pathways at much lower densities, approaching an N_c of 0 for truly 1-D objects. This fact will have consequences in forming composites, as well as in using nanotubes in applications where transparency is needed. Unlike the critical density N_c, the critical exponent (α) depends solely on the dimension of the percolating space; theory predicts a value of 1.33 for two dimensions and 1.94 for three dimensions [13] for objects of arbitrary geometry. This exponent is precisely accurate only when the density approaches N_c from above; for densities greater than N_c, the exponent in Eq. 10.1 is density dependent and approaches 1 for films where $N >> N_c$. Experimental results on percolating nanotube networks in two dimensions have confirmed these results [10] (see Fig. 10.2e).

10.2.3.2 Temperature Dependent Conductivity

The temperature dependence of the resistivity of films of various densities has been examined in detail [14–16]. At high densities, well above the percolation threshold, the resistivity is weakly temperature dependent, mimicking the behavior of a "bad" metal. In contrast, for smaller network densities – in particular near to the percolation threshold – the conductivity depends more strongly on the temperature, giving evidence that temperature-driven charge transport processes dominate the conductivity. The overall temperature and concentration dependence, including both low-density and high-density networks, can be qualitatively understood based on what has been discussed above, with one additional factor: at low densities the conducting networks most likely have at least some semiconducting SWNTs, while at higher densities all-metallic SWNT pathways are feasible. This can be described with a model that includes both semiconducting and metallic pathways. These notions have been elaborated upon by various groups, often with detailed fits to theories [14–16]. All of these are based on the inherent randomness and temperature-driven charge transport across random barriers. What is important for applications is that carbon NT films can serve as a semiconducting channel (essential for transistor operation) when the density is close to the percolation threshold and also as a "metallic" interconnect or sheet of conductor for high network densities.

10.2.3.3 Frequency Dependence and Optical Conductivity

As expected for a random network where charge transport is limited by large resistances between the tubes, the conductivity is also frequency dependent. At low frequencies:

$$\sigma(\omega) = \sigma_{dc} + A\omega^\beta \qquad (10.3)$$

where ω is the angular frequency, σ_{dc} is the dc conductivity, and $\beta \leq 1$ (Zhou and Gruner, unpublished). The coefficient A in Eq. 10.3 is weakly density dependent, while σ_{dc} is a strong function of the density, with $\sigma_{dc} \rightarrow 0$ at the percolation threshold. Thus, the conductivity is increasingly dependent on the frequency as the network density approaches N_c from above. For densities well exceeding N_c, the increase in the conductivity with increasing frequency sets in only at high frequencies in the millimeter wave optical range.

The conductivity has been evaluated in a broad spectral range [17, 18], the most recent results displayed in Fig. 10.3. As discussed before, with increasing frequency an increase of $\sigma(\omega)$ is first observed, due to the progressively reduced role of the intertube barriers. This is followed by the typical Drude roll-off in the infrared spectral range. Among the additional features observed at higher frequencies, a strong

Fig. 10.3 Conductivity versus frequency over a broad spectral range shows the onset of a plasma frequency, followed by the first two band gaps

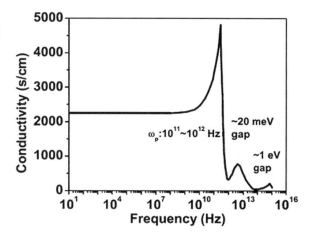

interband absorption in the visible range is the feature most relevant to applications of the films as a transparent and conducting material.

10.2.3.4 Geometric Factors

The measured intratube resistance along individual SWNTs is a strong function of the tube length L_T, decreasing linearly with L_T for tubes longer than the carrier mean free path as expected for diffusive transport, and crossing over to a length-independent resistance for smaller length scales (see Fig. 10.4) [19–22]. The limiting resistance ($R = 6$ kΩ) corresponds to the quantum of resistance for a wire

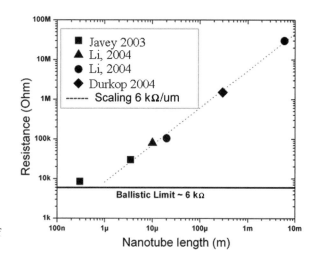

Fig. 10.4 Compilation of resistance values for single-walled nanotubes of varying length. Nanotube resistance approaches the ballistic limit of ∼6 kΩ for tubes <1 μm in length. Above 1 μm the resistance scales with length at a rate of 6 kΩ/μm

with ballistic transport and boundary conditions [23]; one can conclude, therefore, that the mean free path λ of the charge carriers is about 1 μm.

Measurements on the intertube resistance associated with crossed SWNTs show that the junction resistance between SWNTs is about 200–400 kΩ for a metal–metal junction, 1 MΩ for a semiconducting–semiconducting SWNT junction, and 100 MΩ for a metal–semiconducting junction (at low-bias voltage) [11]. As discussed in the preceding paragraph, nanotubes on the order of 1 μm long have a tube resistance of about 10–20 kΩ. Therefore, NT films consisting of a random network of 1 μm long tubes represent a regime where the junction resistance between tubes is several orders of magnitude higher than the resistance along the tube. In this regime, one expects the network conductivity to increase as a power law with increasing nanotube length [24], as each charge carrier has to cross fewer and fewer junctions to traverse the sample. This has been confirmed by experimental values, where NT films were formed by first using high-powered sonication to create tubes (or small bundles of tubes) of controllable length (see Fig. 10.5a, b); these tubes were then deposited as networks. Figure 10.5c shows the conductivity of NT films as a function of the average tube length in the network. While there are no firm predictions of how the overall conductivity should scale with parameters such as nanotube length or the diameter of the bundles that are formed, it has been experimentally established that the conductivity varies as $\sigma_{dc} \sim L_T^{1.46}$ for networks where the average tube length is less than 10 μm [25]. The average bundle size in the network will also play a role here, as some evidence has shown that NT networks consisting of smaller bundles have higher conductivity [26]. This is likely due to the fact that the characteristic length for the current to diffuse through a bundle is longer than the average length between bundles; therefore, most of the current flows at the surface of the NT bundles, causing nonconducting voids toward the center of the bundles.

10.2.4 Doping and Chemical Functionalization

Experiments on individual (semiconducting) tubes give evidence for p-type behavior. Various experiments, conducted at different temperatures and at different oxygen concentrations, confirm that O_2 species in direct contact with the nanotubes is responsible for this p-type behavior [27]. A broad range of doping experiments have also been conducted, and a qualitative picture has emerged. Substantial doping of the carbon structure leads to a drastic reduction of the mobility and mean free path due to the large scattering potential that cannot be avoided by the charge carriers. A noncovalent doping through (weak) attachment of the dopant molecules to the surface of the tubes has shown to lead to changes of the carrier number, often with relatively minor degradation of the mobility of the network. Electron withdrawing species like Br, NO_2, and $TCNQF_4$ [28–31] lead to increasing carrier numbers for the p-type materials, and electron donating materials such as NH_3 or the polymer poly(ethyleneimine) (PEI) lead to an n-type material. For all molecular dopants, the binding energy

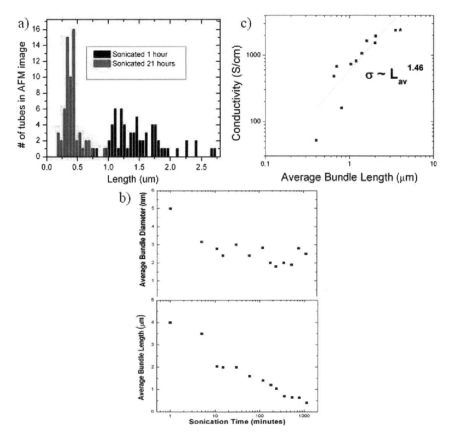

Fig. 10.5 (**a**) Histogram of nanotube lengths (obtained from AFM images) for tubes after 1 hour (*black*) and 24 hours (*white*) of sonication. (**b**) Average tube diameter (*top*) and length (*bottom*) after various sonication times. It is clear that the effect of sonication on nanotube solutions is to both shorten the tubes, as well as to debundle the tubes. (**c**) Conductivity of nanotube networks versus average tube length in the network. Due to the high tube–tube junction resistance, the conductivity shows a power law scaling with nanotube length

between the dopant and the nanotubes is small and thus the doping is not stable; as of today, molecules that lead to stable *p*- or *n*-doping have not been found.

10.3 Fabrication Technologies

Room-temperature fabrication of NT films (in contrast to direct growth on silicon or quartz that require temperatures exceeding 300°C) ensures compatibility with a variety of surfaces that are used for printed electronics on flexible substrates. Solubilization and deposition are the cornerstones of the technology, and careful optimization of both is needed.

10.3.1 Solubilization

Nanotubes, owing to their large aspect ratio, are subject to large intertube Van der Waals forces; these forces cause the tubes to stick together in a solvent, forming large bundles. Therefore, one of the major challenges to fabricating an NT film is to separate the tubes in solution without using covalent chemistry that tends to decrease the tube conductivity [4]. Researchers have examined several different methods for separating tubes in solution, using a variety of solubilization agents including surfactants [32, 33], polymers [34], and DNA [35]. In all cases, the binding of the solubilization agent to the nanotube forms a protective coating, preventing agglomeration between nanotubes. The same solubilization agents, however, prevent direct nanotube–nanotube contact in the film that is subsequently formed through deposition of the tubes onto a surface, and thus have to be removed after (or during) deposition. Of course, those materials that are best at separating the nanotubes are typically the ones that interact most strongly with the tubes and are therefore the ones that are most difficult to remove. Water soluble surfactants, such as sodium dodecyl sulfate (SDS), have been found to best combine good nanotube dispersion with the ability to be removed or rinsed from the films post deposition and are the agents of choice for making the devices discussed in this chapter. Centrifugation and sonication – well-established methods of dispersion of particles in a liquid – are used to create well-dispersed NT solutions. However, additional factors must be considered because sonication and centrifugation tend to shorten the tube length in solution, which, as discussed in Section 10.2.3.4, leads to films of lower conductivity. Factors such as sonication time and power, as well as centrifuge time and speed, must be optimized to obtain solutions leading to films of highest conductivity.

10.3.2 Deposition

Once a stable solution of well-dispersed tubes is obtained, transfer of these tubes uniformly from solution to a given substrate proves a significant challenge. Tube reaggregation during the drying process causes the "coffee ring" effect, where there exists an extreme density gradient in the as-deposited film. However, a number of room-temperature deposition techniques have been developed and are briefly discussed below. Note that several deposition methods are compatible with roll-to-roll technology, the method of fabrication preferred by industry. There are several problems associated with most deposition techniques. First, bundling of the tubes during deposition, along with incomplete removal of the surfactant from the deposited film lead to films of lower conductivity. Also, nonuniformity of the deposited NT film remains a problem for many deposition techniques. Last, due to the current high costs per gram of NT material, the yield of the deposition process is an important factor to consider.

10.3.2.1 Spraying

The simplest deposition technique is a process involving spraying the solution of well-dispersed tubes onto a substrate. Nanotube solutions are sprayed using an artist spray brush (Fig. 10.6a) onto a substrate that is heated to about 80–100°C. Heating the substrate ensures that the small droplets that hit the surface are evaporated quickly, so the tubes in solution do not have time to aggregate before depositing. Figure 10.6b–d shows AFM images of a sprayed NT film along with a trace that indicates an average bundle size of ~5–10 nm. Rinsing the films in water, either after the deposition or several times during deposition, is important to remove the surfactant from the film surface, and it tends to improve the conductivity almost an order of magnitude (Hecht, 2006, unpublished). Spray technology may be especially useful in coating irregularly shaped objects such as fabrics [36] or very large objects such as vehicles or airplanes requiring electrostatic discharge. However, it

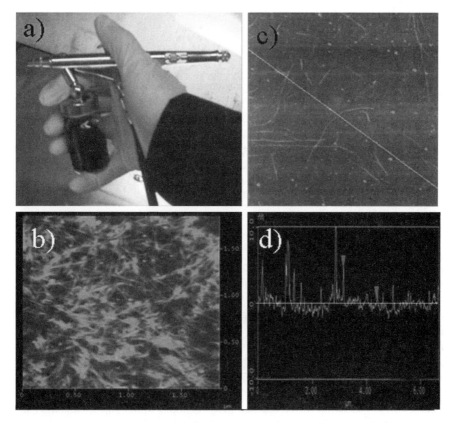

Fig. 10.6 (**a**) Photograph of nanotube spray coating set up using an artists spray brush. (**b**) AFM image of a sprayed nanotube film. (**c,d**) AFM image of a sprayed nanotube film showing a profile scan indicating a nanotube bundle size between 5 and 10 nm

is difficult to obtain high degrees of surface uniformity using a spray deposition method, due to the nonuniformity of the spray mist from the nozzle. Also, the yield of a spray deposition process is a concern, as some of the sprayed droplets have a tendency to either bounce off of the substrate surface or miss the surface entirely. Due to the high current costs of nanotube material, a low yield in the deposition process would make applications cost prohibitive.

10.3.2.2 Slot Coating

Slot coating is a process where a thin film (5–30 µm) of solution is applied evenly to the substrate surface through a slot die. This method yields films of high uniformity and high yield, and is a well known commercial process used in roll-to-roll manufacturing. The conductivity of films deposited by this method is limited by surfactant molecules remaining in the film structure after deposition; therefore, methods to thoroughly rinse away or otherwise remove all unwanted solubilization agents must be explored.

10.3.2.3 Spin Coating

Spin coating of aqueous-based surfactant-wrapped SWNT solutions is difficult on most hydrophobic substrates such as silicon wafers. One successful method that was reported involves the simultaneous spin coating of a stream of methanol along with the nanotube suspension. The stream removes the nanotube from its surfactant-wrapped protection immediately before hitting the substrate, and allows the tube to deposit on the surface [37]. This method, however, leads to highly nonuniform films and is clearly not compatible with roll-to-roll manufacturing.

10.3.2.4 Filtration/Stamping

The deposition method that leads to NT films of highest conductivity involves first depositing the films on a filter and then transferring the film from the filter to the substrate of interest. To deposit the film, a dilute suspension of nanotubes is quickly vacuum filtered onto a porous alumina filter (pore size 100 nm). As the solvent falls through the pores, the SWNTs are trapped on the filter surface, forming an interconnected surface. After forming the film, large amounts of water or other solvents can be washed through the film, rinsing away any residual surfactants. Since the porous alumina filter is not transparent or flexible, the network has to be subsequently transferred onto a transparent surface, such as glass or polyethylene terephthalate (PET). A printing method using a poly(dimethylsiloxane) (PDMS) "stamp" is illustrated in Fig. 10.7a. The stamp is first touched to the nanotube–filter surface and the nanotubes are lifted off of the filter onto the PDMS stamp. Then, the stamp is pressed against a PET surface and heated to about 60°C for 1 h, transferring the tubes from the stamp to the PET surface. This method can also be used to produce patterned films, by simply patterning the PDMS "stamp" before transferring the nanotube

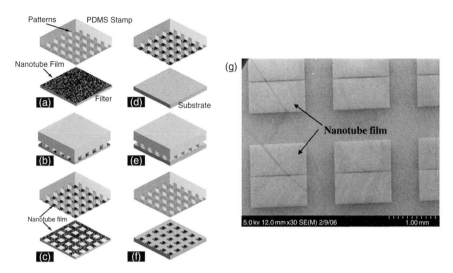

Fig. 10.7 Method to pattern nanotube films using a poly(dimethylsiloxane) (PDMS) based stamp. The stamp is first patterned (**a**), contacted to the nanotube network surface (**b**), and lifted off (**c**), to transfer the nanotubes to the PDMS surface. The stamp is then brought into contact with the substrate of interest (**d–e**), and mild heat (80°C) is applied to transfer the pattern to the substrate (**f**). (**g**) SEM image of such patterned nanotube films showing a gap size of ~20 μm

film. Figure 10.7b shows a scanning electron microscope (SEM) image of a film patterned with this method, with a feature size of about 50 μm [38].

10.4 Carbon NT Films as Conducting and Optically Transparent Material

Due to the combination of the optical properties of individual tubes, and their hollow/1-D nature, NT films at appropriate densities display both high conductivity and high optical transparency.

10.4.1 Network Properties: Sheet Conductance and Optical Transparency

Early experiments have established that the reflectivity of these networks is low in the visible spectral range, and thus the transmission is mainly determined by the optical absorption of the network [18]. It has been also shown that, in contrast to the dc conductivity, the optical conductivity in the visible spectral range is due to local excitations associated with the constituent carbon atoms. Thus, the absorption of the NT film is proportional to the density of the network and is not strongly influenced by effects that influence the dc conductivity, such as the intertube resistances, NT length, and doping.

The electrical and optical properties of the NT films can be parameterized by the measured values of the visible light transmission (VLT) and the sheet resistance (R_s), where the latter quantity is defined as:

$$R_s = \frac{1}{\sigma_{dc}d} \tag{10.4}$$

Here, σ_{dc} is the dc conductivity of the material and d is the material thickness. Both the VLT and the R_s of an NT film are a function of the network density: as the networks become denser, both the sheet resistance and the optical transparency of the films decrease. Figure 10.8 shows R_s as a function of network density; by increasing the film thickness one can tailor the films resistance over several orders of magnitude. Most applications that use a transparent electrode require a VLT between 70 and 95%, and so belong to region II in Fig. 10.8 (the less dense networks in region I are needed for thin-film transistors (TFTs), and more dense networks in region III are optimal for use as supercapacitors or fuel cells).

Figure 10.9 shows the sheet resistance of the films plotted versus VLT (taken at 550 nm, where the human eye is the most sensitive), for films of varying density. Several data points are shown for films of varying thickness, which exhibit a general trend toward lower sheet resistance and lower VLT at higher network densities. The data have been fit to the equation relating VLT and R_s in a thin-metallic film [39]:

$$VLT = \left(1 + \frac{1}{2R_S}\sqrt{\frac{\mu_0}{\varepsilon_0}\frac{\sigma_{op}}{\sigma_{dc}}}\right)^{-2} = \left(1 + \frac{188(\Omega)}{R_S}\frac{200(S/cm)}{\sigma_{dc}}\right)^{-2} \tag{10.5}$$

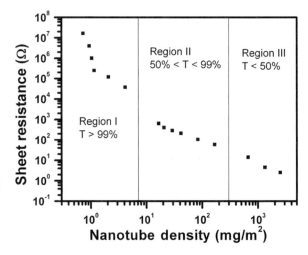

Fig. 10.8 Sheet resistance of carbon nanotube networks of varying density. Notice that the sheet resistance can be easily tailored over seven orders of magnitude. The three regions indicated by the vertical lines loosely indicate the densities needed for various applications: Region I for transistors, region II for transparent electrodes, and region III for supercapacitors and batteries

Fig. 10.9 Plot of the sheet resistance versus transmittance at 550 nm for nanotube films of varying density, with a fit to Eq. 10.4. The corresponding conductivity of these films is 2,100 Scm^{-1}

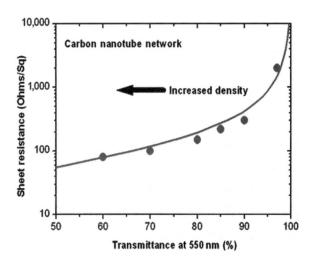

Here, the free space permeability (μ_0) is $4\pi \times 10^{-7}$ NA^{-2}, the free space permittivity (ε_0) is 8.854×10^{-12} C^2 N^{-1}m^{-2}, and σ_{op} and σ_{dc} are the optical and dc conductivities, respectively. This formula is valid for films where the absorptivity is much greater than the reflectivity, which has been established to be true for NT films [18].

The wavelength dependence of the transmittance in the visible spectral range for two films of different densities is shown in Fig. 10.10a. The transmission spectrum

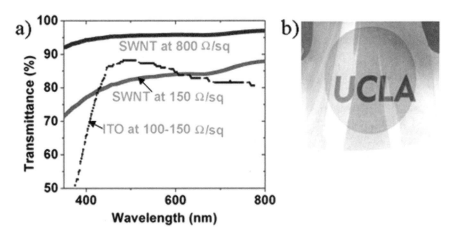

Fig. 10.10 (**a**) Transmission spectra through nanotube films of two different densities (thicknesses) in the visible frequency range. Plotted for comparison is the transmission spectra for an ITO film. Notice the peak in the ITO film which gives it a "tint" relative to the more neutral (*gray*) color of the nanotube film. (**b**) Optical photograph of a nanotube film at 80% transmission on a flexible PET substrate showing its neutral color. Note: UCLA print is behind the film

has no strong absorption peaks in the visible region, leading to films that have a neutral, gray color to the human eye (see Fig. 10.10b), which is desirable for many applications. For comparison purposes, the transmission spectrum through a film of indium tin oxide (ITO) is also shown; the peak in the spectrum at 550 nm gives ITO a yellowish appearance.

10.4.2 Applications: ITO Replacement

Because of the high conductivity and optical transparency, together with the possibility of room-temperature fabrication, NT films are promising candidates to replace ITO in a variety of applications where transparent coatings or electrodes are required. These applications include touch screens, smart windows [40], solar cells [41–43], light emitting diodes [44–46], LCD's flexible EPDs. They all use NT films as either one (solar cells and organic light-emitting diodes (OLEDs)) or both (touch screens and smart windows) electrodes. The performance of the devices is close to that of devices where ITO has been used. These developments have been summarized in a recent review [47]; therefore, these devices will not be described and only some comments are in order here. Several parameters beyond VLT and R_s are of importance for such applications. The surface roughness of the films has to be small in order to allow integration with other layers, layers where charge generation (e.g., solar cells) or photon creation (e.g., OLEDs) occur. For some applications, the work function of the material is a critical parameter; carbon NT films have a work function that is suitable for their use as anodes in either solar cells or OLEDs. Mechanical flexibility and robustness are most likely not critical issues for the adoption of films for a variety of devices, as such films have already demonstrated excellent performance when compared to the more brittle ITO films [26, 48].

10.4.3 Challenges and the Path Forward

The parameters (R_s and VLT) displayed on Fig. 10.9 are already adequate for the use of NT films in touch screens, as well as for electroluminescent lamps. However, some additional factors such as long-term stability and resistance of the films to environmental factors such as humidity have to be tested and addressed if necessary. Lower sheet resistance (for the same optical transparency) is required for OLED and solar cells; this can be accomplished by several routes including increasing the purity/length of the tubes, better dispersion, and also doping. In addition, the surface roughness has to be below some (application dependent) critical value in order to eliminate, or reduce, leakage current between the device electrodes. Integration with other (polymer or organic) layers must also be explored together with different patterning methods. Manufacturing and price issues remain a challenge. Roll-to-roll and ink-jet depositions are the preferred deposition routes due to their low cost and, in the case of ink-jet deposition, ease of patterning. Such challenges notwithstanding, these initial results suggest that with further optimization such networks may offer a direct alternative to ITO and other transparent conducting oxides, in particular under circumstances where mechanical flexibility is also required.

10.5 TFTs with Carbon Nanotube Conducting Channels

Carbon nanotube-based thin-film transistors (NTTFTs) have a short, but substantial, history. The notion that carbon nanotube networks at low density can resemble the properties of silicon, and serve as the backbone of transistors, emerged about 5 years ago, first in some patent applications and then in research papers [49, 50]. The first NTTFTs were fabricated by growing nanotubes directly on a silicon oxide surface and using doped silicon and metal leads as the gate and source and drain, respectively. Such devices were also used to evaluate the performance and to compare the response to expressions widely used in CMOS devices. The first devices on a plastic surface were obtained through transfer printing [51]; tests also demonstrated that the performance did not degrade significantly by bending and mechanical deformation. Devices that incorporate nanotube networks of appropriate densities as both the gate and the conducting channel were first made in 2005; such devices were shown to be highly transparent (Fig. 10.11) [52]. Devices in which all conducting

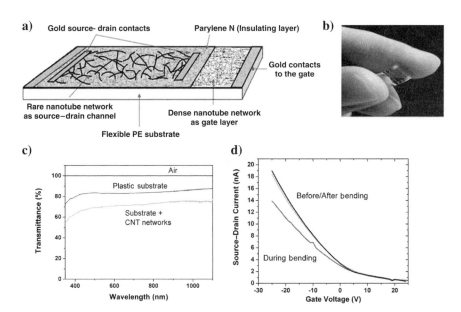

Fig. 10.11 First example of a transparent, flexible transistor. (**a**) Schematic of the device. Carbon NT FILMs form both the conducting channel (rare) and the gate (dense), with evaporated Parylene N forming the dielectric. (**b**) Photograph of device. (**c**) Transmittance through the device in the optical spectral range. (**d**) Transfer characteristics of the device before, during, and after being bent

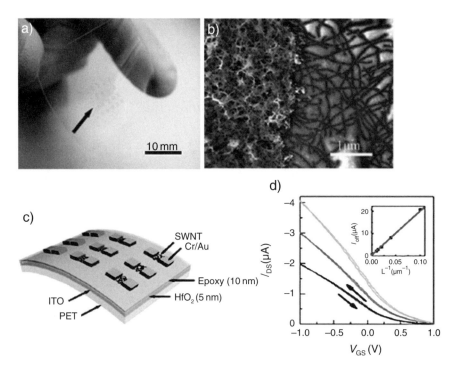

Fig. 10.12 Flexible, transparent transistor with SWNTs as the source, drain, gate, and conducting channel, with an epoxy/PDMS dielectric. (**a**) SEM image of the boundary between the thick printed nanotube source and the thin printed nanotube conducting channel. (**b**) Photograph of the device, which has a mobility of $\sim$30 cm^{2}V^{-1}s^{-1} and an on/off ratio of $\sim$120. The drawback is the large voltage operation range ($\pm$80 V range). (**c**) Device schematic showing a carbon NT FILM with metal contacts on a dielectric consisting of 10 nm of epoxy and 5 nm of HfO$_2$. (**d**) Transfer curve of the device shows operating voltage range from -1 to $+1$ V

components (source and drain contacts, gate, and conducting channel) consisted of carbon nanotubes were also fabricated recently (Fig. 10.12a,b) [53]. Improvements made to the gate dielectric have yielded devices with lower operating voltage ranges (Fig. 10.12c, d) [54]. Further improvements in device performance can be expected as devices enriched in semiconducting tubes are explored.

10.5.1 Device Characteristics

The device characteristics of NTTFTs have been measured by several groups, and substantial variation has been seen in devices fabricated using different sources or having different geometries. The reason for this is that no systematic study, and thorough comparison with theory, has been made as of today. Just like for CMOS transistors, different length scales such as the channel length, the screening length, and the mean free path will determine the detailed behavior of the response to drive

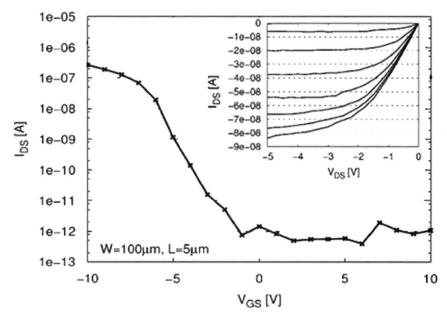

Fig. 10.13 Transfer characteristics (IV_g) for spin coated NT FILM device. The inset shows the output characteristics for a similar device. The mobility for this device is ~ 1 cm^{2}V^{-1}s^{-1}, and the ON/OFF ratio is $\sim 10^5$

voltages. Although some theories have been advanced [55, 56], the measured device characteristics are typically compared with standard formulas that have proven to be useful for CMOS architectures [57]. This approach is followed here.

Figure 10.13 summarizes the device characteristics for a typical solution-deposited NTTFT: The source–drain current (I_D) versus gate voltage (V_G) at constant source–drain voltage (V_D) (transfer characteristics) and the source–drain current versus source–drain voltage at constant gate voltage (output characteristics) (inset). At low source–drain voltage ($V_D < (V_G - V_T)$), the source–drain current increases linearly with V_D (linear regime) and can be approximated by:

$$I_D = \frac{WC_G}{L}\mu(V_G - V_T - \frac{V_D}{2})V_D \quad V_D < (V_G - V_T) \qquad (10.6)$$

Here, μ is the mobility, L is the channel length, W is the channel width, C_g is the gate capacitance per unit area (the calculation of the gate capacitance is expanded on in Section 10.5.2.3), and V_T is the threshold voltage. In particular, the mobility can be calculated in this region from the transconductance (g),

$$g = (\frac{\partial I_D}{\partial V_G})_{V_D=const} = \frac{WC_G}{L}\mu V_D \qquad (10.7)$$

Equation 10.7 can be used to calculate the "device" mobility and is useful to compare device properties. This device mobility is different than the inherent mobility of the material itself and includes effects such as contact resistances at the source–drain electrodes. Using Eq. 10.7, the device mobility for the NTTFT shown in Fig. 10.13 can be calculated using the device parameters of $W = 100$ µm, $L = 5$ µm, $V_D = 0.5$ V, and dielectric thickness of 50 nm of aluminum oxide; this leads to a calculated mobility on the order of 1 $cm^2V^{-1}s^{-1}$ [58].

Notice in Fig. 10.13 (inset) that when the source–drain voltage becomes much larger than the gate voltage, the device current saturates; the saturation source–drain current is given by the equation:

$$I_D = \frac{WC_G}{2L}\mu(V_G - V_T)^2 \quad V_D > (V_G - V_T) \tag{10.8}$$

The transconductance can again be used to calculate the device mobility in the saturation region from:

$$g = \mu \frac{C_{OX}W}{L}(V_{GS} - V_T) \tag{10.9}$$

In both the linear and saturation regions, short channel effects can introduce large error between the device-specific mobility calculated from Eqs. 10.7 and 10.9, and the inherent mobility of the nanotube network. In the linear region, high contact resistances at the metal-semiconductor contact can cause nonlinearities in the I_D versus V_D curve, leading to underestimation of the NT film mobility. For channel lengths down to 10 µm, the on current scales linearly with the reciprocal of the channel length, indicating that the contact resistance at the source and drain are much smaller than the channel resistance (inset Fig. 10.12d) [59]. In the saturation region, when channel lengths are comparable to the dielectric thickness, the I_D versus V_D curve may not saturate, leading to overestimation of the NT film mobility. This is not typically a problem for NTTFTs owing to the long channel lengths required (L > 10 µm) to ensure no metallic nanotubes short the device.

10.5.2 Device Parameters

There have been major inroads made in improving the key characteristics of an NTTFT; substantial progress has been made in understanding and improving the device mobility, ON/OFF ratio, operating voltage range, subthreshold swing, hysteresis, and stability. Each one of these features presents a challenge for NTTFTs, and much progress has already been made toward solving each.

10.5.2.1 Mobility

Figure 10.14 indicates a plot of the progress made since 1984 in field effect device mobility for various organic semiconductors, where incremental improvements have been made by improving the morphological properties of the devices or by synthesizing new organic molecules. However, the device mobility in organic molecule-based devices has begun to reach its saturation between 1 and 10 $cm^2V^{-1}s^{-1}$, where the mobility is fundamentally limited by the weak intermolecular forces (mainly Van der Waals) in organic semiconductors [59, 60]. For comparison, the highest reported mobility for an NTTFT on plastic (with ON/OFF ratio > 10^4) is ~1 $cm^2V^{-1}s^{-1}$ for a random network and ~480 $cm^2V^{-1}s^{-1}$ for an aligned nanotube network (see Fig. 10.14) [58, 61]; the latter mobility exceeds the best organic TFT mobilities by two orders of magnitude and is higher than the mobility of polycrystalline silicon.

It has been shown that the device mobility increases with increasing network density [50]; a further analysis of this ensues. From Eq. 10.6, and recognizing that $\frac{\partial I_D}{\partial V_D} = R_D^{-1} = G_D$(source–drain conductance), the mobility of an NTTFT in the linear region is given by the expression:

$$\mu = \frac{G_D L}{C_G W V_D} \tag{10.10}$$

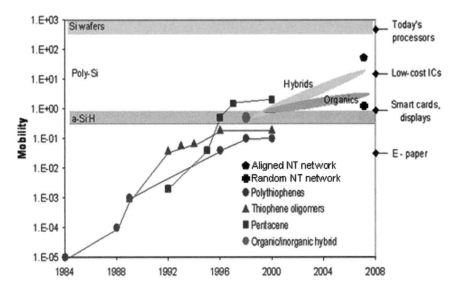

Fig. 10.14 The evolution of the mobility of organic semiconductor TFT devices over the last two decades. The mobility of carbon NT FILM FETs is indicated by the blue cross, as well as the current mobility of various forms of silicon. Some application barriers are also indicated on the right side of the figure

In Eq. 10.9, both the source–drain conductance and the gate capacitance depend on the network density. For densities above the percolation threshold, a NT film can be modeled as a percolating system composed of a random distribution of 1-D sticks spanning a 2-D surface. Standard percolation theory predicts that, for this system, the channel conductance (G_D) is given by $G_D \sim (N\text{-}N_c)^{1.33}$ (see Section 10.2.3.1), where N is the device density (tubes/area) and N_c is the critical density where a percolating path first forms. This scaling law is precisely accurate only when the density approaches N_c from above; for densities greater than N_c, the exponent in Eq. 10.10 is density dependent and approaches 1 for films where $N \gg N_c$. However, experiment has shown that Eq. 10.10 is approximately valid for network densities up to 10 times N_c [10]. The gate capacitance also depends on the network density. For low-density networks near the percolation threshold, the gate capacitance is equal to the product of the capacitance of a single isolated tube and the number of tubes [62], or $C_G \sim N$. Putting it together, one expects the mobility of NTTFTs above the percolation threshold to scale as:

$$\mu \propto \frac{(N - N_C)^{1.33}}{N} \text{(close to percolation)} \tag{10.11}$$

At higher densities, ($N \gg N_c$), the gate capacitance saturates; therefore, the mobility increases linearly with N, even as the network channels become thicker than one monolayer. Unfortunately, as the network density increases, the ON/OFF ratio of the NTTFTs decreases exponentially, as the presence of metallic tubes in the network short the source–drain electrodes.

10.5.2.2 ON/OFF Ratio

Semiconducting SWNT field-effect transistors (FETs) have ON/OFF ratios above 10^5 [63, 64], while nanotube network-based devices report ON/OFF ratios that vary from as high as 10^5 to as low as ~ 1 [58, 65]. Typically, ON/OFF ratios approaching those found in single tube devices are found only in NTTFTs using conducting channels with densities just above the percolation threshold; as the density is increased above this value, the ON/OFF ratio falls off exponentially. This behavior is due to the presence of metallic tubes in the source–drain channel, which short the device, providing a source–drain current pathway when the device is in its "off" state. Figure 10.15 summarizes the fundamental relationships between device ON/OFF ratio and mobility; there is a general trend toward lower ON/OFF ratio for devices with higher mobility (density) [50, 52, 53, 54, 57, 58, 80, 81]. The solid line should serve as a guide to the eye, separating results obtained from solution-deposited NTTFTs from results obtained from chemical vapor deposition (CVD) grown nanotubes. It is clear that NTTFTs deposited from solution exhibit lower mobilities for the same ON/OFF ratios, likely due to larger tube bundles in the network or residual surfactant.

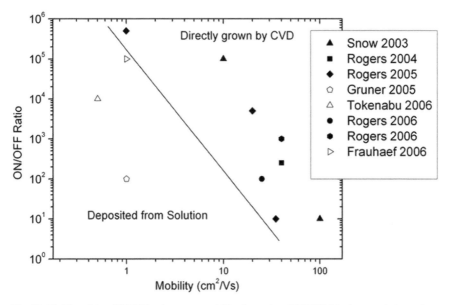

Fig. 10.15 Plot of the ON/OFF ratio versus mobility for carbon NT FILM devices made by various groups over the last four years. The diagonal line coarsely separates the solution deposited devices (*open points*) from the CVD grown devices (*filled points*). Notice the general correlation between higher device mobility and lower ON/OFF ratio

10.5.2.3 Device Capacitance

The capacitance of TFT devices can critically affect device performance, as a higher device capacitance leads to a lower operating voltage range, which is essential for practical applications. At high network densities (such that the average tube–tube spacing, d, is comparable to the dielectric thickness, L_{OX}, see Fig.10.16a), the device capacitance approaches that of a standard parallel plate capacitor. The capacitance per unit area (C_G) of two metal plates separated by a distance L_{OX} by a material with dielectric constant ε_r is:

$$C_G = \frac{\varepsilon_o \varepsilon_r}{L_{OX}} \text{(Parallel Plate Capacitance)} \qquad (10.12)$$

Below this density, the NT film capacitance is more difficult to calculate due to the non-uniform electric field in the dielectric layer. Detailed calculations of the network capacitance at low densities indicate that the capacitive coupling of each tube in the network to the gate is approximately the same as for isolated tubes [62]; with each tube modeled as a metal cylinder of radius (r) embedded in a dielectric above an infinite conducting plane, the capacitance (C_G) is given by:

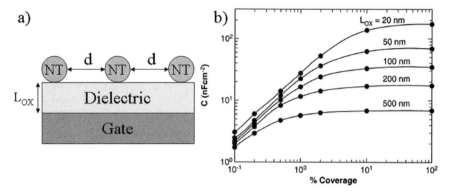

Fig. 10.16 (a) Capacitance of a carbon NT FILM as a function of the density, as obtained by computer modeling and (**b**) schematic of the system. At low densities, when the average tube spacing (*d*) is large compared to the dielectric thickness (L_{OX}), the capacitance increase is proportional to the tube density; this behavior saturates at higher densities to a value given by the standard parallel plate capacitance

$$C_G = \frac{2\pi\varepsilon_o\varepsilon_r L_T N}{\ln(\frac{2L_{OX}}{r})} \tag{10.13}$$

The behavior of the capacitance as a function of density is summarized in Fig. 10.16b. As the above formulas dictate, there are two major avenues for increasing the capacitance of an NTTFT: use dielectric materials that have higher dielectric constants or decrease the thickness of the dielectric layer. It has been demonstrated that using a thin layer of the high-κ gate dielectric Hafnium Oxide (HfO_2) can lead to a gate capacitance approaching the 1-D quantum capacitance limit of 4 pFcm^{-1} [54]. In addition to having a high dielectric constant, an ideal dielectric layer for NTTFTs must form pinhole free layers upon deposition, to prevent device shorting. Also, as TFTs push toward roll-to-roll manufacturing processes, a dielectric that can be printed directly from solution will have to replace more exotic (and costlier) techniques such as atomic layer deposition. Some companies are already making significant efforts toward the realization of printable dielectric materials.

10.5.2.4 Operating Voltage

The required voltage range required to switch a TFT device from the off to the on state is an important device parameter, especially with the continued push toward low cost, lower power consumption devices. Today's active matrix devices for TFT backplanes operate between −20 V and +5 V, and RFID electronics require ∼5 V for operation. Early NTTFT devices fabricated on silicon are operated between ±10 V, while devices on flexible substrates required ±100 V [66]. However, improvements in the gate dielectric have subsequently produced NTTFTs on plastic substrates, which operate in voltage ranges as low as ±1 V (see Fig. 10.12d) [54].

10.5.2.5 Threshold Voltage

The threshold voltage is the value of the gate voltage where the TFT device begins to switch from the off state to the on state; it can be determined by extrapolating the linear region of the transfer curve to the gate voltage axis. The threshold voltage in NTTFTs is determined by the doping level of the nanotube conducting channel, as well as the gate capacitance. It has been found that the magnitude of the threshold voltage can be modified by p- or n-type doping of the NT films with various molecules or biological species; this has the effect of shifting the transfer curve toward positive/negative gate voltages for p/n-type dopants [67, 68]. Although this property of NTTFTs makes them excellent sensors [69], for electronics applications it is imperative that the threshold voltage be constant over repeated device cycles. Therefore, NTTFTs may require a protective, or encapsulation, layer to render them insensitive to the environment. Section 10.5.2.8 will elaborate further on methods to maintain the device stability.

10.5.2.6 Subthreshold Swing

The subthreshold swing (S) is an important device parameter because it corresponds to the voltage range required to switch the device between its off and on states. The subthreshold swing is determined by the inverse ratio of the gate capacitance (C_g) to the capacitance due to interfacial traps (C_{IT}):

$$S = \frac{k_b T}{e} \ln 10(1 + \frac{C_{IT}}{C_g}) \tag{10.14}$$

The room-temperature limit ($C_{IT} \ll C_g$) is 60 mV/decade and has been achieved by single tube devices with high-κ dielectrics [70] or by electrolyte gated devices [71]. In NTTFT devices, however, S has been reported in a range from 280 to 2,500 mVdec^{-1} [54]. The reason for the larger subthreshold swing measured in network devices is not completely understood; one theory is that conduction through the metallic tubes in the network (which does not modulate with gate voltage) is the most likely cause for the nonideal behavior of the subthreshold swing in NTTFT devices. Evidence supporting this can be seen in Fig. 10.17, which shows S decreasing with increasing device ON/OFF ratios (decreasing number of metallic pathways).

10.5.2.7 Hysteresis

One undesirable feature present in NTTFT devices is what is known as hysteresis, which manifests as a difference in the source–drain current for the increasing and decreasing gate voltage sweeps. Experiments have shown that the magnitude of the hysteresis increases as the gate sweep frequency decreases, indicating that slowly moving species cause the hysteretic behavior. Furthermore, the introduction of an electrolytic polymer coating to the NTTFT, as well as exposure of the device to a

Fig. 10.17 Subthreshold Swing of a carbon NT FILM device as a function of the ON/OFF ratio. The increase in subthreshold swing with decreasing ON/OFF ratio (increasing density) is likely due to the parasitic effects of the metallic tubes

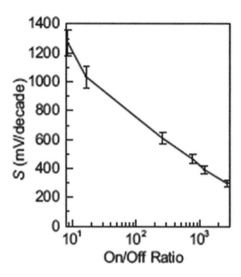

humid environment, has been shown to increase the device hysteresis [66]. These results indicate that the main source of device hysteresis is mobile surface ions, possibly associated with a hydration layer on the substrate. This is substantiated by subsequent experiments that show that the inclusion of a hydrophobic, self-assembled monolayer on the device substrate reduces device hysteresis [72].

10.5.2.8 Device Stability

Carbon nanotubes, due to their unique hollow, 1-D geometry, are extremely sensitive to changes in their environment. Although this property is ideal for making sensors, it is a distinct drawback for using this material for TFT applications requiring long-term stability. As discussed in Section 10.5.2.5, NTTFTs exhibit a change in the device characteristics in the presence of oxidizing or reducing molecules. Exacerbating the problem of stability is the fact that nanotubes are not intrinsically p-type materials; they behave that way because they are hole-doped by oxygen in the air. However, the binding energy between an O_2 molecule and a SWNT is 0.25 eV [73], which enables the molecules to readily desorb from the NTTFT surface by applying heat, ultraviolet radiation, or vacuum [74]. One route toward solving this particular stability issue is to find dopant molecules that have a higher binding energy to carbon nanotubes. One candidate, tetrafluoro-tetracyanoquinodimethane, has shown to be stable after 60 hours in vacuum [75]. Another method to increase the device stability is to coat the NTTFT device with an appropriate encapsulation layer to passivate the device from the environment.

10.5.2.9 Doping and Logic Elements

As discussed in Section 10.2.4, NTTFTs behave as a *p*-type material due to the accidental doping of the carbon atoms by adsorbed oxygen molecules from the surrounding air. To make active logic elements, it is imperative to have both *p*- and *n*-type materials. Luckily, certain polymers, when noncovalently bound to nanotube networks, act as electron donating species; this hybrid nanotube–polymer structure can act as an *n*-type material. The most prevalent example of such a polymer is PEI, which, when spin coated onto a nanotube network, acts as a stable, *n*-type circuit element (see Fig. 10.18a). One can combine the *p*- and *n*-type nanotube materials to form simple circuits, the simplest of which is an inverter; this has been demonstrated for both single tube [63] and network [54] devices. A typical inverter (gain $= 8.2$) from a nanotube network device is shown in Fig. 10.18b.

10.5.3 Challenges and the Path Forward

The early results discussed in this chapter indicate that NTTFTs could play a significant role in printed electronics, in particular when high operating speed and long lifetime are required. Some improvements are necessary, however, before these devices can pose significant challenges to transistors fabricated using polymer, organic molecule, or amorphous silicon-based conducting channels.

The mobility of the devices already exceeds that of devices with amorphous silicon, conducting polymer, or organic molecule conducting channels. Mobilities surpassing $10 \, \mathrm{cm^2V^{-1}s^{-1}}$ are readily attainable; this represents (together with stability and environmental robustness) the most significant value proposition of the devices for printed electronics.

The ON/OFF ratio is currently smaller than what is required for applications both for displays and RFID tags, two applications that have been identified as having significant potential. This parameter is currently limited by the presence of metallic

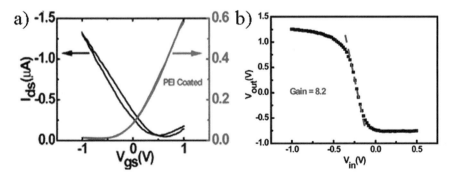

Fig. 10.18 (a) Carbon NT FILM transistor before (*left*) and after (*right*) doping with the polymer PEI. (**b**) Electrical characteristics of an inverter fabricated using both bare (*p*-type) and PEI doped (*n*-type) material

tubes in the network. Such tubes lead to finite conduction even if a gate voltage with appropriate polarity removes the contribution I_{sd} coming from the conducting pathways that consist of semiconducting nanotubes from the conducting channel. Elimination of the metallic tubes either before the fabrication of the TFTs or after the fabrication process is, therefore, the main route for enhancing this parameter. While several groups have made attempts in separating metallic from semiconducting tubes, the only methods that have been demonstrated to lead to a significant improvement of the ON/OFF ratio is that of density differentiation [76] and selective gas-phase etching [77]. An improvement of about a factor of 10^3 clearly indicates the potential of this route. Chemical routes to eliminate the metallic tubes, such as treatment with diazonium [78], together with a selective burnout of the metallic pathways [79], have also been experimented with to partial success.

The operating voltage range of most of the devices fabricated to date exceeds the range that is required of applications. Display applications require a voltage range of ~25 V, while smaller operating voltages of about 5 V are needed for electronic devices, such as RFID tags. The solution to reducing the operating voltages lies in the dielectric layer. For all other device characteristics the same voltage range is inversely proportional to the device capacitance. Thus, as described in Section 10.5.2.3, a thin layer, preferably composed of a high dielectric constant material is required. The dielectric layer cannot be decreased to arbitrary thinness; too thin and "pinholes" in the layer will lead to leakage currents between the gate and the conducting channel of the nanotube network, especially when using "rougher" plastic substrates.

The hysteresis observed is due to mobile ions at the interface between the conducting channel and the underlying (dielectric) substrate. This has also been borne out by experiments where the device characteristics were examined at different driving speeds, and also by changing the humidity, and thus the hydration layer on the surface of the devices. The conclusions have been confirmed by modeling the influence of the surface ions. Not surprisingly the hysteretic behavior can be largely removed by self-assembled hydrophobic monolayers and protective encapsulation by suitable fluoropolymers that prevents the absorption of moisture [64].

10.6 Conclusions

The early progress described in this paper demonstrates that carbon films of carbon nanotubes have achieved a competitive position in the area of printed electronics. This derives from several factors. First, the films already meet the performance for certain applications, and there is reason to believe that further progress will be made on this front. Second proof-of-concept devices have demonstrated the usefulness for a variety of applications. Third, room-temperature fabrication avenues – still subject to perfection – make integration into devices in the plastic/organic technology arena straightforward. Modification of the tubes so that they have increased performance as a metallic electrode, or a semiconducting material (needed for the conducting channel in NTTFTs) is also advancing at a rapid pace, and methods that lead to separation of the metallic and semiconducting tubes are also emerging.

It is also expected that applications beyond what is currently pursued will also emerge, such as electrodes for supercapacitors, printed batteries, and fuel cells. In these cases, transparency is not an issue, and the attribute that can be exploited is the high surface area of the tubes, and consequently of the films as well.

Finally, other nanoscale carbon materials will continue to emerge, with graphene already showing considerable promise.

By considering these advances and opportunities, it is only a question of time when NT films will find their application in a variety of areas that require flexible, conductive, and/or transparent coatings, both in a patterned or nonpatterned form.

Acknowledgements   We would like to thank Derek Hecht and Liangbing Hu for useful discussions and textual improvements.

References

1. Curl RF, Smalley RE (1988) Probing C_{60}. Science 242:1017–1022
2. Iijima A (1991) Helical microtubles of graphitic carbon. Nature 354:56–58
3. Novoselov KS, Geim AK, Morozov SV, Jiang D, Zhang Y, Dubonos SV, Grigorieva IV, Firsov AA (2004) Electric field effects in atomically thin carbon films. Science 306:666–669
4. Dresselhaus MS, Dresselhaus G, Avouris P (2001) Carbon nanotubes: Synthesis, structure, properties, and applications, Vol 80. Springer, Berlin
5. Durkop T, Getty SA, Cobas E, Fuhrer MS (2004) Extraordinary mobility in semiconducting carbon nanotubes. Nano Lett 4:35–39
6. Saito R, Dresselhaus G, Dresselhaus MS (1998) Physical properties of carbon nanotubes. Imperial College Press, London
7. Gao R, Pan Z, Wang ZL (2001) Work function at the tips of multiwalled carbon nanotubes. Appl Phys Lett 78:1757–1759
8. Sun JP, Zhang ZX, Hou SM, Gu ZN, Zhao XY, Liu WM, Xue ZQ (2004) Work function of single-walled carbon nanotubes determined by field emission microscopy. Appl Phys A: Mater Sci Process 75:479–483
9. Wu Z, Chen Z, Du X, Logan J, Sippel J, Nikolou M, Kamaras K, Reynolds J, Tanner D, Hebard A, Rinzler AG (2004) Transparent, conductive carbon nanotube films. Science 305:1273–1276
10. Hu L, Hecht DS, Gruner G (2004) Percolation in transparent and conducting carbon nanotube networks. Nano Lett 4:2513–2517
11. Fuhrer MS, Nygard J, Shih L, Forero M, Yoon Y, Mazzoni MSC, Choi HJ, Ihm J, Louie S, Zettl A, McEuen PL (2000) Crossed nanotube junctions. Science 288:494–497
12. Pike GE, Seager CH (1973) Percolation and conductivity: A computer study. Phys Rev B 10:1421
13. Stauffer G (1985) Introduction to percolation theory. Taylor and Francis, London
14. Kaiser AB, Dusberg G, Roth S (1998) Heterogeneous model for conduction in carbon nanotubes. Phys Rev B 57
15. Hone J, Llaguno C, Nemes NM, Johnson AT, Fischer JE, Walters DA, Casavant MJ, Schmidt J, Smalley RE (2000) Electrical and thermal transport properties of magnetically aligned single wall carbon nanotube films. Appl Phys Lett 77:666–668
16. Bekyarova E, Itkis ME, Cabrera N, Zhao B, Yu A, Gao J, Haddon RC (2004) Electronic properties of single-walled carbon nanotube networks. J Am Chem Soc 127: 5990–5995

17. Xu H, Anlag SM, Hu L, Gruner G (2007) Microwave shielding of transparent and conducting single-walled carbon nanotube films. Appl Phys Lett 90:183119
18. Ruzicka B, Degiorgi L, Gaal R, Thien L, Bacsa R, Salvetat JP, Forro L (2000) Optical and dc conductivity study of potassium-doped single-walled carbon nanotube films. Phys Rev B 61:R2469
19. Durkop T, Getty SA, Cobas E, Fuhrer MS (2004) Extraordinary mobility in semiconducting carbon nanotubes. Nano Lett 4:55–59
20. Li S, Yu Z, Rutherglen C, Burke PJ (2004a) Electrical properties of 0.4 cm long single-walled carbon nanotubes. Nano Lett 4:2003–2007
21. Li S, Yu Z, Yen SF, Tang WC, Burke PJ (2004b) Carbon nanotube transistor operation at 2.6 GHz. Nano Lett 4:753–756
22. Javey A, Guo J, Wang Q, Lundstrom M, Dai HJ (2003) Ballistic carbon nanotube field-effect transistors. Nature 424:654–657
23. Wharam DA, Thornton TJ, Newbury R, Pepper M, Ahmed H, Frost JEF, Hasko DG, Peacock DC, Ritchie DA, Jones GAC (1988) One-dimensional transport and the quantization of the ballistic resistance. J Phys C: Solid State Phys 21:L209–L214
24. Balberg I, Binenbaum M, Anderson CH (1983) Computer study of the percolation threshold in a two-dimension anisotropic system of conducting sticks. Phys Rev B 51:1605
25. Hecht DS, Hu L, Gruner G (2006a) Conductivity scaling with bundle length and diameter in single walled carbon nanotube networks. Appl Phys Lett 89:13312
26. Kaempgen M, Duesberg GS, Roth S (2005) Transparent carbon nanotube coatings. Appl Surf Sci 252:425–429
27. Collins PG, Bradley K, Ishigami M, Zettl A (2000) Extreme oxygen sensitivity of electronic properties of carbon nanotubes. Science 287:1801–1804
28. Star A, Bradley K, Gabriel JP, Gruner G (2003a) Nano-electronic sensors: Chemical detection using carbon nanotubes. Polymeric Mater: Sci Eng 89:204
29. Lee RS, Kim HJ, Fischer JE, Thess A, Smalley RE (1997) Conductivity enhancement in single-walled carbon nanotube bundles doped with K and Br. Nature 388:255
30. O'connell MJ, Eibergen EE, Doorn SK (2005) Chiral selectivity in the charge-transfer bleaching of single-walled carbon-nanotube spectra. Nat Mater 4:412
31. Takenobu T, Kanbara T, Akima N, Takahashi T, Shiraishi M, Tsukagoshi K, Kataura H, Aoyagi Y, Iwasa Y (2005) Control of carrier density by a solution method in carbon-nanotube devices. Adv Mater 17:2430–2434
32. Islam MF, Rojas E, Johnson AT, Yodh AG (2003) High weight fraction surfactant solubilization of single-wall carbon nanotubes in water. Nano Lett 3:269–273
33. Matarredona O, Rhoads H, Li Z, Harwell JH, Balzano L, Resasco DE (2003) Dispersion of single-walled carbon nanotubes in aqueous solutions of the anionic surfactant NaDDBS. J Phys Chem B 107:13357–13367
34. Star A, Joshi V, Han T, Virginia M, Altoe P, Gruner G, Stoddart JF (2004) Electronic detection of the enzymatic degradation of starch. Organic Lett 6:2089–2092
35. Zheng M, Jagota A, Semke ED, Diner BA, Mclean RS, Lustig SR, Richardson RE, Tassi NG (2003) DNA-assisted dispersion and separation of carbon nanotubes. Nat Mater 2:338–342
36. Hecht DS, Hu L, Gruner G (2007) Electronic properties of carbon nanotube/fabric composites. Curr Appl Phys 7:60–63
37. Meitl M, Zhou Y, Gaur A, Jeon S, Usrey ML, Strano MS, Rogers JA (2004) Solution casting and transfer printing single-walled carbon nanotube films. Nano Lett 4:1643–1647
38. Zhou Y, Hu L, Gruner G. (2006) A method of printing carbon nanotube thin films. Appl Phys Lett 88:123109
39. Dressel M, Gruner G (2002) Electrodynamics of solids: Optical properties of electrons in mattter. Cambridge University Press, Cambridge
40. Hu L, Gruner G, Li D, Kaner RB, Cech J (2007) Patternable transparent carbon nanotube films for electrochromic devices. J Appl Phys 101:016102

41. Pasquire AD, Unalan HE, Kanwal A, Miller S, Chhowalla M (2005) Conducting and transparent single-wall carbon nanotube electrodes for polymer-fullerene solar cells. Appl Phys Lett 87:203511

42. Lagemaat J, Barnes TM, Rumbles G, Shaheen SE, Coutts TJ, Weeks C, Levitsky I, Peltola J, Glatkowski P (2006) Organic solar cells with carbon nanotubes replacing In_2O_3:Sn as the transparent electrode. Appl Phys Lett 88:233503

43. Rowell MW, Topinka MA, McGehee MD, Prall J, Dennler G, Sariciftci NS, Hu L, Gruner G (2006) Organic solar cells with carbon nanotube network electrodes. Appl Phys Lett 88:233506

44. Li J, Hu L, Wang L, Zhou Y, Gruner G, Marks TJ (2006) Organic light-emitting diodes having carbon nanotube anodes. Nano Lett 6:2472–2477

45. Aguirre CM, Auvray S, Pigeon S, Izquierdo R, Desjardins P, Martel R (2006) Carbon nanotube sheets as electrodes in organic light-emitting diodes. Appl Phys Lett 88:183104

46. Zhang G, Qi P, Wang X, Lu Y, Li X, Tu R, Bangsaruntip S, Mann D, Zhang L, Dai H (2006a) Selective etching of metallic carbon nanotubes by gas-phase reaction. Science 314: 974–977

47. Gruner G (2006) Carbon nanotube films for transparent and plastic electronics. J Mater Chem 16:3533

48. Saran N, Parikh K, Suh D, Munoz E, Kolla H, Manohar SK (2004) Fabrication and characterization of thin films of single-walled carbon nanotube bundles on flexible plastic substrates. J Am Chem Soc 126:4462–4463

49. Gabriel JC (2003) Large scale production of carbon nanotube transistors. Mater Res Soc Symp Proc 776:271–277

50. Snow ES, Novak JP, Campbell PM, Park D (2003) Random networks of carbon nanotubes as an electronic material. Appl Phys Lett 82:2145–2147

51. Bradley K, Cumings J, Star A, Gabriel JP, Gruner G (2003a) Influence of mobile ions on nanotube based FET devices. Nano Lett 3:639–641

52. Artukovic E, Kaempgen M, Hecht DS, Roth S, Gruner G (2005) Transparent and flexible carbon nanotube transistors. Nano Lett 5:757–760

53. Cao Q, Hur S, Zhu Z, Sun Y, Wang C, Meitl M, Shim M, Rogers J (2006a) Highly bendable, transparent thin-film transistors that use carbon-nanotube-based conductors and semiconductors with elastomeric dielectrics. Adv Mater 18:304–309

54. Cao Q, Xiz M, Shim M, Rogers J (2006b) Bilayer organic-inorganic gate dielectrics for high-performance low-voltage, single-walled carbon nanotube thin-film transistors, complementary logic gates, and p-n diodes on plastic substrates. Adv Funct Mater 16:2355–2362

55. Appenzeller J, Knoch J, Radosavljevic M, Avouris P (2004) Multimode transport in Schottky-barrier carbon-nanotube field-effect transistors. Phys Rev Lett 92:226802

56. Koswatta SO, Lundstrom MS, Nikonov DE (2007) Band-to-band tunneling in a carbon nanotube metal-oxide-semiconductor field-effect transistor is dominated by phonon-assisted tunneling. Nano Lett 7:1160–1164

57. Neaman DA (1992) Semiconductor physics and devices. Irwin, Chicago

58. Schindler A, Brill J, Fruehauf N, Novak JP, Yaniv Z (2007) Solution-deposited carbon nanotube layers for flexible display applications. Physica E 37:119–123

59. Yoo B, Jung T, Basu D, Dodabalapur A, Jones BA, Facchetti A, Wasielewski MR, Marks TJ (2006) High-mobility bottom-contact n-channel organic transistors and their use in complementary ring oscillators. Appl Phys Lett 88:082104

60. Dimitrakopolous CD, Mascaro DJ (2001) Organic thin-film transistors: A review of recent advances. IBM J Res Dev 45:11–27

61. Kang SJ, Kocabas C, Ozel T, Shim M, Pimparkar N, Alam MA, Rotkin SV, Rogers JA (2007) High-performance electronics using dense, perfectly aligned arrays of single-walled carbon nanotubes. Nat Nanotechnol 2:230–236

62. Cao Q, Xia M, Kocabas C, Shim M, Rogers JA, Rotkin S (2007) Gate capacitance coupling of singled-walled carbon nanotube thin-film transistors. Appl Phys Lett 90:023516

63. Bachtold A, Hadley P, Nakanishi T, Dekker C (2001) Logic circuits with carbon nanotube transistors. Science 294:1317–1320
64. McGill SA, Rao SG, Manandhar P, Xiong P, Hong S (2006) High-performance, hysteresis-free carbon nanotube field-effect transistors via directed assembly. Appl Phys Lett 89:163123
65. Unalan HE, Fanchini G, Kanwal A, Pasquier AD, Chhowalla M (2006) Design criteria for transparent single-wall carbon nanotube thin-film transistors. Nano Lett 6:677–682
66. Bradley K, Gabriel JP, Gruner G (2003b) Flexible nanotube electronics. Nano Lett 3: 1353–1355
67. Bradley K, Briman M, Star A, Gruner G (2004) Charge transfer from adsorbed proteins. Nano Lett 4:253–256
68. Hecht DS, Ramirez RJ, Briman M, Artukovic E, Chichak KS, Stoddart JF, Gruner G (2006) Bio-inspired detection of light using porphyrin-sensitized single-wall carbon nanotube FETs. Nano Lett 6:2031–2036
69. Novak JP, Snow ES, Houser EJ, Park D, Stepnowski JL, McGill RA (2003) Nerve agent detection using networks of single-walled carbon nanotubes. Appl Phys Lett 83:4026–4028
70. Javey A et al (2002) High-κ dielectrics for advanced carbon-nanotube transistors and logic gates. Nat Mater 1:241
71. Rosenblatt S, Yaish Y, Park J, Gore J, Sazonova V, McEuen PL (2002) High performance electrolyte-gated carbon nanotube transistors. Nano Lett 2:869
72. McGill SA, Rao SG, Manandhar P, Xiong P (2006) High-performance, hysteresis-free carbon nanotube field-effect transistors via directed assembly. Appl Phys Lett 89:163123
73. Jhi S, Louie SG, Cohen ML (2000) Electronic properties of oxidized carbon nanotubes. Phys Rev Lett 85:1710–1713
74. Chen RJ, Franklin NR, Kong J, Cao J, Tombler TW, Zhang Y, Dai H (2001) Molecular photodesorption from single-walled carbon nanotubes. Appl Phys Lett 79:2258–2260
75. Shiraishi M, Nakamura S, Fukao T, Takenobu T, Kataura H, Iwasa Y (2005) Control of injected carriers in tetracyano-p-quinodimethane encapsulated carbon nanotube transistors. Appl Phys Lett 87:093107
76. Arnold MS, Green AA, Hulvat JF, Stupp SI, Hersam MC (2001) Sorting carbon nanotubes by electronic structure using density differentiation. Nat Nanotechnol 1:60–65
77. Zhang D, Ryu K, Liu X, Polikarpov E, Ly J, Tompson ME, Zhou C (2006b) Transparent, conductive, and flexible carbon nanotube films and their application in organic light-emitting diodes. Nano Lett 6:1880–1886
78. Strano MS, Dyke CA, Usrey ML, Barone PW, Allen MJ, Shan H, Kittrell C, Hauge RH, Tour JM, Smalley R (2003) Electronic structure control of single-walled carbon nanotube functionalization. Science 301:1519–1522
79. Collins PG, Arnold MS, Avouris P (2001) Engineering carbon nanotubes and nanotube circuits using electrical breakdown. Science 292:706–709
80. Austing DG, Lefebvre J, Bond J, Finnie P (2007) Carbon contacted nanotube field effect transistors. Appl Phys Lett 90:103112
81. Takenobu T, Takahashi T, Takayoshi K, Tsukagoshi K, Aoyagi Y, Iwasa Y (2006) High-performance transparent flexible transistors using carbon nanotube films. Appl Phys Lett 88:033511
82. Hur S, Yoon M, Gaur A, Shim M, Facchetti A, Marks TJ, Rogers JA (2005) Organic nanodielectrics for low voltage carbon nanotube thin film transistors and complementary logic gates. J Am Chem Soc 127:13808–13809

Chapter 11
Physics and Materials Issues of Organic Photovoltaics

Shawn R. Scully and Michael D. McGehee

11.1 Introduction

Organic materials hold promise for use in photovoltaic (PV) devices because of their potential to reduce the cost of electricity per kWh ultimately to levels below that of electricity produced by coal-fired power plants. Deposition of organics by techniques such as screen printing, doctor blading, inkjet printing, spray deposition, and thermal evaporation lends itself to incorporation in high-throughput low-cost roll-to-roll coating systems. These are low-temperature deposition techniques which allow the organics to be deposited on plastic substrates such that flexible devices can easily be made. In addition to the inherent economics of high-throughput manufacturing, lightweight and flexibility are qualities claimed to offer a simple way to reduce the price of PV panels by reducing installation costs. Flexible PVs also open niche markets like portable power generation and aesthetic-PV in building design. This chapter reviews the current state-of-the-art in making efficient organic and hybrid inorganic–organic PV devices. We discuss the basic physics of operation in a systematic way and also discuss current material limitations and identify areas that need improvement. Special emphasis is given to materials design and materials and device characterization. This chapter should serve as a guide to researchers in the field who plan to develop better material systems and optimize devices to push organic PV power conversion efficiencies above 10%.

11.2 Basic Operation

The efficient conversion of a photon absorbed by an organic chromophore to an electron–hole pair extracted and driven through an external circuit involves many steps. Upon photon absorption a strongly bound electron–hole pair, known as an

S.R. Scully (✉)
Department of Materials Science and Engineering, Stanford University, Stanford, CA, USA
e-mail: mmcgehee@stanford.edu

W.S. Wong, A. Salleo (eds.), *Flexible Electronics: Materials and Applications*, Electronic Materials: Science & Technology, DOI 10.1007/978-0-387-74363-9_11, © Springer Science+Business Media, LLC 2009 exciton, is formed [1]. The exciton must migrate to an interface where there is a sufficient electrochemical potential drop to drive exciton dissociation into an electron–hole pair that spans the interface across the *donor* (material with low electron affinity) and *acceptor* (material with high electron affinity). This *geminate pair* is still coulombically bound [2–6] and consequently must be dissociated. After dissociation, each charge must be transported through the device, avoiding trapping or bimolecular recombination, to the appropriate contact where the excess potential energy left over after all the above processes can be used to drive a load in an external circuit.

One of the primary ways to characterize a solar cell is by measuring current as a function of applied voltage in the dark and under illumination. The figures-of-merit are the short-circuit current (J_{sc}), which is the current at zero bias, the open circuit voltage (V_{oc}), defined as the voltage at which the total current is zero, and the fill factor (FF):

$$FF \doteq \frac{P_{max}}{J_{sc} \cdot V_{oc}},$$

where P_{max} is the maximum product of current times voltage along the current–voltage curve and, of course, the total efficiency, which is the percentage of P_{max} relative to the incident power.

Figure 11.1 shows representative I–V curves for the dark and the photocurrent for an organic solar cell. The total current, J_{light}, is the sum of the dark current and the photocurrent, assuming the principle of superposition which simply implies that the dark current does not change with illumination and consequently that the dark current and photocurrent may be added to obtain the total current. In this case, the V_{oc} is the potential at which the dark current is equal and opposite to the photocurrent. Also shown is the total current for an ideal solar cell in which the photocurrent

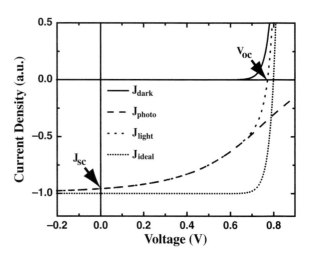

Fig. 11.1 Normalized current density as a function of voltage is shown for dark current (*solid*), light current (*dotted*) and photocurrent (*dashed*) representative of many organic solar cells, and the ideal current (*fine dotted*) corresponding to a voltage-independent photocurrent

is independent of voltage, as is the case for many inorganic solar cells. Unlike the champion inorganic devices, most organic solar cells show a strong voltage dependence of the photocurrent, which can lower the J_{sc} such that saturation (in this case, to $J = -1.0$) requires an extra driving force (i.e., a negative applied bias). The V_{oc} is also reduced simply because the photocurrent near the V_{oc} is lowered; however, as seen in Fig. 11.1, the most dramatic effect of a voltage-dependent photocurrent is a lowering of the *FF*. This has consequences for the maximum efficiency obtainable using organics. Obtaining higher efficiency organic PVs requires knowledge of the many factors that can lead to the strong bias-dependent photocurrent often seen in organics such that novel materials and architectures can be designed to alleviate these efficiency limiting factors.

11.2.1 Photocurrent

The external quantum efficiency (EQE) is the efficiency of charge pairs extracted per photons absorbed. A useful way to write the EQE is as a product of factors corresponding to each of the processes involved in photon-to-electron conversion:

$$\eta_{EQE} = \eta_{abs} \cdot \eta_{exharvest} \cdot \eta_{CT} \cdot \eta_{GS} \cdot \eta_{CC} \quad (11.1)$$

The last two terms in Eq. (11.1) corresponding to the efficiency of geminate separation (η_{GS}) and of charge collection (η_{CC}) are the factors that lead to a voltage-dependent photocurrent, whereas absorption (η_{abs}), exciton harvesting ($\eta_{exharvest}$), and charge transfer (η_{CT}) are considered to be voltage-independent. The photocurrent produced by a solar cell with incident photon flux, $SR(\lambda)$, is related to the EQE through an integral over the spectrum,

$$J_{photo}(V) = e \cdot \int_{all\lambda} \eta_{EQE}(\lambda, V) \cdot SR(\lambda) d\lambda \quad (11.2)$$

The spectrum most relevant for rooftop applications is the AM1.5G solar spectrum shown as a flux in Fig. 11.2. Integration over all wavelengths gives an incident power of 1,000 W/m^2.

11.2.2 Dark Current

Dark current is the current that would flow through the device under a bias in the dark in the opposite direction as the photocurrent. This current can be dominated by injection from the contacts, transport of the fastest carrier, recombination at the electrodes, or recombination in the bulk of the device. In inorganic p–n junction cells, this current is most often written as:

Fig. 11.2 AM1.5G photon
flux is shown as a function of
wavelength and energy. The
overall shape roughly
follows a blackbody
spectrum corresponding to
the sun's temperature of
about 5,700°C. The notches
arise from molecular gas
absorption in the earth's
atmosphere

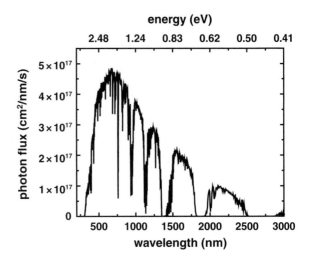

$$J_{dark} = J_o \exp\left(-\frac{e\,(V - V_{bi})}{nk_B T}\right) \tag{11.3}$$

where J_o is the reverse saturation current and can be related to the intrinsic carrier
density and their mobilities, V_{bi} is the built-in potential, and n is the "ideality factor"
which can vary between 1 and 2 depending on whether recombination in the depletion region or minority carrier diffusion to/from junction dominates the current [7].
The dark current–voltage curves of many organic devices can be experimentally fit
by such an equation even though the fundamental device physics are likely different
from that of an inorganic p–n junction. Currently, the efficiencies of most organic
architectures are limited by the low magnitude and voltage dependence of the photocurrent rather than the dominance of the dark current. However, understanding
how to limit the dark current will become increasingly important as problems with
the photocurrent are solved and these devices are made more ideal.

11.3 Organic and Hybrid Solar Cell Architectures

Before discussing the device mechanisms in greater depth, it is worth reviewing
the cell architectures shown in Fig. 11.3. The simplest organic PV architecture is a
single layer material sandwiched between a transparent electrode and a reflecting
metallic electrode. Light absorption, exciton splitting, and electron and hole transport all occur in the same material. These devices typically have quantum efficiencies of <1%, due to poor exciton dissociation because high fields (>10^6 V/cm), are
necessary to overcome the binding energy. Most of the excitons created naturally
decay rather than dissociate free carriers.

Fig. 11.3 Architectures of typical organic solar cells. Planar bilayer heterojunction (*left*), disordered bulk heterojunction (*center*) and ordered bulk heterojunction (*right*)

A simple improvement upon this architecture involves adding a second material, either organic or inorganic, which has HOMO and LUMO energy levels offset relative to the original material shown in Fig. 11.3a. Excitons that migrate to the heterojunction are dissociated into a geminate charge pair which spans the interface. An additional benefit of using two materials is that the photoresponse of the device can be made broader by choosing two materials that have absorption spectra spanning different parts of the solar spectrum. These devices when properly optically engineered have attained quantum efficiencies as high as 60% [8] and power conversion efficiencies of over 3% [9, 10].

While planar heterojunction devices show vast improvements over single layer devices, less than half of the incident photons over the absorption spectrum are absorbed and ultimately converted to free electrons and holes. The primary reason for this is low exciton harvesting. As we will discuss, excitons typically migrate distances of only 3–8 nm before decaying [11–16]. Consequently, most excitons created in the organic films do not arrive at the heterojunction before decaying.

Exciton harvesting was increased through the invention of the bulk heterojunction (Fig. 11.3b), which is an intimately mixed blend of two materials that has a high internal junction interfacial area. This random blend can be composed of small domains such that an exciton can easily migrate to a heterojunction before decaying. In such cases, all excitons can be harvested. In many bulk heterojunctions, improved exciton harvesting comes at the expense of poor charge collection [17]. The random nature of the two networks creates highly circuitous pathways that impede transport. Randomness can also increase energetic and positional disorder, which reduces electronic coupling. In spite of these issues, recent efficiencies have been reported to be as high as 5% [18–20] and nearly 6% [21] in an optically engineered bulk heterojunction device.

Looking forward to obtaining even higher efficiencies, however, motivated the design of the ordered bulk heterojunction [22] in Fig. 11.3c. This architecture is engineered such that domains are small so that all excitons can be harvested and ordered direct pathways within each phase would maximize charge collection. Using an inorganic scaffold to form the nanostructure has the added benefit that the interface could be modified before subsequent deposition of the primary organic. This is, to some, the ideal architecture for a bulk heterojunction PV [22, 23].

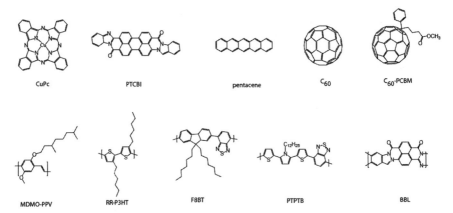

Fig. 11.4 Chemical structures of common organic PV materials. CuPc, copper phthalocyanine; PTCBI, 3,4,9,10-perylenetetracarboxylic bis-benzimidazole; pentacene, C_{60}, C_{60}-PCBM, (6,6)-phenyl-C61-butyric acid methyl ester; MDMO-PPV, poly(2-methoxy-5-(3′,7′-dimethyloctyloxy)-1,4-phenylene-vinylene); RR-P3HT, regioregular poly(3-hexylthiophene); F8BT, [poly(9,9′-dioctylfluorene-co-benzothiadiazole)]; PTPTB, poly(N-dodecyl-2,5-bis(2′-thienyl)pyrrole-2,1,3-benzothiadiazole); BBL, poly(benzimidazobenzophenanthroline ladder)

11.4 Materials

Figure 11.4 shows the chemical structure of many common organic PV materials used in organic and hybrid inorganic/organic solar cells.

11.5 Light Absorption

Good light absorption is essential to making a high efficiency solar cell. Organics are attractive because they can have high absorption coefficients ($\sim 2 \times 10^5$/cm) near their absorption maximum, which is synthetically tunable. Figure 11.5 shows the absorption coefficient versus wavelength for some example materials.

In efficiency calculations, assumptions are often made about the absorption of the material to predict the maximum current achievable under AM1.5G irradiance. Figure 11.5 is a collection of representative absorption data. Figure 11.6a shows the current density found by integrating the solar spectrum below various absorption onsets.

The maximum current producible from converting all solar photons below 4,000 nm (above $\sim$0.3 eV) into electrons yields $\sim$70 mA/cm^2, as shown in the inset of Fig. 11.6a. Converting all photons below the absorption onset for silicon of $\sim$1,100 nm gives just over 45 mA/cm^2. The above calculated currents assume 100% absorption above the onset. Real materials have finite bandwidths and absorption coefficients. Consequently, it is important to consider the thickness dependence of the current for a given absorption spectrum.

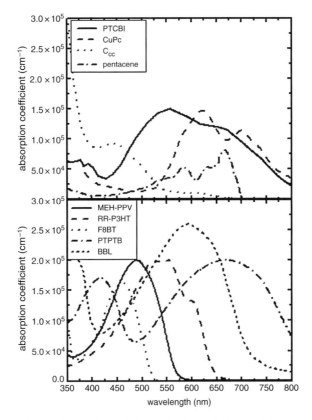

Fig. 11.5 Absorption coefficients for typical organic photovoltaic materials

Figure 11.6b shows representative absorption spectra for RR-P3HT and P3HT-like materials. The spectra for the P3HT-like materials are obtained, assuming the same density of states (DOS) and transition strengths, by shifting the spectrum of RR-P3HT by $-0.5, 0.5$, and 1 eV. Figure 11.6c shows that the absorption bandwidth increases with film thickness for the spectrum, which is redshifted by 0.5 eV relative to RR-P3HT. Also shown are the currents found by integrating the absorption response with the solar spectrum. This spectrum has an onset at $\sim$920 nm. This calculation assumes Beer's law absorption. As the thickness of the film increases, the spectrum broadens. It is important to remember that absorption is not complete when only the peak absorbs 90+% of the light. Ideally, we would engineer devices to absorb nearly all light above the absorption onset, as is done for efficient inorganic solar cells. An extension of this calculation is shown in Fig. 11.6a. The top line, as mentioned, is the maximum current, assuming 100% absorption above the onset. The curves that fall below this maximum correspond various finite film thicknesses with P3HT-like absorption spectra. This further shows that for maximum absorption, devices near 500 nm should be made. This also shows that at this thickness

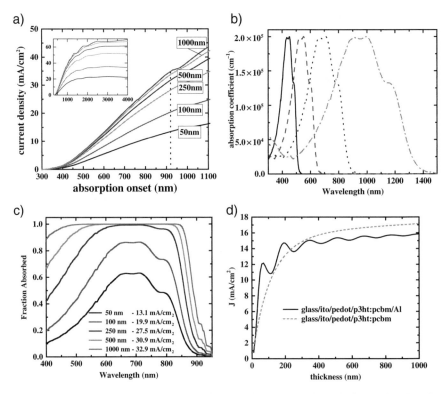

Fig. 11.6 (**a**) Integrated current density as a function of wavelength found by integrating the solar flux (Fig. 11.2) with absorption response for the case of complete absorption above the band edge as well as for P3HT-like absorption spectra, assuming different film thicknesses. (**b**) Representative absorption spectra for RR-P3HT and P3HT-like materials. The P3HT-like spectra are RR-P3HT spectra shifted in energy. (**c**) Calculated fraction of light absorbed as a function of film thickness is shown for a P3HT-like material with band edge shifted by 0.5 eV toward the *red*. Also shown are the current densities found when integrating over the solar spectrum for each film thickness. This effect corresponds to the *dotted vertical line* in panel (**a**). (**d**) Current density as a function of film thickness for P3HT/PCBM solar cells calculated using a transfer matrix approach and all optical properties of each material in stack. Also shown is the prediction assuming Beer's law absorption

one material has a sufficiently broad absorption spectrum to absorb nearly all photons above the band edge.

One aspect of absorption we have ignored in these calculations is the effects of optical interference and reflection off the substrate. Figure 11.6d compares the integrated current for a P3HT:PCBM blend as a function of blend layer thickness found by performing a full optical model, taking into account coherent reflections and transmissions at each layer interface in the device [16, 24], similar to what was recently reported [25, 26], to calculations assuming Beer's law absorption. The optical properties used in the calculation correspond to solvent annealed films, which show enhanced absorption [18] in the red relative to films processed from chlorobenzene and dried quickly. Optical interference effects give rise to the oscillatory behavior of the current with blend layer thickness. These effects are strongest for thin films (<200 nm). In fact, comparing the currents for a 50-nm blend layer thickness, we see nearly a factor of 2 increase when including interference effects relative to Beer's law absorption. We see for films thicker than this the current calculated for the full optical model remains below that predicted by Beer's law. This reduction, of $\sim$1.3 mA/cm^2 or $\sim$8% off the maximum, is primarily due to reflection off the stack. Future work should eventually involve minimizing this reflection by light trapping, antireflective coatings, or surface texturing, as is currently done for silicon [27].

Besides increasing thicknesses to maximize absorption, designing materials with low bandgaps to more effectively cover the solar spectrum is an obvious route to increase efficiencies by increasing the maximum possible photocurrent as shown in Fig. 11.6a. Numerous excellent reviews on the design and synthesis of low bandgap organic materials exist [28, 29] and absorption onsets as low as 1,000 nm have been reported recently [30–33]. Three synthetic approaches are commonly used to make low bandgap materials: extending the conjugation, minimizing bond-length alternation, and the push–pull approach.

The first of these methods for lowering bandgaps involves designing molecules with larger extended pi systems. Anthony et al. has worked with the acene family and stabilized derivatives of hexacene and heptacene driving the absorption onset up to 900 nm (1.38 eV) [30]. Calculations show that infinite polyacene chains could have an absorption onset above 3,500 nm and some have predicted this to be a potential candidate for a zero gap material [29]. While intuitive, this approach tends to be synthetically challenging because the molecules become insoluble due to the reduction in configurational entropy in the system.

In some cases, extending the conjugation is not possible because of bond-length alternation or breaks in conjugation due to twisting. Consequently, many materials have bandgaps that saturate with molecular size at or above 2 eV [34]. This brings up the second route to achieve lower bandgap materials, that is to minimize bond alternation. This is sometimes described as making the ground state have more "quinoid character" [29]. Bond-length alternation opens the bandgap of most materials just as it makes intrinsic polyacetylene a semiconductor rather than a metal [35].

The third option to achieve low bandgap materials involves introducing alternating electron-rich and electron-poor units onto the conjugated backbone in the push–pull or donor–acceptor approach.

Figure 11.7 shows how resonance interactions between the units create a composite gap that arises primarily from the HOMO of the electron-rich block and the LUMO of the electron-poor block. This has been hugely successful in tuning bandgaps due to the ease of synthesis [31, 32, 36–38]. Unfortunately, many of these materials suffer from poor charge carrier mobilities [33, 36] or large-scale phase separation [32] when used in polymer/PCBM devices. Further research is needed to probe if low mobilities are intrinsic to the push–pull design.

As a last note of caution, one should remember that the shape of the absorption spectrum is not the only important aspect of a prospective material, absolute absorption coefficients are equally important. Since optical transitions occur

Fig. 11.7 Schematic showing the formation of a small optical gap as a resonance between donor and acceptor energy levels

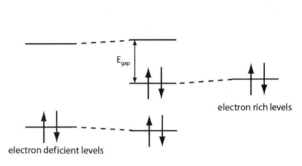

electron rich levels

E_{gap}

electron deficient levels

hybridized push–pull levels

between donor and acceptor units in the push–pull systems, it is not hard to imagine that the weaker the coupling between donor and acceptor, the weaker the absorption. Low absorption coefficients demand even more stringent requirements for exciton harvesting, geminate separation, and charge extraction because films must be made thicker to absorb all the photons. Future work will most certainly involve developing new low bandgap materials with high absorption coefficients ($\sim 2 \times 10^5$/cm) and moderately high ($>10^{-4}$ cm^2 V^{-1}s^{-1}) mobilities.

11.6 Exciton Harvesting

The absorption of a photon creates a bound singlet exciton which must migrate to a donor–acceptor interface, where the exciton may be dissociated into a geminate charge pair. Consequently, exciton migration and harvesting are enormously important for obtaining high quantum efficiencies since the necessary domain sizes are set by the distance excitons can migrate. In planar architectures, exciton harvesting is the primary limiting factor for obtaining high quantum efficiencies since exciton harvesting can only be efficient for film thicknesses less than the value of the diffusion length. Bulk heterojunction cells attempt to overcome this by intimately blending two materials such that the excitons need to migrate minimally to reach a heterointerface. In spite of this, there are numerous examples where the domain size is still too large to efficiently harvest all excitons [39–43]. Hence, understanding the limiting factors for exciton migration is immensely important. In this section, we present background theory and experiment to explain what factors limit the distance excitons migrate in relevant PV materials, methods to measure exciton migration and harvesting, and finally some example strategies to overcome small diffusion lengths.

Singlet excitons can migrate via coulomb coupling (Förster transfer in the point-dipole approximation [44–46]) or by electron exchange [47]. For longer range hops, dipole–dipole coupling dominates since this coupling decreases as $1/r^6$, whereas exchange decreases exponentially with distance and multipole coupling also decreases quickly with distance. On the other hand, short-range nearest-neighbor hops can involve both exchange and coulomb coupling. The figure-of-merit most often quoted for exciton migration is the *exciton diffusion length*. This is often written as:

$$L_D = \sqrt{D \cdot \tau} \qquad (11.4)$$

However, we will show that in most relevant organic materials, singlet exciton migration is not a true random walk and consequently Eq. (11.4) is technically invalid. The deviation from random-walk theory is due to the fact that exciton migration is greatly influenced by disorder which effectively leads to a time-dependent diffusion coefficient [48]. In spite of this complication, it is useful to first consider how large the exciton diffusion length could be in the absence of energetic and positional disorder and use this foundation to discuss the relevant limiting factors for exciton migration.

In Eq. (11.4), D is the exciton diffusivity and τ is the natural lifetime of the exciton. From random-walk theory the diffusivity can be written in terms of a hopping rate, Γ, hopping distance, a, and dimensionality, d, as:

$$D = \frac{\Gamma \cdot a^2}{2 \cdot d} \qquad (11.5)$$

As a lower bound for the diffusion length, we can assume the hopping rate is equivalent to the rate of Förster transfer between nearest neighbor chromophores in the point-dipole approximation. Förster theory [44] predicts that the rate of exciton transfer between two chromophores separated by a distance, a, with natural lifetime, τ, is

$$\Gamma = \frac{1}{\tau} \left(\frac{R_o}{a} \right)^6 \qquad (11.6)$$

The Förster radius, R_o, is the characteristic distance for transfer. Substituting Eqs. (11.5) and (11.6) into Eq. (11.4) gives

$$L_D = \frac{R_o^3}{a^2 \sqrt{2 \cdot d}} \qquad (11.7)$$

Note that the lifetime has dropped out and the diffusion length is determined by the hopping distance, the dimensionality of migration, and the Förster radius, which is simple to calculate from spectroscopic properties. As an example, we can take zinc phthalocyanine (ZnPc). The Förster radius is primarily determined by the photoluminescence quantum efficiency of the chromophores and the overlap between absorption and emission spectra.

Fig. 11.8 Normalized
absorption and emission
spectra of ZnPc. The
molecular structure is shown
in the inset

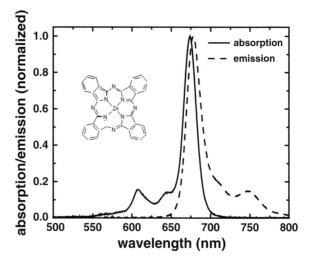

Figure 11.8 shows this overlap for ZnPc in solution [49]. Assuming randomly oriented transition dipoles, using a refractive index of 2 and the measured emission efficiency of 30% [50], we calculate a Förster radius of 5.5 nm. The molecular density of most phthalocyanines is near 1.7×10^{21} cm^{-3} [51], which corresponds to an average intermolecular separation of 0.84 nm. Using these numbers yields a predicted exciton diffusion length of 97–168 nm. We emphasize that this should be a lower bound for the diffusion length as higher hopping rates would result by including exchange coupling. It is also the case that singlet transfer by coulomb coupling deviates from the point-dipole description for exciton hops in close proximity due to the molecular shape, nonzero size of the molecule, and contributions from higher multipole coupling [52, 53]. These deviations would generally predict an increased hopping rate relative to what we assumed above. Finally, correlation of the orientation of the molecular transition dipole moments should also increase transfer rates. Experiments on phthalocyanines, however, report diffusion lengths of only 10–30 nm [16, 54, 55], nearly an order of magnitude less than predicted.

11.6.1 Effects of Disorder

As mentioned earlier, the answer to this discrepancy between the values of predicted (based on random walk theory) and measured diffusion lengths likely involves a number of factors – the most important is that exciton migration is not a true random walk but in fact is dominated by downhill migration to low-energy sites where subsequent trapping occurs [56]. In many materials, energetic and positional disorder reduce the distance excitons can migrate during their natural lifetime to <10 nm [11, 12, 16]. Figure 11.9 schematically shows the energetic landscape both for the

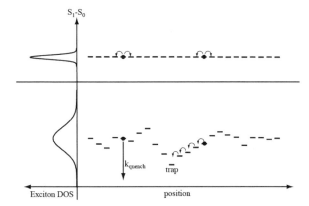

Fig. 11.9 Schematic representing the energetic landscape for a highly ordered material (*top*) and a realistic material with disorder (*bottom*). The exciton DOS is shown for each case. In the ordered material, all hopping rates are equivalent which leads to a random walk of the exciton. In the disordered material, downhill hops are favored which lead to a funneling of excitons to low energy sites that act as traps. Also shown is the possibility of an efficient quenching mechanism due to impurities or defects, shown as the *vertical arrow*

case that there is minimal disorder, leading to a true random walk, and for a realistic case with a large degree of disorder.

This figure shows a 1-D representation of the exciton levels, corresponding to each of the chromophores, as a function of position in a thin-film solid. The corresponding schematic of the exciton DOS is shown to the left of each energetic landscape. The top panel shows the case corresponding to a true random walk. The energy levels are practically the same for all chromophores and therefore the associated DOS is narrow. Consequently, the rate of exciton hopping is the same between any two chromophores. This is in great contrast to a more realistic energy landscape shown in the bottom panel of Fig. 11.9. In this case, there is a large degree of disorder and downhill exciton hopping in energy will be much faster than uphill energy transfer. In this way, excitons funnel toward local energetic minima in the film. A typical migration pathway is shown schematically by the series of arrows leading from the representative exciton to a low-energy site labeled as a trap. Once the exciton arrives at a site whose nearest neighbor chromophores are all significantly higher in energy, the exciton becomes trapped for the remainder of its lifetime. Significant evidence supports this description, most notably time-resolved photoluminescence measurements [12, 57]. Figure 11.10 shows time-resolved photoluminescence spectra for a polythiophene derivative.

The peak of the luminescence spectrum is near 2.1 eV at early times and progressively redshifts until it stabilizes below 2.0 eV after 50 ps. This phenomena is known as *spectral diffusion* and indicates that the average exciton has lost about 0.1 eV during its lifetime. Both thermalization of the exciton via internal conversion and electronic and structural relaxation associated with the Franck–Condon shift typically occur on a subpicosecond timescale [1, 58], which supports the idea that

Fig. 11.10 PEOPT
luminescence spectra as a
function of time following an
ultrafast excitation pulse.
The spectra redshift with
time which indicates exciton
migration to low energy sites
in the film. APS Copyright
2000. http://link.aps.org/
abstract/PRB/v61/p12957

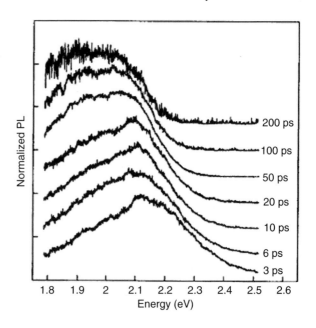

the longer timescale (picosecond to nanosecond) relaxation seen above must be due
to exciton migration to lower energy sites [59].

Another enlightening experiment is to compare time-resolved luminescence
spectra of pristine organic films to steady-state luminescence spectra of organic
films which include amounts of an electron accepting quenching molecule [57].
Figure 11.11 shows the shift of the luminescence maximum as a function of time (in
open circles) for a PPV derivative that has been excited by an ultrafast laser pulse
[57]. As expected, the maximum progressively redshifts with time due to exciton
migration as mentioned. Also shown are the values of the maxima of the steady-
state spectra (filled circles) for each of the organic films that have been blended with
a high electron affinity molecule which quenches the exciton by charge transfer. As
the volume percentage of quenching molecules increases, the luminescence spectra
of the organic film progressively blueshifts. At high quenching molecule volume
percentages, the steady-state spectrum of the film resembles the time-resolved spectrum at early times (<1 ps). This is because most excitons are quenched quickly
so the observed PL at steady state only comes from excitons that have decayed at
early times (<1 ps). This is in contrast to the PL of films with small volume percentages of the quencher. In this case, the steady-state spectrum is identical to the
time-resolved spectrum at long times. This demonstrates that the quenchers intercept the excitons on their migration pathways toward the low energy sites in the
film. Higher quencher concentrations quench excitons faster such that the excitons
do not have time to migrate to the tail of the DOS. The data in Fig. 11.11 were com-
prehensively fit to extract the amount of energetic disorder and it was found that a
Gaussian with width 0.125 eV described the data well [57].

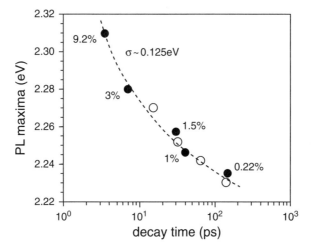

Fig. 11.11 Maxima (in energy) of the luminescence spectrum of a PPV derivative as a function of time following an ultrafast excitation. Also shown are the luminescence maxima of steady state spectra for various concentrations of a high electron affinity quenching molecule randomly blended in the PPV derivative, which demonstrates that the quenching molecules intercept excitons on their migration paths to low energy sites. The *dotted curve* was generated from a microscopic Monte Carlo model of exciton migration and quenching in an inhomogeneously broadened DOS with width 0.125 eV [57]. Copyright [2002], American Institute of Physics

We can now understand how energetic disorder affects the rate of exciton hopping and how a time-dependent diffusivity arises. The experiments suggest a picture like that in Fig. 11.12 where we show representative photoluminescence spectra for early, mid, and late times following absorption by an ultrafast pulse as well as the average absorption spectrum of all chromophores. At early times, there is strong

Fig. 11.12 Schematic showing normalized absorption and luminescence spectra following an ultrafast excitation pulse. The figure shows that the overlap between absorption and emission gets smaller as the excitons migrate to the low energy sites in the film. In this way the diffusivity becomes time dependent being high at early times and low at long times

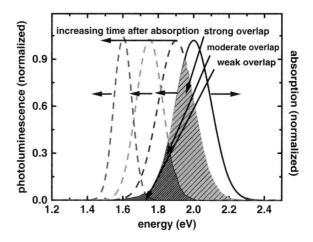

overlap between absorption and emission spectra, which gives rise to strong coupling and consequently fast exciton hopping. At mid times, the exciton population has migrated to lower energy sites with reduced density. Accordingly, the overlap between emission and absorption has lessened and exciton hopping has slowed. After long times, the exciton population has migrated to the tail of the exciton DOS, which gives rise to extremely weak overlap and slow hopping. At these times, the excitons are trapped because the hopping rate becomes long compared to the natural lifetime ($\sim$1 ns). This simple model explains how the inherent disorder in polycrystalline and amorphous films is primarily responsibly for a time-dependent exciton diffusivity and is the major reason excitons migrate much shorter distances during their lifetime than one would predict from basic random walk theory.

11.6.2 Extrinsic Defects

There are other factors that can limit exciton migration lengths. For example, reports exist on the effect of extrinsic quenching sites which effectively reduce the lifetime of the exciton [60–63]. This effect is represented in Fig. 11.8 schematically as the vertical arrow in the lower panel. This mainly affects exciton migration by reducing the average lifetime of the exciton in the same way as the quenching molecules described in Fig. 11.11. An exciton can arrive at a site where deactivation, by charge transfer for example, is fast compared to exciton hopping. O_2 incorporation in films of C_{60} and metal-free phthalocyanine have been shown to act as efficient quenching centers [61, 63]. If O_2 or other chemical impurities create the defect, purification is crucial to obtaining materials with high diffusion lengths. In polymers, the carbonyl defect, commonly formed in PPV derivatives upon exposure to oxygen, acts as an efficient quencher as well [60].

More recently, exciton migration was studied in three different PPV derivatives with various degrees of energetic disorder [13], as determined by charge transport measurements. The measured exciton diffusivity varied by a factor of 10 between the derivatives which correlated well with the degree of disorder seen through charge transport measurements. As expected, exciton migration was faster in the materials with a lower degree of disorder. However, the materials with faster diffusion also had shorter lifetimes. Consequently, the three materials, while having different diffusivities and lifetimes, all had practically the same diffusion length [13]. Unfortunately, increasing the diffusivity simply resulted in faster diffusion to quenching centers in the film.

11.6.3 Measuring Exciton Harvesting

As emphasized, Eq. (11.4) is technically invalid due to the fact that the diffusivity is time-dependent because of energetic disorder. In spite of this deviation, steady-state experiments are still remarkably well fit by a model assuming a random walk and an *empirical effective diffusion length* (hereafter referred to as "diffusion length") [11, 12, 16, 64]. Consequently, the length extracted from common experiments does represent a characteristic distance excitons migrate, but one should always remember the underlying microscopic physics are quite different from a true random walk in spite of what the name "exciton diffusion length" implies. With that said, knowing the value of this length is invaluable for considering device design and so we should discuss what the diffusion length empirically corresponds to and how to measure it.

For large thickness ($t_{film} >> L_D$) films, the diffusion length is the perpendicular distance an exciton will diffuse with probability equal to 1/e. Usually, more relevant for solar cells is the case that the film thickness is equal to the diffusion length. In this case, assuming uniform generation of excitons and perfect reflecting and quenching boundaries on either side of the organic film, approximately three-fourth of all excitons created in the film will be harvested [11].

For the accurate determination of the diffusion length, steady-state or time-resolved photoluminescence quenching studies are often used [11, 12, 16, 65, 66]. In a planar geometry, the films of the organic of interest are deposited to form a heterojunction with a material with staggered energy levels to drive exciton dissociation at the interface. To prevent overestimation of the diffusion length, the other material in the heterojunction must be chosen such that only electron transfer rather than energy transfer will occur [11]. Wide bandgap materials such as TiO_2 or ZnO are ideal quencher materials. Photoluminescence measurements are then performed on the organic materials in the heterojunction as well as neat layers on glass or quartz substrates for a variety of organic film thicknesses. Figure 11.13 shows the results of a typical experiment for MDMO-PPV when using titania as a quenching substrate [11].

The luminescence of the organic films on nonquenching substrates increases linearly with thickness for thin films (where the light is negligibly attenuated due to

Fig. 11.13 Integrated photoluminescence as a function of MDMO-PPV thickness on glass (*squares*) and titania (*circles*). The luminescence is reduced on titania where excitons can migrate and dissociate at the heterointerface. The *solid lines* were produced simulating exciton migration and harvesting assuming a 6 nm exciton diffusion length. Reprinted with permission from [11]. Copyright [2006], American Institute of Physics

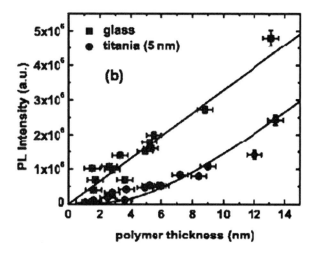

absorption by the film) and relative to this the luminescence of the organic on the quenching substrate is reduced. The reduction in luminescence on the quenching substrates is due to exciton migration and subsequent charge transfer at the quenching interface. By solving the continuity equation for exciton migration with appropriate boundary conditions, one can fit experimental data and extract a diffusion length:

$$\frac{\partial n}{\partial t} = D\frac{\partial^2 n}{\partial x^2} - \frac{n}{\tau} + G(x) \tag{11.8}$$

The first term on the right corresponds to exciton diffusion, the second to natural decay, and the third to exciton photogeneration. Assuming the heterojunction is a perfect exciton splitter and the organic/gas and organic/glass interfaces are perfect exciton reflectors, one finds by solving Eq. (11.8) at steady state:

$$PL_{nonquench}(d) = \gamma I_o d \tag{11.9}$$

$$PL_{quench}(d) = \gamma I_o \left(d - L_d \tanh\left(\frac{d}{L_d}\right) \right) \tag{11.10}$$

where I_o is the incident flux, d is the film thickness, and γ is a multiplicative constant. To obtain accurate measurements, one must ensure a well-defined heterojunction exists with no interdiffusion [66] and low roughnesses of all interfaces, accurately model and understand dominating interference effects in these experiments, and finally ensure excitons can only be harvested by migration and subsequent electron transfer and not by other means such as energy transfer. Following these guidelines, we have measured a diffusion length for MDMO-PPV and MEH-PPV of 6±1 nm [11]. We emphasize the importance of interference effects. To properly measure the diffusion length in these materials, we used thin (<5 nm) TiO$_2$ films. If thick films (>30 nm) were used, models incorporating diffusion lengths of 15–20 nm could fit the data because interference effects can lead to dramatic changes in the optical structures tested [11].

Performing the above quenching experiment in a time-resolved fashion and tracking the exciton lifetime instead of the integrated steady state, PL intensity has the advantage of insensitivity to optical interference effects and can be used to decouple the diffusivity, D, and the lifetime, τ, for a more complete picture of migration [67]. Exciton–exciton annihilation measurements can also be performed and have the benefit that only a single sample needs to be used [14, 15, 47]. However, uncertainty in the annihilation radius can make the results difficult to interpret. In spite of this concern, fair agreement exists between techniques. Finally, exciton diffusion lengths can also be extracted by accurately modeling the thickness dependence of the quantum efficiency in planar heterojunction cells by incorporating optical interference effects if the charge collection efficiency can be made perfect (e.g., by applying a reverse bias if necessary) [8, 16, 24].

In bulk heterojunction cells, a simple litmus test to determine the exciton harvesting efficiency is to compare luminescence, normalized by absorption, of a donor–acceptor blend to that of a neat organic film [68–71]. The fraction of luminescence quenched directly gives an upper bound for the quantum efficiency. While an extremely simple test, care must still be taken as optical interference effects, scattering, and changes in morphology can invalidate this measurement.

11.6.4 Approaches to Overcome Small Diffusion Lengths

We have already discussed bulk heterojunction cells which were originally designed for the specific purpose of harvesting more excitons. For example, PCBM blended with the polymer, MDMO-PPV, quenches (>95%) the luminescence of the polymer at only 1% PCBM loading. At 50% PCBM loading, the photoinduced absorption signature of radical anions which are formed after charge transfer in the polymer form in under 50 femtoseconds [72]. Such fast kinetics imply the exciton must already be formed adjacent to one or more PCBM molecules. In this case, minimal migration, if any, is necessary for efficient exciton harvesting. Another extreme example where exciton harvesting is near perfect is the dye-sensitized solar cell [73]. In this architecture, the exciton is created in a Ru-dye which is anchored to the electron acceptor titania. Consequently, the excitons are all created right next to the titania. This is surely the reason why ultrafast electron transfer from the dye to titania has also been observed in this system [74].

While some bulk heterojunction systems have high exciton harvesting efficiencies, others have domains too large for efficient exciton harvesting. As mentioned earlier, disordered heterojunctions can have other problems such as poor charge transport. It is therefore often believed that a more ordered structure is better. In both planar heterojunctions and current ordered bulk heterojunctions, exciton harvesting is far from perfect. Two major approaches have been used to increase exciton harvesting.

The first approach is to employ triplets by choosing materials with high intersystem crossing rates. Triplet excitons are long lived (~100 ns–1 ms) relative to singlet excitons (0.1–10 ns). Consequently, they have more time to migrate and detrap if stuck at low energy sites. A diffusion length of 40 nm was reported for C_{60} and attributed to the formation of triplets [16]. Others have used metal-centered complexes to ensure high intersystem crossing rates due to spin–orbit coupling. Enhanced harvesting efficiencies were reported when using the common phosphorescent OLED material PtOEP and a diffusion length near 20 nm was qualitatively extracted [75]. More recently, a novel white light OLED architecture was designed and utilizes long-range triplet migration [76]. Also of interest is the technique used to measure the triplet diffusion length of materials incorporated in the OLED [77]. Since these triplet excitons (like many) were nonemissive, a phosphorescent antenna was incorporated that accepts excitons and phosphoresces upon exciton capture. In this way, the same thickness dependence study as described above can be performed, but the time-resolved phosphorescence of the antenna layer is monitored as a function of the triplet material thickness rather than tracking the luminescence of the triplet material itself.

While employing triplet diffusion can enhance exciton harvesting, this can come at the expense of a loss in energy when converting the singlet to triplet. The singlet–triplet splitting of energy levels, due to electron exchange, is often 0.5–0.8 eV [78]. This loss reduces the maximum voltage obtainable using these materials. C_{60} has been reported to have an exchange loss of only 0.15 eV [79, 80]. An extremely novel approach would be to use exciton fission of the singlet exciton to two triplet excitons [81]. Using this scheme, the exchange energy would not be lost and would instead be stored in a second exciton. Only if the exchange energy can be minimized (as in C_{60}) or utilized (through exciton fission for example) will it be practical to utilize triplets to achieve high efficiencies.

A novel approach our group has been pursuing is to use hetero-energy transfer to enhance exciton harvesting [82, 83]. By incorporating a thin highly absorbing layer which has an absorption spectrum overlapping the emission spectrum of the donor material at the heterojunction, singlet excitons can be harvested over 30 nm away. Figure 11.14 shows exciton harvesting measurements for our model system of DOW Red (as energy donor) and PTPTB (as energy acceptor).

The inset shows the strong overlap in emission and absorption spectra necessary for efficient energy transfer. This high exciton harvesting at very large distances is due to the increased dimensionality of the system. The rate of energy transfer goes as $1/x^3$ [84, 85] instead of $1/r^6$ [44] because the total rate of energy transfer is determined by the sum of all excited chromophore–acceptor pairs. Here, x is chosen to emphasize that the distance is the perpendicular one from the exciton (excited donor) to acceptor sheet rather than a radial distance, r, as it is between two point

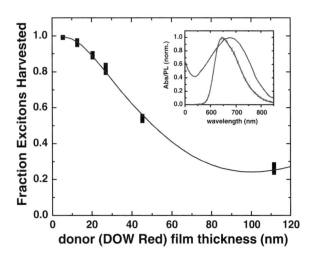

Fig. 11.14 Fraction of excitons photogenerated in DOW Red film on a thin film of the energy acceptor PTPTB as a function of DOW Red film thickness. The inset shows the overlap between emission of DOW Red and absorption of PTPTB. The *solid line* corresponds to an effective diffusion length of 27 nm

dipoles. This approach looks extremely promising due to the effectiveness in long-range exciton harvesting and straightforward design criteria for materials choice.

With this said, abundant opportunities exist to increase exciton harvesting by controlling the nanomorphology of bulk heterojunctions, through the design of better ordered bulk heterojunction architectures, by employing triplet materials with small exchange losses, using resonant energy transfer, or by designing materials with higher intrinsic diffusion lengths. For the latter, reducing energetic disorder, minimizing relaxation processes by choosing stiff materials, and using syntheses and materials processing which minimize contaminants that are responsible for forming quenching centers are all necessary to obtain high diffusion lengths and consequently high quantum efficiencies.

11.7 Exciton Dissociation

Following exciton migration, exciton splitting must occur to obtain extractable electrons and holes. Paramount to this discussion is the issue of the exciton binding energy. While formerly controversial, there appears to be a consensus that absorption of a photon creates a bound exciton in most organic materials [86–89]. Reports of the binding energy have ranged between 0.3 and 1.4 eV depending on the material [88–90]. This binding energy is often associated with the attractive coulombic energy between the photoexcited electron and hole. As an example, two opposite point charges separated 1.2 nm in a medium of dielectric constant $\varepsilon \sim 4$ corresponds to a coulombic binding energy of 0.3 eV, whereas a separation of 5 Angstroms corresponds to a binding energy of 0.72 eV. More precise calculations involving discrete partial charges have shown these qualitative predictions to be close to reality [89].

Comparison of photoelectron spectroscopy, which gives the energy levels associated with the positively and negatively charged molecules, with the optical spectra offers a way to measure this energy difference. UPS and IPES have been employed to identify the energy levels associated with the free hole and electron, respectively. The gap between these levels is the so-called *transport gap*. Similar measurements of the transport gap can be performed using scanning tunneling microscopy [89] as well as cyclic voltametry. Comparison between this gap and the optical bandgap always shows the optical gap to be smaller.

This is easily understood by recognizing that the exciton is composed of an electron coulombically attracted to a hole and hence stabilized relative to the case involving the free, separated (infinitely apart), pair. Simply subtracting the optical gap from the transport gap gives the binding energy.

The exciton binding energy is often associated with the energy drop necessary at a donor–acceptor interface to split the exciton. We note a few discrepancies with this interpretation. The first, as will be discussed in greater detail in the following section, is that after splitting the exciton, the geminate charge pair is often still strongly bound due to the close proximity of the charges to each other. In other

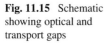

Fig. 11.15 Schematic showing optical and transport gaps

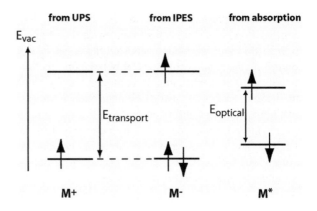

words, splitting the exciton *does not* create free charges. Usually, the coulomb binding energy is reduced by charge transfer, but it is by no means zero. Consequently, a field or other electrochemical potential bias is necessary to free these bound charges. However, using a simple conservation of energy argument, it does appear to be the case that, as defined, the exciton binding energy represents the energy necessarily lost between photon absorption and charge extraction. We note, however, that there are additional losses due to the structural relaxation associated with the Franck–Condon shift and from exciton migration to low energy sites in the film as described in the previous section. In some materials systems, each of these loss mechanisms can easily account for 0.1 eV energy loss. The losses associated with exciton relaxation which corresponds to the difference in energy between absorption and emission spectra certainly impose restrictions on the energy levels necessary for exciton splitting.

Disorder in the energy levels at the interface has also been shown to greatly influence electron transfer rates [91]. Marcus nonadiabatic electron transfer theory also gives guidelines for the driving force necessary for fast electron transfer. In this case, the driving force which maximizes the rate of electron transfer is that which equals the reorganization energy of the system.

While the exact value of the energy offset necessary for exciton splitting still needs further investigation, luminescence quenching and photoinduced absorption can be a good way to determine whether exciton splitting is occurring. If one can ensure intimate random mixing, then simply correlating luminescence quenching and formation of radical-ion photoinduced absorption as a function of energy offset should offer a straightforward means to "observe" exciton splitting. Unfortunately this can be complicated by exciton diffusion limitations. However, numerous measurements suggest charge transfer can occur with a drop in some systems of only 0.2–0.5 eV [92]. Key to future studies which probe this important question will be to develop higher resolution energy level measurements. Currently, both cyclic voltammetry and photoelectron spectroscopy have errors of ±0.3 eV.

11.8 Dissociating Geminate Pairs

While early in the field it was thought that charges were free immediately following exciton splitting, a large consensus has grown that this is not the case. Immediately following exciton splitting, a geminate charge pair spans the donor–acceptor interface and is separated by ~1–3 nm [2–4, 6]. This necessarily implies the charge pairs are strongly coulombically bound due to the low dielectric constants of the materials and the small extent of the electronic wavefunctions associated with these charges. For efficient PV operation, these charge pairs must be separated efficiently.

The charge transfer process at a donor–acceptor heterojunction has been explained to be analogous to the autoionization and subsequent thermalization processes of "hot excitons" in molecular crystals [1, 3]. Upon charge transfer, the excess energy of the transferred charge due to the band offsets is lost in discrete scattering events, which convert electrochemical potential energy of the charges into local molecular vibrational energy. The charges thermalize through these events and come to rest some distance away from the interface and from each other. This characteristic distance, known as the *thermalization radius*, then determines the coulomb binding energy of the geminate pair via the coulomb potential:

$$E_{gp} = \frac{1}{4\pi\varepsilon_o \cdot \varepsilon_r} \frac{e}{r_{th}} \qquad (11.11)$$

Geminate recombination has come to be viewed as one of the major loss mechanisms in organic PVs. In the "single photon regime" (i.e., at low light intensity), separation of geminate pairs will be the dominant process that determines the shape of the photocurrent versus the voltage curve. In this regime at low light intensity bimolecular recombination and space charge can be made inconsequential. The quantum efficiency is therefore only limited by the separation of the geminate pair, assuming efficient photon and exciton harvesting. *FF* in small molecule heterojunctions and polymer/PCBM cells have been shown to be limited by geminate separation [3, 4]. Normalized photocurrent versus applied field directly measures the probability of separation versus field.

Onsager-type models of separation have been invoked to explain the probability of separating the geminate pairs that span a donor–acceptor interface following exciton dissociation [2–4, 6]. General to all models is that the charges are coulombically attracted to each other and, through stochastic motion, undergo a biased random walk. Field, mobility, disorder, dielectric constant, binding energy, back-transfer time, and geometry are all important factors that influence the likelihood of pair separation. To first order, separation occurs when the carriers are outside of their Onsager radius:

$$r_o = \frac{e}{4\pi\varepsilon_o\varepsilon} \frac{1}{k_B T} \qquad (11.12)$$

which is the separation distance at which the coulomb binding energy of the pair becomes approximately equal to the thermal energy, that is $\sim k_B T$.

Geminate separation in polymer/PCBM devices has been modeled using an average continuum model that reproduces the field and temperature dependence of this particular device [4]. Key to the model is that a distribution of thermalization distances and consequently a distribution of binding energies, due to inherent disorder in these systems, are necessary to explain the observed current–voltage characteristics. Spatially averaged mobilities and dielectric constant are used and a back-transfer time, corresponding to the rate of recombination when the electron and hole reside on neighboring molecules, must be assumed.

This last point is more delicate than it might be assumed at first since most recombination measurements (transient photovoltage or transient photoinduced absorption) involve measuring the full kinetics of the recombination process (biased random walk plus back transfer) rather than only the interfacial back-transfer time. Molecular dyads in solution have been studied and their back-transfer kinetics can serve as useful guides; however, there are potential large differences between these dynamics in solution and in the solid state [52, 93–97].

Monte Carlo simulations have also been used to predict the charge separation dynamics on the microscopic scale [3, 6]. The models only assume average dielectric properties and all other parameters are treated microscopically. Geometry is an important parameter in these dynamics as the electrons and holes are energetically confined to their own phase such that a donor–acceptor interface increases the likelihood of charge pair separation [3]. Asymmetry in the dynamics of charge hopping has also been suggested to lead to high separation efficiencies because the likelihood of the electron and hole meeting on neighboring sites is reduced [3].

Common to all three reported models is that the ratio of charge carrier hopping rate to back-transfer rate is an important parameter since this determines the probability of survival when the carriers arrive on adjacent molecules. For example, a mobility of 10^{-4} cm^2 V^{-1}s^{-1} corresponds to a hopping rate of about 5 hops/ns assuming the Einstein relation, 3-D diffusion, and a hopping distance of 5 Angstroms. Each hop takes $\sim$200 ps. If the back-transfer time were also 200 ps, each time a geminate pair sits next to each other they would have about a 50% probability of recombining and a 50% probability of hopping away from each other. Were the back-transfer time instead 2 ns, the probability of recombination would be reduced to 10%. Some fullerene–porphyrin dyads have shown back-transfer times as long as 200 μs [98]. If this were representative of back transfer in a bulk heterojunction, the likelihood of recombination would be extremely reduced. The longer the back-transfer time is, the more chances the geminate pairs will have to climb out of their coulomb potential wells.

Another crucial but poorly understood parameter which influences geminate separation is the thermalization distance as this determines the pair's binding energy. It has been suggested that the thermalization distance should be related to the band offset and ultimately to the excess energy between relaxed excitons and relaxed geminate charge pairs however no thorough study exists that probes this hypothesis

[3]. The excess electrochemical potential energy is partially converted to help the geminate pairs climb out of their mutual coulomb potential wells. It is unknown how efficient this conversion is; however, evidence suggests that it is likely *in*efficient because exciton splitting across heterojunctions with band offsets near 1 eV still produce geminate pairs with binding energies of 0.2–0.3 eV rather than free carriers [3, 4, 6, 99]. This implies that much of the excess potential energy is lost to heat in the form of molecular vibrations.

One of the most recent studies probing these physics reported on the dynamics of transient photoconductivity and transient photoinduced absorption for blends of MDMO-PPV/PCBM and methyl-substituted ladder-type poly(p-phenylene) (MeLPPP)/PCBM [100]. MeLPPP is more rigid and is energetically more ordered than MDMO-PPV. MeLPPP also has a high bandgap relative to MDMO-PPV and consequently a larger energy offset at the polymer/PCBM interface. It was reported that the geminate pairs in the MDMO-PPV/PCBM blend have a higher probability of separating than in MeLPPP/PCBM blends and this was attributed to a larger thermalization radius and lower binding energy of the geminate pairs in the MDMO-PPV blend compared to the MeLPPP blend. This is despite a lower band offset in the MDMO-PPV/PCBM blend and implies a more complicated relationship between band offset and thermalization distance.

Charge separation following exciton splitting can also be inhibited through the formation of an exciplex state which if significantly more stable than the charge separated state will trap the charges until back transfer occurs [5, 101]. Figure 11.16 shows a schematic describing the energetics of this situation.

Two blend systems were recently investigated: PFB:F8BT and TFB:F8BT [5]. In both blends, photopumping the F8BT results in the formation of a new luminescence spectrum that is not a simple linear combination of the neat layer luminescence for the polymers that comprise the blend. The new spectrum is broad and redshifted relative to the spectra of the pure materials and has a characteristic lifetime of its own. The interpretation of this phenomena was that the spectrum arises from an *inter*molecular excited state, termed an *exciplex*, composed of an electron on an acceptor molecule and a hole on a neighboring donor molecule. Diode structures were fabricated to investigate the field dependence of the luminescence dynamics of the excimer. The *generation* of the exciplex was found to be highly field-dependent but the luminescence decay corresponding to the exciplex lifetime was found to be field-independent. Potential energy curves are shown for the ground state system, the state with excited acceptor (F8BT), and the state which has the electron residing on the acceptor and the hole residing on the donor. The field dependence of the exciplex generation is explained by suggesting a "spectroscopically dark" geminate pair is always created following exciton dissociation. This pair can be dissociated by a field or can collapse into a highly stable exciplex whereafter it can regenerate bulk F8BT excitons with a thermal activation energy ranging between 0.1 and 0.3 eV. These findings are relevant for geminate splitting as collapse to an exciplex offers an alternate route to back transfer during the dynamics of geminate migration and separation.

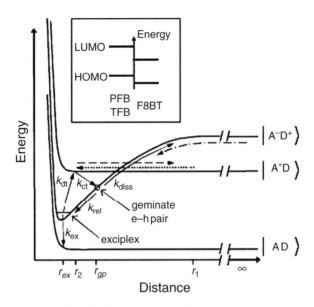

Fig. 11.16 Schematic showing the free energy for a donor–acceptor system, with energy levels shown schematically in the inset, as a function of distance (electron–hole or intermolecular separation) for the ground state of donor and acceptor ($|$AD) distance is intermolecular separation), for the case that the exciton resides on the acceptor ($|$A*D) distance is intermolecular separation), and for the case that the electron and hole reside on the acceptor and donor respectively $\left(|A^-D^+\right\rangle$ distance is electron–hole separation). This system shows that an exciplex is simply a tightly bound geminate pair. The arrows represent different transitions/pathways for the electron–hole pair. APS Copyright 2004. http://link.aps.org/abstract/PRL/v92/e247402

Following this work, the triplet energy levels of the donor and acceptor relative to the exciplex and the charge separated state were shown to be equally important in determining the fate of the collapsed geminate pair [101].

Blend systems of MDMO-PPV and PCNEPV were investigated. Figure 11.17 summarizes the spectroscopic data found for this system. In these polymer blends, the energy level of the MDMO-PPV triplet is lower than the triplet energy of the exciplex. Following dissociation of photoexcited excitons in MDMO-PPV, the geminate pair can survive long enough for spin dephasing of the two carriers, which occurs on a microsecond timescale. After spin dephasing, the geminate pair can collapse into a triplet exciplex with a probability dictated by spin statistics (3/4 to first-order). Decay from the exciplex triplet to the MDMO-PPV triplet via triplet energy transfer is then spin allowed. Since the triplet level in MDMO-PPV is significantly lower than the triplet exciplex, the exciton remains trapped until decay to the ground state whereby the energy is lost. This process represents an important potential loss mechanism of charges in organic PVs. The triplet level is often ~0.5–0.8 eV below the singlet [78]. We may estimate the charge separated state energy as the effective HOMO–LUMO gap arising from the HOMO of the donor and the LUMO of the acceptor. This relationship might suggest the effective gap

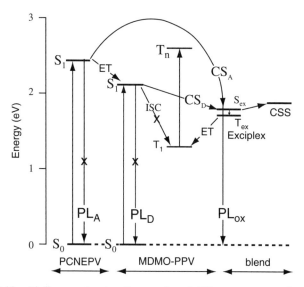

Fig. 11.17 Jablonski diagram showing the energies of different spectroscopic species for the MDMO-PPV/PCNEPV system studied by Janssen and coworkers. *Arrows* indicate transition pathways. In this system, it is shown that the separated geminate pair can collapse back to a triplet exciplex whereafter triplet energy transfer to the triplet state of MDMO-PPV can occur and effectively trap photogenerated charges. APS Copyright 2005. http://link.aps.org/abstract/PRB/v72/e045213

must be less than the relaxed singlet energy minus the exchange energy to prevent the geminate pair from collapsing to a triplet level of one of the polymers. Others have offered similar warnings on blends of polythiophene derivatives and PCBM [102]. Understanding the dynamics of charge separation will remain an active area of research. Future findings will have potentially large implications on the maximum efficiency these cells can achieve.

11.9 Heterojunction Energy Offsets

These offsets are extremely important because they affect the maximum V_{oc} that can be obtained since the quasi-Fermi levels will never practically be pushed beyond the band edges because the DOS is high. The energy difference between the absorbed photon and the effective gap of the heterojunction will always be consequently lost to heat.

Multiple groups have shown a strong correlation between the effective gap and the maximum obtainable V_{oc} [103–106]. Figure 11.18b shows an energy level diagram for an ideal material as well as for one where a Gaussian DOS is superimposed over the energy levels.

A recent review [106] shows the linear correlation between polymer HOMO level (and consequently effective gap) and V_{oc} for over 20 different polymer/PCBM combinations. In spite of the strong correlation, the V_{oc} is often less than the effective

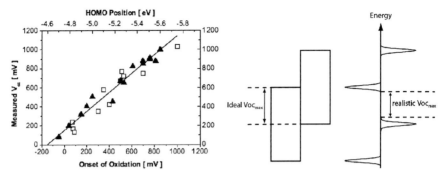

Fig. 11.18 (**a**) V_{oc} as a function of polymer HOMO level for polymer/PCBM solar cels for a series of polymers. The *solid line* has a slope of 1 indicating a clear dependence of V_{oc} on polymer HOMO level. (**b**) Shematic showing idealized energy level diagram as well as a more realistic case where the rigid energy levels are replaced gaussians. Disorder which gives rise to the Gaussian tail states can then be easily understood to pin the Voc at less than would be predicted from the idealized model

gap by 0.2–0.4 V. This can be due to an early turn-on of dark current, pinning of the cathode and/or anode workfunctions to be within the gap relative to the HOMO of the donor and the LUMO of the acceptor respectively [107]. It is also possible that measurements of the HOMO and LUMO levels are just not accurate enough to conclude within about 0.2 eV where they lay. In fact, just as important as the average energy levels are is where the tail states are relative to these average levels. Having less than 1% of the energy states to be within the gap 0.2 eV below the average levels could easily explain this discrepancy because these states would pin the quasi-Fermi levels as shown in Fig. 11.18b. This pinning is the same issue that keeps amorphous silicon solar cells from achieving higher V_{oc}s. The tail states pin the quasi-Fermi levels such that the V_{oc} can only be as high as 0.7 V even though the bandgap is nearly 1.4 eV. Future work will show if we must always lose this energy or if it is specific to the material system used. Perhaps using more ordered materials will reduce this drop.

Further research is necessary to fully explain the energy offsets necessary for efficient exciton splitting at the heterojunction. With that said, the required offsets are likely to be less than current offsets in state-of-the art metal oxide/polymer, P3HT/PCBM, and CuPc/C60 systems of nearly 1 eV.

A novel approach to tune the energy level offset was recently reported [108] where a percentage of a magnesium oxide (MgO) precursor was incorporated in solution of a zinc oxide (ZnO) precursor to form a ternary $Zn_{1-x}Mg_xO$ compound.

Figure 11.19 shows V_{oc} and J_{sc} as a function of Mg content for planar solar cells, incorporating P3HT and the ternary compound. As shown, the V_{oc} of solar cells was increased by nearly 0.4 V with incorporation of about 35% Mg. Unfortunately, this came at the expense of the J_{sc} which dropped by about 30% at this Mg concentration relative to the case without Mg. However, an increase of about 0.3 V was obtained while maintaining a high J_{sc}. Further studies combined with more accurate energy level measurements and spectroscopy will be needed to explore how small an energy

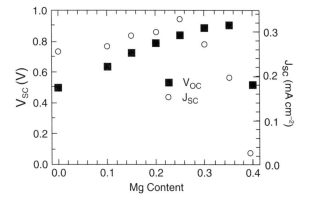

Fig. 11.19 V_{oc} and J_{sc} as a function of Mg content in the ternary $Zn_{1-x}Mg_xO$ compound. Increasing Mg content raises the conduction band edge of the compound which gives rise to a larger V_{oc}

level drop is needed to maintain high currents and fill factors. Pushing efficiencies beyond about 7% will require minimizing wasted energy by reducing unnecessary junction offsets.

11.10 Charge Transport and Recombination

Following photon absorption, exciton harvesting, charge transfer, and geminate separation leaves only charge extraction as the remaining mechanism to understand and optimize for high quantum efficiencies. Two major efficiency-limiting mechanisms must be understood and solved to engineer efficient devices. These are the formation of space charge and the mechanism of bimolecular recombination. These two effects are what prevent achieving fill factors as high as those seen in inorganic solar cells.

Space charge arises when there is a mismatch in carrier mobilities which leads to a buildup of net charge in the device. This charge consequently leads to a redistribution of the electric field, which hinders dissociation of geminate charge pairs and the collection of dissociated charges. The space-charge limited photocurrent for a bulk heterojunction is given by [109]:

$$J_{SCLPC} = e \left(\frac{9\varepsilon_o \varepsilon \mu_{low}}{8e} \right)^{1/4} G^{3/4} V^{1/2} \qquad (11.13)$$

where μ_{low} is the lowest of the two carrier mobilities, ε_o is the permittivity of free space, ε is the dielectric constant of the medium, G is the generation rate of charge pairs, e is the electric charge, and V is the internal voltage of the cell. It was quite convincingly shown that polymer:PCBM devices showed all the characteristics of space-charge limited photocurrent in blends that had a high mismatch in mobilities (2–3 orders of magnitude difference) [109]. Figure 11.20 summarizes these findings.

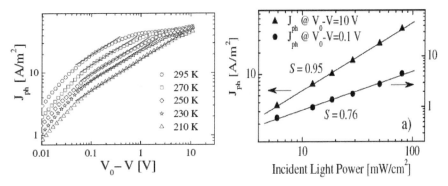

Fig. 11.20 (**a**) Photocurrent as a function of internal voltage demonstrating the occurrence of space-charge limited photocurrent. The *straight lines* indicate the regions the photocurrent is space-charge limited. As the temperature is lowered, the mobilities are lowered and the mobility mismatch increases such that the region over which the photocurrent is space-charge limited expands. However, as the internal field is increased photocurrent becomes limited by absorption rather than space charge. (**b**) Photocurrent is shown as a function of light intensity for two different internal voltages. For 0.1 V, the photocurrent is highly space-charge limited and hence follows a three-quarter dependence on power, whereas for 10 V, the photocurrent scales nearly linearly with light intensity

Figure 11.20a shows the current–voltage curves for different temperatures. The solid lines show the square root dependence on voltage for the regime that the photocurrent is space-charge limited. The electron and hole mobilities in the above study had different activation energies. Consequently, by lowering the temperature, the mobility mismatch was increased from 100 to 1,000 which made the voltage range over which the photocurrent was space-charge limited also increase. Figure 11.20b shows the photocurrent as a function of light intensity for two characteristic voltages. For voltages within the square root regime, the photocurrent follows a near three-quarter power dependence on light intensity as predicted by Eq. (11.13). In the saturation regime, at strong reverse bias the photocurrent follows a near linearly relationship with light intensity. It has since been shown that the photocurrent versus light intensity can fall in a regime where it is limited to some extent, but not fully, by space charge [110]. The extent to which the current is space-charge limited can be measured by the exact dependence of photocurrent on light intensity. It was concluded that for many of the polymer/PCBM cells, the efficiency was primarily limited by space charge which in turn limits the fill factor [110]. Hence, given low mobilities $<10^{-4}$ cm^2 V^{-1}s^{-1}), it is important to balance the mobilities to prevent space charge effects.

We would like to point out that the situation is markedly different in the case of planar heterojunctions because all charge is being created at the planar heterojunction rather than throughout the bulk of the film. In this case, there is always net charge on either side of the junction since the carriers are physically separated. This is analogous to the space-charge limited diode where the carriers are instead photoinjected. Hence, neglecting diffusion, the space-charge limited photocurrent should simply follow Child's law:

$$J_{SCLC} = \frac{9}{8} \varepsilon \cdot \varepsilon_o \mu_{low} \frac{V^2}{L_{low}^3} \qquad (11.14)$$

where μ_{low} and L_{low} are the mobility and the layer thickness, respectively, of the low mobility material. In this case, space charge can become a problem even if the mobilities are matched.

For the case that the mobilities are matched and high enough, the only other mechanism by which charge carriers can be lost is through bimolecular recombination. This involves nongeminate electrons which recombine with holes as the carriers traverse the device to their respective electrodes. Usually in organics with electron density, n, and hole density, p, the recombination rate (in # of carriers cm^{-3} s^{-1}) is given by:

$$R = -\beta np \qquad (11.15)$$

11.10.1 Diffusion-Limited Recombination

As has been reported [2, 111, 112], recombination in most organic cells appears to follow Langevin-like kinetics. In this case, β is given as:

$$\beta = \frac{\langle \mu_h + \mu_e \rangle}{e \varepsilon_o \langle \varepsilon \rangle} \qquad (11.16)$$

where the brackets indicate "spatial average". This corresponds to a "diffusion-limited" scenario wherein the rate of recombination is limited by the rate an electron and hole meet.

Langevin recombination has been successful in modeling the recombination current in single layer OLEDs. Recently, however, a deviation from Langevin kinetics in bulk heterojunction cells was reported [112] and it was explained that the recombination rate is limited by the slowest carrier, rather than the fastest as above, because the carriers are confined to their own phase – electrons in fullerene and holes in polymer. Replacing the spatial average of the mobility with the mobility of the slowest carrier reproduced the current–voltage curves. Understanding the role of morphology and geometry on recombination will surely be important in the near future.

While recombination is rarely unimportant in most solar cell architectures the effects of space charge and geminate separation have largely overshadowed the importance of bimolecular recombination in organic bulk heterojunctions. However, as these architectures are made more efficient by choosing materials which yield efficient geminate separation and high mobilities (which consequently eliminate space charge), bimolecular recombination will then be left as the top culprit for limiting efficiencies.

Already in 100-nm-thick MDMO-PPV:PCBM blends, it is estimated that at the maximum power point nearly 25% of the photogenerated charge carriers bimolecularly recombine [113]. It has been shown that the importance of bimolecular recombination will increase at greater device thicknesses because the carriers must be transported further and hence will be more likely to recombine. It is not obvious just from Eq. (11.16) whether an increase in mobility would lead to an increase or decrease of bimolecular recombination. Increasing mobility would lead to faster extraction, but the recombination coefficient would also increase in proportion to the mobility. However, it is important to recognize that the steady-state carrier density also varies with mobility. For similar generation rates, increasing the mobility leads to a reduction in the steady-state carrier concentration. This effect outweighs the increase in bimolecular rate constant such that for higher mobilities bimolecular recombination is lessened.

11.10.2 Interface-Limited (Back Transfer Limited) Recombination

A few other options exist for increasing the FF which in the case of efficient geminate separation and in the absence of space charge would be limited by recombination. Designing materials with higher dielectric constants may also be a route to reduce recombination. Otherwise, to exceed these efficiency limits we need to design devices that have recombination which is not diffusion limited. Diffusion-limited recombination implies the interfacial back-transfer rate, corresponding to the electron and hole sitting on adjacent molecules, is not limiting. Therefore, if a system can be designed such that the back-transfer rate becomes limiting, and we are in an *interface limited regime*, recombination rates can be slowed to even less than that predicted by Langevin theory. This concept has been investigated extensively in the dye-sensitized architecture. One recent study [93] is summarized in Fig. 11.21.

By adjusting, through molecular design, the spatial location of the HOMO wavefunction of the dye relative to the surface titania, the recombination rate from photoinjected electrons in the titania through back transfer to the dye can be changed by five orders of magnitude. These findings clearly indicate that the recombination rate is interface rather than diffusion limited in this system. The solid line in Fig. 2.21 is an exponential fit and can be understood by considering the basic nonadiabatic Marcus electron transfer equation:

$$k_{ET} \propto \exp\left(-\beta \cdot r\right) \exp\left(-\frac{(\Delta G_o + \lambda)^2}{4\lambda k_B T}\right) \qquad (11.17)$$

The first exponential represents the electronic coupling between the dye wavefunctions and the titania wavefunctions and contains the distance dependence. The

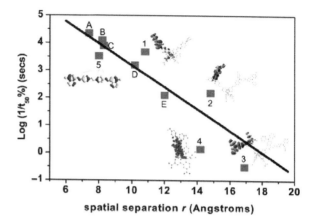

Fig. 11.21 Recombination time as a function of calculated spatial separation of the HOMO wavefunctions for a series of dyes from the titania surface. The *straight line* is an exponential fit and indicates that the recombination process is limited by tunneling from the titania to the dye HOMO level. [93] – Reproduced by permission of The Royal Society of Chemistry

second exponential contains the information relating to the thermal activation barrier given as:

$$\Delta G = \frac{(\Delta G_o + \lambda)^2}{4\lambda} \tag{11.18}$$

where ΔG_o is the free energy change between initial (electron in titania and hole on dye) and final (electron back on dye in the ground state) states, and λ is the so-called "reorganization energy," which is related to the polarization and relaxation in the surrounding medium.

Marcus theory has been used to explain electron transfer between single donor and acceptor molecules and has also been extended to describe transfer between discrete and continuous states by integrating over the DOS to account for all possible transfer pathways [93, 114].

This simplistic equation gives two parameters to tune the rates of forward and back electron transfer. As mentioned, the electronic coupling can be reduced through molecular design. Because the rate of forward electron transfer in many systems is orders of magnitude larger than necessary, there is some leeway to reduce the electronic coupling for both forward and back electron transfer without sacrificing the yield of electron transfer. For example, in the MDMO-PPV:PCBM system formation of the photoinduced charge separated state through electron transfer has been shown to occur faster than 50 fs, whereas the excited state lifetime of MDMO-PPV is near 200 ps [72]. For 90% charge transfer, an electron transfer rate of only 20 ps is necessary. In this case, the forward transfer is 5,000 times higher than is necessary for efficient photogeneration. This system is a good example where the electronic coupling could likely be reduced without the expense of a reduction in electron transfer yield. The reduction in electronic coupling would likely also reduce, to some extent, the rate of back transfer.

Fig. 11.22 Simplified
energy level diagram
indicating changes in free
energy for forward and back
electron transfer

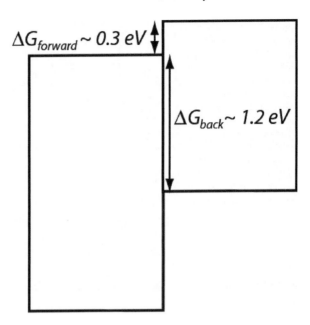

The second parameter we, as chemists and engineers, have to control is the change in free energy of the systems which effectively amounts to changing energy levels. Appropriate design can also lead to asymmetry in the system such that forward electron transfer is barrierless and back transfer has a large activation energy. To first-order, this is already the case in donor–acceptor systems with small energy offsets. Figure 11.22 shows this case.

In this case, assuming a value for the reorganization energy equal to the change in free energy associated with forward electron transfer, we would predict the forward transfer to be barrierless but the back transfer to be thermally activated with a barrier equal to 0.68 eV. Consequently, the back transfer would be slowed by over 11 orders of magnitude relative to forward transfer. Of course this example is overly simplistic and complications such as differences in electronic coupling and reorganization energy between forward and back transfer as well as the effects of disorder would most certainly change the details of this picture. It is also possible that recombination would proceed through back transfer to the LUMO of the donor which should only have an activation barrier of 0.3 eV followed by direct recombination within the donor. With these considerations in mind, slowing back-transfer rates such that recombination is interface limited rather than diffusion limited offers a straightforward route, by introducing asymmetry in the electronic coupling or activation energies for back and forward electron transfer, to reduce the probability charge carriers recombine and consequently achieve higher efficiencies. Ordered bulk heterojunctions which utilize an inorganic scaffold are the ideal architectures to tune recombination kinetics because the interface can be easily modified. Furthermore one can engineer systems such that molecules attach in a preferential geometry to the scaffold, thereby better controlling the electronic coupling.

11.10.3 Measurements Relevant for Extracting Charge

One of the most important and informative measurements one can do to investigate charge extraction issues is to measure the charge carrier mobilities. Because of the high degree of disorder and consequent field-dependent mobilities, measurements of mobility corresponding to the carrier densities found in working solar cells under one sun are most relevant. The mobility can vary by four orders of magnitude simply because of the differences in carrier density [115, 116]. Mobilities at lower densities are usually lower because low energy sites are not occupied and consequently tend to trap charges as they traverse the films, whereas at high carrier density traps can be filled which leads to higher measured mobilities. Time-of-flight (TOF) or space-charge limited diode (SCLD) measurements are the best options for obtaining meaningful numbers because the carrier densities in these experiments are similar to densities in solar cells under one sun illumination (10^{15}/cm^3). Thin-film transistor (TFT) mobilities are less relevant because the charge carrier density is higher (10^{19}/cm^3).

Some effort has been made to focus on mobility studies relevant for solar cells. Using TOF, it was shown that slowing the kinetics of film formation of RR-P3HT:PCBM films led to a balancing of electron and hole mobilities [18]. Using similar processing, the hole mobility in the RR-P3HT, measured in SCLDs, was shown to increase by a factor of 33 to 5×10^{-3} cm^2 V^{-1}s^{-1} [117]. We have shown that aligning polymer chains in anodic alumina can increase the mobility by about this amount as well and found that, due to the increase in dielectric constant, the space-charge limitations were reduced which led to higher currents through the structures than the SCLC of neat films of similar thickness even though less volume was occupied by polymer [118]. Development of high mobility materials tested under relevant conditions is ongoing and still needed.

Recombination times are most commonly measured by transient photoinduced absorption [74, 93], transient photovoltage [119, 120], or impedence spectroscopy [121]. Transient microwave photoconductivity [122], charge extraction in linearly increasing voltage (CELIV), modified TOF [111], as well as time-delayed collection field [123] experiments offer other alternatives to measure back-transfer times. Key to these experiments is understanding the recombination mechanisms to properly interpret the data. Transient photoinduced absorption monitors the kinetics of the total photoinduced charge population, whereas transient photovoltage appears only to measure the mobile population which reach the electrodes. All of these experiments can likely mask the interfacial back-transfer kinetics since the diffusion-limited kinetics often occur on a longer timescale. In this regard, transient photoinduced absorption experiments that involve molecular dyads in solution are ideal to measure back-transfer times [93–96, 98], in spite of the fact that the kinetics can dramatically change with environment. With this potential complication in mind, we need to plod ahead because back-transfer times are so crucial to understanding the kinetics of geminate separation and bimolecular recombination.

11.11 Nanostructures

In this section, we review progress on nanostructures and discuss the ideal architecture. Most agree the ideal nanostructure involves straight vertical pores or pillars. For porous structures, the pore diameter should be near the exciton diffusion length in size such that most excitons are harvested. If a wide-bandgap scaffold is used, like titania or ZnO, the scaffold volume should be as small as possible to maximize light absorption (since the scaffold absorbs no visible light) and high aspect ratios are needed such that devices upward of 500 nm thick can be made.

A first attempt at this structure was made by our group using a block copolymer self-assembly route [42]. A structuring directing agent, P123, was mixed with a sol–gel titania precursor. Upon evaporation of solvent, the P123 self-assembles into hydrophobic and hydrophilic domains where the hydrophilic domains collect the polar titania precursor. Upon calcination, the titania crystallizes and the block copolymer is removed. Figure 11.23 shows SEM pictures of the corresponding structures and SAXs data suggested uniform pore size near 6.5 nm in calcined films.

Over 60% of the pore volume could be filled with RR-P3HT upon a thermal treatment. Unfortunately, these structures were later shown to have a body-centered cubic structure such that the pathways through the pores were actually quite tortuous rather than straight and vertical.

Later, we developed a method to emboss porous titania structures [124]. By infiltrating a stiff crosslinkable polymer into a porous anodic alumina mold, we formed

Fig. 11.23 SEM pictures of mesoporous titania fabricated via block copolymer self-assembly

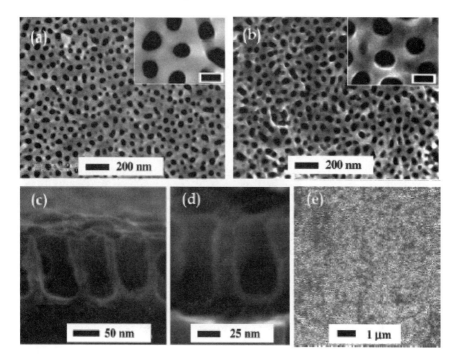

Fig. 11.24 SEM pictures of nanoembossed titania films showing straight pores

a stamp that replicated the anodic alumina structure when pressed onto a substrate covered in a titania sol-gel precursor. This resulted in the titanium dioxide architecture with vertical pores shown in Fig. 11.24.

This technique allowed a straightforward way to tune pore depths and diameters by controlling the anodization conditions of the alumina master; however, the pore size could not be reduced to less than about 40 nm because the polymer pillars of the stamp would begin to stick together due to the flexibility of and the strong van der Waals interactions between the pillars.

A very promising approach was demonstrated using a templated hydrothermal route to grow ZnO nanowires [125, 126]. Figure 11.25 illustrates SEM pictures showing the dense, high aspect ratio nanowires.

This growth route was also shown to be reproducible on large wafer scale. And these structures have been used in liquid dye-sensitized cells where it was shown that the nanowires have superior transport compared to nanocrystalline colloidal titania films.

A novel route was presented where ordered organic nanostructures were created by proper choice of the growth conditions when using organic vapor phase deposition [127]. Figure 11.26 shows SEM and AFM images of CuPc structures.

By adjusting the growth parameters, the α-CuPc phase, a highly anisotropic phase which grows predominately along the z-axis due to the columnar stacking of CuPc, can be encouraged to grow as evidenced by the SEM photos and XRD

Fig. 11.25 SEM pictures of
oriented ZnO nanowires
produced from oriented
seeded nucleation layers

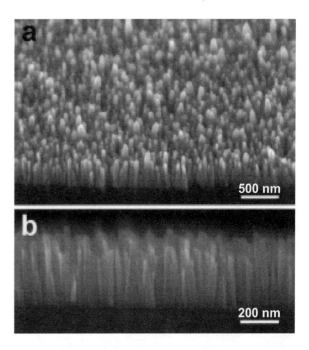

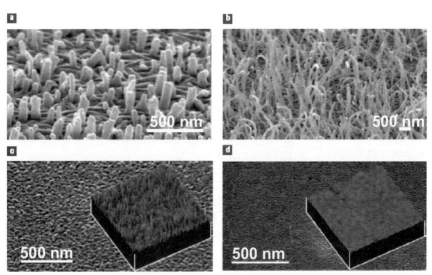

Fig. 11.26 SEM pictures of OVPD grown CuPc films on silicon (**a**) and ITO (**b–c**) and a thermally
evaporated CuPc film on ITO (**d**). The different morphologies arise from different deposition conditions and kinetics (i.e., surface energy, substrate temperature, molecular vapor delivery rate, and
pressure). Insets in (**c**) and (**d**) are AFM images of the surface which show the OVPD deposited film
to have an average peak-to-valley height of 35 nm compared to just 3 nm for the thermally evaporated film. Reprinted by permission from Macmillan Publishers Ltd: [Nature Materials] [127],
copyright [2005]

data. By adjusting the substrate temperature, the density of nucleation sites was controlled. This structure could be effectively filled by a second deposition of the acceptor material, PTCBI.

11.12 Efficiency Limits and Outlook

Numerous attempts to estimate the practical efficiency limits of organic solar cell architectures have been made [23, 106, 128]. All agree that the major improvement upon current materials systems through new materials design will be reduction of the heterojunction offset, which currently limits V_{oc}s to ~0.6 V or less for TiO_2/P3HT, P3HT/PCBM, and CuPc/C60 systems. The consequences of reducing this drop on other device characteristics (*FF*, photocurrent, dark current, etc.) are still unknown, but we are confident that through the proper engineering of architectures additional losses can be kept at a minimum. Understanding the effects of exciton binding energy, geminate pair thermalization, and other interface energetics will be crucial to design more efficient devices. Additional understanding of what practically limits the V_{oc} will be required, which may be tied to more quantitatively understanding the tail states in the organics as well as understanding the dark current. Some estimates have assumed that EQE would practically be limited to 65%. We believe we can engineer devices such that EQEs are at the absorption limit as this has actually already been demonstrated. By using light trapping and antireflection coatings, we may raise EQEs over the entire absorption range of the organic to above 90%. Reducing bandgaps will become important to raise the photocurrent by absorbing more light, though the exact value of the optimum bandgap depends strongly upon the heterojunction offset and losses from the ideal V_{oc}. Finally, investigation of the ultimate limitations on the *FF* (geminate separation, bimolecular recombination, dark current) through intentional device design will determine just how high the efficiency of this exciting technology can be, though we are confident that devices will be designed to nearly eliminate the voltage dependence of the photocurrent. Numerous opportunities exist for the chemist and device physicist, and we hope this chapter will serve as a guide to better design materials and architectures necessary to rapidly improve the efficiency of these devices. Transferring the technology of highly efficient organic and hybrid inorganic–organic PVs to flexible substrates will most certainly require better understanding of the underlying device physics since materials processing can have such a dramatic influence on the structure of organic films, the materials' electronic properties, and consequently the ultimate PV efficiencies achievable using exciting next-generation processing associated with flexible devices.

References

1. Pope M, Swenberg CE (1982) Electronic processes in organic crystals. Clarendon Press, Oxford
2. Barker JA, Ramsdale CM, Greenham NC (2003) Phys Rev B (Condensed Matter and Materials Physics) 67:75205

3. Peumans P, Forrest SR (2004) Chem Phys Lett 398:27
4. Mihailetchi VD, Koster LJA, Hummelen JC, Blom PWM (2004) Phys Rev Lett 93:216601
5. Morteani AC, Sreearunothai P, Herz LM, Friend RH, Silva C (2004) Phys Rev Lett 92:247402
6. Offermans T, Meskers SCJ, Janssen RAJ (2005) Chem Phys 308:125
7. Sze SM (1981) Physics of semiconductor devices. Wiley, New York
8. Yoo S, Domercq B, Kippelen B (2004) Appl Phys Lett 85:5427
9. Peumans P, Forrest SR (2001) Appl Phys Lett 79:126
10. Schulze K, Uhrich C, Schüppel R, Leo K, Pfeiffer M, Brier E, Reinold E, Bäuerle P (2006) Adv Mater 18:2872
11. Scully SR, McGehee MD (2006) J Appl Phys 100:034907
12. Theander M, Yartsev A, Zigmantas D, Sundström V, Mammo W, Anderson MR, Inganäs O (2000) Phys Rev B 61:12957
13. Markov DE, Tanase C, Blom PWM, Wildeman J (2005) Phys Rev B 72:045217
14. Shaw PE, Lewis AJ, Ruseckas A, Samuel IDW (2006) Proceedings of SPIE – The International Society for Optical Engineering, Bellingham, 6334
15. Lewis AJ, Ruseckas A, Gaudin OPM, Webster GR, Burn PL, Samuel IDW (2006) Org Electron: Phys Mater Appl 7:452
16. Peumans P, Yakimov A, Forrest S (2003) J Appl Phys 93:3693
17. Peumans P, Uchida S, Forrest SR (2003) Nature 425:158
18. Li G, Shrotriya V, Huang JS, Yao Y, Moriarty T, Emery K, Yang Y (2005) Nat Mater 4:864
19. Ma W, Yang C, Gong X, Lee K, Heeger AJ (2005) Adv Func Mater 15:1617
20. Reyes-Reyes M, Kim K, Carroll DL (2005) Appl Phys Lett 87:083506
21. Xue J, Uchida S, Rand BP, Forrest SR (2004) Appl Phys Lett 85:5757
22. Coakley KM, Liu YX, Goh C, McGehee MD (2005) MRS Bull 30:37
23. Coakley KM, McGehee MD (2004) Chem Mater 16:4533
24. Pettersson LAA, Roman LS, Inganas O (1999) J Appl Phys 86:487
25. Hoppe H, Shokhovets S, Gobsch G (2007) Phys Status Solidi – Rapid Res. Lett. (RRL) 1:R40
26. Monestier F, Simon JJ, Torchio P, Escoubas L, Flory F, Bailly S, de Bettignies R, Guillerez S, Defranoux C (2007) Sol Energy Mater Sol Cells 91:405
27. Campbell P, Green MA (1987) J Appl Phys 62:243
28. Winder C, Sariciftci NS (2004) J Mat Chem 14:1077
29. van Mullekom HAM, Vekemans JAJM, Havinga EE, Meijer EW (2001) Mater Sci Eng: R: Reports 32:1
30. Payne MM, Parkin SR, Anthony JE (2005) J Am Chem S 127:8028
31. Wang EPX, Oswald F, Langa F, Admassie S, Andersson MR, Inganäs O (2005) Adv Funct Mater 15:1665
32. Wienk MM, Turbiez MGR, Struijk MP, Fonrodona M, Janssen RAJ (2006) Appl Phys Lett 88:153511
33. Mühlbacher MSD, Morana M, Zhu Z, Waller D, Gaudiana R, Brabec C (2006) Adv Mater 18:2884
34. Gierschner JCJ, Egelhaaf HJ (2007) Adv Mater 19:173
35. Hoffmann R, Janiak C, Kollmar C (1991) Macromolecules 24:3725
36. Brabec CJ, Winder C, Sariciftci NS, Hummelen JC, Dhanabalan A, van Hal PA, Janssen RAJ (2002) Adv Funct Mater 12:709
37. Thompson BC, Kim YG, Reynolds JR (2005) Macromolecules 38:5359
38. Young-Gi K, Thompson BC, Ananthakrishnan N, Padmanaban G, Ramakrishnan S, Reynolds JR (2005) J Mater Res 20:3188
39. Coffey DC, Ginger DS (2006) Nat Mater 5:735
40. Snaith HJ, Arias AC, Morteani AC, Silva C, Friend RH (2002) Nano Lett 2:1353
41. Olson DC, Piris J, Collins RT, Shaheen SE, Ginley DS (2006) Thin Solid Films 496:26

42. Coakley KM, Liu YX, McGehee MD, Frindell KL, Stucky GD (2003) Adv Funct Mater 13:301
43. Dexter DL (1953) J Chem Phys 21:836
44. Forster T (1959) Discussions Faraday Soc 27:7
45. Madigan C, Bulovic V (2006) Phys Rev Lett 96:046404
46. Scheidler M, Lemmer U, Kersting R, Karg S, Riess W, Cleve B, Mahrt RF, Kurz H, Bassler H, Goebel EO, Thomas P (1996) Phys Rev B 54:5536
47. Namdas EB, Ruseckas A, Samuel IDW, Lo S-C, Burn PL (2005) Appl Phys Lett 86:91104
48. Burlakov VM, Kawata K, Assender HE, Briggs GAD, Ruseckas A, Samuel IDW (2005) Phys Rev B (Condens Matter Mater Phys) 72:75206
49. Du H, Fuh RCA, Li JZ, Corkan LA, Lindsey JS (1998) Photochem Photobiol 68:141
50. Seybold PG, Gouterman M (1969) J Mol Spectroscopy 31:1
51. Hammond RB, Roberts KJ, Docherty R, Edmondson M, Gairns R (1996) J Chem Soc-Perkin Trans 2:1527
52. Hukka TI, Toivonen T, Hennebicq E, Bredas JL, Janssen RAJ, Beljonne D (2006) Adv Mater 18:1301
53. Hennebicq E, Pourtois G, Scholes GD, Herz LM, Russell DM, Silva C, Setayesh S, Grimsdale AC, Mullen K, Bredas JL, Beljonne D (2005) J Am Chem S 127:4744
54. Kerp HR, Donker H, Koehorst RBM, Schaafsma TJ, van Faassen EE (1998) Chem Phys Lett 298:302
55. Stubinger T, Brutting W (2001) J Appl Phys 90:3632
56. Yu J, Hu DH, Barbara PF (2000) Science 289:1327
57. Im C, Lupton JM, Schouwink P, Heun S,. Becker H, Bassler H (2002) J Chem Phys 117:1395
58. Turro NJ (1978) Modern Molecular Photochemistry. Benjamin/Cummings, Menlo Park, CA
59. Westenhoff S, Daniel C, Friend RH, Silva C, Sundstrom V, Yartsev A (2005) J Chem Phys 122:094903
60. Yan M, Rothberg LJ, Papadimitrakopoulos F, Galvin ME, Miller TM (1994) Phys Rev Lett 73:744
61. Blasse G, Dirksen GJ, Meijerink A, Vanderpol JF, Neeleman E, Drenth W (1989) Chem Phys Lett 154:420
62. Kerp HR, van Faassen EE (2000) Chem Phys Lett 332:5
63. Murata K, Ito S, Takahashi K, Hoffman BM (1996) Appl Phys Lett 68:427
64. Kroeze JE, Savenije TJ, Vermeulen MJW, Warman JM (2003) J Phys Chem B 107:7696
65. Gregg BA, Sprague J, Peterson MW (1997) J Phys Chem B 101:5362
66. Markov DE, Amsterdam E, Blom PWM, Sieval AB, Hummelen JC (2005) J Phys Chem A 109:5266
67. Markov DE, Hummelen JC, Blom PWM (2005) Phys Rev B 72:045216
68. Arkhipov VI, Emelianova EV, Bassler H (2004) Phys Rev B 70:085413
69. Haugeneder A, Neges M, Kallinger C, Spirkl W, Lemmer U, Feldmann J, Scherf U, Harth E, Gugel A, Mullen K (1999) Phys Rev B Condens Matter (USA) 59:15346
70. Choong V, Park Y, Gao Y, Wehrmeister T, Mullen K, Hsieh BR, Tang CW (1996) Appl Phys Lett 69:1492
71. Markov DE, Blom PWM (2006) Phys Rev B (Condens Matter Mater Phys) 74:085206
72. Brabec CJ, Zerza G, Cerullo G, De Silvestri S, Luzzati S, Hummelen JC, Sariciftci S (2001) Chem Phys Lett 340:232
73. O'Regan B, Gratzel M (1991) Nature 353:737
74. Haque SA, Palomares E, Cho BM, Green ANM, Hirata N, Klug DR, Durrant JR (2005) J Am Chem S 127:3456
75. Shao Y, Yang Y (2005) Adv Mater 17:2841
76. Sun YR, Giebink NC, Kanno H, Ma BW, Thompson ME, Forrest SR (2006) Nature 440:908
77. Giebink NC, Sun Y, Forrest SR (2006) Org Electron: Phys Mater Appl 7:375
78. Kohler A, Beljonne D (2004) Adv Funct Mater 14:11

79. Arbogast JW, Darmanyan AP, Foote CS, Rubin Y, Diederich FN, Alvarez MM, Anz SJ, Whetten RL (1991) J Phys Chem (USA) 95:11
80. Capozzi V, Casamassima G, Lorusso GF, Minafra A, Piccolo R, Trovato T, Valentini A (1996) Sol State Commun 98:853
81. Paci I, Johnson JC, Chen X, Rana G, Popovic D, David DE, Nozik AJ, Ratner MA, Michl J (2006) J Am Chem Soc 128:16546
82. Liu YX, Summers MA, Edder C, Fréchet JMJ, McGehee MD (2005) Adv Mater 17:2960
83. Scully SR, Armstrong PB, Edder C, Frechet JMJ, Mcgehee MD (2007) Adv Mat 19(19): 2961–2966
84. Kuhn H (1970) J Chem Phys 53:101
85. Haynes DR, Tokmakoff A, George SM (1994) J Chem Phys 100:1968
86. Gomes da Costa P, Conwell EM (1993) Phys Rev B 48:1993
87. Bredas J-L, Cornil J, Heeger AJ (1996) Adv Mater 8:447
88. Alvarado SF, Seidler PF, Lidzey DG, Bradley DDC (1998) Phys Rev Lett 81:1082
89. Hill IG, Kahn A, Soos ZG, Pascal RA (2000) Chem Phys Lett 327:181
90. Barth S, Bassler H (1997) Phys Rev Lett 79:4445
91. Haque SA, Park T, Holmes AB, Durrant JR (2003) ChemPhysChem 4:89
92. Halls JJM, Cornil J, dos Santos DA, Silbey R, Hwang DH, Holmes AB, Bredas JL, Friend RH (1999) Phys Rev B Condens Matter (USA) 60:5721
93. Durrant JR, Haque SA, Palomares E (2006) Chem Commun 3279
94. Handa S, Giacalone F, Haque SA, Palomares E, Martin N, Durrant JR (2005) Chem – A Eur J 11:7440
95. van Hal PA, Meskers SCJ, Janssen RAJ (2004) Appl Phys A: Mater Sci Process 79:41
96. van Hal PA, Beckers EHA, Meskers SCJ, Janssen RAJ, Jousselme B, Blanchard P, Roncali J (2002) Chem – A Eur J 8:5415
97. Peeters E, van Hal PA, Knol J, Brabec CJ, Sariciftci NS, Hummelen JC, Janssen RAJ (2000) J. Phys Chem B 104:10174
98. Harriman A (2004) Angewandte Chemie International Edition 43:4985
99. Mihailetchi VD, Xie HX, de Boer B, Koster LJA, Blom PWM (2006) Adv Funct Mater 16:699
100. Muller JG, Lupton JM, Feldmann J, Lemmer U, Scharber MC, Sariciftci NS, Brabec CJ, Scherf U (2005) Phys Rev B (Condens Matter Mater Phys) 72:195208
101. Offermans T, van Hal PA, Meskers SCJ, Koetse MM, Janssen RAJ (2005) Phys Rev B (Condens Matter Mater Phys) 72:45213
102. Ohkita H, Cook S, Astuti Y, Duffy W, Heeney M, Tierney S, McCulloch I, Bradley DDC, Durrant JR (2006) Chem Commun 3939
103. Liu Y, Scully SR, McGehee MD, Jinsong L, Luscombe CK, Frechet JMJ, Shaheen SE, Ginley DS (2006) J Phys Chem B 110:3257
104. Ramsdale CM, Barker JA, Arias AC, MacKenzie JD, Friend RH, Greenham NC (2002) J Appl Phys 92:4266
105. Brabec CJ, Cravino A, Meissner D, Sariciftci NS, Fromherz T, Rispens MT, Sanchez L, Hummelen JC (2001) Adv Func Mater 11:374
106. Scharber MC, Wuhlbacher D, Koppe M, Denk P, Waldauf C, Heeger AJ, Brabec CL (2006) Adv Mater 18:789
107. Mihailetchi VD, Blom PWM, Hummelen JC, Rispens MT (2003) J Appl Phys 94:6849
108. Olson DC, Shaheen SE, White MS, Mitchell WJ, van Hest M, Collins RT, Ginley DS (2007) Adv Funct Mater 17:264
109. Mihailetchi VD, Wildeman J, Blom PWM (2005) Phys Rev Lett 94:126602
110. Koster LJA, Mihailetchi VD, Xie H, Blom PWM (2005) Appl Phys Lett 87:1

111. Pivrikas A, Juska G, Mozer AJ, Scharber M, Arlauskas K, Sariciftci NS, Stubb H, Oster-backa R (2005) Phys Rev Lett 94:1
112. Koster LJA, Mihailetchi VD, Blom PWM (2006) Appl Phys Lett 88:52104
113. Koster LJA, Smits ECP, Mihailetchi VD, Blom PWM (2005) Phys Rev B (Condens Matter Mater Phys) 72:85205
114. Tachibana Y, Haque SA, Mercer IP, Moser JE, Klug DR, Durrant JR (2001) J Phys Chem B 105:7424
115. Tanase C, Blom PWM, de Leeuw DM, Meijer EJ (2004) Phys Status Solidi A – Appl Res 201:1236
116. Tanase C, Meijer EJ, Blom PWM, de Leeuw DM (2003) Phys Rev Lett 91:216601
117. Mihailetchi VD, Xie HX, de Boer B, Popescu LM, Hummelen JC, Blom PWM, Koster LJA (2006) Appl Phys Lett 89:012107
118. Coakley KM, Srinivasan BS, Ziebarth JM, Goh C, Liu YX, McGehee MD (2005) Adv Funct Mater 15:1927
119. O'Regan BC, Scully S, Mayer AC, Palomares E, Durrant J (2005) J Phys Chem B 109:4616
120. O'Regan BC, Lenzmann F (2004) J Phys Chem B 108:4342
121. Bisquert J, Vikhrenko VS (2004) J Phys Chem B 108:2313
122. Savenije TJ, Kroeze JE, Yang XN, Loos J (2005) Adv Funct Mater 15:1260
123. Offermans T, Meskers SCJ, Janssen RAJ (2006) J Appl Phys 100:074509
124. Goh C, Coakley KM, McGehee MD (2005) Nano Lett 5:1545
125. Greene LE, Law M, Goldberger J, Kim F, Johnson JC, Zhang Y, Saykally RJ, Yang P (2003) Angew Chem Int Ed 42:3021
126. Greene LE, Law M, Tan DH, Montano M, Goldberger J, Somorjai G, Yang PD (2005) Nano Lett 5:1231
127. Yang F, Shtein M, Forrest SR (2005) Nat Mater 4:37
128. Koster LJA, Mihailetchi VD, Blom PWM (2006) Appl Phys Lett 88:093511

Chapter 12
Bulk Heterojunction Solar Cells for Large-Area PV Fabrication on Flexible Substrates

C. Waldauf, G. Dennler, P. Schilinsky, and C. J. Brabec

12.1 Introduction and Motivation

12.1.1 Photovoltaics

Photovoltaics (PV) is the harvesting of energy from sunlight and conversion into electrical power. This technology is being increasingly recognized as one of the key components in our future global energy scenario. The limited resources of fossil fuel and the steadily increasing costs to supply fossil fuels to the market give a natural limit when renewable energies will begin to kick in as major energy suppliers. In addition, the detrimental long-term effects of CO_2 and other emissions from burning fossil fuels into our atmosphere underscore the multiple benefits of developing renewable energy resources and commercializing them.

Certainly, PV technology is only one of many alternative renewable energies like wind, biomass, and water. Though all of these technologies are expected to give significant contributions to the world's energy supply in the next century, PV has three key properties that make it unique:

- it directly generates electricity (without the need of generators);
- it is an outstanding flexible technology, supplying electrical power in form of portable modules in the mW scale toward PV power plants with peak capacities in the multiple MW regime;
- it is the only renewable energy which can be customized, i.e., handled by individuals.

Though it is not surprising that PV becomes increasingly recognized as part of the solution to the growing energy challenge and an essential component of future global

C.J. Brabec (✉)

Konarka Technologies GmbH, Altenbergerstrasse 69, A-4040 Linz, Austria; Konarka Technologies GmbH, Landgrabenstrasse 94, D-90443 Nürnberg, Germany

e-mail: cbrabec@konarka.com

W.S. Wong, A. Salleo (eds.), *Flexible Electronics: Materials and Applications*, Electronic Materials: Science & Technology, DOI 10.1007/978-0-387-74363-9_12,

© Springer Science+Business Media, LLC 2009

energy production, the big drawback of the current PV technologies is their economy of energy production. Currently, electricity from PV is about 10 times more expensive than energy from fossil fuels and about three times more expensive than other renewable energies. Provided that PVs can be made truly economically competitive with fossil fuels, large-scale manufacturing of these devices offers a pathway to a sustainable energy source that can supply at least 20% of our energy needs.

12.1.2 Technology Overview

To address the economics of PV technology, Martin Green from University of New South Wales has originally developed the concept of third-generation PVs (see Fig. 12.1).

The third generation is currently separated into two categories. In the first, IIIa are novel approaches that strive to achieve very high efficiencies. Key words for IIIa technologies are hot carriers, multiple electron–hole pair creation, concentrator cells, and thermophotonics. All these concepts have theoretical maximum efficiencies well in excess of the Shockley–Queisser limit of ~31% for single junction devices [1]. Certainly, these high-efficiency cells can allow higher costs and still show a favorable €/Wp balance. In the second type of third-generation device, IIIb, the goal is opposite. A low €/Wp balance shall be achieved via moderate efficiencies (15–20%), but at very low costs. This will require inexpensive materials for the semiconductors, the packaging, the production process, low-temperature atmospheric processing, high fabrication throughput, low or no investment into the production facility, and a production on demand scenario. Novel PV technologies, such as organic-based PVs, with their potential to be manufactured by well-established printing techniques in a roll-to-roll process, are found in this latter category. Such low-temperature and therefore low-energy consuming techniques require less capital investment than fabrication techniques for Si-based devices.

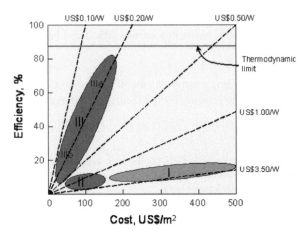

Fig. 12.1 Efficiency versus cost of the different PV technologies currently followed. First generation are thick film Si devices, and second generation are inorganic thin-film technologies. Third-generation technologies are described in more detail in the text

Today, we have several promising technologies for this third-generation low-cost PVs, summarized under the key word bulk heterojunction (BHJ) PVs. They all have in common that at least one of the key functionalities, typically the absorber layer for the PV energy conversion, consists of a blend of organic or hybrid semiconductors processed from solution.

The most prominent and mature technology are the dye sensitized solar cells (DSSCs). DSSCs use an organic dye to absorb light and undergo a rapid electron transfer to a nanostructured oxide such as anatase TiO_2. The mesoscopic structure of the TiO_2 allows processing rather thick nanoporous films. At an active layer thickness of several microns, light is entirely absorbed and these devices reach external quantum efficiencies in excess of 80%. The hole transport is handled by a redox couple, such as iodide/triiodide ($I^-/I3^-$). There is much interest in replacing the liquid electrolyte by a solid-state hole transporter; however, current progress is limited by the transport properties of the solid-state system.

Another interesting technology, which this chapter will focus on, consists of employing a conjugated polymer such as poly(3-hexylthiophene) as the donor and a fullerene derivative as the acceptor.

A close hybrid alternative to the fully organic BHJ are organic–inorganic composites that combine a light-absorbing conjugated polymer as the donor and hole transporter with a nanostructured, inorganic semiconductor such as CdSe, CdS, PbS, TiO_2, or ZnO as the acceptor and/or electron transporter. Depending on the choice of the semiconductor, both components can efficiently absorb light, and the bandgap of the nanocrystals can be tuned by growing them to different sizes.

One step further are hybrid BHJ, where the electron acceptor and electron transporter are grown in self-organized structures on a substrate, filled up with conjugated polymers as hole transporters. These work similarly to the organic ones; however, the gross morphology of the mixture is determined by that of the nanostructured oxide. One has to mention that this approach is relatively close to the solid-state DSSC devices.

Though all these technologies are under development and none of them has received full commercialization, comparison of their current performance and roadmaps is not always fair, and certainly not easy. Such roadmaps may only reflect the current situation, which frequently is a mirror of the person years and development dollars spent on a technology program. Nevertheless, it is sometimes helpful to use these roadmaps as an orientation from where a technology has come, as depicted in Fig. 12.2.

12.1.3 Motivation for Large-Area, Solution-Processable Photovoltaics

The biggest argument for the third-generation organic and hybrid technologies is certainly their promised ultralow costs. As a result, the thought of PV elements based on thin plastic carriers, manufactured by printing and coating techniques from reel to reel and packaged by lamination techniques, is not only intriguing but highly

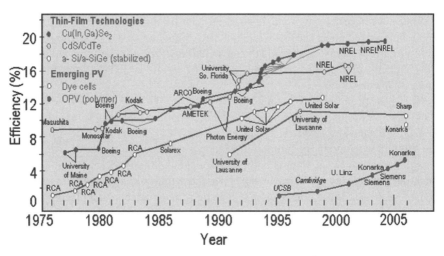

Fig. 12.2 Roadmap for prominent thin film PV technologies. Organic BHJ and DSSC data till 2007 were added to this roadmap initiated by the National Renewable Energies Laboratory (NREL)

attractive from a cost standpoint. In order to fulfil these requirements, high volume production technologies for large-area coating must be applied to a low-cost material class. Solution-processable organic and inorganic semiconductors have a high potential to satisfy these requirements. Flexible chemical tailoring allows the design of organic semiconductors with the desired properties, and printing or coating techniques like screen, ink-jet, offset, gravure, or flexo printing are being established for semiconducting polymers today, driven by display or general electronic demands. Altogether, the BHJ technology offers many attractive features for large-area electronic applications among them

- the potential to be **flexible and semitransparent**
- the potential to be manufactured in a **continuous printing and coating process**
- **large-area** coating
- **easy integration** in different devices
- **significant cost reduction compared to** traditional solutions
- substantial **ecological and economic advantages**

These features are beneficial for commercialization; however, as their classical counterparts, these new technologies have to fulfil the basic requirements for renewable energy production. In the energy market, the competitive position of every solar technology is mainly determined by these factors: efficiency, lifetime, and costs (per Wp). The potential of organic PV has to be judged by these key figures as well (Fig. 12.3).

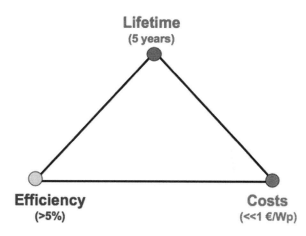

Fig. 12.3 The critical triangle for PVs. BHJ solar cells have to fulfil all requirements simultaneously: lifetime, efficiency, and costs; otherwise, they will be limited to a niche market. Additional criteria like semitransparency, flexibility, or very short *break even times* are beneficial but cannot make up for a significant deficiency in one of the triangle's corners

12.2 The Concept of Bulk Heterojunction Solar Cells

12.2.1 Basics of Organic Solar Cell Materials

Organic semiconductors distinguish themselves significantly from inorganic semi-conductors by their structural and energetic disorder, which result in a different nature of the primary photoexcited state [2]. Most inorganic semiconductors have a large dielectric constant and are thus characterized by a quite small exciton binding energy, which leads to a direct photogeneration as the primary event after photoexcitation. In organic semiconductors, on the other hand, photoexcitation preferentially leads to an excitonic state dominated by the Coulomb binding energy. The precise nature of the exciton as well as the exciton binding energy is still under scientific discussion [3], but most groups generally agree that the exciton binding energies are in the range of 0.3–0.5 eV, while extreme values as low as 0.05 eV and as high as 1 eV [4] were reported.

Excitonic materials with large binding energies are not useful for PV applications, and various strategies to overcome this limitation have been reported over the last decade. Among the most successful concepts so far, the so-called BHJ composites were discussed [5–8]. The BHJ relies on the concept to use two different materials with a relative shift in their energetic levels (donor–acceptor) sufficient to break down an exciton at their interface. In the BHJ, these donor–acceptor interfaces are distributed all over the bulk of the photoactive layer, justifying the name BHJ compared to a planar bilayer heterojunction. For a given exciton diffusion length and for a given carrier mobility, a simple design rule lays out how to create an efficient BHJ composite.

Design Rule: After photogeneration, the exciton needs to diffuse to the nearest charge separating interface. Since there is no driving force toward these randomly distributed interfaces, the diffusion will follow in first approximation the random walk theory well described by the Onsager theory [9, 10] The typical exciton diffusion length for most organic semiconductors lies between 5 and 15 nm [11–13] before they decay radiatively or nonradiatively. Therefore, the average donor–acceptor separation should be of the order of the exciton diffusion length. A more intimate mixture leads to recombination losses of charge carriers, while a more coarse mixing leads to loss of excitons before charge generation.

Other strategies to separate excitons, like trap dissociation, [2], electrical field dissociation [14], or dissociation of "hot exciton" [15] states with excess energies currently have no relevance for the organic PV field.

One very efficient embodiment of a BHJ system was independently reported by Wang et al. [16], the Santa Barbara group [17], and the Osaka group [18]. Their experiments clearly evidenced that a photoexcited exciton on a conjugated polymer is broken down into free carriers at an interface with a C_{60}-containing molecule (acceptor), leading to a photoinduced charge transfer system from the polymer to the fullerene. The forward electron transfer was observed to happen within few tens of femtoseconds [19], which is much faster than any competing relaxation processes. The internal quantum efficiency of this process is $\sim$100%.

12.2.2 Fundamentals of Photovoltaics

12.2.2.1 Solar Radiation

According to Planck, a universal dependence of the energy density per photon energy interval has been found for the so-called blackbody radiation. Any body, which has an absorptivity of $a(\hbar\omega) = 1$ for photons with energy $\hbar\omega$, will emit radiation proportionally to its temperature. Although the sun consists mainly of protons, alpha particles, and electrons, its absorptivity is $a(\hbar\omega) = 1$ for all photon energies $\hbar\omega$ by virtue of its enormous size. There is a relatively thin layer of a few hundred km at the surface of the sun, in which the temperature is constant and all incident photons are totally absorbed. The solar spectrum, observed just outside of the earth's atmosphere, agrees well with Planck's law for a temperature of $T_S = 5,800$ K.

Light passing through the atmosphere of the earth gets modified by scattering and absorption and it is particularly attenuated in the ultraviolet and in the infrared regime, by an amount proportional to the path length through the atmosphere. This path length depends on the incident angle of the sunlight and the light intensity as well as the spectral distribution of sunlight at the earth's surface. As a standard spectrum, for which solar cell efficiencies are rated, a distance of 1.5 times the atmosphere's thickness is chosen and the spectrum is designated AM1.5 (air mass 1.5). The solar spectrum outside the atmosphere is accordingly AM0. The total energy current density obtained by integrating over the spectrum amounts to 1.35 kW/m^2 for the AM0-spectrum and to 1.0 kW/m^2 for the AM1.5-spectrum.

For solar cells, the photon current density is much more important than the energy current density. Neglecting nonlinear processes like impact ionization effects or multiple carrier generation, the excitation of one electron requires at least one absorbed photon. The photon current density, j_{ph}, is derived from Planck's blackbody equation:

$$\frac{d j_{ph,ideal}}{d\hbar\omega} = \frac{2\Omega}{\hbar^3 c^2} \frac{(\hbar\omega)^2}{\exp\left(\frac{\hbar\omega}{kT}\right) - 1}$$ (12.1)

with Ω as the solid angle perpendicular to the emission front. For radiators with an absorptivity smaller than 1 (i.e., part of the photons are reflected), Kirchoff found that the photon current density is proportional to the absorptivity:

$$\frac{d j_{ph}}{d\hbar\omega} = a\,(\hbar\omega) \frac{2\Omega}{\hbar^3 c^2} \frac{(\hbar\omega)^2}{\exp\left(\frac{\hbar\omega}{kT}\right) - 1}$$ (12.2)

12.2.2.2 Band Considerations of a Two-Level System

A two-level system is the most general electronic model of a solar cell. It allows a simple and rigorous treatment of optical transitions and the formulation of the excitation state. A precise description of the formal treatment of the fermionic two-level system can be found in [20]. Such a system has a lower energy level at ε_1 and an upper energy level at ε_2. In classical semiconductor terminology, the highest mostly occupied energy range is called valence band, and the next higher mostly unoccupied energy range is called conduction band. In organic materials, the highest occupied states are called HOMO (Highest Occupied Molecular Orbital) and in the next higher energy range states are called LUMO (Lowest Unoccupied Molecular Orbital). The difference between the HOMO and the LUMO is the band energy gap ε_G of the semiconductor, and the carrier statistics is given by the Fermi distribution for the individual energy levels.

Specifically for BHJ-type solar cells, it is necessary to define more precisely the relevant energy levels. In the previous section, we have outlined that pristine organic semiconductors have a low quantum efficiency for charge generation. The quantum efficiency for charge generation is significantly increased by designing a so-called heterojunction, in which a donor-type material is blended with an acceptor-type material. A BHJ composite is therefore a four-level system, with a $HOMO_{donor}$, a $HOMO_{acceptor}$, a $LUMO_{donor}$, and a $LUMO_{acceptor}$, having two separate bandgaps, ε_{donor} and $\varepsilon_{acceptor}$. Specifically for organic BHJ systems, it was demonstrated recently [21] that the four-level system can be reduced to a two-level system by assuming that the majority carrier concentrations will reside either in the $HOMO_{donor}$ or in the $LUMO_{acceptor}$. Furthermore, the effective electrical bandgap (not to be confused with the optical bandgap) is given by $\varepsilon_{gap} = HOMO_{donor} - LUMO_{acceptor}$. With this set of assumptions, we will be able to calculate the carrier density in the respective bands according to well-established equations of semiconductor physics.

According to Fermi statistics, the probability to find an electron in a state in which it has the energy, ε_C, is given by:

$$f(\varepsilon_C) = \frac{1}{\exp\left(\frac{\varepsilon_c - \varepsilon_F}{kT}\right) + 1} \tag{12.3}$$

If the electron density is significantly smaller than the density of states (i.e., in the Boltzmann limit), the "+1" can be omitted. Assuming a two-level system, the total number of electrons per volume in the LUMO (respectively conduction band) is found by integrating the density of states per energy interval multiplied by the Fermi distribution over the energy range of the conduction band:

$$n_e = \int_{\varepsilon_C} f(\varepsilon_e) D(\varepsilon_e) \, d\varepsilon_e \tag{12.4}$$

where ε_C designates the bottom of the conduction band. Since the Fermi function decreases exponentially with increasing energy, if ε_C exceeds the Fermi energy by some kT, only states in the lower part of the conduction band contribute to the integral in Eq. (12.4). As a result, the total number of electrons is in good approximation:

$$n_e = N_C \exp\left(-\frac{\varepsilon_C - \varepsilon_F}{kT}\right) \tag{12.5}$$

where N_C is the effective density of states, which depends on the temperature and the effective electron mass. The effective density of states strongly depends on the nature of the semiconductor, while Eq. (12.5) is quite independent from the nature of a material. The calculation of the total concentration of holes n_h in the valence band is analogous to Eq. (12.5) and yields:

$$n_h = N_V \exp\left(\frac{\varepsilon_V - \varepsilon_F}{kT}\right) \tag{12.6}$$

where N_V is the effective density of states in the valence band. The Fermi energy ε_F, in between the conduction and the valence band, follows from the condition of charge conservation.

Doping of the semiconductor shifts the position of the Fermi level relative within the band. This is easily rationalized by substituting the carrier density n_e by $n_e + N_D$. In this case, the increased electron density shifts the Fermi level closer to the conduction band.

For a semiconductor in equilibrium under a constant temperature, the product $n_e n_h$ of electron density in the conduction band and hole density in the valence band is independent of doping and is a constant. It has the same value for doped as well as undoped semiconductors, when the electron and hole densities are equal and are called intrinsic density n_i:

$$n_e n_h = N_C N_V \exp\left(-\frac{\varepsilon_C - \varepsilon_V}{kT}\right) = n_i^2 \tag{12.7}$$

A semiconductor under illumination or with a current flow is considered to be in equilibrium, though the electrons and holes will be in a dynamic equilibrium. That equilibrium, however, is a thermodynamic equilibrium, in which the blackbody background radiation is kept in balance with its inverse, a recombination process. Deviations from this detailed balance result in a perturbation. Due to the additional charge carrier generation, however, the electrons in the conduction band are not in detailed balance equilibrium with the holes in the valence band. This perturbation may not be intermixed with the Coulomb neutrality of the charge carriers, which is always valid. A more detailed discussion on the equilibrium conditions of a semiconductor under illumination can be found in [20]. As a consequence, the occupation probability for holes in the valence band and for electrons in the conduction band with electrons has different Fermi energies:

$$n_e = N_C \exp\left(-\frac{\varepsilon_C - \varepsilon_{F,C}}{kT}\right) \tag{12.8}$$

and

$$n_h = N_V \exp\left(\frac{\varepsilon_V - \varepsilon_{F,V}}{kT}\right) \tag{12.9}$$

The product of electron and hole densities is now:

$$n_e n_h = n_i^2 \exp\left(\frac{\varepsilon_{F,C} - \varepsilon_{F,V}}{kT}\right) \tag{12.10}$$

and can exceed n_i significantly.

12.2.2.3 Transport Phenomena

The charge carrier transport in solar cells follows the classical transport equation, irrespective of the nature of the semiconductor. Solutions to the transport equations for organic solar cells have been published recently, and allow a great insight into the balance between charge carrier generation and charge carrier recombination. [21–23]. The transport equations are a set of four equations: the Poisson equation, the continuity equation, and the electron and the hole transport equations that are summarized below.

- The Poisson equation

$$\frac{d^2}{dx^2}\psi(x) = \frac{q}{\varepsilon . \varepsilon_0}[n(x) - p(x)] \tag{12.11}$$

where $\psi(x)$ is the electric potential in the device as a function of position, and $n(x)$ and $p(x)$, the electron and hole densities, respectively.

- The current continuity equation

$$\frac{d}{dx}J_n(x) = \frac{d}{dx}J_p(x) = q\left[G(x) - R(x)\right] \qquad (12.12)$$

where $J_n(x)$ and $J_p(x)$ are the electron and hole current densities, respectively, $G(x)$ the generation rate of charge carriers, and $R(x)$ the recombination rate of charge carriers.

- The current transport equations for electrons and holes

$$J_n(x) = -q.n(x).\mu_n.\frac{d}{dx}\psi + q.D_n\frac{d}{dx}n(x) \qquad (12.13)$$

$$J_p(x) = -q.p(x).\mu_p.\frac{d}{dx}\psi + q.D_p\frac{d}{dx}p(x) \qquad (12.14)$$

where μ_n and μ_p are the electron and the hole mobilities, respectively. D_n and D_p, the electron and hole diffusion coefficients, are assumed to obey the Einstein relation:

$$D_{n,p} = \mu_{n,p}.\frac{k.T}{q} \qquad (12.15)$$

A specific property of disordered semiconductor is the field and temperature dependence of the mobility. Various microscopical models have been used to describe mobility in a highly disordered system. Among them are the Marcus-type jump theory [24], the Miller-Abrahams form to describe hopping rates between adjacent sites, [25] an empirical equation [26] reflecting explicitly the field dependence of the mobility, and a general formula within the framework of the disorder formalism:

$$\mu(T, E) = \mu_0 \exp\left[-\frac{2}{3}\left(\frac{\sigma}{kT}\right)^2\right] \exp\left\{C\left[\left(\frac{\sigma}{kT}\right)^2 - \Sigma^2\right]E^{1/2}\right\} \qquad (12.16)$$

where σ is the width of the Gaussian density of states in a disordered semiconductor, Σ is a parameter characterizing the positional disorder, μ_0 is a prefactor mobility in the energetically disorder-free system, E is the electric field, and C is a fitting parameter. This formalism attributes the temperature and the electric field dependence of the mobility to two main parameters, namely the energetic and the positional disorder. The former one is related to the width σ of the Gaussian density of states normalized by kT. The latter one, on the other hand, arises from fluctuation of the intersite coupling due to either the variation of intersite distance between the charge transport sites or simply the variation of the overlap between corresponding electronic orbital. Equation (12.16) predicts a non-Arrhenius-type activation of the mobility extrapolated to zero electric field following [27, 28]:

$$\ln\left[\mu(0, T)\right] \propto \frac{1}{T^2} \qquad (12.17)$$

and a potentially negative field dependence of the mobility, experimentally observed by Mozer et al. in the case of P3HT [29].

The lifetime τ of carriers is the second important parameter besides the mobility to consider for loss analysis. It is the $\mu\tau$-product of a semiconductor, which determines the volume of the semiconductor contributing to the external circuitry. The kinetics of charge carrier recombination can be of two different natures, namely monomolecular (first-order), or bimolecular (second or higher order). It is important to keep in mind that monomolecular does not mean that a carrier can recombine with itself and violate the Coulomb neutrality. Rather, monomolecular recombination means that the species under investigation recombines at a rate that is linearly proportional to its population density. The light intensity dependence of the photocurrent is frequently looked at in order to distinguish between first- and second-order recombination processes [30, 31], but only a few studies have focused on determining the photoinduced carrier lifetime of BHJ composites. Among them, transient absorption spectroscopy and transient photocurrent/photovoltage spectroscopy [32–34], flash photolysis time-resolved microwave conductivity (FP-TRMC) [35, 36], and photoinduced charge carrier extraction by linearly increasing voltage (photo-CELIV) [37–40] technique studies all yielded a second-order-type recombination behavior with lifetimes of the mobile carriers in the microsecond regime, whereas the lifetime of trapped carriers can extend into the millisecond regime. Depending on the type of polymer, the bimolecular recombination was found to be of Langevin type ($\beta_L = e.\mu/\varepsilon.\varepsilon_0$) for more amorphous semiconductors and of non Langevin-type for more ordered or semi-crystalline semiconductors [40, 41].

12.2.2.4 Current–Voltage Characteristic

The most simple way to discuss and describe a two-level system is by assuming that a semipermeable membrane at the interface of the two-level system with the external circuitry is selectively allowing only carriers with a distinct energy to pass. In case of an electrical current flowing through the system, independently of whether the carriers are injected or photogenerated and extracted, this membrane does only allow holes with an energy ε_1 and electrons with an energy ε_2 through. From Eq. (12.12), the continuity equation for electrons or holes we have:

$$\frac{d}{dx}J(x) = q\left[G(x) - R(x)\right]$$

For the qualitative derivation of the current (I)–voltage (V) relation, we now follow the thermodynamic approach and redefine the generation and recombination rates for photons instead of carriers. The generation $G(x)$ remains, as defined above, the generation of charged carriers. But it is important to understand that these carriers are generated radiatively. Therefore, the number of absorbed photons needs to equal the number of emitted photons, whereas recombination of carriers is regarded to lead to radiative emission. Nonradiative recombination, i.e., recombination without emission of visible light, generates heat and this heat will be emitted as IR photons. This allows us to give an analytical expression for $R(x)$, the radiative recombination constant [20], with R^0 as the spontaneous emission rate of the material in equilibrium with its environment:

$$R(x) = -eR^0 \left[\exp\left(\frac{\varepsilon_{F2} - \varepsilon_{F1}}{kT} \right) - 1 \right] \tag{12.18}$$

By integrating over the thickness of the two-level system, the total current density j_Q can be calculated as follows:

$$j_Q = e \int R^0 \left[\exp\left(\frac{\varepsilon_{F2} - \varepsilon_{F1}}{kT} \right) - 1 \right] dx - e \int G(x)dx \tag{12.19}$$

We further assume that our semipermeable membranes give an ideal unipolar ohmic contact so that the Fermi energy of the higher level is identical to the Fermi energy of the electron membrane, and the Fermi energy of the lower level becomes identical to the Fermi energy of the hole membrane. The difference of the Fermi energies of the two membranes is then related to the voltage V between the contacts to the membranes, $eV = \varepsilon_{F2} - \varepsilon_{F1}$, and by introducing the dark saturation current density j_0 and the photocurrent density under short-circuit conditions j_{SC}, we can rewrite Eq. (12.19) as Eq. (12.22), which is the familiar one-diode equation for semiconductors with two distinct energy levels for electrons and holes (plotted in Fig. 12.4):

$$j_0 = e \int R^0 dx \tag{12.20}$$

$$j_{SC} = -e \int G(x)dx \tag{12.21}$$

$$j_Q = j_{rev} \left[\exp\left(\frac{eV}{kT} \right) - 1 \right] + j_{sc} \tag{12.22}$$

12.2.2.5 The V_{oc}

The nature of the V_{oc} in BHJ solar cells has long been debated. Due to the similarity of the device architecture to that of organic light-emitting diodes (OLEDs), it was intitially suspected that BHJ solar cells can be described by a MIM model (metal–intrinsic–metal) [42], where the built-in voltage is controlled by the difference in the

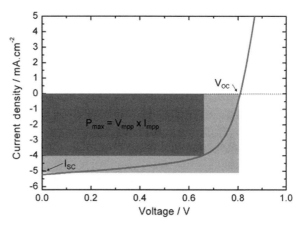

Fig. 12.4 Photocurrent density of a two-level system versus the energy of the electron–hole pairs (i.e., eV). The maximum usable photocurrent density is found when absorbed photons and reemitted photons are balanced and is called the maximum power point (P_{max}). At the energy eV_{OC}, all carriers recombine (i.e., all absorbed photons are reemitted) and the total current density becomes zero

work function of the two metal electrodes. Systematically, varying the work function of the electrodes [43–47] or the HOMO/LUMO positions of the semiconductors showed quite clearly that the MIM picture is wrong. By today, it is commonly accepted that the appearance of the V_{oc} is induced by the splitting of the electron and hole pseudo-Fermi levels [48], whereas the role of the electrodes is to provide semipermeable membranes at the interface of the semiconductor. Fermi level pinning occurs at the interfaces. Most metals irrespective of their work functions can get pinned to surface states of fullerenes [44], while at the polymer interface only the metals with work functions outside the polaron bands get pinned to the energetic position of the polaron [49].

12.2.3 Understanding and Optimization of BHJ Composites

The printing and coating of large-area semiconductor devices is a completely new challenge for the thin-film electronics industry. Traditional printing is well understood as the art to generate visual pictures with low defects. Quality control of printed patterns is easily done by various optical and microscopical methods. The printing of semiconductors requires a level of control of the printed or coated films which goes well beyond the traditional printing picture. Printing of semiconductors implies the printing of a functionality. This functionality can be of electrical (OFET), magnetic (storage), electro-optical (OLED, OPV), electrochemical (OFET, storage, sensors) or of other nature. Traditional printing does not handle or control these disciplines so far. Modeling the printed devices will be a first and important step to introduce quality control for printed semiconductor layers. A successful device model can give insight into the printed film properties which goes beyond the traditional quality control of printing. Expectations are that such a model will answer questions on (1) the correlation between processing parameters and device performance, (2) quality and homogeneity of the printed layers,

(3) morphology problems during solid-state film forming, and (4) electrical and/or mechanical defects of the films.

12.2.3.1 The One-Diode Model for Organic Solar Cell

In order to optimize solar cell devices and to explain the dependence of the fill factor (*FF*) and V_{oc} on light intensity, models are needed to describe the light-dependent processes in BHJ solar cells. In this section, an extension of the standard one diode model is introduced, which allows the simulation of *I–V* curves measured under different light intensities.

It was shown recently that based on the self-consistent solutions to the transport equations, the electrical field and the carrier concentration in the active layer of the BHJ solar cells can be calculated (Fig. 12.5a). These simulations showed that a BHJ composite can be successfully described by a single intrinsic semiconductor layer with the energy levels of the valence and conduction bands as the HOMO of the polymer and the LUMO of the fullerene, respectively [50, 57]. The electric field is found to be constant over the active layer and no or negligible deviation from this behavior is observed under intense illumination (Fig. 12.5a,b).

Based on this finding, a macroscopic replacement circuit is suggested, capable of describing the *I–V* characteristics of BHJ solar cells under different illumination densities with a single set of parameters. Such a replacement circuit has to take the field dependence of the photocurrent into account. Before describing the expanded replacement circuit, the deficiencies of the standard one-diode model [51]

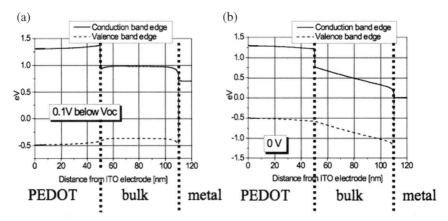

Fig. 12.5 Dependence of the electrical potentials on the applied voltage as calculated by solving self-consistently the transport equations (Eqs. 12.11–12.14). Shown are the conduction band edge (*full line*) and the valence band edge (*dotted line*) as a function of distance from the ITO electrode. The PEDOT is up to 50 nm thick and assumed to be a p-type semiconductor. The BHJ spans a 50- to 100-nm distance from the ITO electrode and assumed to be a single ambipolar semiconductor. The metal contact is more than 110 nm away from the ITO electrode. The potential within the bulk of the solar cell is nearly flat for voltages near V_{oc} (**a**) and shows a constant potential decrease, which can be seen clearly for 0 V (**b**) without band bending

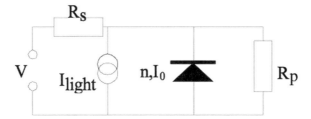

Fig. 12.6 Equivalent circuit diagram of the macroscopic model for describing a solar cell with one diode. The total current for a given voltage V is the sum of the single currents through the diode, which is represented by the parameters n and I_0, the shunt resistance R_p, the limiting series resistance R_s, and the photoinduced current I_{light}

need to be analyzed. The model is given by Eq. (12.21) and is schematically depicted in Fig. 12.6.

Here, the photoinduced current j_{light} is added as a current source:

$$j - j_0 \times \left(e^{\frac{e \times (V - I \times Rs)}{nk_B T}} - 1 \right) - \frac{V - j \times Rs}{Rp} + j_{light} = 0 \qquad (12.23)$$

where j is the current density through the diode, j_0 is the saturation current density, k_B is the Boltzmann constant, T is the temperature in kelvin, V is the external bias applied to the diode, n is the ideality factor of the diode, R_s is the series resistance, and R_p is the parallel resistance. The dependence of the parallel resistance R_p on the illumination intensity is taken into account by adding a second photoconductive parallel resistor $R_{photoshunt}$, where the photoconductivity is proportional to the light intensity P_{light}. Figure 12.7 shows typical I–V characteristics of BHJ solar cells, measured under different illumination intensities together with the best fit to Eq. (12.23).

The fit parameters are collected in Table 12.1, where R_p is calculated with $R_{p,dark}$ and $R_{photoshunt}$ via the light intensity. Experimental data have been taken on devices using a poly(3-hexylthiophene) (P3HT) as donor and a [6,6]-phenyl-C_{61}-butyric acid methyl ester (the fullerene PCBM) as acceptor.

The general quality of the fits is satisfying. The shunt (characteristic in the voltage regime between -0.2 V and 0.2 V) and the diode turn-on (characteristic in the voltage regime between 0.3 and 0.6 V) of the dark I–V curve are correctly reproduced with a meaningful set of parameters. Please note, that the slope of the j–V characteristic in the diode turn-on regime is governed by the ideality factor n. Also the illuminated j–V curves can be correctly reproduced; however, in order to properly describe the open circuit voltage under increasing illumination density, a continuous increase of the saturation current j_0 is necessary. The variation of j_0 by a factor of 2 is significant and gives an equivalent shift of the open-circuit voltage by nearly 50 mV. However, the variation of dark diode parameters under illumination cannot be justified, since it would correspond to a change of the dark diode itself under illumination. This is in conflict with Eq. (12.23), where j_{light} is the only

Fig. 12.7 I–V curves measured under various light intensities (*boxes*) with fitted curves using standard one-diode model (*line*). Values for the parameters n and I_0 are given for each light intensity. The light intensity is given by the value of optical density of the various gray filters and the light intensity of 110 mW/cm² @ OD0

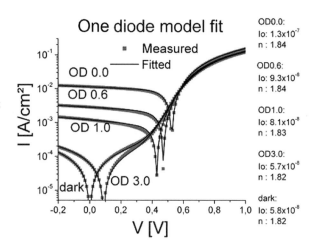

One diode model fit

■ Measured
— Fitted

OD 0.0
OD 0.6
OD 1.0
OD 3.0
dark
OD 3.0

ODO.0:
Io: 1.3×10^{-7}
n : 1.84

OD0.6:
Io: 9.3×10^{-8}
n : 1.84

OD1.0:
Io: 8.1×10^{-8}
n : 1.83

OD3.0:
Io: 5.7×10^{-8}
n : 1.82

dark:
Io: 5.8×10^{-8}
n : 1.82

Table 12.1 Fit parameters n, j_0, and R_s for I–V curves measured under various light intensities, fitted using standard one-diode model. The light intensity is given by the value of optical density (OD), the various gray filters and the light intensity of 110 mW/cm² @ OD0. R_p is calculated with the dark shunt $R_{p,dark} = 1,540~\Omega cm^2$ parallel to a photo conductive resistor with the photo conductivity $1/R_{photoshunt} = 5.3 \times 10^{-5}~(\Omega cm^2~mW)^{-1}$

OD filter	n	Saturation current I_0 [A/cm²]	R_s [Ω cm²]
0.0	1.84	1.3×10^{-7}	2.2
0.6	1.84	9.3×10^{-8}	2.5
1.0	1.81	8.1×10^{-8}	2.5
3.0	1.82	5.7×10^{-8}	2.6
dark	1.82	5.7×10^{-8}	2.4

parameter sensitive to illumination. A fixed set of parameters j_0 and n do not correctly reproduce the illumination dependency of V_{oc} (see Fig. 12.8): the higher the light intensity, the larger the error in V_{oc}. Also a two-diode model, where the second diode is implemented as a Shockley–Read-Hall (SRH) – recombination diode [52] – does not explain the illumination intensity of the V_{oc}, independent of the ideality factors of these two diodes.

The discrepancy discussed above brings up the deficiencies of the one-diode model given in Eq. (12.23). In order to remove this discrepancy, a more sophisticated model was introduced. It is well known that the photoinduced current of BHJ solar cells is strongly driven by the internal electrical field. This is not surprising for devices with active layers in the 100-nm regime, and a constant electrical field (as seen in Fig. 12.8) further supports these findings. Applying now an external bias close to the built-in voltage V_{bi} will minimize this field. Consequently, a reduction

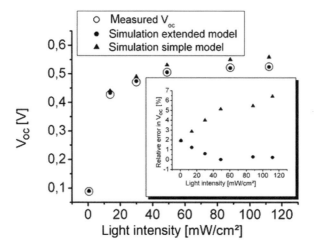

Fig. 12.8 The open-circuit voltage (V_{oc}) over the light intensity as measured (*open circles* ○), fitted with the standard one-diode model (*closed triangle* ▲), and fitted with the extended model (*closed circle* ●). The inset shows the relative error of the V_{oc} value of the standard model and the extended model to the measured value

of the electrical field is expected to reduce the average mean carrier distance, which can be given as follows:

$$\text{mean distance} = \tau \times \mu \times \frac{(V - Vbi)}{L} \qquad (12.24)$$

where τ is the average lifetime of the carriers, μ the carrier mobility, and L the thickness of the sample. Equation (12.24) suggests that j_{light} should not be regarded as constant with voltage. If the average mean carrier distance is smaller than the thickness of the sample, just the fraction (mean distance/layer thickness) of carriers will reach the contacts and can be extracted from the solar cell. Therefore in the expansion of the one-diode model, the dependence of j_{light} on the voltage will be taken into account. Similar models have been used to describe j–V curves of amorphous silicon (a-Si) solar cells [53], when transport is limited by the field-driven average mean carrier distance.

First, the photogenerated charge carrier flux density is calculated. Since the solar cells discussed in this chapter show no or only negligible carrier losses (an internal quantum efficiency of ~100% and an external quantum efficiency of ~80%), the carrier flux density can be estimated directly from the short circuit current j_{sc}, which is equivalent to j_{light} (0 V). The photogenerated charge carriers are now driven by the potential difference between the internal built-in voltage V_{bi} [54] and the applied voltage V. The number of carriers which can be extracted from the device equals the total number of carriers multiplied by the ratio of the mean distance to the thickness of the sample, where this ratio should be in the range 0–1

Table 12.2 Fit parameters n, j_0, R_s, $R_{p,dark}$, $1/R_{photoshunt}$, μ, τ, and V_{bi} for j–V curves measured under various light intensities, fitted using expanded one-diode-model with nonconstant j_{light}

Parameter	Value
n	1.79
j_0 (A/cm^2)	4.8×10^{-8}
R_s (Ω cm^2)	2.1
$R_{p,dark}$ (Ω cm^2)	1,540
$1/R_{photoshunt}$ (Ω cm^2 mW)$^{-1}$	5.3×10^{-5}
μ (cm^{2}V^{-1}s^{-1})	0.001
τ (s)	7.1×10^{-6}
V_{bi} (V)	0.61

The overall photogenerated current j_{light} can be rewritten as:

$$
j_{light}(V) = \begin{cases} -|j_{sc}| & \text{if } \mu \times \tau \times (-V + V_{bi})/L > L \\ |j_{sc}| & \text{if } \mu \times \tau \times (V - V_{bi})/L > L \\ |j_{sc}| \times \mu \times \tau \times (-V + V_{bi})/L^2 & \text{else} \end{cases}
$$

(12.25)

Reversal of the photocurrent between the fourth and the first quadrant is a consequence of a field-driven light-generated current. In the forward direction, the photocurrent contributes to the total current, e.g., due to photoconductivity; Eq. (12.25) does not contain direct carrier recombination. Please note that in the case of recombination losses, j_{sc} in Eq. (12.25) has to be replaced by j_{sc0}, the maximum possible (optically limited) primary photocurrent, which can be extracted under large reverse bias. For comparison, more complex models, regarding the influence of direct recombination processes as discussed for a-Si devices [53], were applied to fit the BHJ data presented in this section. The result was that direct recombination processes are not relevant for most of the presented BHJ data. However, we cannot exclude that these more sophisticated models are relevant for other BHJ materials or materials combinations.

This extended one-diode model composed of the standard one-diode model and the field-depended light current was tested using the experimental data set already shown in Fig. 12.8. The fitting of I–V curves is done by iteratively minimizing the mean least error χ^2, which is given by the sum over the differences between the measured and the fitted currents at each voltage:

$$
\chi^2 = \sum_V \left[I_{measured}(V) - I_{fitted}(V) \right]^2 + \left[\log(I_{measured}(V)) - \log(I_{fitted}(V)) \right]^2
$$

(12.26)

With this iterative method, dark diode parameters j_0, n, R_s, and $R_{p,dark}$ are determined first. Second, keeping the dark diode parameters constant, the mobility μ and the carrier life time τ are determined by fitting the illuminated j–V curves. The set of parameters are listed in Table 12.2 for the light–current Eq. (12.25). The mobility is kept constant at 10^{-3} cm^{2}V^{-1}s^{-1} in agreement with literature values.

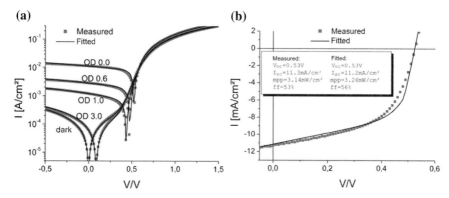

Fig. 12.9 (a) I–V curves measured under various light intensities (*boxes*) and fitted curves using one-diode model with nonconstant light current I_{light} (*line*). Values for the model parameters are given in Table 12.2 and are the same for each light intensity. The light intensity is given by the value of optical density of the various gray filters and the light intensity of 110 mW/cm² @ OD0. (b) The fourth quadrant of measured (*boxes*) and fitted (*line*) I–V curves in a linear scale for 110 mW/cm². The values for the short circuit current (I_{sc}), open-circuit voltage (V_{oc}), maximum power point (mpp), and fill factor (FF) are given for both fitted and measured I–V curves

Figure 12.9 shows again the experimental data compared to the fitted data as calculated from Table 12.2. Again, excellent agreement is observed; but in contrast with the simple one-diode model, the expanded model fits the experimental data with a high degree of accuracy with one and the same set of parameters. Most importantly, the open-circuit voltage is fitted correctly over more than three orders in illumination with one set of parameters. Plotting the I–V curve on a linear scale reveals a small deviation in the fourth quadrant just before the open-circuit voltage is observed (Fig. 12.9). This deviation is supposed to originate from the rather crude assumption in the model that all photogenerated carriers have identical mobility and lifetime. Because we neglect the statistical distribution of mobility and lifetime values, the photocurrent is slightly overestimated at the opening of the diode, and therefore, the FF is also slightly overestimated.

Another advantage of the expanded model is the estimate of a mean carrier lifetime τ, which is elaborate to obtain otherwise. The value of τ determined by fitting experimental data with the expanded diode model (several microseconds) is similar to the carrier lifetimes determined from transient absorption measurements on polymer BHJ films [55]. The value of the fitted built-in voltage of 0.61 V is also in good agreement with the measured open-circuit voltage. Summarizing it is necessary to expand the one-diode model by introducing the field dependence of the photocurrent. Efforts to use this slightly more complex model are rewarded by the accurate estimate of the built-in voltage V_{bi} and the mean carrier lifetime τ or of the $\mu\tau$-product if the mobility μ of the tested material is unknown. Small deviations in the fourth quadrant are explained by the fact that the model does not take into account the statistical distribution of the mobility and lifetime.

12.2.3.2 Consequences of the One-Diode Model

In the previous section, a simplified model for BHJ solar cells was introduced, allowing to deviate from the misleading concept of a MIM picture as the charge-generating principle in the solar cell. Clearly, the intimate mixture of electron-accepting and hole-accepting materials is responsible for the direct charge generation, independently of the choice of metal contacts. Since it is agreed that the charges are generated everywhere within the bulk, the proper picture for the organic bulk heterojunction (oBHJ) solar cells is a homogenously distributed p–n junction, with the p–n junction formed between the donor and the acceptor molecules. In contrast to other models [56, 57], this picture allows to explain the light intensity and temperature dependence of solar cells without the introduction of microscopic transport parameters.

The fundamental question for any PV technology is the origin of the diode rectification, which is the energetic barrier which physically separates the charge carriers from each other and prevents their recombination after separation. This energetic barrier is the origin of the diode I–V characteristics which was discussed in Section 12.2.2.4. Addition of a parallel and a series resistance to the diode resulted in the expanded equivalent circuit with a field-dependent photocurrent. In case of no illumination ($j_{light}=0$), assuming $R_p >> R_s$, the electrical response of the solar cell is expected to consist of three distinct regimes:

- A linear regime at negative voltages and low positive voltages where the current is limited by R_p.
- An exponential behavior at intermediate positive voltages where the current is controlled by the diode.
- A second linear regime at high voltages where the current is limited by the series resistance.

Typical j–V characteristics of nonilluminated BHJ devices are presented in Fig. 12.10. The three regimes can be easily identified. Differently processed solar cells differ in the ohmic response in regimes (1) and (3), but have a common positive slope between 0.5 V and 0.75 V. The linearity of this j–V regime in the semilogarithmic representation indicates an exponential correlation between current density and voltage. Exponential behavior is usually attributed to a clearly defined interface between two materials with chemical potentials shifted relative to each other, as present in bilayer or Schottky devices. The relation between current density j and voltage V therefore is analytically best described by Eqs. (12.23) and (12.25).

For devices consisting of an intimate mixture of two materials, a well-defined interface for charge separation, such as a p–n junction, suggests an arrangement of the two blended semiconductor materials as shown schematically in Fig. 12. 11.

The concept of organic BHJ devices was initiated by the idea of an increased interface for charge separation in order to overcome the limits due to low exciton diffusion lengths [58, 59] in organic materials. For several reasons, it is sound to

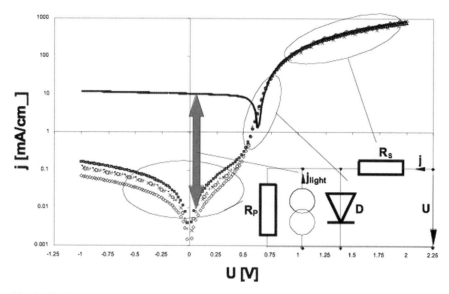

Fig. 12.10 j–V characteristics of seven oBHJ devices($\triangle$,$\square$,+, , , , −) of one production batch. The inset shows the standard replacement circuit. The characteristic of one device under incident light of one sun is shown as a *solid line*

accept that this interface is responsible for the charge generation and also for the charge separation.

The charge transfer from polymers to derivatives of C_{60} is known to occur on an ultrafast timescale [60], while the back transfer is strongly hindered due to energetic reasons [61, 62]. Once the carriers are separated, selective contacts are still needed to extract them. The choice of nonselective contacts may lead to surface recombination at the metal–semiconductor interface. These losses, namely the effect of the type of metal contacts on solar cell parameters, were frequently misinterpreted as an indication of the MIM picture.

Due to the clearly defined interface, its 3-D expansion throughout the bulk and the selective transport of the appropriate charge carriers along phases of the respective material it makes sense to regard BHJ solar cells as "expanded p–n junction" compared to the classical "planar p–n junction." With that model developed, it is much easier to work out the physical differences to the MIM picture: in a p–n picture, the diode properties – i.e., the ideality factor n and the reverse saturation current j_0 – will depend on the two semiconductor materials.

The ideality factor reflects the bias dependence of the energetic barrier between the two semiconductors, controlling injection as well as recombination. It is a measure of the morphology of the p–n junctions and will vary with the processing conditions [63], which is clearly reflected in different ideality factors for different solutions.

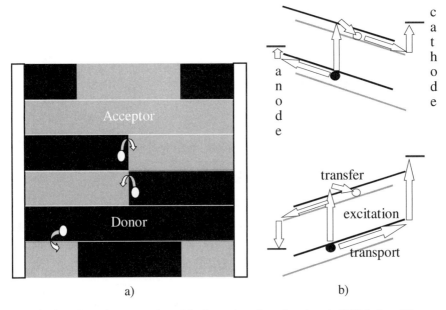

a)   b)

Fig. 12.11 Schematic presentation of the formation of p–n junctions in BHJ devices: The two materials arrange as a bilayer, the electrodes are connected by a pure phase, or one material is completely enclosed by the other (**a**). After excitation, undirected charge transfer takes place across the donor–acceptor interface subsequently followed by field-driven transport along pure materials phases (**b**)

The saturation current density reflects the majority type carriers which can over-come the barrier, independently of the applied bias. In conventional p–n junctions, the carrier density consists of a temperature independent (intrinsic) part reflecting the purity of the materials j_i and a thermally activated part. The latter shows an exponential dependence on the temperature T and on the height of the barrier ϕ.

Analysis of temperature-dependent j–V measurements on oBHJ devices clearly reveals $j_0 = j_i e^{q\phi/nkt}$ (Fig. 12.12), with the barrier height ϕ being identical to the difference to the LUMO$_{PCBM}$ and the HOMO$_{POLYMER}$ (see Table 12.3). This correlation is somewhat the ultimate proof that the MIM picture can not be correct.

It is worthwhile to compare the barrier values calculated from the temperature dependence of the saturation current to either V_{oc} or better V_{bi} values extracted from the j–V characteristics. Theory for p–n junction predicts that these values should be more or less identical. The values for the V_{bi} were calculated according to Section 12.2.3.1 and compared in Table 12.3 and excellent agreement is found. A different model was recently suggested by Koster et al. [56], whereas this model needs to redefine the specific material parameters for each new material system.

An interesting consequence of the extended p–n junction picture is the photode-pendence of the parallel resistance. In the extended p–n junction, pure phases of the

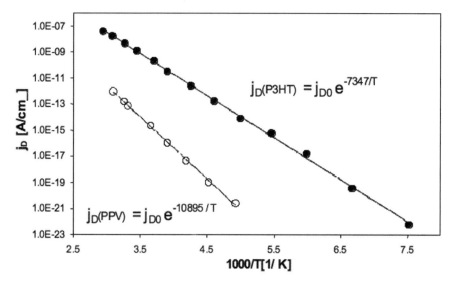

Fig. 12.12 Temperature dependence of the reverse saturation current of P3HT:PCBM (*closed circles*) and PPV-based (*open circles*) oBHJ devices. The *closed circles* are values obtained by fitting the exponential part of the j–V characteristics in the dark. The *line* is the best fit to the experimental data

Table 12.3 Values for the built-in potential V_{bi} obtained from intensity-dependent j–V data compared to the energetic barrier ϕ responsible for the temperature dependence of the saturation current j_0. Both values are in strong correlation with the difference between the LUMO level of the acceptor and the HOMO level of the donor which were determined from cyclovoltammetric measurements. The ideality factor n is listed as well. The second batch of MDMO-PPV:PCBM devices was not measured under different temperatures; therefore, no energetic barrier is indicated

Materials/ batch	V_{bi} (V) determined via $j_{light} = 0$	ϕ (eV) determined via j_0 (T)	$E_{LUMO} - E_{HOMO}$ (eV)	n
MDMO-PPV:PCBM/1	1.05 ± 0.1	1.25 ± 0.05	1.1 ± 0.1	1.38 ± 0.02
MDMO-PPV:PCBM/2	1.05 ± 0.1	–	1.1 ± 0.1	1.23 ± 0.02
P3HT:PCBM/1	0.85 ± 0.1	0.92 ± 0.05	0.95 ± 0.1	1.6 ± 0.02
P3HT:PCBM/2	0.85 ± 0.1	0.93 ± 0.05	0.95 ± 0.1	1.72 ± 0.02

two materials can form continuous transport paths between the anode and the cathode, thereby generating a parallel shunt resistance. The dark parallel shunt resistance reflects the intrinsic conductivity of the materials, which for most organic semiconductors is quite low. Under illumination, light-induced charge generation ("photodoping"), caused by the charge transfer between donor and acceptor, is expected to reduce the shunt dramatically. The conductivity σ of materials is given by $\sigma = (nq)\mu$, where n is the charge density, q the charge, and μ the mobility, which is assumed to be independent of illumination. Under illumination, the photogenerated carriers N_{ph} increase the semiconductor conductivity via $\sigma \approx (n + N_{ph})q\mu$.

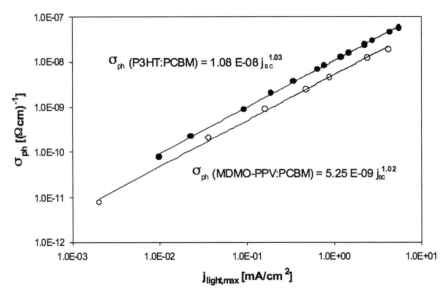

Fig. 12.13 Relation between the light generated photoconductivity σ_{ph} and the light generated current density j_{light} for P3HT:PCBM-devices (*closed circles*) and MDMO-PPV:PCBM devices (*open circles*). $\sigma_{ph} = \sigma_{ill} - \sigma_0 = f(j_{light})$. The lines are the best power law fits to the data

Since the photogenerated carriers are proportional to the photocurrent, one would expect the shunt to scale with photocurrent density. The experimental data for this correlation are shown in Fig. 12.13 for P3HT:PCBM as well as MDMO-PPV:PCBM devices.

Summarizing this section, BHJ solar cells are characterized by a p–n type junction between the donor and the acceptor phases. The nature of the junction is observed in the temperature dependence of the dark current, built-in potential and photoshunt of various material systems, including P3HT:PCBM, MDMO-PPV:PCBM, and PF:PCBM devices. Device analysis based on the expanded p–n junction picture is a powerful tool to analyze performance losses of the solar cells. This aspect will be discussed in more detail in the next section.

12.2.3.3 Application to Large-Area Solution-Processed Solar Cells

The establishment of a well-working simulation model for solar cells opens up the possibility to analyze the interplay between processing and performance. Simply looking at the device performance is most of the times insufficient to understand why certain process conditions give a fully working device, while another set of process conditions may cause failure. A successful device analysis consists of two parts: (1) a semiconductor bulk analysis needs to determine the density long living photocharges, while (2) the interface analysis needs to determine the charge extraction efficiency to the electrodes. It is generally accepted that deficiencies or defects of the bulk semiconductor dominantly cause lower current densities, while interface deficiencies lead to FF and V_{oc} losses. However, there are situations where different bulk semiconductor morphologies significantly influence FF or V_{oc}. These atypical scenarios are difficult to identify. A cost and time efficient alternative to spectroscopic investigations is the simulation of j–V curves.

Processing of organic solar cells via printing or coating technologies is the essential advantage for large-area electronics applications. A true low-cost technology demands low processing temperatures, compatibility to existing coating or printing equipment with reasonable low resolution and raw materials in abundance. However, with a few exceptions, most of the low-cost, solution-processed solar cells reported in literature have been processed by spin coating technologies. Spin coating is a suitable technology for the precision coating of photoresists, for instance, resulting in outstanding smooth and homogenous surfaces. It is at least questionable whether a printing technology can give films with comparable specifications. Another complexity is the impact of drying on the semiconductors morphology. Especially for two-component BHJ composites, drying is known to majorly influence the solar cell performance. Slow drying gives more time to two components to organize and arrange themselves into the equilibrium configuration. A few reports in literature demonstrated high power efficiencies for coated solar cells [64–66], but a systematic investigation on the correlation between production technology and device performance is missing so far and cannot be settled within the scope of this chapter. Nevertheless, the performance of fairly efficient BHJ solar cells produced by a reel-to-reel compatible coating technique will be analyzed and compared to a reference, namely devices produced by a discrete spin coating technique.

Figure 12.14 shows a typical j–V curve of an efficient P3HT:PCBM solar cell produced by large-area coating of the semiconductor. This particular device had a short-circuit density I_{sc} of 11.5 mA/cm^2, an open-circuit voltage V_{oc} of 615 mV, and

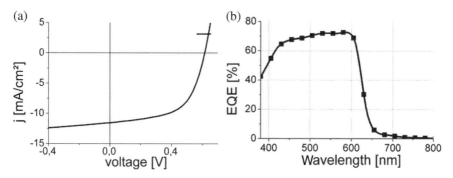

Fig. 12.14 (a) j–V curve of a P3HT:PCBM BHJ solar cell under 100 mW/cm^2 white light illumination. This particular device has a AM1.5 power efficiency of more than 4%. (b) Spectrally resolved photocurrent (EQE) of a doctor-bladed device. By convoluting the external quantum efficiency with the AM1.5G solar spectrum, a short circuit current density of over 11 mA/cm^2 is derived. From comparison to the measured value and by utilizing outdoor measurements, a mismatch factor for the I_{sc} of 0.75–0.8 is calculated. All values in this chapter have been calculated by a mismatch factor of 0.75

a *FF* of 58% under AM1.5 illumination with 100 mW/cm². This corresponds to a power efficiency of over 4%. More than 45 devices were produced and compared for this study, and the performance data varied between 7.5 and 12 mA/cm² for I_{sc}, between 600 and 645 mV for V_{oc}, between 56 and 65% for the FF, corresponding to power efficiencies between ~3% and well beyond 4%. It is important to say that the current densities were calculated by using a mismatch factor which was derived by convoluting the EQE spectrum with the AM1.5G spectrum. Together with the small size of the devices, which may favor parasitic effects from the edges, the current densities should be accurate within a 10–15% tolerance.

Printing techniques are a natural choice for applying and fixing a fluid in a structured way onto a flexible substrate. Shaheen et al. recently presented screen-printed organic solar cells [67]; other printing and coating techniques for organic semiconductors under investigation are, for instance, ink-jet printing [68]. However, organic solar cells have the advantage that the semiconductor layer does not necessarily have to be structured by printing because the dimensional requirements for the lateral resolution or registration are relaxed. This situation favors the use of large-area coating techniques such as doctor blading or slot extrusion, methods which are known to give lateral resolutions in the submillimeter regime. Though it is still widely believed that spin coating is superior to other coating technologies, mainly due to homogeneity of the film thickness, reduced particles, and reduced number of pinholes, it appears obvious that conventional printing and coating techniques, such as solution extrusion or doctor blading, are the more proper tools for production of organic PV devices. This is demonstrated in Fig. 12.15 where the typical perfor-

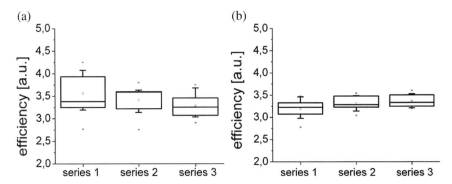

Fig. 12.15 Comparison of the power efficiency of (**a**) coated versus (**b**) spin-coated devices. Data are presented in box plots. The *horizontal lines* in the box denote the 25th, 50th, and 75th percentile values. The *error bars* denote the 5th and 95th percentile values. The two symbols below and above the 5th/95th percentile error bar denote the highest and the lowest observed values, respectively. The *open square* inside the box denotes the mean value. The height of the box is the measure for the tolerance. Devices were illuminated with 100 mW/cm² from a solar simulator. Efficiencies were corrected for the mismatch factor of the illumination source and represent AM1.5 values

mance of spin-coated versus doctor-bladed organic solar cells is compared. Three series of 16 solar cells were compared to evaluate the quality of the doctor blading process. For these batches, it was found that the average performance of doctor-bladed solar cells is even slightly higher than that of batch-processed, spin-coated solar cells from the reference experiment. Further, the tolerance of the doctor blading process is comparable to that of the spin coating process. The slight increase in the variation of the efficiency for the bladed devices is mainly due to a gradient in the film thickness of the bladed layers, which is frequently observed for coating or printing of small areas.

The correlation between the morphology of a BHJ composite and its PV performance is under intense scientific discussion [69]. One approach to this complex situation is a more in-depth analysis of the I–V behavior of BHJ solar cells based on the above introduced equivalent circuit for BHJ solar cells.

One reason for the higher efficiency of the coated devices is a slightly higher I_{sc}, by ~10%, which is attributed to a slightly increased film thickness. EQE measurements evidenced that the quantum efficiency for charge generation or charge separation is comparable to doctor-bladed and spin-coated films. Layers produced by each of these techniques are therefore suggested to have comparable morphology in the 5–30 nm scale, i.e., in the relevant dimensions for charge generation and transport. In parallel, it is observed that doctor-bladed devices have a higher open-circuit voltage (Fig. 12.16a), which somehow contradicts the hypothesis of identical morphologies. Statistical analysis reveals that the V_{oc} of coated devices is ~30–40 mV higher than that of spin-coated devices, corresponding to a 6–7% increase. The diode parameters of the equivalent circuit for both types of solar cells were analyzed as described above and are depicted in Fig. 12.16b.

The focus of the evaluation is on two-diode parameters: the ideality factor n and the reverse saturation current density j_0. These two parameters determine the low voltage shape of the j–V curve in the absence of shunts. The saturation current j_0 determines the absolute size of the diode current around zero bias, while the exponential behavior is dominated by the diode ideality factor n. Following PV theory [70, 71, 72], the physics of a solar cell, namely the separation of the photogenerated charges, is described by these two parameters. This concept is well developed and understood for inorganic solar cells.

The ideality factor reflects the "opening behavior" of the diode with the applied voltage, that is its recombination behavior. Recombination will take place at interfaces where carriers with opposite charges can meet. The diode ideality factor n therefore reflects the density of such interfaces and is representative of the morphology between the conjugated polymers and the fullerenes [73]. Different solvents alter the morphology of pristine as well as of blended films [74], and this is usually reflected in different ideality factors of solar cells processed from different solutions [72].

The second parameter that affects the exponential part of the j–V characteristic is the saturation current, which reflects the number of charges able to overcome the energetic barrier under reverse (zero) bias, $j_0 \approx j_i e^{q\phi/kT}$. Also, the dark current

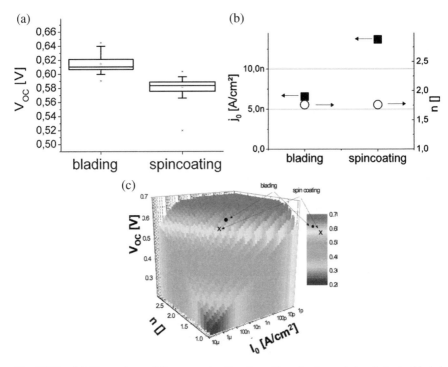

Fig. 12.16 (**a**) Comparison of the open-circuit voltage V_{oc} of spin coated and doctor-bladed devices. The experimental data of 16 devices are presented in the box diagrams. (**b**) Diode ideality factor n (*open circles*) and saturation current density j_0 (*closed squares*) of spin-coated and coated devices according to the analysis with an equivalent circuit. (**c**) Dependence of the V_{oc} on n and j_0. The *cross* marks the parameter set for spin-coated devices, and the *dot* for coated devices

density is representative of the loss mechanisms as luminescent recombination. As demonstrated above, the height of the respective energetic barrier ϕ correlates to the difference between the conjugated polymer's HOMO and the fullerene's LUMO.

For the investigated samples, analysis of spin-coated versus doctor-bladed cells reveals differences in the saturation current density j_0. The diode ideality factor n is unchanged as it can be expected for devices processed from the same solvent. It is this reduction of j_0 by approximately a factor of 3 which is responsible for the V_{oc} enhancement. For BHJ devices, j_0 correlates with the probability that carriers are generated across a barrier. This barrier originates from the difference of the donor's HOMO relative to acceptor's LUMO. As such, there is a distinct difference between n and j_0. The former parameter correlates with the number of interfaces and the latter parameter correlates with the quality of the interface with respect to recombination and generation processes. Different processing techniques yield different morphologies of the composite exerting a strong influence on the absolute value of j_0, since j_0 may be governed by the particular combination of grain boundaries, crystalline regions, mixed amorphous regions between the two semiconductors and especially to 2-D and 3-D polymer aggregates typical of polyhexylthiophenes [75]. Formally, a reduction of j_0 can also be explained by a reduced doping of one of the components. It is unclear how different processing could lead to different background doping.

Based on the results from the equivalent circuit analysis, one can analyze the potential for further improvements. For the given materials and their absorption spectrum, the current densities are already close to the optimum. An additional small enhancement of I_{sc} is expected with better light incoupling, e.g., antireflection coating, more transparent electrodes, and electrode layers or light trapping. The bigger potential for improvement is in the fill factor FF and especially in the open-circuit voltage V_{oc}. To understand how the open-circuit voltage V_{oc} can be influenced by the diode parameters, it is important to recall that the V_{oc} is the bias where the photocurrent and the dark injection current are matching, resulting in an overall zero current densitiy. Though the maximum value of V_{oc} is the difference of the donor's HOMO and the acceptor's LUMO, also referred to as the built-in potential (V_{bi}), dark current contributions usually result in V_{oc} values significantly below V_{bi}.

Figure 12.16c summarizes the correlation of the V_{oc}, the diode ideality factor n, and the dark saturation current density j_0. A further reduction of j_0 by one order of magnitude is needed to increase the V_{oc} of P3HT:PCBM solar cells above 650 mV and close to the 700-mV level, which is a further relative efficiency increase of $\sim 10\%$.

The development of proper coating techniques for organic semiconductors is imperative for large-area reel-to-reel processing. The difference between differently processed (i.e., coated and spin-coated devices) is, according to the analysis based on the extended one-diode replacement circuit, negligible. More freedom to deposit a semiconductor layer with the optimized film thickness together with improved diode parameters are additional strong arguments that BHJ solar cells are compatible with large-area electronics production via printing and coating technologies.

12.3 Challenges for Large-Area Processing

12.3.1 Production Scheme

As illustrated on Fig. 12.17, BHJ-based devices consist in a quite simple structure made of one photoactive organic layer sandwiched between conductive electrodes. Up to now, the most widely used bottom transparent electrode is indium tin oxide (ITO) deposited by reactive sputtering on either rigid glass or flexible polymer foils like PET. In order to compensate for the roughness of this electrode, a layer of poly(3,4-ethylenedioxythiophene) (PEDOT) highly doped with poly(styrenesulfonate) (PSS) can be applied. On the top of this layer, the active donor–acceptor blend is deposited. Finally, the back electrode is evaporated after having deposited a small layer of typically low work function metals like Ca, Ba, Al, or combinations of LiF/Al, all of them known to enhance cell performance.

Fig. 12.17 Schematic
drawing for a BHJ solar cell

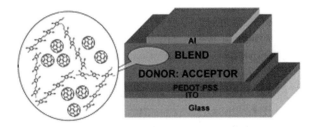

The obvious advantage of plastic solar cells compared to their inorganic counterparts comes from the possibility to produce large-area flexible devices with processing technologies employed in the very mature printing industry [76]. While for
small samples, spin casting technique is commonly carried out, large-scale devices
call upon other approaches like ink-jet printing [77] or doctor blading [78]. Shaheen et al. showed even the applicability of screen printing. Though the application
of screen printing for the deposition of a few 100-nm layers is under discussion,
Shaheen's work certainly paved the route to potentially larger scale and faster roll-
to-roll processing [67]. Figure 12.18 summarizes the schematics of doctor blading
and screen printing.

It has to be mentioned here that an optimization of the device design is manda-
tory to achieve high efficiencies over large areas. Indeed, the bottom transparent
electrodes usually possess sheet resistance of about 10 Ω/square on rigid substrates,
while flexible ITO substrates typically are in the 20–40 Ω/square regime. Therefore, the series resistance induced by this part of the device can dramatically reduce
charge collection. A solution to overcome this problem is to limit the path of the
charge carriers in this electrode, that is, to design individual cells that have large
width but short length, similarly to the case of transistor channels. Furthermore,
since BHJ organic solar cells usually possess a V_{mpp} of several 100 mV, the realization of modules delivering several volts requires efficient series connections of
a large number of individual cells. Figure 12.19 shows the cross section of a BHJ
module. The individual cells are interconnected in a so-called Z-interconnect, where
the top electrode (cathode) of one adjacent cell makes an electrical interconnect to
the anode of the neighboring cell. Such connections, already used to produce inorganic thin-film solar cell modules for a long time, have to ensure minimized surface
area losses and optimized connections of the anode of one cell to the cathode of the

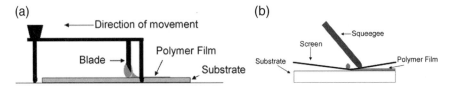

Fig. 12.18 Schematics of the (**a**) doctor blading and the (**b**) screen printing techniques

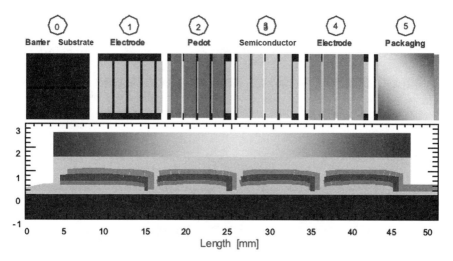

Fig. 12.19 Schematic layout of a BHJ solar cell

next cell. To achieve such integration, the cells should be precisely aligned and their size perfectly controlled. Moreover, such a process should be compatible to roll-to-roll production at high speed, thus avoiding any masking steps. Different strategies to pattern organic devices have been proposed. They mostly rely on ink-jet printing [77], photolithography [79], soft lithography [80], and laser etching [81].

Looking at the layout of a BHJ module, it becomes obvious why coating and printing technologies are expected to be the ideal production technologies. However, recently a very attractive alternative to printing was suggested: the combination of coating technologies and laser patterning [82]. It was shown that a patterning technique based on Nd:YAG laser etching can allow to easily shape self-aligned series-connected solar cells into large-scale modules (Fig. 12.20). This technique, inspired by the state-of-the-art processing used in a-Si thin-film solar cell production [83], relies on selective etching of the photoactive material and the electrode of the cell. When comparing the absorption spectra of ITO MDMO-PPV:PCBM blend to the emission lines of the fundamental and the second harmonic of a Nd:YAG laser, one realizes that ITO can be etched by the 1,064-nm line, while the organic photoactive material is much more sensitive to the 532-nm line. This selective absorption in principle allows to access complicated patterns with one single laser without any mask or lithographic operation. Figure 12.21 shows the successive steps for the production of large modules made of series-connected cells:

- Deposition of ITO and PEDOT:PSS all over the substrate.
- Structuring of the ITO and PEDOT:PSS by ablation with the 1,064-nm Nd:YAG laser line to separate the cells.
- Deposition of a blanket layer of the photoactive material.

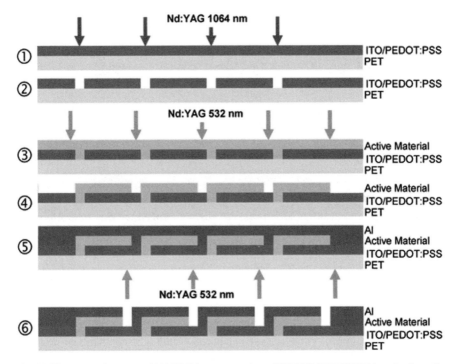

Fig. 12.20 Successive steps of Nd:YAG laser patterning of MDMO-PPV:PCBM based solar cells connected in series (reproduced with permission from [82])

- Structuring of the MDMO-PPV:PCBM material by ablation with the 532-nm Nd:YAG laser line to open vias to the underlying ITO electrodes.
- Deposition of a blanket layer of the top electrode material over the surface of the device.
- Structuring of the top electrode by lift-off due to ablation of the MDMO-PPV:PCBM blend with the 532-nm Nd:YAG laser to separate the individual cells.

Using a 30-kHz pulsed laser, etching speed up to 300 mm.s^{-1} has been achieved, suggesting that this facile way of creating tightly packed series-connected cells might be entirely compatible with roll-to-roll process [82].

12.3.2 Encapsulation of Flexible Solar Cells

Conjugated polymers such as poly(p-phenylene vinylene) (PPV) are known to be rather unstable in air [84], being particularly susceptible to photodegradation induced by oxygen and moisture [85, 86]. The mechanism involves the binding of oxygen atoms to vinyl bonds, which breaks the conjugation and leads to the formation of carbonyl groups [87, 88]. Spectroscopic ellipsometry studies showed

Fig. 12.21 OTR versus WVTR of commercially available polymer films, food packaging requirements, and organic electronics requirements

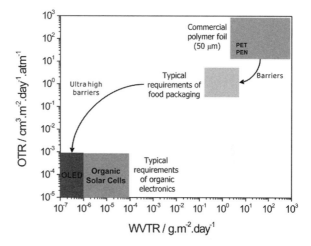

that during simultaneous exposure to air and light, the thickness of the active layer increases, while its refractive index and absorption coefficient drastically decrease [89]. Moreover, the material for the low work function electrode is usually chosen from metals like Al and Ca, in order to maximize the open-circuit voltage of the solar cells, as explained above. These metals rapidly undergo oxidation when exposed to air. This leads to the formation of thin insulating oxide barriers [90], hindering electric conduction, and collection of the charge carriers. Moreover, it has been reported that water can affect the interface between the metallic contact and the organic semiconductors by an electrochemical process that causes delamination of the electrode [91]. This effect is directly connected to the formation of dark spots in OLEDs.

While all these phenomena drastically complicate the usage of OLEDs in air, conjugated polymer:fullerene blends have been shown to be slightly less sensitive than pure conjugated polymer. Neugebauer et al. studied the degradation of MDMO-PPV, PCBM, and blend of the two materials by attenuated total reflection fourier transform infrared (ATR-FTIR) [93]. The authors observed that the MDMO-PPV signal decreases by 50% within 5 hours, under illumination and pure oxygen atmosphere. Though these numbers are relative and mainly reflect the experimental setup, when mixed with an electron acceptor, the stability of the MDMO-PPV is drastically increased relative to the pristine material. There may be two distinct explanations of this effect. First, since the electrochemical potential of the photoexcited state of conjugated polymer is relatively high, it might immediately lead to direct electrochemical interactions with oxygen and water vapor present in its vicinity. The ultrafast electron transfer from the LUMO of the conjugated polymer to the LUMO of the acceptor empties the excited state of the polymer and lowers the electrochemical energy of the excited electron. Second, the intersystem crossing of the polymer can produce triplet excited states that may create singlet oxygen by energy transfer [92]. This highly reactive form of oxygen is expected to react with the polymer backbone creating carbonyl-type defects. As shown above, the ultrafast photoinduced electron transfer does quench the inert system crossing to the triplet state, by hindering this degradation route.

Several studies focused on the lifetime investigation of conjugated polymer:fullerene solar cells. In the case of MDMO-PPV:PCBM cells encapsulated between glass plates, Neugebauer et al. reported a decrease of 40% of the V_{oc} and 60% of the I_{sc} within 3,000 hours, whereas unprotected cells lasted only couple of hours [93]. Accelerated lifetime test showed that the degradation was about 10 times faster at 80°C than at 25°C [94]. Moreover, the degradation was found to be influenced by some morphological components. Indeed, Jeranko et al. investigated the aging effect by photocurrent imaging, which revealed the formation of island of higher efficiency [95]. Finally, deterioration of the electron collecting electrode–polymer interface has been proposed as a determining factor [96, 97]. Several attempts have been tried to reduce the degradation of the active materials, such as replacing the easily oxidized Al by more stable Au [98] or blending the photoactive polymer with a potentially protective host matrix [99]. However, efficient encapsulation solutions are required in order to ensure extended lifetime to this type of device. In order to fully take advantage of the opportunities offered by soluble materials like conjugated polymers, these solution have to be mechanically flexible.

The ability of oxygen and moisture to cross an encapsulating membrane is quantified by the oxygen transmission rate (OTR) and the water vapor transmission rate (WVTR). It is generally accepted that the lifetime of OLED devices above 10,000 hours requires WVTR and OTR of about 10^{-6} g.m^{-2}.day^{-1} and 10^{-4}cm^3.m^{-2}.day^{-1}.atm^{-1}, respectively [100]. As illustrated on Fig. 12.21, these values are about three to five orders of magnitude lower than the corresponding values of commercially available polymer films. However, since conjugated polymer:fullerene-based solar cells are slightly less sensitive to oxidative agents, it is commonly considered that WVTR around 10^{-4} g.m^{-2}.day^{-1} might be sufficient for this type of device. It has to be noted here that WVTR and OTR are inversely proportional to the thickness of a homogeneous membrane. This result arises from the fact that these parameters are experimentally evaluated by placing the membrane in between a fully saturated and a fully depleted atmosphere (under constant flux of inert drain gas). The concentration of diffusing species decreases linearly across the thickness of the membrane. Thus, in order to decrease by four orders of magnitude, the OTR of a 50-μm-thick PET substrate, one would have to increase its thickness by the same factor. Luckily, and for the sake of flexibility, another solution consists of using inorganic gas barrier thin-film deposited on thin polymer substrates.

Transparent thin-film barrier coatings resistant to permeation of gases and vapors deposited onto flexible polymer substrates have been intensively studied for applications in food and pharmaceutical packaging, where improving the barrier capabilities of the bare plastic films by 1–3 orders of magnitude is usually considered sufficient (Fig. 12.21). It has been shown that silicon-based dielectric coatings deposited by plasma enhanced chemical vapor deposition (PECVD) on plastic films can be used as single layer permeation barriers yielding barrier improvement factors (BIF) up to 3 orders of magnitude [101, 102]. Permeation through those barrier materials has been proven to be a phenomenon controlled by nanometer-size structural defects present in the barrier coating, mainly caused by intrinsic or extrinsic surface roughness [101, 103, 104]. Moreover, theoretical calculations have shown that the total permeation rate through many small pinholes is much higher than that corresponding to the same total pinhole area combined in a few larger defects [105, 106]. This can be explained by lateral diffusion which is of crucial importance when the diameters of the defects are small compared with the thickness of the substrate. Thus, BIF beyond three orders of magnitude are unlikely to be achieved with single inorganic layers.

The most common technique used to achieve ultrahigh barrier properties is based on alternating organic–inorganic multilayers; sandwiching inorganic barriers between polymeric buffers has been reported to reduce the number of pinholes significantly [107]. This can be explained in terms of smoothing of the coated surface, reduction of mechanical damage, and increase of thermal stability [100] Moreover, repeating the alternating process yields stacked structures that allow the organic layers to "decouple" the defects from neighboring inorganic layers. Polyacrylate/Al_2O_3 alternating coatings produced in a multistep process [108] have been used to encapsulate OLEDs. The organic layer typically is produced by flash evaporation of an acrylic monomer that is subsequently cured by UV light, while the inorganic coating

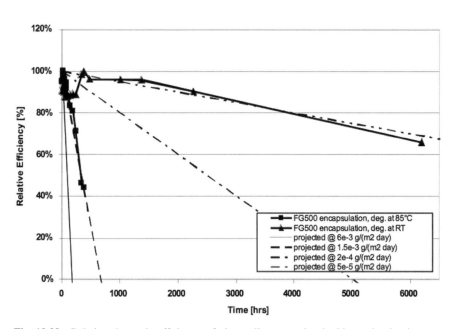

Fig. 12.22 Relative change in efficiency, of glass cells encapsulated with an ultra barrier, as a function of time. The change in efficiency can be modeled using a constant WVTR rate and the volume of Ca. Degradation is ca. 4× slower than predicted with the WVTR rates measured with a Ca sensor

is usually deposited by reactive sputtering. WVTR of about 2×10^{-6} g.m^{-2}.day^{-1} has been reported [109], yielding 5-mm^2 OLED lifetime of 2,500 hours [110].

In the case of conjugated polymer:fullerene solar cells, the usage of another type of ultrahigh barrier was reported. This barrier, entirely produced by PECVD of alternating organic/SiO$_x$ on 175-μm-thick PET substrate, succeeded accelerated "calcium test" (50°C, 85% relative humidity) for more than 1,000 hours, which corresponds to at least 10,000 hours under normal ambient conditions [111]. Dennler et al. showed that MDMO-PPV:PCBM cells fabricated on such substrates and subsequently encapsulated exhibit shelf lifetime of about 3,000 hours in spite of their appreciable mechanical flexibility. The correlation between the device lifetime and the barrier property was recently investigated by Hauch et al. (Fig. 12.22). Figure 12.22 correlates the lifetime of single cells with the WVTR of the packaging material. For a variety of WVTRs, the device lifetime correlated quite well with the WVTR of the top package. Thus, solar cells with sensitive electrodes like Ca do have an environmental stability issue. Careful engineering of the devices was necessary to achieve longer lifetimes. Operational lifetimes beyond 1,000 hours under one sun and elevated temperatures were reported for glass as well as flexible substrates [112–114]. These observations reveal that the encapsulation may not be the limiting factor and the combined usage of more stable conjugated polymers, better defined interfaces in combination with reasonable high barrier coatings, will lead to very long lifetimes [95].

12.4 Conclusions

BHJ solar cells made impressive progress over the last couple of years. Starting with roughly 1% efficient devices on a few millimeter square active areas, processed by sublimation of compounds such as merocyanines [115] and pthalocyanines [116] in the late 1980s, today's efficiency is over 6%, on large areas and on flexible substrates with impressive lifetime. The rapid progress in this field is mostly based on the progress in the materials science and production technologies. Today's BHJ solar cells can be processed on glass as well as on flexible substrates with relative ease and comparable performance. From today's perspective, it seems only a question of time when devices with 10% power conversion efficiency on flexible substrates and large areas will be demonstrated. One key development allowing such high efficiencies certainly will be on the materials side. The second, as important development, is to gain improved control over the solid-state morphology of donor–acceptor composites. The near future will show exciting developments, and some of them can be anticipated already today, like self-assembled materials, nanosized device geometries release the requirements on the proper morphology, metallic nanoparticles and plasmon absorption and highly complex molecules which fulfil all functions in distinct parts of the molecule. In parallel to these exciting new fields, we will see a fast development of the solution-based techniques such as ink-jet, screen, or flexographic printing (a form of rotary web letterpress using flexible relief plates) for the production of BHJ solar cells. Other important economic factors addressing the balance-of-system costs are novel semiconducting electrodes to replace ITO, low-cost barriers against water and oxygen permeation as well as flexible electronic components to be combined with flexible PV power supply. The BHJ technologies will see an exciting nearby future in the application of PVs.

References

1. Shockley W, Queisser HJ (1961) Detailed Balance Limit of Efficiency of p-n Junction Solar Cells J Appl Phys 32:510
2. Gregg BA, Hanna MC (2003) J Appl Phys 93:3605
3. Sariciftci NS (ed) (1997) Primary Photoexcitation in Conjugated Polymers: Molecular Exciton Versus Semiconductor Band Model. World Scientific, Singapore
4. Gregg BA (2003) J Phys Chem B 107:4688
5. O'Regan B, Graetzel M (1991) Nature 353:737
6. Yu G, Gao J, Hummelen JC, Wudl F, Heeger AJ (1995) Science 270:1789
7. Drechsel J, Männig B, Kozlowski F, Pfeiffer M, Leo K, Hoppe H (2005) Appl Phys Lett 86:244101
8. Huynh WU, Dittmer JJ, Alivisatos AP (2002) Science 295:2425
9. Onsager L (1938) Phys Rev 54:554
10. Schweitzer B, Bässler H (2000) Synth Met 109:1
11. Haugeneder A, Neges M, Kallinger C, Spirkl W, Lemmer U, Feldmann J, Scherf U, Harth E, Gügel A, Müllen K (1999) Phys Rev B 59:15346
12. Vacar D, Maniloff ES, McBranch DW, Heeger AJ (1997) Phys Rev B 56:4573
13. Theander M, Yartsev A, Zigmantas D, Sundström V, Mammo W, Andersson MR, Inganäs O (2000) Phys Rev B 61:12957
14. Popovic ZD, Hor A-M, Loutfy RO (1988) Chem Phys 127:451
15. Arkhipov VI, Emelianova EV, Bässler H (1999) Phys Rev Lett 82:1321
16. Wang Y (1992) Nature 356:585
17. Sariciftci NS, Smilowitz L, Heeger AJ, Wudl F (1992) Science 258:1474
18. Morita S, Zakhidov AA, Yoshino K (1992) Solid State Commun 82:249
19. Brabec CJ, Zerza G, Cerullo G, De Silvestri S, Luzzati S, Hummelen JC, Sariciftci NS (2001) Chem Phys Lett 340:232
20. Würfel P, Bauer G (2003) Quantum solar energy conversion in organic solar cells. In: Brabec CJ, Dyakonov V, Parisi J, Sariciftci S (eds) Organic Photovoltaics, Springer, Berlin, pp 118–158
21. Waldauf C, Schilinsky P, Hauch J, Brabec CJ (2004) Thin Solid Films 451–452:503
22. Koster LJA, Smits ECP, Mihailetchi VD, Blom PWM (2005) Phys Rev B 72:085205
23. Mihailetchi VD, Koster LJA, Hummelen JC, Blom PWM (2005) Phys Rev Lett 93:216601
24. Marcus RA (1993) Rev Mod Phys 65:599
25. Miller A, Abrahams E (1960) Phys Rev 120:745
26. Pai DM (1970) J Chem Phys 52:2285
27. Mozer AJ, Dennler G, Sarificti NS, Westerling M, Pivrikas A, Österbacka R, Juska G (2005) Phys Rev B 72:035217
28. Rakhmanova SV, Conwell EM (2001) Synth Met 116:389
29. Mozer AJ, Sariciftci NS, Pivrikas A, Österkacka R, Juska G, Brassat L, Bässler H (2005) Phys Rev B 72:035214
30. Riedel I, Parisi J, Dyakonov V, Lutsen L, Vanderzande D, Hummelen JC (2004) Adv Funct Mater 14:38
31. Gommans HHP, Kemerink M, Kramer JM, Janssen RAJ (2005) Appl Phys Lett 87:122104

32. Montanari I, Nohueira AF, Nelson J, Durrant J, Winder C, Loi MA, Sariciftci NS, Brabec CJ (2002) Appl Phys Lett 81:3001
33. O'Regan BC, Scully S, Mayer AC, Palomares E, Durrant J (2005) J Phys Chem B 109:4616
34. Meskers SCJ, van Hal PA, Spiering AJH, Hummelen JC, Janssen RAJ (2000) Phys Rev B 61:9917
35. Kroeze JE, Savenije TJ, Vermeulen MJW, Warman JM (2003) J Phys Chem B 107:7696
36. Savenije TJ, Kreoze JE, Wienk MM, Kroon JM, Warman JM (2004) Phys Rev B 69:155205
37. Juška G, Arlauskas K, Viliūnas M, Kočka J (2000) Phys Rev Lett 84:4946
38. Juska G, Arlauskas K, Viliunas M, Genevicius K, Österbacka R, Stubb H (2000) Phys Rev B 62:16235
39. Langevin P (1903) Ann Chem Phys 28:289
40. Pivrikas A, Juska G, Mozer AJ, Scharber M, Arlauskas K, Sariciftci NS, Stubb H, Osterbacka R (2005) Phys Rev Lett 94:176806
41. Mihailetchi VD, Koster LJA, Blom PWM, Melzer C, de Boer B, van Duren JKJ, Janssen RAJ (2005) Adv Funct Mater 15:795
42. Parker ID (1994) J Appl Phys 75:1656
43. Brabec CJ, Cravino A, Meissner D, Sariciftci NS, Fromherz T, Rispens MT, Sanchez L, Hummelen JC (2001) Adv Funct Mater 11:374
44. Brabec CJ, Cravino A, Meissner D, Sariciftci NS, Rispens MT, Sanchez L, Hummelen JC, Fromherz T (2002) Thin Solid Films 403–404:368
45. Kim H, Jin S-H, Suh H, Lee K (2004) Proc. SPIE 5215:111
46. Gadisa A, Svensson M, Andersson MR, Inganäs O (2004) Appl Phys Lett 84:1609
47. Frohne H, Shaheen SE, Brabec CJ, Mueller D, Sariciftci NS, Meerholz K. (2002) Chem Phys Chem 9:795
48. Würfel P. (2004) Physics of Solar Cells. Wiley VCH, Weinheim
49. Crispin A, Crispin X, Fahlman M, Berggren M, Salaneck, W (2006) Appl Phys Lett 89:213503
50. Waldauf C, Schilinsky P, Hauch J, Brabec CJ (2004) Thin Solid Films 451–452:503
51. Hall RN (1952) Phys Rev 87:387
52. Shockley W, Read WT (1952) Phys Rev 87:835–842
53. Merten J, Asensi JM, Voz C, Shah AV, Platz R, Andreu J (1998) IEEE Transactions on Electron Devices 45:423
54. Brabec CJ, Cravino A, Meissner D, Sariciftci NS, Fromherz T, Rispens MT, Sanchez L, Hummelen JC (2001) Adv Funct Mater 11(5):374–380
55. Montanari I, Nogueira AF, Nelson J, Durrant JR, Winder C, Loi MA, Sariciftci NS, Brabec C (2002) Appl Phys Lett 81:3001
56. Koster LJA, Mihailetchi VD, Ramaker R, Blom PWM (2005) Appl Phys Lett 86:123509
57. Mihailetchi VD, Koster LJA, Blom PWM (2004) Appl Phys Lett 85(6):971
58. Haugeneder A, Neges M, Kallinger C, Spirkl W, Lemmer U, Feldmann J, Scherf U, Harth E, Gugel A, Mullen K (1999) Phys Rev B Condens Matter 59(23):15346
59. Kankan B, Castelino K, Majumdar A (2003) Nano Lett 3(12):1729
60. Brabec CJ, Zerza G, Cerillo G, DeSilvestris S, Luzatti S, Hummelen JC, Sariciftci NS (2001) Chem Phys Lett 340:232
61. Morteani AC, Sreearunothai P, Herz LM, Friend RH, Silva C (2004) Phys Rev Lett 92(24):247402–247413
62. Halls JJM, Cornil J, dosSantos DA, Silbey R, Hwang DH, Holmes AB, Bredas JL, Friend RH (1999) Phys Rev B Condens Matter 60(8):5721
63. Fujii A, Mizukami H, Umeda T, Shirakawa T, Hashimoto Y, Yoshino K (2004) Jpn J Appl Phys 43(12):8312–8315
64. Schilinsky P, Waldauf C, Brabec CJ (2002) Appl Phys Lett 81:3885
65. Waldauf C, Schilinsky P, Hauch J, Brabec CJ (2004) Thin Solid Films 451–452:503
66. NREL certificate for Konarka solar cell E8-6 from July, 18th, 2005
67. Shaheen S, Radspinner R, Peyghambarian N, Jabbour GE (2001) Appl Phys Lett 79:2996

68. Sirringhaus H, Kawase T, Friend RH (2001) MRS Bulletin 26(7):539
69. Duren KJ, Yang X, Loos J, Bulle-Lieuwana CWT, Sieval AB, Hummelen JC, Janssen RAJ (2004) Adv Func Mat 14:425
70. Sze SM (1985) Physics of Semiconductor Devices, 2nd edn. John Wiley & Sons, New York
71. Lewerenz HJ, Jungblut H (1995) Photovoltaik. Springer, Berlin
72. Waldauf C, Scharber MC, Schilinsky P, Hauch JA, Brabec CJ (2006) J Appl Phys 99:104503
73. Waldauf C, Schilinsky P, Hauch J, Brabec CJ (2004) Thin Solid Films 451–452:503
74. Chang J-F, Sun B, Breiby DW, Nielsen MM, Solling T, Giles M, McCulloch I, Sirringhaus H (2004) Chem Mater 16:4772
75. Osterbacka R, An CP, Jiang XM, Vardeny ZV (2000) Science 287:839
76. Brabec CJ, Hummelen JC, Janssen RAJ, Sariciftci NS (1999) Synth Met 102:861
77. Sirringhaus H, Kawase T, Friend RH, Shimoda T, Inbasekaran M, Wu W, Woo EP (2000) Science 290:2123
78. Brabec CJ, Sariciftci NS, Hummelen JC (2001) Adv Funct Mater 11:15
79. Mäkelä T, Pienimaa S, Jussila S (1999) Synth Met 101:705
80. Granlund T, Nyberg T, Roman LS, Svensson M, Inganäs O (2000) Adv Mater 12:269
81. Tak Y-H, Kim C-N, Kim M-S, Kim K-B, Lee M-H, Kim S-T (2003) Synth Met 138:497
82. Dennler G, Lungenschmied C, Neugebauer H, Sariciftci NS, Labouret A (2005) J Mater Res 20:3224
83. Akimasa U, Susumu K (2001) Method of Fabricating Integrated Thin Film Solar Cells. Patent US6168968
84. Morgado J, Friend RH, Cacialli F (2000) Synth Met 114:189
85. Sutherland DGJ, Carlisle JA, Elliker P, Fox G, Hagler TW, Jimenez I, Lee HW, Pakbaz K, Terminello LJ, Williams SC, Himpsel FJ, Shuh DK, Tong WM, Lia JJ, Callcott TA, Ederer DL (1996) Appl Phys Lett 68:2046
86. Low HY (2002) Thin Solid Films 413:160
87. Scurlock RD, Wang B, Ogilby PR, Sheats JR, Clough RL (1995) J Am Chem Soc 117:10194
88. Cumpston BH, Parker ID, Jensen KF (1997) J Appl Phys 81:3716
89. Kumar S, Biswas AK, Shukla VK, Awasthi A, Anand RS, Narain J (2003) Synth Met 139:751
90. Do LM, Han EM, Nidome Y, Fujihira M, Kanno T, Yoshida S, Maeda A, Ikushima AJ (1994) J Appl Phys 76:5118
91. Schaer M, Nüesch F, Berner D, Leo W, Zuppiroli L (2001) Adv Func Mater 11:116
92. Hale GD, Oldenburg SJ, Halas NJ (1997) Phys Rev B 55:16069
93. Neugebauer H, Brabec CJ, Hummelen JC, Sariciftci NS (2000) Sol Ener Mater Sol Cells 61:35
94. Schuller S, Schilinsky P, Hauch J, Brabec CJ (2004) Appl Phys A 79:37
95. Jeranko T, Tributsch H, Sariciftci NS, Hummelen JC (2004) Sol Ener Mater Sol Cells 83:247
96. Krebs FC, Carlé JE, Cruys-Bagger N, Andersen M, Lilliedal MR, Hammond MA, Hvidt S (2005) Sol Ener Mater Sol Cells 86:499
97. Kroon JM, Wienk MM, Verhees WJH, Hummelen JC (2002) Thin Solid Films 403–404:223
98. Sahin Y, Alem S, De Bettignies R, Nunzi J-M (2005) Thin Solid Films 476:340
99. Brabec CJ, Padinger F, Sarificti NS, Hummelen JC (1999) J Appl Phys 85:6866
100. Lewis JS, Weaver MS (2004) IEEE. J Select Topics Quantum Elect 10:45
101. da Silva Sobrinho AS, Latreche M, Czeremuszkin G, Klemberg-Sapieha JE, Wertheimer MR (1998) J Vac Sci Technol A 16:3190
102. Leterrier Y (2003) Prog Mater Sci 48:1
103. da Silva Sobrinho AS, Czeremuszkin G, Latreche M, Dennler G, Wertheimer MR (1999) Surf Coat Technol 116–119:1204
104. Dennler G, Houdayer A, Ségui Y, Wertheimer MR (2001) J Vac Sci Technol A 19:2320
105. Russi G, Nulman M (1993) J Appl Phys 74:5471
106. da Silva Sobrinho AS, Czeremuszkin G, Latreche M, Wertheimer MR (2000) J Vac Sci Technol A 18:149

107. Affinito JD, Gross ME, Coronado CA, Graff GL, Greenwell EN, Martin PM (1996) Thin Solid Films 290–291:263
108. Affinito JD, Gross ME, Mournier PA, Shi MK, Graff GL (1999) J Vac Sci Technol A 17:1974
109. Weaver MS, Michalski LA, Rajan K, Rothman MA, Silvernail JA, Burrows PE, Graff GL, Gross ME, Martin PM, Hall M, Mast E, Bonham C, Bennett W, Zumhoff M (2002) Appl Phys Lett 81:2929
110. Chwang AB, Rothman MA, Mao SY, Hewitt RH, Weaver MS, Sivernail JA, Rajan K, Hack M, Brown JJ, Chu X, Moro L, Krajewski T, Rutherford N (2003) Appl Phys Lett 83:413
111. Dennler G, Lungenschmied C, Neugebauer H, Sariciftci NS, Latrèche M, Czeremuszkin G, Wertheimer MR (2005) Thin Sol Films 349
112. Kroon JM, Veenstra SC, Slooff LH, Verhees WJH, Koetse MM, Sweelssen J, Schoo HFM, Beek WJE, Wienk MM, Janssen RAJ, Yang X, Loos J, Michailetchi VD, Blom PWM, Knol J, Hummelen JC (2005) Proceedings of the 20th European Photovoltaic Solar Energy Conference and Exhibition, Barcelona, Spain, June 6–10, 2005
113. Schuller S, Schilinsky P, Hauch J, Brabec CJ (2004) Appl Phys A 79:37
114. Brabec CJ, Schilinsky P, Waldauf C, Hauch J (2005) MRS Bulletin January 30, 2005
115. Chamberlain GA (1983) Solar Cells 8:47
116. Simon J, Andre J-J (1985) Molecular Semiconductors. Springer-Verlag, Berlin

Chapter 13
Substrates and Thin-Film Barrier Technology for Flexible Electronics

Ahmet Gün Erlat, Min Yan, and Anil R. Duggal

13.1 Introduction

The term "flexible electronics" encompasses a wide array of applications such as flexible displays, low-cost and/or large-area sensors, conformal lighting, and solar cells to name a few, with one common ingredient: the ability to fabricate electronic and optoelectronic devices on nonrigid substrates such as plastic films, metal foil, or thin glass, without losing the functionality of the devices during operation. Other terms often used to convey this concept include "printable electronics", "macro-electronics," and "organic electronics." The promise of flexible electronics lies in the potential for building large-area electronic devices with much lower cost than possible with conventional silicon-based technology [1]. The technologies being developed to enable this all revolve around building devices using low-cost printing techniques that are compatible with high-volume "roll-to-roll" manufacturing. Example technologies that are potentially compatible with low-cost printing techniques include organic light-emitting devices (OLEDs) [2–3], organic photovoltaic devices [4–6], thin-film transistors (TFTs) and TFT arrays using both organic [7–9] and solution-processible inorganic materials [10, 11], electronic paper [12], wearable electronics [13], flexible batteries [14], RFID tags [15], sensors [16], and more complicated circuitry [17–19].

Fabricating these devices on flexible substrates, particularly on plastic, brings with it new challenges in order to make them commercially viable. One critical challenge is to design a means of hermetic packaging of the organic electronic device because many such devices exhibit a very short shelf lifetime if not protected from the ingress of environmental permeants such as oxygen and water vapor [20]. Such a packaging scheme can be viewed to have two components. First, the substrate on

A.G. Erlat (✉)
General Electric Global Research Center, 1 Research Circle, KWC 331, Niskayuna, NY 12309
e-mail: erlat@research.ge.com

W.S. Wong, A. Salleo (eds.), *Flexible Electronics: Materials and Applications*, Electronic   413
Materials: Science & Technology, DOI 10.1007/978-0-387-74363-9_13,
© Springer Science+Business Media, LLC 2009

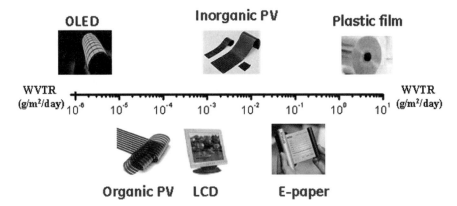

Fig. 13.1 Water vapor barrier requirements for various optoelectronic applications (OLED, organic light-emitting device; PV, photovoltaic; LCD, liquid crystal display; E-paper, electrophoretic paper display)

which the device is built needs to provide a barrier; second, the device and the substrate need to be "capped" with a form of encapsulation to complete the hermetic package. Figure 13.1 demonstrates the barrier levels for water vapor (in terms of water vapor transmission rate, WVTR) required for many of the technologies mentioned above. One can see that plastic materials alone cannot provide a sufficiently hermetic barrier and that an OLED device requires the most stringent level of barrier protection. The focus of this chapter is to present a comprehensive account of the thin-film barrier technologies available to enable flexible electronics. The main emphasis will be on barrier technologies that can be integrated with an OLED device since this represents the most challenging application. Section 13.2 outlines the requirements for hermetic packaging in detail. Section 13.3 describes the engineering challenges in creating a thin-film barrier, measuring its effectiveness, and the various approaches and solutions that are available. Section 13.4 delves into device–barrier integration challenges both for the case where the barrier is part of the device substrate and for when it is used as an encapsulation layer on top of the device.

13.2 Barrier Requirements

Various packaging schemes that have been considered for a flexible OLED are depicted in Fig. 13.2. In each flexible scheme, the device is built up from a roll-processible film substrate which needs to provide adequate barrier properties, and then covered with a flexible top barrier which consists of either a superstrate film (A, B, C) or a thin encapsulating film (D). Schemes A, B, and C utilize an adhesive to attach the superstrate and substrate films together. As shown, the adhesive fills the whole area between the two films, but other variations can be considered

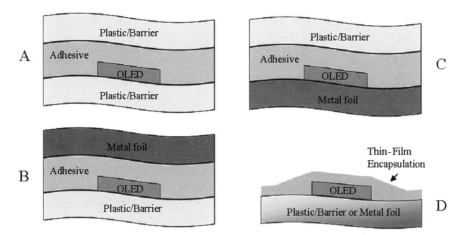

Fig. 13.2 Packaging schemes for flexible OLEDs. (**A**) Barrier plastic film as substrate with barrier plastic film as superstrate, (**B**) Barrier plastic film as substrate with metal foil as superstrate, (**C**) Metal foil as substrate with barrier plastic film as superstrate, and (**D**) Flexible substrate with thin-film encapsulation

where the adhesive fills only the peripheral area outside of the devices. Regardless, whenever an adhesive is utilized, it must be recognized that its thickness and barrier properties are also important since water and oxygen can diffuse into the device from the edge of the adhesive seal. This can be the leading device failure mechanism when the substrate and superstrate provide a high-quality barrier.

OLEDs represent a subset of flexible electronic devices that require optical transmission through either the bottom or the top of the device. The difference between schemes A, B, and C relate simply to how this optical transmission is accomplished. Schemes B and C utilize a metal foil, e.g., stainless steel or aluminum, as either a substrate or a superstrate and a plastic film coated with a thin-film barrier on the opposite side of the device. Metal foils provide essentially a perfect barrier but are of course opaque. Hence, the combination of the plastic film and thin film barrier needs to be transparent at the wavelengths appropriate for the device. Scheme A utilizes a plastic film and thin film barrier as both substrate and superstrate. This scheme enables the possibility of a fully transparent flexible electronic device such as a heads-up display.

Scheme D represents an alternative packaging approach where a thin-film barrier is coated directly onto the device. The potential advantage of this scheme is that there are no longer edge-permeation issues due to the elimination of the need for an adhesive. The primary disadvantage is the extra-imposed constraint that the thin-film barrier deposition process has to be integrated with the device-making process such that the underlying device is not damaged. Given a transparent barrier coating, the underlying substrate can be either transparent or opaque depending on the application.

In the following, the requirements for a barrier coating suitable for an OLED device are described. In particular, first the generic properties that are required for both an encapsulation coating and a substrate coating are described and then the needs that are unique to the substrate coating application are given.

13.2.1 Generic Requirements

13.2.1.1 Barrier Level

The packaging of an OLED device affects mainly its "shelf lifetime," i.e., how the device degrades in the environment even when it is not in use. Shelf lifetime is limited due to several potential degradation mechanisms. OLEDs utilize low work function cathodes that are evaporatively deposited [3, 21]. These materials are extremely reactive with water and can lose their electrical properties rapidly upon exposure to moisture [22]. Also, water vapor permeating through the pinholes in the vacuum-deposited cathode can diffuse into the metal–organic interface and limit electron injection as well as cause delamination [22]. In addition, the commonly used organic materials within the active layers of the device are known to lose their emissive properties due to moisture exposure [22, 23]. As a result of all or some of these mechanisms, degradation presents itself as growing dark spots distributed throughout the active area or as a front of nonemissive area growing in from the edges of the device. Simple calculations, primarily based on the degradation of the cathode material when exposed to water vapor, suggest that in order to achieve a device lifetime of 10,000 hours at room temperature, the substrate must provide a barrier that limits permeation rates to less than 10^{-6} g/m²/day and 10^{-3} cm³/m²/day for water vapor and oxygen, respectively [24, 25].

13.2.1.2 Optical Properties

Any absorption of light in the barrier film through the emissive side of the device results directly in a loss in device efficiency. Hence, any such loss needs to be minimized. In addition, because light from an OLED is emitted from an optical microcavity, the refractive index of the barrier film needs to be understood and controlled [26]. This is particularly important for cases where the barrier coating consists of multiple components with different refractive indices since these can act as dielectric mirrors, which greatly enhance microcavity effects. Depending on device design, these microcavity effects can enhance, suppress, or distort the desired light emission.

13.2.1.3 Mechanical Flexibility

Mechanical flexibility is required for roll-to-roll (R2R) processing and for various end-use applications such as bendable to "roll-up" displays. A typical metric to test the barrier structure for the latter application is to bend the barrier-coated substrate over a 1" diameter 100 or 1,000 times and measure the barrier properties before and after this test. It has been demonstrated that the most advanced barrier films on plastic substrates can pass this test [26–28].

13.2.1.4 Compatibility of the Barrier Coating

It goes without saying that the process conditions used to deposit the barrier must be compatible with the underlying layers. For a device encapsulation application, this can be a demanding requirement – particularly for organic electronic devices where the active organic layers degrade at relatively low temperatures (150–200°C) and are sensitive to UV radiation and ion bombardment. For a barrier coating on plastic substrate application, the same considerations apply but there is less sensitivity to slight degradation in the substrate since it is not one of the active layers of the device.

13.2.2 Substrate-Specific Requirements

13.2.2.1 Substrate Options

When mechanical flexibility is not a requirement, the substrate material of choice for OLED devices is glass as it is transparent, smooth, impermeable to water and oxygen, and chemically and thermally stable. Glass can actually be considered for flexible electronics applications because, when fabricated with a thickness of <200 μm, it is relatively bendable. Currently, glass sheet as thin as 30 microns can be produced using down-draw processing [29]. Unfortunately, the application space open to thin glass is likely to be small because this material is still brittle, difficult to produce in large sheets, and unsuitable for R2R processing.

Metal foils are flexible, have excellent barrier and thermal properties, and are compatible with R2R processing. In fact, at the research stage, stainless steel foil has been successfully integrated into a flexible active-matrix electrophoretic display device by E Ink Corporation [30] and into a flexible active-matrix OLED display device by Universal Display Corporation [31]. Hence, it appears likely that metal foil substrates will play a prominent role in flexible electronics. The main drawback is that applications are limited to the packaging scheme C depicted in Fig. 13.2 and, for optical devices, still require a barrier encapsulation coating or a superstrate. It should be noted that typically the metal foil substrate still requires its own thin-film coating to provide the appropriate combination of surface smoothness, electrical isolation, and chemical resistance required by the application.

For applications where a transparent flexible substrate is required, a plastic film is necessary. In choosing an appropriate plastic film, it is useful to categorize the available options by glass transition temperature (T_g) as this sets the scale for the accessible processing temperatures and time. Another key distinction is whether the underlying structure is amorphous or semicrystalline as this leads to different thermomechanical film properties. Amorphous polymers can be further subdivided into thermoplastic materials, which can be converted into films via an extrusion process, and higher T_g materials, which require solvent casting for film formation.

Semicrystalline polymeric films, such as PET and PEN, are commercially available from DuPont under the names Melinex® and Teonex® respectively. PET has a relatively low T_g around 78°C, while PEN has T_g around 120°C. Due to their semicrystallinity, PET and PEN have relatively low coefficient of thermal expansion

(CTE), which is around 15 ppm/°C and better solvent resistance when compared with amorphous polymeric films. PET and PEN films can be melt cast followed by biaxial stretching and heat setting processes to crystallize the film. The dimensional stability of PET and PEN films can be further enhanced by a heat stabilization process in which the internal stress in the film is relaxed by exposure to high temperature while under minimum line tension [32]. The heat stabilization process dramatically reduces the shrinkage rate of PET and PEN films at temperatures above their glass transition temperatures and thereby extends the accessible process temperatures. For instance, even though PET has a T_g of 78°C, when heat stabilized it can be processed at temperatures above 100°C.

Examples of thermoplastic amorphous polymers include polycarbonates (PCs) and polyethersulfones (PES). The classic bisphenol A (BPA) PC (e.g., GE's Lexan®) has a T_g of ~150°C, while new engineered high heat PCs are now commercially available with T_gs around 220°C (e.g., GE's Lexan® XHT resin). PES, materials (e.g., Sumitomo Bakelite's Sumilite®) also offer T_gs around 220°C. Both PC and PES have a relatively high CTE (50~70 ppm/°C) due to their amorphous structure.

Several high T_g polymers can only be solvent cast due to the fact that their melting points are too high to allow the polymers to be melt processed without significant degradation. This typically results in a higher cost for the resulting film. Examples of solvent cast amorphous polymer films include aromatic fluorine-containing polyarylate (PAR), e.g., Ferranias Arylite®; polycyclic olefin (PCO), e.g., Promerus Appear®; and polyimide (PI), e.g., DuPont Kapton® [32, 33].

Most of the plastic films described above can deliver optical transparency over the visible light range except for PI, which is yellow. PC, in particular, has excellent optical properties (high transmission, low haziness). Polymeric films tend to undergo undesirable change in dimension when heated to high temperatures due to molecular relaxation associated with the increased mobility of polymer chains and relaxation of residual strain within the film. This effect is, in general, less pronounced for semicrystalline polymers than for amorphous polymers. In addition, semicrystalline films are inherently stiffer than amorphous films (~3× greater Young's modulus). Nevertheless, the thermomechanical stability of any plastic film is substantially inferior to glass or metal foil and hence any robust manufacturing process needs to be designed with this in mind.

All these substrate choices have advantages and disadvantages and they most often require additional coatings that can provide functions other than barrier to moisture and oxygen.

13.2.2.2 Chemical Resistance

Most device fabrication processes utilize solvents and chemicals that may dissolve or otherwise damage the plastic substrate. A typical list of the materials that the substrate must be compatible with includes methanol, isopropanol, acetone, tetrahydrofuran, n-methylpyrrolidone, ethylacetate, sulfuric acid, glacial acetic acid, hydrogen peroxide, and sodium hydroxide. Fabricating a thin-film barrier structure that protects the substrate from such chemicals would be ideal but in most cases it is not possible. More often than not, a chemical hard coat, preferably on both sides of the substrate, is necessary [28, 34].

13.2.2.3 High-Temperature Stability

High-temperature stability substrate films are desired for flexible electronics in order to increase the range of process steps accessible for device fabrication. Such steps can range from simple drying steps where higher temperature results in faster processing to TFT material deposition where higher temperature results in higher performance [35]. Specific processes have different temperature requirements and/or tradeoffs with temperature. However, a useful upper-limit target value is 350°C as this is currently a lower-limit processing temperature for conventional high-performance α-Si TFTs [35]. These films also need to have high dimensional stability with low level of shrinkage through thermal process cycles in order to enable precise registration, for example, in active-matrix displays. In fulfilling these requirements, care has to be taken to match the mechanical properties of the barrier coating to the underlying plastic film to as great an extent as possible. The barrier coatings discussed in this chapter all utilize inorganic coatings which have low thermal expansion coefficients relative to most plastic substrates. Hence, the barrier has to be designed to be robust to mismatches in thermal expansion to avoid cracking under thermal cycling conditions.

13.2.2.4 Surface Quality

The total thickness of the active layers of an OLED and most other organic electronic devices is on the order of 100 nm. Hence the surface morphology and roughness of the substrate need to be tightly controlled to avoid adversely affecting device integrity. In particular, the surface needs to be ultrasmooth (<1 nm rms roughness) and free of local surface anomalies such as spikes greater than a few tens of nanometers. Typical plastic substrate films do not meet this requirement [34, 36]. Thus, the barrier coating must be designed to "smooth over" the relatively rough morphology of the plastic film surface. In order to achieve this, most barrier coatings include some form of planarizing or smoothing layer.

13.3 Thin-Film Barrier Technology

13.3.1 Historical Background

Many applications, including food packaging [37, 38], pharmaceuticals [39], medical [40] and electronic devices [41, 42], require some sort of barrier to protect against the permeation of gases and vapors that are harmful to the application.

Barrier packaging can be divided into two categories: rigid and flexible. Rigid packaging utilizes metal or glass and has been around for centuries. Both these materials are still being used today to hermetically seal electronic devices. Flexible packaging also dates back almost to first humankind, in the shape of fruit skins, leaves, and animal skin, driven mostly by the need to preserve food. More recently, paper has been used as a packaging material for two centuries and plastics have been used for about a century [43]. Plastic films offer unique advantages compared to rigid and other flexible packaging forms in that they are thin, light-weight, cost-effective and lend themselves to much broader design-freedom depending on the requirements of the application (e.g., transparency for an OLED) [44]. The down-side, however, is that none of the plastic films available today are able to satisfy the barrier requirements outlined in the previous section.

It was soon realized in the mid-1900s that coating the plastic film with a very thin, single inorganic layer by means of vacuum deposition provided significant improvement in the barrier properties [45, 46]. At first, these films were based on metals such as Al but then, in the 1980s, due to the increased demand for visual access to food products as well as for compatibility of packaging with microwave use, transparent thin barrier coatings based on oxides such as SiO_x and AlO_x were developed. [46–48]. Oxide barrier films are now routinely coated onto plastic films such as PET to provide transparent barrier packaging. Typically, they provide a barrier to water permeation of 10^{-2} g/m^2/day [48].

Flexible electronic applications such as OLED devices require more than four orders of magnitude higher barrier performance than possible with single layer oxide films. In this section, the development of these advanced barrier films is described. One important and often ignored aspect in the development is how to reliably measure such low permeation rates. Hence, this section will start with a description of measurement methods. Following this, the basics and limitations of single layer oxide coatings will be described to provide a basis for the final discussion of today's state-of-the-art advanced barriers.

13.3.2 Permeation Measurement Techniques

The food and medical packaging industries typically measure oxygen and water vapor transmission rates (OTR and WVTR, respectively) using commercially available equipment such as that available from MOCON, Inc. [49, 50]. The principle of measurement for this type of equipment is illustrated in Fig. 13.3a for water vapor. The upstream side of the barrier film is saturated with a certain concentration of water and the downstream side is purged with a moisture-free carrier gas. As the water vapor permeates through the film, the carrier gas transports it to a sensor. Once steady state is reached, WVTR values are reported at a given temperature and relative humidity (RH) level. The sensitivity of this technique is limited primarily by edge leakage and sensor technology to values of 5×10^{-3} g/m^2/day for WVTR and 5×10^{-3} cm^3/m^2/day/atm for OTR at temperatures from 5 to about 50°C [50].

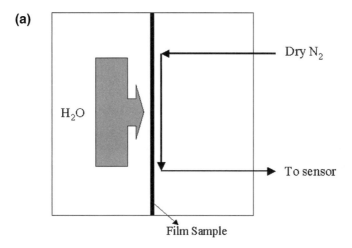

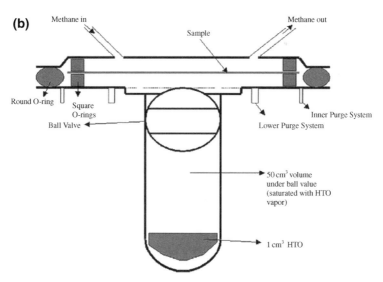

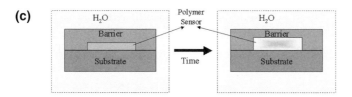

Fig. 13.3 (a) Schematic for water vapor transmission rate measurement system developed by MOCON, Inc., (b) Schematic for tritium water vapor transmission rate measurement system developed by General Atomic, (c) Schematic for water vapor transmission rate measurement method developed by National Institute of Standards and Technology (NIST)

Numerous techniques have been developed over the last few years with better sensitivity, most notably mass spectrometry [51, 52], measurements utilizing X-ray and neutron reflectivity [53], tritium diffusion test [54] and the calcium test [55–61] for water vapor permeation measurement, and residual gas analysis in ultra high vacuum [62] for oxygen permeation measurements.

The principle of tritium diffusion test is shown in Fig. 13.3b [54]. This technique follows the same basic principle as the MOCON test except that tritium-enriched water (HTO-enriched water) permeates through the barrier film and is carried to a beta counter by means of a dry methane gas. In the ionization chamber, radioactivity counts are converted to WVTR values once steady state is reached, taking into account the concentrations and geometry of the setup. In this case, since only HTO contributes to beta decay, the influence of water vapor contamination from other sources such as edge leakage is alleviated. Hence, a theoretical detection limit of 2×10^{-7} g/m^2/day is possible and so far a practical detection limit of 1×10^{-6} g/m^2/day has been reported [54]. This technique meets the barrier testing requirement for an OLED but suffers from two shortcomings: (1) only WVTR spatially averaged over the measurement area can be determined and (2) it produces radioactive waste contaminants.

Another technique introduced to study barrier films with ultralow permeation rates works by measuring film swelling using X-ray and neutron reflectivity measurements as a function of time at given temperature and humidity conditions [53]. The measurement utilizes an undercured polymer film sandwiched between the barrier and the base substrate as shown in Fig. 13.3c. As water vapor permeates through the barrier and reaches the polymer, swelling of the polymer is used as a means of calculating the amount of water vapor that has permeated. This technique, although shown to be a useful tool to understand the water vapor permeation mechanism through multilayer films [53], has not been taken up by the industry as a standard means of measurement, likely due to the need for expensive measurement apparatus and very specific processing required to prepare the polymer film for swelling studies.

The technique that is the most widely used today is the so-called "Calcium test", first introduced by Nisato et al. [55, 56]. The principle is simple: very thin, evaporated Ca metal is sealed between a perfect barrier such as glass and the film to be tested and serves as a well to collect the water vapor that permeates through the barrier film. The test relies on the high level of reactivity of Ca metal with H_2O to create transparent and nonconducting calcium hydroxide as depicted in Reaction (13.1).

$$Ca + 2H_2O \rightarrow Ca(OH)_2 + H_2 \qquad (13.1)$$

It should be noted that calcium also reacts with oxygen but the contribution of oxygen to the degradation of Ca has been shown to be negligible at room temperature [60, 61]. Typically, the rate of Reaction (13.1), and hence the water permeation rate, is monitored by measuring the change in visual appearance and/or optical density (OD) [55–59] or the electrical resistivity [60] of the calcium film as a function of time.

Although the principle of the calcium test is simple, the application can be problematic because the outcome is sensitive to the means of sample preparation, sealing material and data collection methods. In the following, we describe in detail two types of calcium tests for measuring water permeation: a "Total Rate" test which measures the rate over a selected area and a "Defect Imaging" test which is sensitive to local variations in the water permeation rate [58, 59]. The schematic of a calcium cell, the measurement system, and the cell geometry for the Total Rate WVTR case is shown in Fig. 13.4a. The calcium test cell is constructed by first evaporating a 50-nm-thick calcium layer on top of a cleaned, 50 mm × 75 mm glass slide. This glass slide is then encapsulated in a glove box (H_2O and O_2 levels below 1 ppm) with either another clean glass slide without Ca (the reference cell) or with the polymer/barrier substrate using a UV-curable adhesive as a perimeter

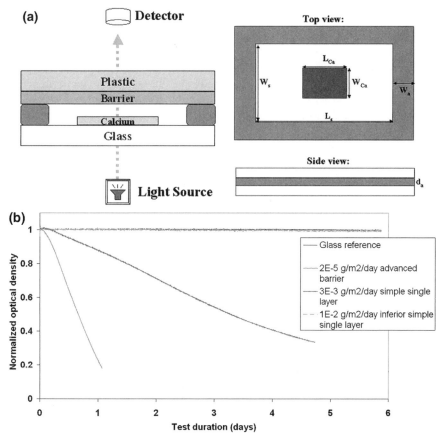

Fig. 13.4 Illustration of the "Total Rate" calcium test (**a**) cell structure and (**b**) measurement capability and standard detection limit

seal. By depositing the calcium onto the glass substrate and not onto the barrier, one is assured that the Ca will corrode uniformly over its area even if the barrier film has nonuniform properties across its area. Hence, the calcium corrosion rate measures the diffusion rate integrated over the barrier area. The Ca cell is placed between a LED light source and a photodetector and is kept in a temperature- and humidity-controlled chamber. As the water vapor permeates through the substrate and/or the edge perimeter seal, or is being released from any internal sources (e.g., residual moisture in adhesive), it reacts with the calcium inside the cell according to Reaction (13.1) above. As water vapor permeation progresses, calcium, corroding uniformly according to Reaction (13.1), becomes thinner and more transparent. The light transmittance (T) of calcium at 880-nm wavelength is measured continuously via the photodetector at an interval of about 1 Hz, and the average value is stored to a file every 5 minutes. As the OD of a film is proportional to its thickness, the WVTR is calculated via the following equation:

$$WVTR = -2A\frac{M[H_2O]}{M[Ca]}\rho_{Ca}\frac{L_{Ca} \cdot W_{CA}}{L_S \cdot W_S}\frac{d(OD)}{dt} \tag{13.1}$$

where A is the scaling factor between calcium thickness and OD; $M[H_2O]$ and $M[Ca]$ are the molar masses of H_2O and Ca with values of 18 and 40.1 amu, respectively; ρ_{Ca} is the density of calcium; L_{Ca} and W_{Ca} are the length and the width of the deposited Ca; L_s and W_s are the length and the width of the permeation area defined by the interior boundary of the edge perimeter seal (Fig. 13.4a), and d(OD)/dt is the slope of the measured optical absorbance.

In order to be able to measure substrate materials with WVTR values on the order of 10^{-6} g/m^2/day, the effective WVTR (WVTR$_{eff}$) from the perimeter edge seal needs to be either significantly below 10^{-6} g/m^2/day or it should have a very low variation so that it can be subtracted from Eq. (13.1) to obtain the true substrate WVTR. One can estimate the steady state WVTR$_{eff}$ for a reference cell with the geometry as shown in Fig. 13.4a with the following equation:

$$WVTR = \frac{P_a}{W_a \cdot L_a \cdot W_s} \cdot d_a \cdot (2L_s + 2W_s) \tag{13.2}$$

where P_a, W_a, L_a, and d_a are the permeability, width, length, and thickness of the adhesive respectively. Figure 13.4b shows OD versus time plots from a Total Rate calcium corrosion test for a single inorganic barrier layer and from an advanced barrier on polymer substrates and glass/glass reference cells. It is clear from this figure that the Total Rate calcium test is a reliable test to measure samples with WVTR in the range of 10^{-2} to 10^{-5} g/m^2/day. With improvements to the sealing technique and seal degassing procedures, it has also been demonstrated that measurements in the 10^{-6} g/m^2/day regime are possible at 23°C/50%RH [63].

One of the shortcomings of the Total Rate calcium test is that it is not possible to assess information about permeation through defects/pinholes. When such information is needed, the Defect Imaging test, originally developed by Nisato [55, 56, 59], is more appropriate. In one implementation of this method [59], ca. 100-nm-thick Ca is thermally evaporated directly onto the barrier coating of a $1'' \times 1''$ plastic substrate in a glove box. This structure is then encapsulated in a "solid-fill" fashion using a UV curable epoxy and a $1''$ diameter glass cover slip. The UV curable epoxy is a lower viscosity material than the one used for the Total Rate method and therefore it is possible to make a seal that is less than 10 μm thick, providing a much lower edge leakage rate of water vapor. An automated imaging system enables imaging and measuring the OD of 24 samples at a time placed on stages built to reduce subsequent handling of the samples. Once the samples are sealed, they are removed from the glove box, imaged and measured for initial OD, and then placed into an environmental chamber. These samples are then periodically imaged using the automated imaging system. As the water vapor permeates through the plastic substrate, mainly through the defects in the UHB and reaches the Ca, it starts reacting with Ca in local regions forming $Ca(OH)_2$ and these local reaction areas expand laterally as a function of time. The schematic of this complete process is shown in Fig. 13.5. The detection limit for this method is more than 1,000 hours at $60°C/90\%RH$. Using this method, it is possible to obtain information on defect related permeation as well as information on effective WVTR through the UHB by analyzing the area fraction of $Ca(OH)_2$ formation as a function of time [56].

With all these techniques facilitating the measurement of barrier performance, the field of barrier films is still lacking a standard technique or procedure to enable comparison of barrier performance for different types of barrier films produced by different institutions. Calcium test is the most common method; however, one-to-one comparison of measurements from variations of calcium test is at best satisfactory, without a common standard. Better understanding will stem from a good understanding of permeation mechanisms for different barrier structures and the next section presents a review of such studies.

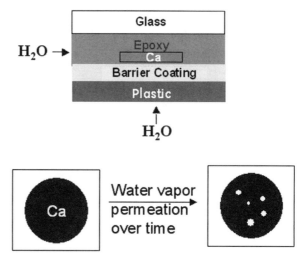

Fig. 13.5 Illustration of the "Defect Imaging" calcium test

13.3.3 Permeation Through Thin-Film Barriers

An extensive amount of research has been put into understanding permeation through single layer barrier films and this forms the basis for more recent developments in advanced barrier structures for flexible electronic applications. This section first addresses key challenges and findings from work on single layer barriers and then discusses advanced barriers and fabrication methods.

13.3.3.1 Simple Single Layer Barrier Films

Single layer thin-film barrier materials are chosen from oxides (SiO_x, AlO_x), nitrides (SiN_x), or oxynitrides (AlO_xN_y), which, in bulk structures, are impermeable to oxygen and water vapor at low temperatures, even at thicknesses on the order of 10 nm [64, 65]. However, when formed on polymer substrates using the most common vacuum deposition techniques such as sputtering, evaporation, and plasma enhanced chemical vapor deposition (PECVD), the barrier properties of the overall composite are found to be limited to around 10^{-2} g/m^2/day for WVTR and 10^{-2} cm^3/m^2/day/atm for OTR [46–48]. Although these values represent two to three orders of magnitude improvement over the OTR and WVTR of an uncoated polymer, they are still many orders of magnitude higher than what is possible with a bulk oxide with the similar thickness [64]. The thickness of the barrier film, defects in the shape of pinholes or cracks formed during deposition, coating density, morphology, and composition are all responsible for this discrepancy in barrier performance between a bulk oxide and a thin-film oxide of similar thickness.

Permeation of gases and vapors through polymer/barrier systems as a function of barrier thickness has been well documented [46, 47, 66, 67] and the relationship is as shown in Fig. 13.6. Three regimes are identified. Permeation rate initially decreases with coating thickness until a critical thickness is achieved (d_c, limit of regime I). Such behavior is linked to the fraction of polymer surface covered by the coating during the initial stages of deposition [66, 67]. Once complete coverage is attained, the permeation rate mostly remains unchanged in regime II with a rate that is much higher than the rate for bulk material. As the thickness is increased further into

Fig. 13.6 Simplified illustration of the change in gas transmission rate of a barrier coated plastic as a function of increasing barrier coating thickness

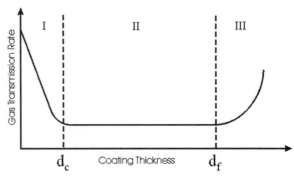

regime III ($> d_f$), the barrier performance degrades mostly due to the internal stress build-up in the glassy coating which leads to the formation of cracks [67, 68]. The determination of the critical thickness (d_c, onset of regime II) is the crucial first step in designing barrier films due to the fact that optimal barrier properties need to be satisfied while avoiding commercially unacceptable processing times (thick films). The value for d_c changes with coating material, deposition method, and to a larger extent the substrate surface [48, 69]. The reported values on polymers are 40–50 nm for PECVD SiO_x [70, 71], ~20 nm for EB-evaporated AlO_x [72], ~20–30 nm for sputtered AlO_x [72, 73], 8–30 nm for PECVD SiN_x coatings [71, 74], ~100 nm for ETP SiN [75], and <25 nm for ALD AlO_x [76, 77].

For the purposes of this chapter, the main interest is in transparent thin-film barriers with thicknesses in regime II. In this regime, thin-film oxides and oxynitrides grown at low substrate temperatures (usually below 100°C) have varying degrees of porosity [69, 78–80]. The permeation channels that create this porosity and hence contribute to varying degrees of barrier performance can be classified as macrochannels (>5 μm), microchannels (1–5 μm), Nanochannels (<1 μm), and the oxide lattice (<0.3 nm). The extent to which these channels are created and influence barrier performance depends on several factors, including handling and cleanliness of the deposition environment, substrate surface roughness and morphology, and deposition technique and process parameters.

Handling and cleanliness (or lack thereof) of the deposition environment is mostly known to contribute to the occurrence of macro- and microchannels. A good example of this was shown by Jamiesson and Windle [45] for their PET/Al barriers where they found that the density and size of pinholes correlate well with the density and size (1–5 μm) of atmospheric dust particles on the polymer surface. In addition, the number density (mm^{-2}) of the channels created correlate well with OTR values (<10 mm^{-2} for ~0.1 cm^3/m^2/day/atm and ~100 mm^{-2} for ~1.3 cm^3/m^2/day/atm). This is a clear indication that proper handling of the barrier coated substrate, implementing effective substrate cleaning procedures, and controlling the particle counts in the deposition environment by means of utilizing "cleanrooms" are the first steps to eliminating macro and some of the microchannels in the barrier film.

High substrate surface roughness is another key factor that adversely affects barrier performance as demonstrated by many researchers [81–83]. Especially for the most widely used commodity polymers such as PET and oriented polypropylene (OPP), it has been shown that the surface may appear very smooth (rms ~ 1–5 nm) when analyzed in μm^2 lateral scale (by AFM), but when studied on a broader lateral range, this roughness can reach tens to hundreds of nanometers [81]. Similar roughness can be created by antiblock, filler particles (i.e., silica) of 1–2 μm diameter [83]. Shaw [84] first reported the application of a flash-evaporated, UV or E-beam curable smoothing coating on such surfaces and through systematic AFM analysis showed that surface anomalies were covered by increasing the thickness of this layer to a few microns [85]. It was further shown that the application of this layer prior to an Al or AlO_x barrier coating improved both the OTR and the WVTR of the composite by more than three orders of magnitude compared to an uncoated polymer [84, 86, 87]. More recently, it has been demonstrated that WVTR values in the range of

10^{-4} to 10^{-5} g/m^2/day, as well as a dramatic reduction in microscale defects, can be achieved by appropriate modification of the polymer surface with a smoothing layer applied either via flash evaporation or simple nonvacuum techniques such as spin coating [88].

Assuming that the deposition environment is kept clean and proper film handling procedures are implemented and if applicable, a smoothing coating is applied, the majority of macro- and microchannels for permeation can be eliminated. Many studies utilizing either analytical characterization [71, 72, 79, 80, 89, 90] or the activated rate theory [70, 79, 90, 91] and several physical "defect" models [45, 69, 78, 92–96] have demonstrated that especially oxygen permeation is dominated by these type of defects. On the other hand, water vapor permeation through polymers coated with SiO$_x$ [70, 80, 97], AlO$_x$ [72, 91, 98, 99] AlO$_x$N$_y$ [99, 100] exhibits less linear dependence on the size and density of larger defects and at times has been shown to be controlled by the nanoscale morphology and composition of the coating (due to its polar nature, water molecules can chemically interact with the coating depending on the material chosen). The remaining engineering challenge then becomes producing a dense thin film as close as possible in structure to its bulk counterpart with good control of composition. The optimal solution depends greatly on the material system chosen, deposition method, and process parameters.

Denser films can be fabricated by supplying high kinetic energy species (atoms of the material to be deposited, ions, neutrals) to the surface. Hence, thermal and E-beam evaporation techniques fall short of producing high-quality barriers, despite exceptionally high deposition rates. Placing an auxiliary plasma between the source and the substrate can improve barrier performance by densifying the films [101]. Magnetron sputtering, another widely used physical vapor deposition technique, is a more preferred and versatile technique to densify films. Since with this technique, there is a dense plasma confined to the cathode surface, acceleration of high energy species to the substrate surface is enhanced [82, 90]. In addition, manipulation of the magnet configuration to leak the plasma toward the substrate, using more advanced power supply schemes such as pulsed DC, has been shown to enable highly densely packed film morphology, which especially aids in improving water vapor barrier performance [102, 103]. Examples have been reported for AlO$_x$ and AlO$_x$N$_y$ on PET [82, 90, 99, 100, 102, 103].

One shortcoming of physical vapor deposition techniques is that they suffer from "line of sight" film deposition, which makes the barrier films very sensitive to even small anomalies on the surface that can create defects. PECVD and ALD are techniques with much reduced directionality of deposition and have been shown to produce superior barrier films for SiN [71, 79, 89, 104] and AlO$_x$ [76, 77], respectively. For PECVD, when the substrate is placed on the powered electrode (RIE mode), additional ion bombardment produces dense SiN$_x$ films with WVTR levels in the 10^{-4} regime. At the same time, compressive stresses can build up in the film due to this high energy process and these may need to be controlled depending on the application [105]. ALD, on the other hand, is perfect for producing highly conformal films and recent developments using plasma-enhanced ALD have produced single layer AlO$_x$ films in the $10^{-4}-10^{-5}$ g/m^2/day regime [76, 77]. Significantly, slower deposition rates for ALD compared to other techniques discussed here, however, make it commercially unattractive for the flexible electronics application.

The materials system and its effect on the permeation of water vapor is a final key aspect to consider. Tropha and Harvey [70] provided the first insight into the mechanism of water vapor permeation by using activated rate theory to calculate an activation energy for H_2O permeation. They deposited a series of SiO_x coatings on PET, polycarbonate(PC), polypropylene (PP), and polystyrene (PS) substrates using PECVD. They show that regardless of the water vapor activation energy (E_A(WVTR)) for permeation through the four uncoated substrates, the activation energy was always the same when each substrate was coated with SiO_x. This is significant because if the mechanism for water permeation were simply diffusion through the constricted pores of the SiO_x coating, then the temperature dependence of permeation should be controlled by the temperature dependence of permeation through the plastic film only. The fact that it is controlled by an activation energy which is characteristic of the coating suggests that some sort of chemical interaction occurs between the water and the coating. Suggested interactions include attachment of water molecules to Si-O or Si-dangling bonds within the SiO_x pores [80]. Other studies on AlO_x and AlO_xN_y coatings have suggested that the level of interaction is coating-material-dependent as might be expected for a chemical process [82, 98–100].

As described in this section, considerable effort has been undertaken to develop simple single layer barrier coatings, but their performance is not adequate for the most demanding flexible electronics applications. The barrier properties are primarily limited by the presence of macro and micro defects, but so far a practical way to eliminate these defects has not been demonstrated. As a result, the technical community began to develop advanced barrier films that can provide a good barrier in spite of the presence of defects. These are described further in the next section.

13.3.3.2 Advanced Barrier Films

The first approach designed to provide a high barrier in spite of the presence of film defects is illustrated in Fig. 13.7. The idea is to have multilayers of inorganic

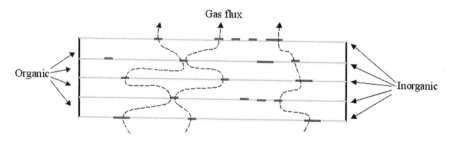

Fig. 13.7 Concept of "defect de-coupling" gas permeation mechanism through multilayer advanced barrier coatings

coatings that are separated by organic coatings [25, 27, 42, 86, 105–110]. The organic coatings themselves have inferior barrier properties to the inorganic coatings, but their purpose is to decouple the growth of a defect from one inorganic coating to the other. In the ideal case, there is then no defect that spans the whole thickness of the coating. Hence, water and oxygen molecules must follow a tortuous path through the defects to get through the whole coating and this presumably leads to a lower permeation rate. More recently, complex single layer coatings which achieve the same kind of decoupling defects through the thickness of the coating have been developed [26, 28, 58, 59]. These achieve the decoupling by grading the composition of the single layer coating in the thickness direction. Note that for both the multilayer and the complex "multi-zone" single layer coatings, the lessons learned in trying to improve simple single layers such as clean deposition environments, proper handling, smoothing the substrate surface, and the choice of suitable material and deposition systems are all still critical to success. In this section, we discuss the approaches introduced by Batelle/Vitex [25, 27, 42, 86, 106], GE [26, 28], and Dow Corning [105, 108] for achieving advanced barriers with an emphasis on coating structure and properties, deposition techniques, and application to batch and/or R2R processing.

Multilayer inorganic–organic films were first developed for capacitors at GE in the early 1980s to realize a significant reduction in pinhole density in Al [84, 86, 111]. This method of depositing alternating layers of organic and inorganic layers on polymer substrates has since been applied to barrier technology by Batelle and Vitex and has been shown to dramatically improve barrier quality compared to simple single layer barrier coatings [25, 27, 42, 86, 106]. The organic layer is composed of a polyacrylate (PA) material and the inorganic "barrier" layer is chosen from evaporated or sputtered Al or AlO_x. PA is deposited using polymer multilayer deposition (PML) [84, 86]. In PML, a desired choice of acrylic monomer fluid is first degassed and then sprayed into an evaporator where it is converted into a molecular vapour (flash evaporation). The evaporator is hot enough to vaporize the liquid but not sufficiently hot to crack or polymerize it. The vapors are then drawn from the evaporator into the lower pressure vacuum chamber and cryocondense on the substrate as a continuous liquid film. This film is subsequently irradiated by an electron beam or a UV source to crosslink and form a smooth and uniform polymer film. The assembly for PML deposition can be inline or incorporated into a conventional vacuum web coater, which can also contain sputtering facilities, thereby facilitating the deposition of alternating layers of PA and AlO_x [86]. The first organic layer, up to several microns, serves as a high-quality smoothing layer, able to coat over surface features larger than 1.5 μm. The subsequent inorganic (AlO_x in most cases) therefore already contains a low density of defects. This full structure in some cases is capped off with another PA topcoat, which serves to mechanically protect the underlying fragile layers [84, 86, 112]. It has also been shown that this polymeric topcoat sometimes improves barrier performance. This has been attributed to possible chemical interactions of this material with the walls of pores in barrier coatings, thereby providing a synergistic effect [113, 114]. Water vapor transmission in the low 10^{-6} g/m^2/day has been reported for 3–5 PA–AlO_x pairs (called "dyads") on polymer substrates using Calcium tests [24, 27, 42, 55, 56] and OLED shelf lifetime tests [106]. However, Kapoor et al. [115] found that processes developed in batch tools were not directly transferable to a R2R coating process. The barrier structure (inorganic barrier layer thickness, number of dyads) was reoptimized, modifications to the R2R coater and the R2R coating processes were implemented, and choice of substrate material was reconsidered in order to minimize defects, improve barrier performance, and improve coating uniformity. Similar considerations for batch to R2R transition are applicable to all types of advanced barriers.

An alternative to the multilayer approach is the graded single layer approach developed by GE. Here, the composition is graded between inorganic-like materials and organic-like materials in a coating less than 1 μm thick using a parallel plate capacitively coupled PECVD reactor [26, 28]. This was first developed in a batch mode process capable of fabricating a complete OLED barrier substrate package based on an experimental high heat Lexan® PC film which has a T_g around 240°C. A schematic diagram of the batch process to make a functional substrate package is shown in Fig. 13.8. The process starts with cutting and heat treating a film sheet at elevated temperature to improve its dimension stability, and then mounting it onto an aluminum frame [116]. The film/frame assembly is then cleaned with high pressure DI water spray and isopropanol rinse, followed by baking in vacuum oven to dry out moisture in the film. The next process step for this dried film/frame assembly is spin coating an epoxy-based smoothing layer. This unique spin-on smoothing layer provided an average surface roughness of 0.6 nm over 200 μm × 250 μm area. The surface morphology of this layer, which is clearly superior to that of an

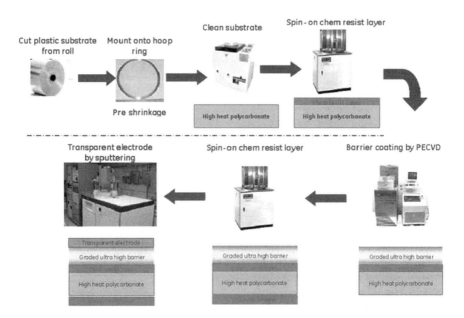

Fig. 13.8 Schematic of the plastic substrate fabrication processes in batch mode

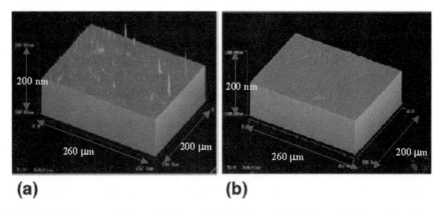

Fig. 13.9 Surface morphology of (**a**) bare high heat Lexan® PC film and (**b**) smoothing layer coated high heat Lexan® PC film. The surface morphology was studied using optical profilometry

uncoated plastic substrate (Fig. 13.9), greatly reduces the number of defects in the barrier coating. This concept is believed to be responsible for the several orders of magnitude improvement in WVTR demonstrated in Fig. 13.10 for GE's batch-mode graded barrier, when deposited on the smoothing layer rather than on a bare plastic substrate.

The graded barrier was coated with plasma enhanced chemical vapor deposition (PECVD) and the barrier consists of SiN_x or SiO_xN_y inorganic and SiO_xC_y organic zones without a discrete interface. In this barrier structure, the organic materials effectively decouple defects growing in the thickness direction but, instead of

Fig. 13.10 WVTR for GE's graded barrier coating on bare high heat Lexan® PC film and on smoothing layer coated high heat Lexan® PC film

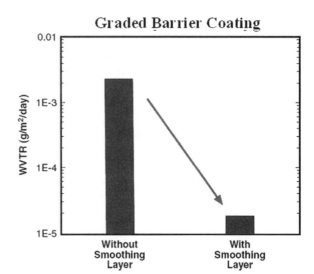

having a sharp interface between inorganic and organic materials, there are "transitional" zones where the coating composition varies continuously from inorganic to organic and vice versa. These "transitional" zones bridge inorganic and organic materials which results in a single layer structure with improved mechanical stability and stress relaxation. The inorganic process utilizes a combination of silane, ammonia, and on occasion, oxygen gases to create a material composition ranging between SiN_x and SiO_x. The organic process includes a combination of a Si-containing organic precursor and Ar gases to create a Si-containing organic material. The inorganic and the organic processes are tailored such that the resulting materials have hardness (inorganic material: 10–15 GPa, organic material: 1 GPa) and elastic modulus (inorganic material: 50–100 GPa, organic material: 10 GPa) similar to those of glass-like materials and thermoplastics, respectively. The graded barrier structure is obtained by gradually mixing the inorganic and the organic processes. At constant pressure and radio frequency (RF) power, each mass flow controller (MFC) for each individual process gas is programmed to achieve continuous compositional changes, while the plasma remains on, in order to achieve a gradual change in the coating composition from inorganic to organic materials and vice versa. The thickness of the "transitional" zone is determined by the time to change the precursor gas composition from the inorganic process to the organic process and vice versa. Bulk WVTR values as low as 4×10^{-6} g/m^2/day at 23°C/50%RH have been demonstrated with this UHB structure on high heat PC substrate with an epoxy smoothing layer [58, 63]. Using the "defect imaging" type Calcium test described in Section 13.3.2, barrier performance as good as glass encapsulation has also been demonstrated at accelerated conditions with occasional defect-related corrosion, mostly due to occasional presence of particles or other surface anomalies. In addition to good barrier performance, the continuous transition from inorganic to organic zones and vice versa ensures excellent adhesion and the ability to tune the refractive indices of the inorganic and organic zones by varying PECVD conditions. The latter capability enables optical tuning without the need to optimize zone thicknesses. For PC substrates, it is necessary to add a chemical resistance coating to the backside of the substrate. This is accomplished by taking the film/frame assembly back to the spin coater to coat the backside of film with same smoothing layer as on the frontside. Finally, an ITO coating is put down on top of the barrier using DC magnetron sputtering. As described in Section 13.4, this complete substrate package is compatible with typical OLED fabrication processes.

GE has also reported successful transition of this batch process to a R2R type process using a prototype R2R PECVD reactor (Fig. 3.11). The major components of the reactor system are the web drive system, unique deposition stations, and the PECVD electrodes. The web drive system is capable of handling an 8″-wide plastic web with highly precise web speed and tension control. To minimize surface damage, the web drive components are designed not to touch the plastic surface where the coating will be deposited. In addition, a bottom-up deposition process, in which the coating is deposited on the plastic web from the bottom electrode, prevents particles from falling onto the web surface, minimizing the possibility of creating defects in the barrier coating. This proof-of-concept R2R reactor has two

Fig. 13.11 GE's
proof-of-concept R2R
PECVD coater for graded
barrier coating on plastic film

substations in one deposition chamber separated by a baffle which has an adjustable opening that allows controlled mixing of reactive gases and thereby controlled grading in the barrier structure. A single pass of web through deposition stations can lay down one inorganic zone and one organic zone along with a transition zone between them. A graded single layer barrier with more zones can be constructed by continuously passing the plastic web back and forth between the two substations.

Using this reactor, the feasibility of both realizing a graded coating structure and achieving high barrier performance using continuous R2R processing has been demonstrated [117]. To date, WVTR values of $\sim 1 \times 10^{-5}$ g/m^2/day have been demonstrated for a three-zone R2R graded barrier coated PET film. Figure 3.12 shows a representative cross section TEM image for a three-zone R2R barrier coated Si sample that was taped to plastic web. This clearly reveals a continuous contrast change between an inorganic SiO_xN_y zone and an organic SiO_xC_y zone. Grading was further evidenced by Energy Dispersive Spectroscopy (EDS) study. EDS was used to study the composition at five locations throughout the thickness of R2R three-zone barrier (Fig. 13.12). One can see that the carbon/oxygen (C/O) ratio is low in the SiO_xN_y-rich zone (Location 1), and increases continuously when moving through Location (2) to a maximum at Location (3), which is the SiO_xC_y-rich zone. The ratio then decreases continuously from Location (3) to Location (5). From Fig. 13.11, it is obvious that a continuous composition change between inorganic SiO_xN_y and organic SiO_xC_y has been achieved. With this structure, GE recently reported a water vapor barrier performance equivalent to that achieved with a batch mode process using both types of Ca test described in Section 13.3.2 [117].

A slightly different PECVD approach has been developed by Dow Corning. They have varied the process conditions to obtain a SiOC stress buffer layer on top of which they deposit SiC barrier films to form a low stress/high stress coating pair [105, 108]. They reported 10^{-4} g/m^2/day for multilayers of these high and low stress films made in batch mode on PET substrates. In order to transition to R2R processing, Dow Corning built a pilot scale R2R PECVD reactor using a high-density Penning Discharge Plasma (PDP) source, which is capable of coating PET films with a

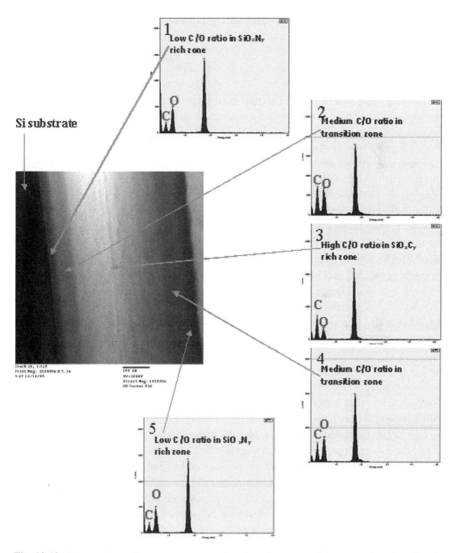

Fig. 13.12 Energy dispersive spectroscopy study of coating compositions at various locations in graded barrier coating. Locations 1 and 5 are SiO_xN_y zones, Locations 2 and 4 are transition zones, and Location 3 is SiO_xC_y zone

SiC:H barrier coating. The PDP source is a magnetically confined plasma source in which the configuration of cathode and anode, in relation to the magnetic and electric field, is similar to a Penning Cell [118]. In Dow Corning's R2R PECVD reactor, the PDP sources are substantially covered by the plastic web, which substantially reduces coating build-up on the electrodes and eliminates the need for frequent etchback plasma clean and other maintenance. Dow Corning's SiC:H coating is based on a graded stack consisting of a stress buffer layer followed by a barrier layer [105].

The stress buffer layer is used to mitigate the high mechanical stress of the barrier layer to the interface with the substrate. The SiC:H coating could be put down at deposition rate above 125 nm·m/min and the R2R coated SiC:H barrier achieved a WVTR below the 10^{-3} g/m^2/day MOCON detection limit [119].

At present, the use of a combination of inorganic and organic materials remains the primary approach to achieving high barrier performance. The details of the permeation mechanism of water vapor through these structures are not well understood. Several modeling approaches [75, 120] and a survey of Calcium test data of various groups suggest that the mechanism of "defect de-coupling" mechanism is dominant. Since the majority of advanced barriers contain either AlO$_x$ or SiN$_x$ as the inorganic component, the two inorganic films shown to be the least reactive with permeating water molecules (see Section 13.3.3.1), it is plausible that this mechanism is dominant. However, recently the question has arisen as to whether the reported values of WVTR (mostly) by Ca test represent the true steady state permeation rate. In particular, Graff and coworkers [25, 121] developed a model for multilayer films based on Fickian mechanics and used it to argue that the structure in Fig. 13.7 promotes a significant increase in lag times for water vapor to "fill up" the structure, but only a modest decrease in the actual steady state transmission rate (at best from 10^{-2} to 10^{-3} for structures with the least number of defects. In other words, it is argued that the improvements in barrier performance shown for advanced barrier films over single layer ones are due to an extended transient period because of the highly increased diffusion path (Fig. 13.7). This argument is supported by Vogt et al. [53] in their X-ray and neutron reflectivity measurements of such multilayer structures (using an undercured PA in dyads to ensure swelling – see Section 13.3.2). They conclude that water accumulation at the oxide–polymer interface, characterized as strong adsorption of water to AlO$_x$, creates a desiccant effect and this effect is responsible for the long transient "fill-up" times they observe for such structures. While these studies provide useful insights to better understand the water vapor permeation mechanisms, it should be noted that for similar barrier structures, steady state, as opposed to transient, consumption of Ca has been reported [55–57, 59] leading to WVTR values of around 10^{-6} g/m^2/day. In addition, for PECVD graded barrier films, tritium diffusion tests where lag times are clearly discernable from steady state behavior, WVTR values of 10^{-5} g/m^2/day have been reported after a transient period of several hundred hours [54]. In general, the complexity of water vapor permeation compared to other nonreactive permeants makes it difficult to pin down one exact mechanism for moisture permeation at present and, as is the case for simple single layer films (see Section 13.3.3.1), the materials used to construct UHB structures can play a significant role in how water vapor permeates.

As clear from the discussion above, there have been significant strides in advanced barrier films with barrier performance necessary for OLED packaging and in developing R2R processes to make the films commercially viable. These efforts have also been accompanied by developments in integrating OLEDs onto polymer/barrier structures and integrating UHB coatings as a top encapsulation for OLEDs. The final section discusses the challenges and developments in these two critical device–barrier integration areas.

13.4 Barrier–Device Integration

13.4.1 Substrate and Barrier Compatibility with OLEDs

The unique thin device structure of OLEDs has made them a natural choice for flexible display applications. Flexible OLED display prototypes have already been demonstrated on barrier-coated plastic substrates. A group from Pioneer Corporation built a 3″ full-color passive-matrix OLED device ($160 \times RGB \times 120$ pixels) on a single layer SiO_xN_y barrier coated plastic substrate [122]. The first proof-of-concept of a flexible OLED on an advanced barrier coating was a 128×64 passive matrix (PM) monochrome display fabricated on a multilayer barrier coated heat stabilized PET made by Vitex [24]. Quantitative tests of the compatibility of the Vitex multilayer barrier as both a substrate and a top encapsulation layer for an OLED have also been performed. In particular, Weaver et al. [106] built an OLED test device on PML barrier coated heat stabilized PET with glass lid encapsulation and achieved a half life of 3,800 hours from an initial luminance of 425 cd/m^2 measured at room temperature environment. Chwang et al. [123] demonstrated a 5-mm^2 OLED test pixel and a passive-matrix flexible OLED device on a multilayer barrier coated heat stabilized PET substrate with transmissive multilayer barrier coating direct encapsulation and reached a room-temperature half life (initial luminance 600 cd/m^2) for a test pixel of 2,500 hours.

In the following, the details of experiments performed to ensure the compatibility of GE's graded barrier with OLED devices will be described. Compatibility is best demonstrated by actual device fabrication since many unanticipated issues such as the presence of particles, handling methods, resistance to a certain set of solvents, stability during high-temperature steps, the effect of mechanical stress due to sealing, performance of the transparent conductor, and shelf life tend to surface at this stage. In order to test for these, two-layer polymer OLED devices consisting of PEDOT/PSS [poly(ethylenedioxythiopene)/poly(sulfonated styrene)] as a hole transport layer and a polyfluorene-based light-emitting polymer (LEP) layer were fabricated onto an ITO and graded barrier-coated high temperature PC substrate [28]. A series-connected architecture described elsewhere [124, 125] was employed to enable a large emitting area. In particular, 10 devices consisting of eight series-connected elements, each with area 1.95 cm^2, were fabricated on a 15×15 cm^2 substrate. Fabrication was accomplished using the fairly standard procedure described below.

In order to enable device fabrication using conventional batch processing tools, the substrate films were first affixed to 3 mm wide, 15×15 cm square titanium frames by means of double-sided Kapton polyimide tape. The affixed film was cleaned with isopropanol, then given a 10-minute dwell in an aqueous detergent (Alconox) with ultrasonic agitation, followed by water rinsing and drying.

ITO patterning was accomplished using a photoresist and immersion etching. The photoresist (AZ1512) was applied by means of spin coating and baked for 10 min at 110°C, producing a film 1 micron thick. The positive photoresist was imaged through a glass mask using a UV collimated light source for 15 sec. The

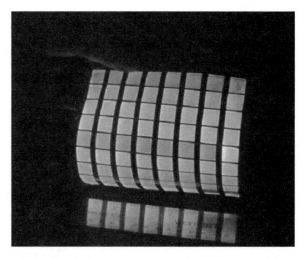

Fig. 13.13 15-cm square flexible OLED on high heat Lexan® PC substrate with graded barrier

image was developed at room temperature for 1 min using OCG 809 (2:1). Following a water rinse and dry step, the exposed ITO was removed by immersion for 5 min in a 45°C solution of 10/10/1 hydrochloric acid/water/nitric acid. The sample was rinsed with water, dried, and the photoresist removed by immersion in acetone. Residual materials were removed by a 10-min dwell in an aqueous detergent (Alconox) with ultrasonic agitation, followed by water rinsing and drying.

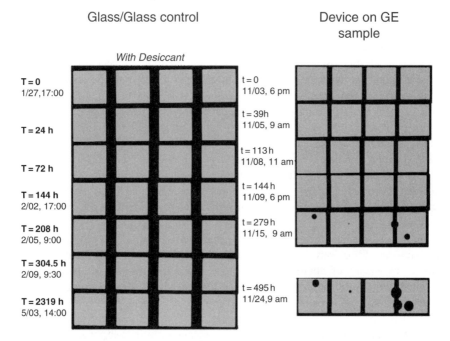

Fig. 13.14 Accelerated shelf lifetime test (60°C/85%RH) of UDC OLED test pixels on (**a**) glass substrate and (**b**) GE graded barrier coated high heat Lexan® PC. All test pixels were encapsulated with glass lid using epoxy edge seal with desiccant incorporated

The active device layers were deposited by means of spin coating. In particular, an aqueous coating of PEDOT/PSS (Baytron P VP CH 8000 from H.C. Starck) was spin coated and baked at 110°C to produce a film of 50 nm thickness. The LEP layer was fabricated using a solution of LUMATION* 1304 (*Trademark of The Dow Chemical Company) in xylene. After spin coating and a 10-min bake on a hot plate, a 70-nm-thick LEP layer was produced. A bilayer cathode made up of NaF and Al was then evaporated onto the films through a stainless steel mask. Finally, the devices were encapsulated with a 100-micron-thick foil of aluminum coated with acrylic adhesive inside a glove box. The samples were then removed from the glove box for testing.

Figure 13.13 shows a picture of the resulting substrate being bent with all devices turned on at a brightness of ~1,000 cd/m². At this brightness, the efficiency, as measured in an integrating sphere, is 8 cd/A. This is comparable to that expected for this LEP material [126]. It thus appears that the PC/graded barrier substrate package is compatible with a polymer-based OLED device fabrication process.

The compatibility of GE's graded barrier coated high heat PC film with OLED fabrication processes was further confirmed by Universal Display Corporation (UDC) for small molecule OLED devices. Small molecule OLED test pixels were first fabricated on GE's graded barrier sample using UDC's standard OLED fabrication processes and then encapsulated with a glass lid using epoxy edge seal with desiccant incorporated inside the test device. For comparison, glass/glass test devices were also fabricated using the same methods and tested side-by-side with devices on the graded barrier PC. Figure 13.14 shows the images of a test device on the barrier film and glass/glass control devices after undergoing accelerated shelf lifetime test at 60°C/85%RH. At 500 hours, though there were some black spots

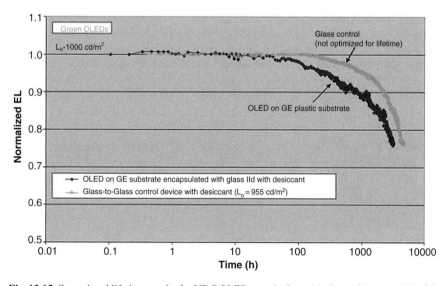

Fig. 13.15 Operational lifetime results for UDC OLED test pixels on (a) glass substrate and (b) GE graded barrier coated high heat Lexan® PC. All test pixels were encapsulated with glass lid using epoxy edge seal with desiccant incorporated. Test was carried out at room temperature and 50% RH

forming, most pixel areas on the graded barrier substrate were still intact. The room temperature operational lifetime test results were illustrated in Fig. 13.15 and it was estimated that the OLED device on the barrier film sample has a half life 80% of that of glass/glass control.

These test results show that barrier coated plastic films can potentially be used in place of a glass substrate for both polymer and small molecule OLED devices.

13.4.2 Thin-Film Encapsulation

Another technology for realizing flexible OLEDs and other organic electronic devices is thin-film encapsulation (TFE) (Fig. 13.1, scheme D). Currently, commercial OLED displays are encapsulated with a thin metal or a glass cap sealed to the base OLED substrate with an epoxy seal [20, 127, 128]. In addition, getter materials such as CaO or BaO are used inside this package to prolong the shelf lifetime of the device [20, 128, 129]. This method of encapsulation limits the ability to produce a fully flexible device. In addition, it is a slow and high-cost process and the standard seal thickness is microns thick and therefore cannot provide adequate barrier without the addition of a desiccant. It is the combination of these disadvantages of current encapsulation technologies that have led researchers and companies to put effort into developing TFE methods that can be commercially viable. Reduction of edge seal thickness, reduction in processing tact time, and thereby reduction in cost along with the potential for enabling full flexibility are the core advantages of this technology. On the other hand, development of TFEs presents unique challenges compared to those outlined for plastic substrates. The substrate is now the OLED structure itself, which is an inorganic/organic multilayer structure with multiple terminating surfaces depending on the OLED architecture (metal cathode, organic layer, contacts, etc.). The barrier structure used for encapsulation therefore needs to be compatible with the underlying OLED and to exhibit low stress and good adhesion. Especially for passive-matrix OLED displays, where cathode partition structures can be microns high and have negative slopes, conformality of the TFE layer is important to ensure continuous coverage of all active areas. Avoiding physical or chemical damage to the OLED during deposition is also crucial since most active organic materials tend to degrade at high processing temperatures and when exposed to high energy species. Finally, an important aspect not generally noted is the ability to properly encapsulate contact leads to the outside of the package to avoid any premature device failure. All these requirements should be fulfilled while maintaining excellent barrier properties for TFE and maintaining electro-optical function of the OLED after prolonged shelf life tests.

Studies to date have explored different TFE deposition methods with different structures on a variety of OLED architectures for both solution-processed polymer and evaporated small molecule devices [105, 123, 130–142]. There are two basic OLED device architectures: "bottom-emitting" and "top-emitting". In the "bottom-emitting" configuration, active organic layers are deposited on top of a transparent substrate with a transparent conductive contact and then capped with a nontransparent low work function metal cathode, such as Ca capped with Al. In the "top-emitting" configuration, the active organic layers are deposited on top of a opaque or reflective substrate with a metallic contact and then capped with a transparent conductive electrode. OLED display devices can be further subdivided into passive-matrix (PMOLED) or active-matrix (AMOLED) depending on how the individual pixels are driven. Combinations of these architectures can be fabricated on glass, metal foil, or plastic substrates.

On glass substrates, most TFE studies have been reported on simple test pixels for bottom-emitting small molecule devices. Here, the PML-type multilayer approach [27, 130] and the PECVD graded single layer approach [131] demonstrate the longest shelf lifetimes with less than 10% pixel degradation at 60°C/90%RH after 500 hours for the former and less than 10% pixel degradation at the same conditions after 1,000 hours for the latter. The ALD method of Ghosh [136] is also promising, with a report of negligible degradation after 1,000 hours at 85°C/85%RH for top-emission small molecule OLEDs.

For passive-matrix display applications, promising results are achieved so far by the deposition of a thick organic layer first to cover the cathode partition topology and subsequent multilayer film deposition [27, 130]. However, recently a very conformal PECVD SiN process coupled with either SiO$_x$ or a topcoat [133, 134] with ink-jet printed organic layers to make up a multilayer barrier [135] has been demonstrated to be able to cover the cathode partition type of topology. In this case, shelf lifetimes of 500 hours with some dark spot formation and edge leakage were reported at accelerated environmental conditions. The primary cause of degradation in these studies is cited as relating to large particles of micron size, which means there is still more work to be done to enable commercial viability. In addition, as a general trend, the operational lifetimes of encapsulated OLEDs are inferior to their glass encapsulated counterparts. This degradation is, in most cases, due to dominance of fast-growing dark spots caused by moisture ingress through defects in the barrier. Weidner et al. [105], however, showed in an experiment where they glass encapsulated PECVD TFE-coated OLED and compared to simple glass-encapsulated OLED, operational lifetime performance of the OLED/TFE was similar and in cases better than glass-encapsulated samples. This result indicates that PECVD TFE does not adversely affect the operational performance of an OLED. Several groups have also demonstrated TFE on OLEDs fabricated on plastic substrates with or without barrier. Chwang et al. [123] applied PML-type multilayer coatings onto OLEDs fabricated on PET substrates with a similar barrier stack. When compared at equivalent conditions, the all plastic structure displayed a lifetime of 2,500 hours, whereas the same TFE encapsulated device on a glass substrate performed to 3,700 hours. The traditional all glass package decayed in 9,000 hours. The same group demonstrated TFE encapsulation of an all plastic 80 dpi, passive matrix display with an operational lifetime of 110 hours, with initial luminance at 110 cd/m^2. Recently, Yoshida et al. have demonstrated a flexible OLED display on PET, encapsulated with a 3-μm thick PECVD SiN layer with a projected operational lifetime (to reach 50% initial luminance of 1,000 cd/m^2) of over 5,000 hours.

The same group also showed that, with identical encapsulation, negligible change in OLED performance for a full-color OLED display (5.2 in. 1/4 VGA) after 500 hours at accelerated environmental testing conditions [143].

Significant ground has been covered over the last five years or so in making TFE a commercially viable technology. Improving both operational and shelf lifetime remains a challenge. Also, there is surprisingly little work on TFE of active matrix displays, which are the most promising candidates for future, high information content, high-resolution displays; so more work in this field is necessary. Last but not least, particle control, yield, and tact time, irrespective of the deposition method for advanced barrier systems remain the primary challenges standing between TFE being an excellent research level effort and TFE being a commercially viable solution for encapsulation of flexible organic electronic devices.

13.5 Concluding Remarks

Hermetic packaging of organic electronic devices is one of the most important challenges that must be overcome before flexible electronic devices can become a commercial reality. In this regard, thin-film barrier technology, especially on plastic substrates and also directly on the device, is a key enabling technology. From this review, it is clear that there are now multiple advanced barrier technologies that, at least in the research lab, can provide the right order of magnitude hermeticity. However, system integration challenges such as substrate deformation during device processing, thermomechanical stability, and compatibility with the device require more work and further improvements. Challenges also remain to demonstrate similar barrier performance using commercially viable R2R fabrication processes. Nevertheless, given the increasing rate of progress and increasing effort over the last decade, it seems likely that flexible thin-film barrier technology will mature in time to enable flexible electronics applications.

References

1. Forrest SR (2004) The path to ubiquitous and low-cost organic electronic appliances on plastic. Nature 428:911–918
2. Friend RH, Gymer RW, Holmes AB, Burroughes JH, Marks RN, Taliani C, Bradley DDC, Dos Santos DA, Bredas JL, Logdlund M, Salaneck WR (1999) Electroluminescence in conjugated polymers. Nature 397:121–128
3. Forrest SR (2000) Active optoelectronics using thin-film organic semiconductors. IEEE J Select Topics Quantum Electron 6:1072–1083
4. Tang CW (1986) Two-layer organic photovoltaic cell. Appl Phys Lett 48:183–185
5. Yu G, Gao J, Hummelen J, Wudl F, Heeger AJ (1995) Polymer photovoltaic cells: Enhanced efficiencies via a network of internal donor–acceptor heterojunctions. Science 270: 1789–1791
6. Halls JJM, Walsh CA, Greenham NC, Marseglia EA, Friend RH, Moratti SC, Holmes AB (1995) Efficient photodiodes from interpenetrating polymer networks. Nature 376: 498–500

7. Lovinger AJ, Rothberg LJ (1996) Electrically active organic and polymeric materials for thin-film-transistor technologies. J Mater Res 11:1581–1592

8. Rogers JA, Bao Z, Baldwin K, Dodabalapur A, Crone B, Raju VR, Kuck V, Katz H, Amundson K, Ewing J, Drzaic P (2001) Paper-like electronic displays: Large-area, rubber-stamped plastic sheets of electronics and microencapsulated electrophoretic inks. Proc Natl Acad Sci 98:4835–4840

9. Snow ES, Novak JP, Campbell PM, Park D (2003) Random networks of carbon nanotubes as an electronic material. Appl Phys Lett 82:2145–2147

10. Duan X, Niu C, Sahi V, Chen J, Parce JW, Empedocles S, Goldman JL (2003) High-performance thin-film transistors using semiconductor nanowires and nanoribbons. Nature 425:274–278

11. Mitzi DB, Kosbar LL, Murray CE, Copel M, Afzali A (2004) High-mobility ultrathin semiconducting films prepared by spin coating. Nature 428:299–302

12. Kazlas PT, McCreary MD (2002) Paperlike microencapsulated electrophoretic materials and displays. MRS Bulletin 27:894–897

13. Park S, Jayaraman S (2003) Smart textiles: Wearable electronic systems. MRS Bulletin 28:585–591

14. Jansen AN, Amine K, Newman AE, Vissers DR, Henriksen GL (2002) Low-cost, flexible battery packaging materials. JOM 54:29–54

15. Robshaw MJB (2006) An overview of RFID tags and new cryptographic developments. Inf Secur Tech Rep 11:82–88

16. Tung S, Witherspoon SR, Roe LA, Silano A, Maynard DP, Ferraro N (2001) A MEMS-based flexible sensor and actuator system for space inflatable structures. Smart Mater Struct 10:1230–1239

17. Crone BK, Dodabalapur A, Sarpeshkar R, Gelperin A, Katz HE, Bao Z (2002) Organic oscillator and adaptive amplifier circuits for chemical vapor sensing. J Appl Phys 91: 140–146

18. Fix W, Ullmann A, Ficker J, Clemens W (2002) Fast polymer integrated circuits. Appl Phys Lett 81:1735–1737

19. Gelinck G, Huitema HEA, Veenendaal EV, Cantatore E, Schrijnemakers L, Van Der Putten JBPH, Geuns TCT, Beenhakkers M, Giesbers JB, Huisman B, Meijer EJ, Benito EM, Touwslager FJ, Marsman AW, Van Rens BJE, De Leeuw DM (2004) Flexible active-matrix displays and shift registers based on solution-processed organic transistors. Nat Mater 3:106–110

20. Lewis JS, Weaver MS (2004) Thin-film permeation-barrier technology for flexible organic light-emitting devices. IEEE J Select Topics Quantum Electron 10:45–57

21. Parthasarathy G, Liu J, Duggal AR (2003) Organic light emitting devices from displays to lighting. The Electrochem Soc Interface 12:42–47

22. Schaer M, Nüesch F, Berner D, Leo W, Zuppiroli L (2001) Water vapor and oxygen degradation mechanisms in organic light emitting diodes. Adv Funct Mater 11: 116–121

23. McElvain J, Antoniadis H, Hueschen MR, Miller JN, Roitman DM, Sheats JR, Moon L (1996) Formation and growth of black spots in organic light-emitting diodes. J Appl Phys 80:6002–6007

24. Burrows PE, Graff GL, Gross ME, Martin PM, Hall M, Mast E, Bonham C, Bennett W, Michalski L, Weaver M, Brown JJ, Fogarty D, Sapochak LS (2001) Gas permeation and lifetime tests on polymer-based barrier coatings. Proc SPIE 4105:75–83

25. Graff GL, Burrows PE, Williford RE, Praino RF (2005) Barrier layer technology for flexible displays. In: Crawford GP (ed) Flexible Flat Panel Displays. John Wiley & Sons Ltd, West Succex, pp 57–78

26. Kim TW, Yan M, Erlat AG, McConnelee PA, Pellow M, Deluca J, Feist TP, Duggal AR, Schaepkens M (2005) Transparent hybrid inorganic/organic barrier coatings for plastic organic light-emitting diode substrates. J Vac Sci Technol 23:971–977

27. Rutherford NM, Moro L, Chu X, Visser RJ (2004) Plastic barrier substrate and thin film encapsulation: Progress toward manufacturability. Third Annual Flexible Displays and Microelectronics Conference
28. Yan M, Kim TW, Erlat AG, Pellow M, Foust D, Liu J, Schaepkens M, Heller CM, McConnelee PA, Feist T, Duggal A (2005) A transparent, high barrier, and high heat substrate for organic electronics. Proc IEEE 93:1468–1477
29. Plichta A, Habeck A, Knoche S, Kruse A, Weber A, Hildebrand N (2005) Flexible glass substrate. In: Crawford GP (ed) Flexible Flat Panel Displays. John Wiley & Sons Ltd, West Sussex, pp 35–55
30. Chen Y, Au J, Kazlas P, Ritenour A, Gates H, McCreary M (2003) Flexible active-matrix electronic ink display. Nature 423:136
31. Ma RQ, Hewitt R, Rajan K, Silvernail J, Urbanik K, Hack M, Brown J (2007) Flexible low power consumption active-matrix OLED displays. International Display Manufacturing Conference and FPD Expo – Proceedings, Taipei, Taiwan 305–308
32. MacDonald WA (2004) Engineered films for display technologies. J Mater Chem 14:4–10
33. MacDonald WA, Rollins K, MacKerron D, Rakos K, Eveson R, Hashimoto K, Rustin R (2005) Engineered films for display technologies. In: Crawford GP (ed) Flexible Flat Panel Displays. John Wiley & Sons Ltd, West Sussex, pp 11–33
34. MacDonald WA (2004) Engineered films for display technologies. J Mater Chem 14:4–10
35. Long K, Kattamis AZ, Cheng IC, Gleskova H, Wagner S, Sturm JC (2006) Stability of amorphous-silicon TFTs deposited on clear plastic substrates at 250°C to 280°C. IEEE Electron Device Lett 27:111–113
36. Mahon JK, Brown JJ, Zhou TX, Burrows PE, Forrest SR (1999) Requirements of flexible substrates for organic light emitting devices in flat panel display applications. 42nd Annual Technical Conference Proceedings of the Society of Vacuum Coaters, Chicago, IL, pp 456–459
37. Kadoya T (ed) (1990) Food packaging. Academic Press, London
38. Gavitt IF (1994) Snack food packaging barrier: How much is enough? Proceedings of the Society Of Vacuum Coaters, 37th Annual Technical cConference, pp 127–132
39. Dean DA(1990) Plastics in pharmaceutical packaging: A literature review, Leatherhead, Pira
40. Coxe M (1998) Improved barrier properties of medical packaging. Annual Technical Conference – ANTEC, Conference Proceedings 13:3380–3383
41. Goosey MT (1985) Permeability of coatings and encapsulants for electronic and optoelectronic devices. In: Comyn J (ed) Polymer Permeability. Elsevier Applied Science, London, pp 309–339
42. Burrows PE, Graff GL, Gross ME, Martin PM, Shi MK, Hall M, Mast E, Bonham C, Bennett W, Sullivan MB (2001) Ultra barrier flexible substrates for flat panel displays. Displays 22:65–69
43. Childs ES (1974) The rapid growth of and big potential for plastics in packaging. In: Bruins PF (ed) Packaging with plastics. Gordon and Breach Science Publishers, New York, pp 1–24
44. McCrum NC, Buckley CP, Bucknall CB (1997) Principles of polymer engineering. Oxford University Press, Oxford
45. Jamieson EHH, Windle AH (1983) Structure and oxygen-barrier properties of metallized polymer film. J Mater Sci 18:64–80
46. Krug TG (1990) Transparent barriers for food packaging. Proceedings of the Society of Vacuum Coaters, 33rd Annual Technical Conference, pp 163–169
47. Krug TG, Ludwig R, Steiniger G (1993) New developments in transparent barrier coatings. Proceedings of the Society of Vacuum Coaters, 36th Annual Technical Conference, pp 302–306
48. Chatham H (1996) Oxygen diffusion barrier properties of transparent oxide coatings on polymeric substrates. Surf Coat Technol 78:1–9
49. Maixner RD (2002) Future trends in permeation measurement. Proceedings of the Society of Vacuum Coaters, 45th Annual Technical Conference, pp 461–464

50. Stevens M, Tuomela S, Mayer D (2005) Water vapor permeation testing of ultra-barriers: Limitations of current methods and advancements resulting in increased sensitivity. Proceedings of the Society of Vacuum Coaters, 48th Annual Technical Conference, pp 189–191
51. Norenberg H, Miyamoto T, Tsukahara Y, Smith GDW, Briggs, GAD (1999) Mass spectrometric estimation of gas permeation coefficients for thin polymer membranes. Rev Sci Instrum 70:2414–2420
52. Norenberg H, Burlakov V (2004) Recent developments in measuring permeation through barrier films and understanding of permeation processes. Proceedings of the Society of Vacuum Coaters, 47th Annual Technical Conference, 164–167
53. Vogt BD, Lee HJ, Prabhu VM, Delongchamp DM, Lin EK, Wu WL, Satija SK (2005) X-ray and neutron reflectivity measurements of moisture transport through model multilayered barrier films for flexible displays. J Appl Phys 11:114509-1–114509-7
54. Dunkel R, Bujas R, Klein A, Horndt V (2005) Method of measuring ultralow water vapor permeation for OLED displays. Proc IEEE 93:1478–1482
55. Nisato G, Bouten PCP, Slinkkerveer PJ, Bennet WD, Graff GL, Rutherford N, Wiese L (2001) Evaluating high performance diffusion barriers: The calcium test. Asia Display/IDW'01 Proceedings, pp 1435–1438
56. Nisato G, Kuilder M, Bouten P, Moro L, Philips O, Rutherford N (2003) Thin-film encapsulation for OLEDs: Evaluation of multilayer barriers using the Ca test. SID 03 Digest XXXIV:550–553
57. Burrows PE, Graff GL, Gross ME, Martin PM, Hall M, Mast E, Bonham C, Bennett W, Michalski L, Weaver M, Brown JJ, Fogarty D, Sapochak LS (2001) Gas permeation and lifetime tests on polymer-based barrier coatings. Proc SPIE 4105:75–83
58. Erlat AG, Schaepkens M, Kim TW, Heller CM, Yan M, McConnelee P (2004) Ultra-high barrier coatings on polymer substrates for flexible optoelectronics: Water vapor transport and measurement systems. Proceedings of the Society of Vacuum Coaters, 47th Annual Technical Conference, pp 654–657
59. Erlat AG, Yan M, Kim TW, Schaepkens M, Liu J, Heller CM, McConnelee P, Feist T, Duggal AR (2005) Ultra-high barrier coatings for flexible organic electronics. Proceedings of the Society of Vacuum Coaters, 48th Annual Technical Conference, pp 116–120
60. Paetzold R, Winnacker A, Henseler D, Cesari V, Heuser K (2003) Permeation rate measurements by electrical analysis of calcium corrosion. Rev Sci Instrum 74:5147
61. Cros S, Firon M, Lenfant S, Trouslard P, Beck L (2006) Study of thin calcium electrode degradation by ion beam analysis. Nucl Instrum Methods Phys Res B 251:257–260
62. Lewis J, Grego S, Vick E, Temple D (2004) Performance of thin films for flexible OLED displays. 4th Annual Flexible Displays and Microelectronics Conference
63. Erlat AG, Yan M, Heller CMH, unpublished results, GE Global Research Center
64. Deal BE, Grove AS (1965) General relationship for the thermal oxidation of silicon. J Appl Phys 36:3770–3778
65. Barrer RM (1941) Diffusion in and through solids. Cambridge University Press, New York
66. Felts JT (1991) Thickness effects on thin film gas barriers: Silicon-based coatings. Proceedings of the Society of Vacuum Coaters, 34th Aannual Technical Conference, pp 99–104
67. Felts JT (1993) Transparent barrier coatings update: Flexible substrates. Proceedings of the Society of Vacuum Coaters 36th Annual Technical Conference, pp 324–331
68. Benmalek M, Dunlop HM (1995) Inorganic coatings on polymers. Surf Coat Technol 76–77:821–826
69. Roberts AP, Henry BM, Sutton AP, Grovenor CRM, Briggs GAD, Miyamoto T, Kano M, Tsukahara Y, Yanaka M (2002) Gas permeation in silicon-oxide/polymer (SiOx/PET) barrier films: Role of the oxide lattice, nano-defects and macro-defects. J Membrane Sci 208:75–88
70. Tropsha YG, Harvey NG (1997) Activated rate theory treatment of oxygen and water transport through silicon oxide/poly(ethylene terephthalate) composite barrier structures. J Phys Chem B 101:2239–2266

71. da Silva Sobrinho AS, Latreche M, Czeremuszkin G, Klemberg-Sapieha JE, Wertheimer MR (1998) Transparent barrier coatings on polyethylene terephthalate by single- and dual-frequency plasma-enhanced chemical vapor deposition. J Vac Sci Technol A 16:3190–3198

72. Henry BM, Dinelli F, Zhao KY, Erlat AG, Grovenor CRM, Briggs GAD (1999) Characterization of oxide gas barrier films. Proceedings of the Society of Vacuum Coaters, 42nd Annual Technical Conference, pp 403–407

73. Barker CP, Kochem KH, Revell KM, Kelly RSA, Badyal JPS (1995) Interfacial chemistry of metallized, oxide coated, and nanocomposite coated polymer films. Thin Solid Films 257:77–82

74. Klemberg-Sapieha JE, Martinu L, Küttel OM, Wertheimer MR (1993) Transparent gas barrier coatings produced by dual-frequency PECVD. Proceedings of the Society of Vacuum Coaters, 36th Annual Technical Conference pp 445–449

75. Schaepkens M, Kim TW, Erlat AG, Yan M, Flanagan KW, Heller CM, McConnelee PA (2004) Ultrahigh barrier coating deposition on polycarbonate substrates. J Vac Sci Technol A 22:1716–1722

76. Groner MD, George SM, McLean RS, Carcia PF (2006) Gas diffusion barriers on polymers using Al_2O_3 atomic layer deposition. Appl Phys Lett 88:0517109

77. Carcia PF, McLean RS, Reilly MH, Groner MD, George SM (2006) Ca test of Al_2O_3 gas diffusion barriers grown by atomic layer deposition on polymers. Appl Phys Lett 89:031915

78. Decker W, Henry BM (2002) Basic principles of thin film barrier coatings. Proceedings of the Society of Vacuum Coaters, 45th annual technical conference, pp 493–502

79. da Silva Sobrinho AS, Czeremuszkin G, Latreche M, Wertheimer MR (2000) Defect-permeation correlation for ultrathin transparent barrier coatings on polymers. J Vac Sci Technol A 18:149–157

80. Erlat AG, Spontak RJ, Clarke RP, Robinson TC, Haaland PD, Tropsha Y, Harvey NG, Vogler EA (1999) SiO_x gas barrier coatings on polymer substrates: morphology and gas transport considerations. J Phys Chem B 103:6047–6055

81. Ling JSG, Leggett GJ (1997) Scanning force microscopy of poly(ethylene terephthalate) surfaces: Comparison of SEM with SFM topographical, lateral force and force modulation data. Polymer 38:2617–2625

82. Erlat AG, Henry BM, Ingram JJ, Mountain DB, McGuigan, A, Howson RP, Grovenor, CRM, Briggs GAD, Tsukahara Y (2001) Characterisation of aluminium oxynitride gas barrier films. Thin Solid Films 388:78–86

83. Cros B, Vallat MF, Despaux G (1997) Characterization by acoustic microscopy of adhesion in poly(ethylene terephthalate) films coated by aluminum. Appl Surf Sci 126:159–168

84. Shaw DG, Langlois MG (1994) Use of vapor deposited acrylate coatings to improve the barrier properties of metallized film. Proceedings of the Society of Vacuum Coaters, 45th Annual Technical Conference, pp 240–247

85. Roehrig M, Shaw D, Langlois M, Sheehan C (1997) Use of evaporated acrylate coatings to smooth the surface of polyester and polypropylene film substrates. J Plastic Film & Sheeting 13:235–251

86. Affinito JD, Gross ME, Coronado CA, Graff GL, Greenwell EN, Martin PM (1996) New method for fabricating transparent barrier layers. Thin Solid Films 290–291:63–67

87. Affinito JD, Eufinger S, Gross ME, Graff GL, Martin PM (1997) PML/oxide/PML barrier layer performance differences arising from use of UV or electron beam polymerization of the PML layers. Thin Solid Films 308–309:19–25

88. Yan M, Erlat AG, Duggal AR, GE Global Research, unpublished results

89. da Silva Sobrinho AS, Czeremuszkin G, Latreche M, Dennle G, Wertheimer MR (1999) Study of defects in ultra-thin transparent coatings on polymers. Surf Coat Technol 116–119:1204–1210

90. Henry BM, Dinelli F, Zhao KY, Grovenor CRM, Kolosov OV, Briggs GAD, Roberts AP, Kumar RS, Howson RP (1999) Microstructural study of transparent metal oxide gas barrier films. Thin Solid Films 355–356:500–505

91. Henry BM, Erlat AG, Grovenor CRM, Briggs GAD, Tsukahara T, Miyamoto T, Noguchi N, Niijima T (2000) Proceedings of the Society of Vacuum Coaters, 43rd Annual Technical Conference pp 373–378

92. Prins W, Hermans JJ (1959) Theory of permeation through metal coated polymer films. J Phys Chem 63:716–719

93. Beu TA, Mercea PV (1990) Gas transport through metallized polymer membranes. Mater Chem Phys 26:309–322

94. Rossi G, Nulman M (1993) Effect of local flaws in polymeric permeation reducing barriers. J Appl Phys 74:5471–5475

95. Czeremuszkin G, Latreche M, da Silva Sobrinho AS, Wertheimer MR (1999) Simple model of oxygen permeation through defects in transparent coatings. Proceedings of the Society of Vacuum Coaters, 42nd Annual Technical Conference, pp 176–180

96. Hanika M, Langowski HC, Moosheimer U, Peukert W (2003) Inorganic layers on polymeric films – Influence of defects and morphology on barrier properties. Chem Eng Technol 26:605–614

97. Koo WH, Jeong SM, Choi SH, Kim W, Baik HK, Lee SM, Lee SJ (2005) Effects of the polarizability and packing density of transparent oxide films on water vapor permeation. J Phys Chem B 109:11354–11360

98. Henry BM, Erlat AG, Deng CS, Briggs GAD, Miyamoto T, Noguchi N, Niijima T, Tsukahara Y (2001) The permeation of water vapor through gas barrier films. Proceedings of the Society of Vacuum Coaters, 44th Annual Technical Conference, pp 469–475

99. Erlat AG, Henry BM, Howson RP, Grovenor CRM, Briggs GAD, Tsukahara Y (2001) Growth and characterization of transparent metal oxide and oxynitride gas barrier films on polymer substrates. Proceedings of the Society of Vacuum Coaters, 44th Annual Technical Conference, pp 448–454

100. Erlat AG, Henry BM, Grovenor CRM, Briggs GAD, Chater RJ, Tsukahara Y (2004) Mechanism of water vapor transport through PET/AlO$_x$N$_y$ gas barrier films. J Phys Chem B 108:883–890

101. Schiller S, Neumann M, Morgner H, Schiller N (1995) Reactive aluminum oxide coating of plastic films – possibilities and limitations for high deposition rates. Proceedings of the Society of Vacuum Coaters, 38th Annual Technical Conference, pp 18–27

102. Schiller N, Reschke J, Goedicke K, Neumann M (1996) Application of the magnetron activated deposition process (MAD-process) to coat polymer films with alumina in web coaters. Surf Coat Technol 86–87:776–782

103. Schiller N, Morgner H, Fahland M, Straach S, Räbisch M, Charton C (1999) Transparent oxide coatings on plastic webs for emerging applications. Proceedings of the Society of Vacuum Coaters, 42nd Annual Technical Conference, pp 392–396

104. Langereis E, Creatore M, Heil SBS, van de Sanden MCM, Kessels WMM (2006) Plasma-assisted atomic layer deposition of Al$_2$O$_3$ moisture permeation barriers on polymers. Appl Phys Lett 89:081915

105. Weidner WK, Zambov LM, Shamamian VA, Perz SV, Camilletti RC, Snow SA, Loboda MJ, Cerny GA (2005) PECVD processed silicon carbide coatings for organic luminescent devices. Proceedings of the Society of Vacuum Coaters, 48th Annual Technical Conference, 158–162

106. Weaver MS, Michalski LA, Rajan K, Rothman MA, Silvernail JA, Brown JJ, Burrows PE, Graff GL, Gross ME, Martin PM, Hall M, Mast M, Bonham C, Bennett W, Zumhoff M (2002) Organic light-emitting devices with extended operating lifetimes on plastic substrates. Appl Phys Lett 81:2929–2931

107. Langowski HC, Melzer A, Schubert D (2002) Ultra high barrier layers for technical applications. Proceedings of the Society of Vacuum Coaters, 45th Annual Technical Conference, pp 471–475

108. Zambov L, Weidner K, Shamamian V, Camilletti R, Pernisz U, Loboda M, Cerny G (2006) Advanced chemical vapor deposition silicon carbide barrier technology for ultra-low permeability applications. J Vac Sci Technol 24:1706–1713

109. Henry BM, Howells D, Topping JA, Assender HE, Grovenor CRM, Marras L (2006) Microstructural and barrier properties of multilayered films. Proceedings of the Society of Vacuum Coaters, 49th Annual Technical Conference, pp 654–657

110. Charton C, Schiller N, Fahland M, Hollander A, Wedel A, Noller K (2006) Development of high barrier films on flexible polymer substrates. Thin Solid Films 502:99–103

111. Yializis A, Powers GL, Shaw DG (1990) A new high temperature multilayer capacitor with acrylate dielectrics. IEEE Trans Components, Hybrids Manufact Technol 13:611–616

112. Leterrier Y (2003) Durability of nanosized oxygen-barrier coatings on polymers. Prog Mater Sci 48:1–55

113. Miyamoto T, Mizuno K, Noguchi N, Niijima N (2001) Gas barrier performances of organic–inorganic multilayered films. Proceedings of the Society of Vacuum Coaters, 44th annual Technical Conference, pp 166–171

114. Haas KH, Amberg-Schwab A, Rose K, Schottner G (1999) Functionalized coatings based on inorganic–organic polymers (ORMOCER) and their combination with vapor deposited inorganic thin films. Surf Coat Technol 111:72–79

115. Kapoor S, Seta ME, Visser RJ, Gross ME, Bennett WD, Newton R, Goodwin K, Deng Y (2006) Pilot production of ultra-barrier substrate for flexible displays. Proceedings of the Society of Vacuum Coaters, 49th Annual Technical Conference, pp 625–631

116. Yan M, Duggal A, McConnelee PA, Schaepkens M (2005) Methods and fixtures for facilitating handling of thin films. US 20050016464 A1

117. Yan M, Erlat AG, Zhao R, Smith DJ, Scherer B, Jones C, Foust D, McConnelee PA, Feist TP, Duggal AR (2007) Roll-to-roll ultra-high barrier substrate for organic electronics. 6th Annual Flexible Displays and Microelectronics Conference

118. Madocks J, Rewhinkle J, Barton L (2005) Packaging barrier films deposited on PET by PECVD using a new high density plasma source. Mater Sci Eng B 119:268–273

119. Shamamian V, Zambov L, Pernisz U, Kim SJ, Perz S, Cerny GA (2006) Progress in the development of SiC:H alloy films on flexible substrates for extremely low moisture permeability applications. 5th Flexible Displays and Microelectronics Conference

120. Czeremuszkin G, Latreche M, Wertheimer MR (2001) Permeation through defects in transparent barrier-coated plastic films. Proceedings of the Society of Vacuum Coaters, 49th Annual Technical Conference, pp 178–183

121. Graff GL, Williford RE, Burrows PE (2004) Mechanisms of vapor permeation through multilayer barrier films: Lag time versus equilibrium permeation. J Appl Phys 96: 1840–1849

122. Sugimoto A, Ochi H, Fujimura S, Yoshida A, Miyadera T, Tsuchida M (2004) Flexible OLED displays using plastic substrates. IEEE J Select Top Quantum Electron 10: 107–114

123. Chwang A, Rothman MA, Mao SY, Hewitt RH, Weaver MS, Silvernail JA, Rajan K, Hack M, Brown JJ, Chu X, Moro L, Krajewski T, Rutherford N (2003) Thin film encapsulated flexible organic electroluminescent displays. Appl Phys Lett 83:413–415

124. Foust D, Balch E, Duggal A, Heller C, Guida R, Nealon W, Faircloth T (2006) Series connected OLED structure and fabrication method. US7049757 B2

125. Duggal AR, Foust DF, Nealon WF, Heller CM (2003) Fault-tolerant, scalable organic light-emitting device architecture. Appl Phys Lett 82:2580–2582

126. LUMATION Green 1300 Series LEP Data Sheets, The Dow Chemical Company

127. Burrows PE, Bulovic V, Forrest SR, Sapochak LS, Mccarty DM, Thompson ME (1994) Reliability and degradation of organic light emitting devices. Appl Phys Lett 65:2922–2924

128. Williams DJ, Rajeswaran G (2003) Proc SPIE 5080:166–169

129. Tsuruoka Y, Hieda S, Tanaka S, Takahashi H (2003) Transparent thin film desiccant for OLEDs. Proc Soc Inform Display Symp Dig Tech Papers, pp 860–863

130. Moro L, Krajewski TA, Rutherford NM, Philips O, Visser RJ, Gross M, Bennett WD, Graff G (2004) Process and design of a multilayer thin film encapsulation of passive matrix OLED displays. Proc SPIE 5214:83–93

131. Erlat AG, Yan M, Zhao R, Tandon S, Smith DJ, Scherer B, Jones C, Foust D, McConnelee PA, Feist TP, Duggal AR (2007) Recent developments in ultra-high barrier (UHB) coatings for organic electronics. Presentation at the 50th Annual Technical Conference, Society of Vacuum Coaters

132. Van Assche FJH, Vangheluwe RT, Maes JWC, Mishcke MS, Bijker MD, Dings FC, Evers MF, Kessels WMM, van de Sanden MCM (2004) A thin film encapsulation stack for PLED and OLED displays. SID 04 Digest 35:695–697

133. Lifka H, van Esch HA, Rosink JJWM (2004) Thin film encapsulation of OLEDs with a NONON stack. SID 04 Digest 35:1384–1387

134. Rosink JJWM, Lifka H, Rietjens GH, Pierik A (2005) Ultra-thin encapsulation for large area OLED displays. SID 05 Digest 36:1272–1275

135. van Rens B (2006) Breaking the LCD cost barrier for OLED production using inline concepts. 2nd Plastic Electronics Conference and Showcase

136. Ghosh AP, Gerenser LJ, Jarman CM, Fornalik JE (2005) Thin-film encapsulation of organic light-emitting devices. Appl Phys Lett 86:223503-1–223503-3

137. Park SK, Oh J, Hwang C, Lee J, Yang YS, Chu HY (2005) Ultrathin film encapsulation of an OLED by ALD. Electrochem Solid State Lett 8:H21–H23

138. Tzen CK, Shen WJ, Wang HC, Su CF, Wu CC, Lu CC, Tzen JC, Su CY, Tang SJ (2006) High transparency and low operation voltage top emission organic light emitting devices with thin film encapsulation. SID 06 Digest 37:972–974

139. Hemerik M, van Erven R, Vangheluwe R, Yang J, van Rijswijk T, Winters R, van Rens B (2006) Lifetime of thin-film encapsulation and its impact on OLED device performance. SID 06 Digest 1571–1573

140. Yi CH, LeeYK, Lee SM, Ju SH, Lee EJ, Choo YM, Park KC, Jang J (2004) A hybrid encapsulation method for OLED. SID 04 Digest 715–716

141. Akedo K, Miura A, Fujikawa H, Taga Y (2003) Plasma-CVD SiNx/plasma-polymerized CNx:H multi-layer passivation films for organic light emitting diodes. Proc Soc Inform Display Symp 34:559–561

142. Yoshida A, Fujimura S, Miyake T, Yoshizawa T, Ochi H, Sugimoto A, Kubota H, Miyadera T, Ishizuka S, Tsuchida M, Nakada H (2003) 3-inch Full-color OLED display using a plastic substrate. Proc Soc Inform Display Symp 34:856–859

143. Kubota H, Miyaguchi S, Ishizuka S, Wakimoto T, Funaki J, Fukuda Y, Watanabe T, Ochi H, Sakamoto T, Miyake T, Tsuchida M, Ohshita I, Tohma T (2000) Organic LED full color passive-matrix display. J Luminescence 87–89:56–60

Index